Role of the Canola Genome for Plant Growth, Development, and Stress Responses

Muhammad Waseem
Hainan University, China

IGI Global
Scientific Publishing
Publishing Tomorrow's Research Today

Vice President of Editorial	Melissa Wagner
Director of Acquisitions	Mikaela Felty
Director of Book Development	Jocelynn Hessler
Production Manager	Mike Brehm
Cover Design	Jose Rosado

Published in the United States of America by
IGI Global Scientific Publishing
701 East Chocolate Avenue
Hershey, PA, 17033, USA
Tel: 717-533-8845 | Fax: 717-533-7115
Website: https://www.igi-global.com E-mail: cust@igi-global.com

Library of Congress Cataloging-in-Publication Data

CIP Data Pending
ISBN13: 9798369386033
Isbn13Softcover: 9798369386040
EISBN13: 9798369386057

British Cataloguing in Publication Data
A Cataloguing in Publication record for this book is available from the British Library.

Editorial Advisory Board

Table of Contents

Detailed Table of Contents

health. Its oilseed cake, a byproduct of oil extraction, is used as high-protein animal feed, contributing to sustainable livestock production.In recent years, advancements in biotechnology have further enhanced canola's role in agriculture

Brassica napus is the most cultivated and economically important plant commonly known as rapeseed (2n=4x=38 “AACC”). Globally rapeseed oil ranks third next to soybean and palm. Besides human consumption, it is a versatile crop with wide range of uses in animal feed, lubricants, biodiesel etc. Higher levels of erucic acid controlled by dominant alleles and elevated amounts of glucosinolates governed by polygenes posed challenges for farmers owing to reduced consumer demand due to health concerns. This reduction subsequently imposed stress on breeders to develop agronomically viable market-friendly crops. The development of Low Erucic Acid Rapeseed (LEAR) followed by double zero (00) rapeseed genotypes resulted from continual breeding efforts over the years mainly by cross-breeding followed by selection. On the contrary, High Erucic Acid Rapeseed (HEAR) is bred specifically for industrial applications. This chapter focuses on the genetic architecture dating back to origin of canola and how traditional breeding strategies contributed significantly to the present-day canola plant.

Genome assembly has emerged as a fundamental tool within genomics, facilitating the reconstruction of entire DNA sequences while providing comprehensive insights into the genetic composition of various organisms. This chapter concentrates on the genome assembly of Brassica napus (canola), a species of considerable agricultural and economic significance. The assembly of the canola genome presents significant potential for expediting breeding initiatives, advancing genetic enhancement, and enriching our comprehension of the plant's biological mechanisms. Nevertheless, the intricate characteristics of the canola genome—characterized by polyploidy, repetitive sequences, and elevated heterozygosity—pose substantial challenges. The resolution of these difficulties necessitates the utilization of sophisticated sequencing technologies, such as long-read platforms, alongside refined bioinformatics methodologies specifically tailored for polyploid genomes. The outlook for canola genomics appears optimistic with climate-resilient varieties to bolster global food security.

Canola (Brassica napus L.) is a globally significant healthy vegetable oil that derives its nutritional and economic value primarily from its seed fatty acid levels; omega-3 fatty acids, oleic acid and linolenic acid. Additionally, canola oil serves as a biodiesel source with glycerin byproducts utilized in personal care products. Consequently, enhancing seed oil content (SOC), which is a key breeding objective for this crop, which ranks second worldwide in oil production after soybean. B. napus has 1,132 Mb genome size and researchers have established an association panel for mapping using genotype-by-sequencing. The availability of extensive germplasm resources and reduced sequencing costs has facilitated genome-wide association studies (GWAS) to investigate economically vital traits. The current chapter provides an outline of GWAS platform development, followed by an examination of experimental designs and sequencing strategies employed in GWAS research focused on canola oil content. Additionally, it explores future perspectives in GWAS research and associated challenges.

High-density Single Nucleotide Polymorphism (SNP) arrays have revolutionized genetic research and breeding in canola (Brassica napus L.), offering deep insights into its genetic structure. These arrays enable Genome-Wide Association Studies (GWAS), Quantitative Trait Loci (QTL) mapping, and genomic selection, identifying critical loci for traits like oil quality, yield, stress resilience and disease resistance. This precision enhances Marker-Assisted Selection (MAS), accelerating the development of superior canola varieties. This chapter details SNP array development, including SNP discovery and optimization and addresses the limitations of SNP arrays compared to next-generation technologies like whole genome sequencing (WGS) and Clustered Regularly Interspaced Short Palindromic Repeats (CRISPR). Despite these, ongoing improvements and integration with multi-omics strengthen the role of SNP arrays in advancing canola productivity and sustainability.

The chapter provides an overview of Canola (Brassica napus) is a vital oilseed crop valued for its high oil content, nutritional benefits and potential in sustainable energy. Key agronomic traits, including seed yield, flowering period, plant architecture, and disease resistance, are critical for maximizing production and environmental adaptability. Advances in molecular genetics, particularly QTL mapping and GWAS, have facilitated the identification of genomic regions tied to traits like disease resistance, drought and heat tolerance, and oil content, using markers like SSRs and SNPs. QTL mapping has pinpointed genetic loci for traits such as plant height, root biomass, and resistance to sclerotinia stem rot. Modern gene-editing tools like CRISPR-Cas9 allow precise modifications, improving traits like glucosinolate content and oil quality. Although QTL mapping can be slow and affected by environmental factors, the discovery of significant QTLs and beneficial alleles holds promise for enhanced agronomic performance and resilience in canola breeding.

The canola plant (Brassica napus L.) is a major oilseed crop globally providing high-quality oil and protein feed for livestock. However, there is limited understanding of the genetic and molecular processes that govern the plant's growth, development and response to stress. This project seeks to investigate the genetic makeup of canola with a specific focus on the chloroplast and mitochondrial genomes, as well as the regulation of gene expression through small RNA and DNA methylation. It is with this in mind that we strive to elucidate the complex interaction between these components to explain the underlying mechanisms that impinge on canola productivity and its adaptability. This research shall stand as one of the major steps toward generation of high-yielding stress-tolerant and climate-resilient canola varieties, thereby further supporting the enhancement of global food security as well as sustainable agriculture.

Genetic transformation in canola represents a transformative approach in modern agriculture, leveraging principles of molecular biology to introduce and express new traits that enhance crop performance and functionality. Another widely used method is Particle Bombardment or Biolistics, which involves physically delivering

biomarkers and traits related to stress tolerance and oil quality improvement. Besides that, metabolic engineering could improve agronomical performance and nutritional quality, opening more perspectives toward sustainable crop improvement under changing environmental conditions.

This chapter explores the possibilities of multi-omics technologies in catalyzing canola breeding for sustainable agriculture. From genomics to transcriptomics, proteomics and metabolomics, omics techniques offer a unified view of genetic and molecular regulation of important agronomic traits. Genomic technologies are essential in screening yield related QTL, oil quality and stress tolerance traits. Proteomics and metabolomics are another level up from genomics and transcriptomics as they show intricate protein–protein and metabolic networks that regulate plant growth, development and stress tolerance. Further, functional genomics styles like CRISPR-Cas9, RNA interference provide new approaches targeting genome modification for canola improvement, disease resistance and efficient nutrient utilization under climate change. Further, author presents potential work direction using multi-omics data with advanced platforms; high-throughput phenotyping and artificial intelligence that will culminate to development of improved canola varieties to enhance food security and industrial utility.

Proteomics, or the large-scale study of proteins and their functions, is an effective method for improving our understanding of the canola plant (Brassica napus). This study focusses on the use of proteomics in canola research, namely in growth, development, and responses to biotic and abiotic challenges. Proteomics allows for the identification and measurement of proteins implicated in essential metabolic

Preface

Canola (*Brassica napus* L.), a globally significant oilseed crop, has witnessed remarkable advancements in breeding, biotechnology, and genomics over the past few decades. As we confront the dual challenges of climate change and increasing food demand, the need to enhance the yield potential, nutritional quality, and stress resilience of canola has never been more urgent. Central to this endeavor is a comprehensive understanding of the canola genome and its functional components that govern plant growth, development, and adaptation to diverse environmental pressures.

This book, *Role of the Canola Genome for Plant Growth, Development, and Stress Responses*, brings together a diverse collection of chapters that reflect the latest scientific progress in the genomics and functional biology of canola. Drawing on multidisciplinary expertise, the volume offers an integrative exploration of how classical genetics, genome sequencing, omics technologies, and molecular breeding tools are unlocking the genetic potential of canola to support sustainable agriculture and climate-smart crop production.

The book is structured into fifteen thematic chapters, beginning with the historical evolution and genomic diversity of canola. It advances through essential topics such as classical and molecular breeding, genome assembly, SNP arrays, GWAS, and QTL mapping for economically significant traits. Further chapters explore organelle genomics, epigenetics, transcriptomics, proteomics, and metabolomics—each highlighting the key regulatory networks and molecular pathways underlying plant development and stress responses.

Chapter 1: Traces the evolutionary history of canola from its hybrid origins, highlighting the role of selective breeding and modern genomic innovations such as CRISPR-Cas9, transcriptomics, and multi-omics in enhancing adaptability to environmental stressors.

Chapter 2: Describes canola's agronomic significance, optimal growing conditions, and economic and environmental benefits, emphasizing its role in soil health, sustainable livestock feed, and the global oil market, especially in Canada.

Chapter 3: Explains the genetic makeup and traditional breeding history of canola, detailing the development of low erucic acid and double-zero varieties, and the use of specific high erucic acid types for industrial purposes.

Chapter 4: Outlines the challenges and advances in assembling the complex polyploid genome of canola, emphasizing the importance of long-read sequencing and tailored bioinformatics for breeding and climate resilience.

Chapter 5: Discusses genome-wide association studies (GWAS) for economically important traits, focusing on methods for mapping seed oil content and the potential for improving nutritional and industrial value.

Chapter 6: Details the creation and application of high-density SNP arrays in canola breeding, their role in GWAS, QTL mapping, and marker-assisted selection, along with a comparison to whole genome sequencing and CRISPR technologies.

Chapter 7: Reviews QTL mapping and beneficial allele discovery for quality and agronomic traits such as yield, disease resistance, and drought tolerance, and examines the use of CRISPR-Cas9 to enhance desirable traits.

Chapter 8: Investigates chloroplast and mitochondrial genomes, small RNA, and DNA methylation in canola, aiming to uncover molecular mechanisms that support productivity and climate resilience.

Chapter 9: Explores genetic transformation methods in canola, including Agrobacterium-mediated transformation and particle bombardment, as well as selection and screening techniques for identifying successful genetic modifications.

Chapter 10: Examines canola's molecular and physiological responses to abiotic stresses, the role of candidate genes and transcription factors, and the potential of transgenic breeding for improved tolerance and productivity.

Chapter 11: Presents transcriptomic analyses of canola growth, development, and stress responses, and discusses how transcriptomics informs breeding for improved adaptability and performance.

Chapter 12: Explores canola metabolomics, detailing the biochemical pathways and metabolites linked to yield, quality, stress tolerance, and oil improvement, and highlighting the role of metabolic engineering.

Chapter 13: Envisions the integration of multi-omics approaches—genomics, transcriptomics, proteomics, and metabolomics—with tools like CRISPR, AI, and high-throughput phenotyping to accelerate canola breeding for food security and industrial applications.

Chapter 14: Describes the use of proteomics to study canola's protein networks, metabolic pathways, and stress responses, and how this information supports breeding and genetic engineering efforts.

Special emphasis is placed on abiotic and biotic stress resistance, with discussions on the functional genomics of drought, salinity, and pathogen responses, as well as transformative breeding strategies through genetic engineering and multi-omics integration. The final chapters offer future perspectives for canola improvement in the omics era, emphasizing translational genomics and precision breeding approaches.

This volume is the product of international collaboration among researchers from India, Pakistan, China, Iran, Turkey, and beyond. We are grateful to all contributors for their commitment to advancing scientific knowledge and for providing a valuable reference that blends foundational biology with applied innovation.

We envision this book as a comprehensive resource for plant geneticists, breeders, molecular biologists, agronomists, and students interested in canola improvement. It offers not only scientific insights but also practical tools for researchers and industry stakeholders aiming to enhance crop productivity, nutritional value, and environmental adaptability.

We hope that this book will inspire further research and development efforts and serve as a cornerstone reference in the continued advancement of canola genomics and breeding.

Chapter 1

Historical Evolution of Canola and the Genomic Variations

Ujjwal Agrawal
https://orcid.org/0000-0002-2604-7953
UM-DAE Centre for Excellence in Basic Sciences, Mumbai, India

ABSTRACT

This chapter covers the origins of the development of canola, Brassica napus, and how genomic variations have sculpted its adaptive responses. Long from its origins as a hybrid from species Brassica, selective breeding transformed canola into varieties with greatly improved oil quality and agronomic traits. Recent advances in genomic technology have sped up the discovery of key genetic variations that underlie its adaptive responses to environmental stressors such as drought, disease, and temperature fluctuations. In this chapter, we dig into the molecular mechanisms that made the canola develop and some of these developments through genomic innovations like CRISPR-Cas9, transcriptomics, and multi-omics approaches. We also consider the evolutionary history of canola, from drawings and trees, where great improvements have led to the development of well-performing varieties that correspond to the needs of modern agriculture.

1. INTRODUCTION

1.1 Overview of Canola and Its Evolution

One of the most important oilseed crops in the world is canola or Brassica napus. The majority of the present increase, specially made for very high oil content and

DOI: 10.4018/979-8-3693-8603-3.ch001

extensive usage in food, biofuels, and industrial sectors, originated from rapeseed, a crop first grown for lamp oil and other industrial oils. Varieties of rapeseed with low erucic acid and glucosinolate content were produced through selective breeding in the mid-20th Century, and that is where Canola comes in. The term "canola" emerged from the phrase "Canadian oil, low acid," credited to the origin and quality of the oil compared to common rapeseed oil. Today, it is grown globally in a balanced ratio that contributes positively to the earth's field of agriculture and industry.

The strong genomic complexity derived from the polyploid nature of this species accounts for why Canola is successful in terms of evolutionary phenomena, especially with regard to genetic diversity brought about by hybridization and natural selection. It is a hybrid between Brassica rapa and Brassica oleracea, a polyploid plant carrying various sets of chromosomes. It was through polyploidy that much higher genetic diversity and adaptability than those possible through the simple addition of genomes were allowed, thus guaranteeing Canola more protection against environmental challenges.

Genomic research has gone very well with the agronomic success of canola development. Scientists have decoded the canola genome and identified certain genetic variations associated with traits such as disease resistance, yield improvement, and stress tolerance. These findings resulted in improvements in breeding strategies, and farmers are now able to cultivate Canola in the widest range of agroecological zones. Natural and artificial inputs of genomic variations have provided the driving force through which Canola has emerged as a high-performing crop that can resist biotic as well as abiotic stresses.

1.2 Role of Genomic Variations in Crop Improvement

The most critical elements of evolutionary studies in any species, and then particularly important for crops like Canola, are such genomic variations. These occur as a result of differential DNA sequences among individuals or populations. They can occur as mutations, gene duplications, or hybridization, so they can also be induced artificially through selective breeding, genetic modifications, or even more modern techniques using tools like CRISPR-Cas9.

Genomic variations have been important for Canola's responsiveness to the environment and agriculture. For example, changes in those genes that are linked with disease resistance make it possible for the canola species to adapt with the defense mechanisms against pathogens like the feared Leptosphaeria maculans (blackleg disease) and Plasmodiophora brassicae (clubroot). Genomic variations that are associated with drought tolerance and salinity tolerance have played a role in the production of Canola through low-productive environmental conditions.

Genomic variations at these sites give new insights into Canola adaptive evolution: how changing genes enhance an organism's survival and reproduction abilities within its environment. Indeed, adaptive evolution in Canola is believed to result not only from natural selection, where plants were selected or self-selected as the environmental conditions they were under kept intensifying, but also through artificial selection, where breeders selected and amplified specific traits in the plant that matched the needs and demands of agriculture.

The identification and exploitation of these variations were very specific and possible because of modern genomic tools. Acceleration by technologies such as next-generation sequencing (NGS), genome-wide association studies (GWAS), and quantitative trait loci (QTL) mapping has sped up the finding of beneficial genetic variations in Canola, therefore developing improved varieties with enhanced agronomic traits.

1.3 Chapter Objectives

This chapter will attempt a comprehensive analysis of how genomic variation has challenged the historical evolution and adaptive responses of Canola. More concretely, the chapter addresses the following objectives:

- **Historical Evolution of Canola:** The origins of Canola, from its ancestral species to the present, will form a backbone for which it outlines how hybridization, polyploidy, and natural selection have crafted its genomic complexity.
- **The scope of genomic variation in Canola's adaptive evolution:** The chapter will focus on which genomic changes allowed Canola to adapt to various environmental stresses, such as biotic challenges from pathogens and abiotic stresses like drought and temperature extremes.
- **Breeding and Genetic Engineering:** The chapter shall outline how selection and breeding programs, along with the use of modern genomic technologies, have molded the canola genome, improved its agronomic characteristics, and rendered it stronger.
- **Future Frontiers in Canola Research:** The concluding paragraphs of the chapter will try to outline possibilities of future breeding and genomic innovation, including gene editing technologies such as CRISPR, to make Canola even more adaptable and productive.

1.4 Significance of This Study

Knowing the evolutionary history and genetic variations that have supported adaptation in Canola would be extremely useful for sustainable agriculture. More resilient, high-producing crops like Canola will be needed as the world's population increases and continues to change due to climate changes. It is by tracing the historical evolution of the development of Canola and identifying the genetic mechanisms behind its adaptability that researchers and breeders will find ways to develop new varieties of Canola that are better equipped to meet future agricultural challenges.

Indeed, this study is relevant to other crops within the **Brassicaceae family**, including mustard and cabbage, as well as broader implications for agricultural biotechnology. Indeed, lessons learned by transferring the genomic evolution of Canola to other important crops may enhance food security and promote sustainable agricultural practices.

2. ORIGINS OF CANOLA: HYBRIDIZATION AND POLYPLOIDY

2.1 Ancestry of Canola (Brassica rapa and Brassica oleracea)

Canola is a species of Brassica napus derived from the hybridization event between two ancestral species, namely Brassica rapa and Brassica oleracea. Both species are included in the Brassicaceae, a family that comprises many crops economically important to human beings, such as mustard, cabbage, broccoli, and cauliflower.

The evolution of Canola is a prototypical example of allopolyploidy, in which a new species forms when the genomes of two distinct species are combined into a new species with two complete sets of chromosomes. Brassica rapa (turnip, field mustard) is diploid with 20 chromosomes (AA genome, $2n = 20$), while Brassica oleracea (cabbage, kale) is also diploid with 18 chromosomes (CC genome, $2n = 18$). This hybridization event generated a plant that consists of a mixture of both the parental genomes(AACC) and ultimately resulted in the formation of the Brassica napus species bearing 38 chromosomes and having a polyploid genome. This polyploidy resulted in increased genetic variation that served as a primary source of the adaptability and robustness of the plant.

The hybridization process might have occurred thousands of years ago, but through the hands of humans and selective breeding, this plant became the Canola we have today. The intercrossing of two different gene pools in Brassica napus allowed the expression of some desirable traits from either the parent species, one of the factors that contributed to adaptation to diverse environments and, more importantly, its utilization as a crop.

2.2 Role of Polyploidy in Canola's Genomic Complexity

Polyploidy, having more than two complete sets of chromosomes, plays a significant role as one of the success factors for Canola. The plant, being a polyploid one, shows a general vigor and adaptability and expresses a level of genetic diversity over its diploid relatives, a product largely of redundancy and flexibility afforded by multiple copies of genes. This redundancy allows a polyploid species to acquire mutations without severely debilitating critical functions, as the backup copy of the gene may still execute the original function. Such mutations over time result in new traits that make the plant better suited for survival under several environmental conditions.

Polyploidy has contributed to the genomic complexity that allows Canola to thrive in diverse ecological zones. Access to more extensive genetic material allows the plant to produce more oil, resist diseases, and be more resilient to abiotic stresses, including drought and salinity. In addition to that, it provides breeders with a wider range of opportunities for selection of useful traits since, despite what can be termed as an almost negligible percentage, there are still many duplicated genes in the canola genome, which most likely constituted a key element during the domestication and improvement of the crop.

2.3 Early Agricultural Uses and Domestication

These relatives of Canola were first cultivated for their oil-filled seeds, mainly as a source of lamp fuel and other industrial uses, in the 18th and 19th centuries. Brassica napus and its cousins were not suitable for human consumption in the first place because of their high content of erucic acid and glucosinolates. The former made the plant unappetizing for human consumption, while the latter caused it to be farmed for use in non-food applications.

Domesticating Canola began in the 20th Century but mainly in Canada. Its breeding focused on reducing levels of erucic acid and glucosinolates so that its oil became palatable for consumption. This selective breeding finally materialized the modern crop of Canola, where the content of erucic acid is less than 2% and nearly negligible amounts of glucosinolates. This transformation from rapeseed to Canola not only rendered the crop more marketable but also more beneficial. Thus, it became one of the staple oilseed crops cultivated worldwide.

Domestication in Canola occurred through artificial selection for desirable traits such as quality oil, higher yield, and disease resistance. The suitable polyploid genome assisted in greatly increasing breeding efforts because genetic variation was almost impossible to avoid, which would lead to significant improvements during a relatively short period. The basis for which breeders can draw select desirable traits and simultaneously but progressively eliminate undesirable ones led to the creation of the crop as edible, highly productive, and adaptable to different agricultural environments.

2.4 Hybridization and its Role in Genetic Diversity

The hybridization event that created Brassica napus laid the foundation for the genetic diversity underpinning Canola's success. Hybridization introduces new combinations of alleles, thereby allowing novel phenotypes to arise and increasing the overall genetic diversity of the plant. With Canola, for instance, the possibility of combining two separate species, BRASSICA RAPA, and BRASSICA OLERACEA, that could benefit by combining a large number of genes or improving desirable traits from both contributing species was available.

Of course, this genetic diversity is significant in the context of adaptation because it allows some population members to carry the alleles with beneficial effects of resistance to diseases, tolerance of environmental stressors, and so on. Hybridization has generated raw material in Canola that breeders have leveraged to develop varieties with increased resistance to pests and diseases, including abiotic stresses like drought and salinity. The potential to use that genetic diversity through selection breeding has been central to the crop's evolutionary success.

Hybridization is still also present in modern breeding programs. In recent years, beneficial traits from related species have already been incorporated into Canola through interspecific hybridization, further adding to its genetic diversity and adaptability. The current trend of developments in molecular genetics and genomic technologies alone will likely continue to influence the course of canola evolution in the years to come.

3. EVOLUTIONARY TIMELINE: HISTORICAL DEVELOPMENT OF CANOLA

3.1 Origins of Rapeseed Cultivation

Canola has its roots traced as far back as more than 4,000 years in the history of rapeseed. Original records of rapeseed growing were found in India, China, and Japan. The seeds are used for oil production and are essential to lamp oil production in those areas. Rapeseed in Europe The cultivation of rapeseed originated in the Middle Ages. The crop was initially farmed for industrial purposes, such as producing lubricating fluids, and later on, during the 18th and 19th centuries, there was increased demand for lubricants because more machinery spread and produced more.

Natural rapeseed oil was deemed unsuitable for human consumption due to its level of erucic acid, which was known to pose health risks. Moreover, rapeseed was known to have a significant number of glucosinolates, which resulted in a rather unpleasant bitterness and, thus, only used to a very restricted extent in animal feed. This subsequently restricted the use of rapeseed primarily to non-food purposes, and until the mid-20th Century, the plant did not receive much attention as a valued crop in agriculture.

3.2 Evolution of Canola in the 20th Century

The transformation from rapeseed to Canola has been one of the greatest success stories from an agricultural history perspective, and it took place at least as early as the first half of the 20th Century. The breakthrough in the development of Canola occurred when there was an urgent strain on the demand for lubricants, mainly for marine engines, during World War II. Given its stability and good boiling point, Rapeseed oil proved ideal for applying lubrication in steam engines and ships. These demands added to the further cultivation of rapeseed, particularly in Canada, which has become a major crop producer.

However, high erucic acid and glucosinolates made the oil unsuitable for human consumption and the meal unpalatable to animals. The researchers in Canada saw the potential of using rapeseed as a food oil and began selective breeding to eliminate these undesirables. They achieved it in the 1970s when the breeders produced low-erucic-acid rapeseed (LEAR), which had less than 2% erucic acid and had safety standards for human consumption.

The term "canola" was coined in 1978 and could easily be said to be a landmark in the crop's history. A word from "Canadian oil, low acid" reflects the crop's origin, along with its reduced levels of erucic acid. Besides its low levels of erucic acid, canola varieties bred since the 1970s also have reduced glucosinolate levels, making the meal suitable as livestock feed.

3.3 Impact of Hybrid Breeding and Genetic Engineering

Advances in plant breeding are closely associated with the development of Canola as a crop worldwide. Genetic engineering technology has recently provided new avenues for improving crop production. Traditional selective crossbreeding was an early breeding tool that introduced desirable traits while slowly eliminating undesirable ones. Still, selective breeding is very slow, so it becomes obvious that new approaches were necessary for accelerating the development of canola varieties with improved traits.

Most recently, during the 1980s and 1990s, hybrid breeding methods were devised so hybrid canola varieties could be bred that manifested heterosis or hybrid vigor. Hybrid varieties were discovered to have rather significant yield improvements, enhanced resistance to disease, and tolerance to stress in comparison with the traditional open-pollinated varieties. Hybrid breeding offered increased genetic diversity with the canola populations, thus enabling crop versatility across different environments.

A second major turning point in the history of Canola was the introduction of genetic engineering late in the 20th Century. Genetic engineering opened up the mechanism to introduce into the Canola genome-specific genes that control traits inaccessible through traditional breeding or otherwise simply impossible to accomplish. Probably the first of these successes in GM canola was the creation of herbicide-resistant varieties that allowed farmers to control weed competition more reliably without reducing their crops. Such innovation comes with an increase in canola yields and reduces the amount of chemicals used in removing weeds, giving it a big boost toward more sustainable agriculture.

Today, almost all Canola coming from countries such as Canada, the United States, and Australia are genetically modified varieties of Canola. GM canola varieties have been engineered to include resistance to herbicides, high oil content, and disease resistance, making them stronger and more productive than their non-GM counterparts. These developments affected the world production of Canola as this crop became one of the world's most important oilseeds.

3.4 Expansion of Canola Cultivation Worldwide

Canola was a local crop that became a global product through late 20th-century agricultural production that spread from Canada to the U.S., Europe, Australia, and China. Canada is the leading producer, accounting for 20-25% of the world's production. By far, the other countries taking the lead are China and Australia.

Factors driving its growth:

- Better Crop traits: Genetic advancement increased productivity and disease resistance, making Canola more aggressive in various climates.
- Versatility: Canola oil, with high levels of omega-3 and low saturated fat, opened new vistas for cooking, while canola meal acts as a high protein animal feed.
- Sustainability: Canola is water- and nutrient-efficient, using herbicide- and disease-resistant varieties that cut chemical use.

These traits made it possible for Canola to penetrate drier climates, such as arid regions of the world, as scientists developed more resistance against drought and heat conditions with climate change.

3.5 Challenges and Opportunities in Modern Canola Breeding

Despite the success of Canola, there is still much to improve on this crop in terms of further enhancement in disease resistance, particularly to new pathogens and environmental changes. Blackleg and clubroot are among the significant diseases that impact the production of Canola. Hence, crop breeders remain tasked with developing varieties that, at any particular time, are resistant to these and other pathogens.

Apart from disease resistance, the most pressing need for developing canola varieties resilient to climate change is becoming urgent. Canola is more exposed to elevated temperature, drought, and other abiotic stresses as the climate becomes warmer and more unpredictable. Therefore, breeding programs are oriented more toward producing improved varieties with better stress tolerance to save the crop from climatic changes.

Advances in genomic tools, such as CRISPR-Cas9 gene-editing technology, offer huge and exciting potential for new approaches in canola breeding. The foundational component is the technology, which facilitates the precision modification of the canola genome in unprecedented ways that will introduce or enhance certain traits. Further application of gene editing is needed to improve varieties about increased nutrient content, disease resistance, and stress tolerance.

4. GENOMIC VARIATIONS AND ADAPTIVE EVOLUTION IN CANOLA

4.1 The Role of Polyploidy in Canola's Evolution

One of the critical genomic features of Canola (Brassica napus) is its polyploid. Canola is an amphidiploid derived through homologous hybridization between two diploid species, Brassica rapa and Brassica oleracea. The hybridization has been proved to have occurred approximately 7,500 years ago and resulted in the forming of a new species, with the set of combined chromosomes from the parental species; B. Napus has a chromosome complement of 38 (2n = 38, AACC) and therefore receives 19 chromosomes from one parent, B. rapa (AA) and 19 from the other, B. oleracea (CC).

Polyploidy has played an important role in the evolutionary success of Canola. The genetic diversity it created and elevated to more varied adaptation to distinct environments helped Canola inherit all the desirable traits from both parents, such as cold tolerance associated with B. rapa and pest resistance with B. oleracea. The genetic richness formed a platform for further evolution and selection, leading to the development of Canola, one of the world's most important crops.

Another consequence of polyploidization for the first generation of canola plants was genomic instability through homoeologous recombination, allowing the exchange of genes from the A and C genomes. This enabled some strongly significant evolutionary consequences and further produced new gene combinations that determined the adaptation and diversification of Canola. These evolutionary processes have been driven by time and selection pressure from a diverse range of environments acting upon genetic variations.

4.2 Genetic Variation for Abiotic Stress Tolerance

Abiotic stress tolerance represents one of the primary factors governing adaptive canola evolution, notably drought, salt, and heat stress. Consequently, genetic variation within the canola genome has allowed traits to be selected that would otherwise be impossible to survive under these stressful conditions.

The reason why drought tolerance for Canola may have been selected for genes related to water use efficiency or root architecture. For instance, varieties having a deeper root system are better equipped to tap water from the deeper soil layers; therefore, these varieties are more drought-resistant. Many QTLs that overlap with canola traits related to drought tolerance include those controlling root length and rate of water uptake, among others, like stomatal conductance (Slafer et al., 2020).

In addition to drought, tolerance to salinity has been another important area where Canola has made significant genomic adaptations. Salinity stress causes ionic imbalances and osmotic stress that affects growth in the case of plants; this has created a significant problem, mainly in regions whose soils possess poor quality. Genomic studies have identified several key genes related to ion transport and osmotic adjustment that have contributed to the evolution of salt-tolerant varieties of Canola (Hussain et al., 2019). They maintain ion homeostasis and cellular stability inside the plant due to the high salinity that increases its fitness in saline environments.

Temperature extremes, especially cold and heat stress, have significantly contributed to the evolution of Canola. The process's primary benefit is the cold tolerance that has allowed Canola to thrive in colder environments, particularly in northern regions of the continent, such as Canada. Studies through genetics research have elucidated several genes responsible for the plant's responses to cold. These respond through mechanisms like antifreeze proteins that regulate membrane fluidity, thus serving to safeguard the plant from damage caused by freezing (Wang et al., 2018).

On the other hand, stress generated by heat has favored the creation of heat-tolerant genes, especially in warm climates. Oxidative stress coupled with metabolic disruption can arise from the machinery involved in photosynthesis if this becomes impaired during heat shock. Adaptive evolution into Canola has seen adaptive gene selections that reduce the impacts of such stresses, including those encoding for heat shock proteins and antioxidant pathways (Li et al., 2021).

4.3 Biotic Stress Resistance: Coevolutionary Arms Race with Pathogens

The co-evolutionary arms race between Canola and its pathogens has been a major driving force of genetic variation and adaptation. Centuries of selective pressure from fungal, bacterial, and insect pests have been specifically engineered over time, especially blackleg disease caused by Leptosphaeria maculans. Once Canola's R-gene-based resistance may trigger defense responses, it has been challenged by better virulent pathogen strains over the years. Similarly, clubroot disease (Plasmodiophora brassicae) has also been a reason for the introduction of resistance genes from wild relatives to ensure crop sustainability. Insect pests include the diamondback moth and flea beetles, which have rendered Canola to evolve direct and indirect defense mechanisms.

4.4 Human-Mediated Selection: Breeding and Biotechnology

Although natural selection has been the driving force in the adaptive evolution of Canola, human-mediated selection, via the application of breeding and biotechnology, has accelerated this process and shaped the modern Canola genome. This transformation from the traditional rapeseed that occurred during the 20th Century was thus due to the process of directed selection on characteristics such as smaller amounts of erucic acid and reduced levels of glucosinolates.

Hybrid breeding was introduced to canola improvement in the 1980s, changing Canola into a breed that expresses heterosis or hybrid vigor. This led to the enhancement of yield, disease resistance, and tolerance to abiotic stress, which explains its popularity. Hybrid breeding has also increased the genetic diversity of Canola because breeders have been able to combine the favorable traits from different parental lines to make superior hybrid varieties (Beckie et al., 2011).

With the advent of genetic engineering, hybrid breeding followed in the footsteps of the genetic revolution that came about towards the end of the last Century. Perhaps the most significant innovation in this respect was the Canola resistance to herbicides, which allowed farmers to have better control over weeds without harming their crops. This was achieved with genes confer herbicide resistance to glyphosate and others, making it possible to use no-till farming approaches and reduce soil erosion and chemical inputs (Harker et al., 2012). Genetic engineering also allows the designing of Canola with high oil content, higher disease resistance, and better nutrient use efficiency.

Canola's improvement has had a new light at the end of the tunnel in the last few years with the innovation of CRISPR-Cas9 gene-editing technology. This is in contrast to the traditional method of genetic engineering, which is normally based on foreign gene use. This tool allows for the precise modification of the canola genome to enhance or knock out specific genes involved in important traits. This technology offers much potential in developing future canola breeds; it will allow more efficiency and directed improvements (Sedeek et al., 2019).

Table 1. Significant events and breakthroughs in the history of canola breeding and cultivation.

Ancient Times	**Early Cultivation**	**Cultivation of Brassica rapa and Brassica oleracea for seeds and leaves.**
Roman Era	Hybridization and Early Use	Formation of **Brassica napus** from **Brassica rapa** and **Brassica oleracea**; early use for oil.
1970s	Introduction of Low-Erucic Acid Varieties	Development of canola with reduced erucic acid and glucosinolates by Canadian scientists.
1980s	Expansion of Cultivation	Widespread adoption of canola in North America, Europe, and Australia.
1990s	Genetic Improvement and Biotech Innovations	Introduction of GM canola varieties with herbicide resistance and enhanced oil quality.
2000s	Focus on Sustainability and Disease Resistance	Development of canola with improved resistance to pests and diseases, and enhanced environmental adaptability.
2010s	Advances in Genomics and Precision Breeding	Utilization of genomic technologies for precise trait selection and breeding.
2020s	Integration of Climate Resilience and Nutritional Quality	Development of canola varieties with better drought tolerance, nutrient profiles, and reduced environmental impact.

5. GENOMIC TOOLS AND TECHNOLOGIES IN CANOLA BREEDING

5.1 Genome-Wide Association Studies (GWAS) in Canola

GWASs have become a powerful tool for identifying the genetic variation associated with complex traits in crops such as Canola and Brassica napus. GWAS exploits natural variation occurring across large populations by scanning the entire genome for association between a specific genetic marker and the expression of the phenotypic trait. This has been invaluable for dissecting the genetic architecture of important agronomic traits like yield, oil content, disease resistance, and stress tolerance.

GWAS was utilized intensively in the mapping of Quantitative Trait Loci in Canola. The QTLs represent areas of the genome associated with differences in a specific trait. Traits considered important in canola commercial value, such as seed size and oil content, have been targeted by researchers who have identified QTLs. These studies have shown that QTL controlling these traits tends to cluster in certain genome regions; thus, breeders can target this region of the chromosome during marker-assisted selection (Basunanda et al., 2010).

Apart from yield-related traits, GWAS has also been employed to identify disease-resistance loci in Canola. For instance, some large QTLs associated with blackleg disease resistance have been mapped on several chromosomes, primarily in regions harboring known R genes. The discovery of these loci has led to the development of blackleg-resistant canola cultivars, the variance necessary for maintaining crop productivity in disease-prone areas (Delourme et al., 2014).

Another important application of GWAS in Canola is the study of abiotic stress tolerance. Drought and salinity are key stresses that generally limit the production of Canola, especially in areas characterized by poor water availability or saline soils. Through GWAS, researchers have identified various loci associated with traits such as root architecture, osmotic regulation, and stomatal conductance, among other factors contributing to the enhanced ability of a crop to use water more efficiently and tolerate salt conditions (Zou et al., 2016).

The capability of GWAS to manage high data with thousands of markers makes it specifically powerful in crops like Canola, where complex traits are often controlled by multiple genes with small individual effects. Coupled with high-throughput phenotyping and sequencing, the above technology has accelerated breeding in Canola to a great extent and provided an elaborative map of the genetic basis of important traits.

5.2 QTL Mapping and Marker-Assisted Selection

In contrast to this, GWAS fundamentally focuses on the natural variation in the population. In contrast, QTL mapping narrows down the approach to use controlled crosses between parental lines for identifying regions of the genome that are linked to specific traits of interest. The QTL approach has contributed significantly to canola breeding; yield, oil quality, and resistance to various diseases have been enhanced.

The QTL mapping has elucidated loci responsible for the genetic variation of Canola regarding its erucic acid content, which contributes to a specific fatty acid. Erucic acid may prove dangerous when ingested over time because it is known to cause harmful effects on the human body. The capacity to produce low quantities of erucic acid in Canola was a significant determinant of the development of low-erucic acid varieties used on an industrial level in crop production today.

QTL mapping has also assisted in selecting Canola for improved glucosinolate content. Glucosinolates are credited with plant defense but impart bitterness and decrease the nutritional quality of canola meal when used as animal feed. Through the QTL analysis, the loci associated with lower glucosinolate contents have been identified, hence the production of canola meal that is more palatable and nutritious (Rahman et al., 2018).

One of the major applications of QTL mapping to Canola was improving disease resistance. For example, this is also one of the debilitating diseases caused by fungi, such as Sclerotinia stem rot. The QTL confers partial resistance, which, in combination with other resistance genes, significantly protects the disease (Wei et al., 2014).

Marker-Assisted Selection (MAS) is a breeding technique that accelerates the selection process by using DNA-based markers linked to desirable traits. MAS has gained popularity in canola breeding and introgressing donor lines with herbicide resistance, disease resistance, and improved oil quality to elite breeding lines. The strategy has often helped pyramid several resistance genes into one variety, increasing their overall performance and resilience (Snowdon et al., 2007).

MAS has significantly reduced the breeding time scale and associated cost because it allows the breeder to select desirable traits at the seedling stage, often many generations before the plants express their final phenotypes. This has helped accelerate the development of high-performing canola cultivars suited to specific environmental conditions and market requirements.

5.3 Genomic Selection: The Revolution in Canola Breeding

GS is a new frontier in plant breeding that may one day revolutionize the development of crops such as Canola. This approach differs from traditional breeding methods, which often rely upon phenotypic selection or on a few markers associated with QTLs by using genome-wide marker data to predict the genetic value of individuals in a breeding population.

The key advantage of GS is that it can capture the effects of tens of thousands of small genetic variants distributed over the entire genome. As such, it is highly effective for complex traits like yield, influenced by many genes with small effect sizes. GS can predict the performance of untested lines and, as such, support breeders in selecting lines with the highest potential for further development by training prediction models on a reference population with known genotypes and phenotypes (Crossa et al., 2017).

GS has successfully been applied in canola breeding programs to speed up variety selection in high-yielding lines. Research indicates that GS reduces the breeding cycle by several years since it does not depend on several generations of phenotypic selection. Such traits as yield, resistance to disease, and oil content involve a lot of time in canola breeding. Hence, this finding is valuable in Canola (Heffner et al., 2009).

Another application of GS in Canola is the enhancement of abiotic stress tolerance. By combining genomic information with environmental data, GS models predict how different lines of Canola will perform well under various stress conditions, such as drought or high temperature. This allows breeders to identify lines that are more likely to perform better in adverse environments, thus enhancing crop resilience to climate change (Jarquín et al., 2014).

Improvements in high-throughput and phenotyping technologies have further developed GS in canola breeding. Rapid genotyping of large populations and collection of detailed phenotypic data allows for more accurate prediction models, increased efficiency, and precision in the breeding process.

5.4 Transcriptomics, Metabolomics, and Proteomics in Canola

In addition to genomic tools, -omics technologies such as transcriptomics, metabolomics, and proteomics have shown much insight into the molecular mechanisms that govern the important traits in Canola. These analyses allow scientists to study patterns of gene expression, metabolite profiles, and protein interactions on a global level that defines how distinct genes and pathways are involved in the growth, development, and stress response of Canola.

Transcriptomics studies the totality of RNA transcripts produced by the genome under various conditions. Transcriptomic studies on Canola have already isolated genes that are highly involved in determining stress response; for example, they are upregulated after drought, cold, or heat stress. These will include transcription factors, signaling molecules, and enzymes that characterize pathways in stress response; for example, they can serve as a source of useful breeding targets (Zhang et al., 2017).

Metabolomics has been defined as the profiling of small molecules or metabolites engaging in various biochemical processes. The identified metabolites were useful in metabolomic studies that explained the metabolic pathways controlling oil content and composition in Canola. Such results have been valuable for breeding for improved oil quality, including varieties with increased oleic acid content and decreased saturated fats (Yang et al., 2018).

Proteomics is the study of all proteins encoded by the genome. Proteomic analysis on Canola has identified defense-related proteins against pathogens and abiotic stresses, including heat shock proteins, antioxidant enzymes, and pathogenesis-related proteins. These proteins have a significant role in defense against the environment and pathogens and, therefore, are vital targets for enhancing resistance in Canola (Hu et al., 2015).

Integrating this heterogeneous data from different omics platforms will shed much deeper insight into molecular networks that govern important traits in Canola. A systems-level approach is therefore essential to identify new targets for breeding and biotechnology and understand the interplay between genetic and environmental interactions in shaping the phenotype of the plant.

6. IMPACT OF NATURAL AND ARTIFICIAL SELECTION ON CANOLA GENOMIC VARIATION

The breeding and domestication of Canola (Brassica napus) have been driven by natural and artificial selection. These events have formed the phenotypic and genotypic diversity in Canola. To unravel how Canola has evolved, it is critical to understand these selection forces and utilize that knowledge in future breeding programs.

6.1 Natural Selection: Environmental Pressures and Evolutionary Adaptations

Generally, natural selection would be the primary mechanism through which plants, including Canola, adapt to the conditions of their environments. Over millennia, wild progenitors of domesticated Canola, such as Brassica rapa and Brassica oleracea, evolved in response to climate, soil, and resource availability. These pressures resulted in genetic diversity that favored survival and reproduction given particular conditions and supplied the genetic diversity required for adaptation.

Natural selection would have molded traits like drought tolerance, cold resistance in Canola, and resistance to pests and pathogens. Thus, adaptation to temperate climates in northern Europe would have selected the variants of the B genes, which enhance the extent of tolerance to cold stress. Several cold-responsive genes were identified in B. napus, which is significant in its ability to survive low temperatures; it's one of the key factors to be considered in its cultivation in regions with cold winters (Zhao et al., 2018). It regulates different physiological activities such as cold acclimation, osmotic adjustment, and membrane stabilization, which help the plant survive frost and freezing conditions.

Natural selection also acted on characters that would allow early wild Brassica species to tolerate drought stress, especially when water was already scarce. Genetic variation for water use efficiency, root architecture, and stomatal conductance likely imparted survival benefits under water-limited conditions in some populations. These features have been retained in the gene pool of the cultivated Canola. Since these features exist, breeders can select the drought-resistant varieties most appropriate for dry or semi-arid regions (Guo et al., 2016).

6.2 Artificial Selection: Human Influence on Canola Genetic Variation

Because these diverse genetic materials were established through natural selection, artificial selection is a significant factor in the current form of the canola genome. Artificial selection is the process of deliberate breeding and selection for desirable traits in plants by humans. Thousands of years ago, this process started when human beings began to domesticate wild plants, and in the last Century, it has emerged into a strong form with modern plant breeding techniques.

This is how the wild rapeseed became what we now have as Canola. It constitutes a very good example of the strength of artificial selection in breeding. It was used to make oil for industrial use, not for food, as it had high levels of erucic acid, which was dangerous to humans if ingested. Between the 1960s and the 1970s, traditional breeding by Canadian plant breeders facilitated the production of low-erucic acid varieties. This led to the development of Canola, a crop whose oil is suitable for cooking and food products (Stefansson & Hougen, 1975). This marked the beginning of the industrial crop's evolution from an industrial crop to a highly important crop globally in oilseed.

Another primary concentration of artificial selection has been glucosinolate content in canola meal reduction. Glucosinolates are natural substances useful to plant defense functions but anti-nutritional at high concentrations in animal feeding diets. Success was achieved in reducing the content for canola meal breeding, thereby providing this source of protein to animals. This was achieved by identifying and selecting genetic variants controlling glucosinolate biosynthesis. The outcome was the production of "double-low" canola varieties with low levels of erucic acid and glucosinolates (Rahman et al., 2018).

6.3 Hybrid Breeding and Heterosis in Canola

One of the most important successes of artificial selection in Canola is hybrid variety development based on heterosis or hybrid vigor. Hybrid breeding entails crossing two genetically different lines so that offspring F1 hybrids would be produced with a performance above that of either parent. Canola hybrids have significantly improved yield, disease resistance, and tolerance to other stresses due to the genetic diversity generated by crossing different lines.

Hybrid Canola has been developed using CMS systems. CMS is the natural condition of some plants where the male reproductive organs, pollen, are sterile and cannot lead to self-pollination. Male-sterile lines are crossed with fertile lines by breeders to produce hybrid seeds. Plants resulting from such crosses always possess heterosis, high vigor, and yield compared to either of their parents. Today, hybrid varieties occupy the biggest spot for Canola in many parts of the world, particularly in Europe and North America.

Heterosis has also been exploited to enhance the tolerance of Canola to biotic and abiotic stresses. For example, hybrids that are tolerant of severe diseases like blackleg and clubroot have been developed through allele selection from various parent lines conferring resistance. Similarly, hybrid varieties that are drought- or heat-stress tolerant have been developed through crosses of lines with complementary stress response traits (Snowdon et al., 2007).

6.4 The Role of Biotechnology in Artificial Selection

Artificial selection in Canola entered a new period with biotechnology and genetic engineering. The ability to directly manipulate the genome of Canola through GM has opened new avenues for character improvement, which could not be made very easily or were slow to attain by traditional breeding. Herbicide-resistant varieties of GM canola are one of the best examples of this kind of manipulation.

The first commercial production of herbicide-resistant Canola started in the 1990s. At the same time, the release of this crop variety has since transformed the cultivation of Canola as it is now possible to exercise management over weeds. GM varieties of Canola are genetically manipulated through genes that make them resistant to glyphosate or Roundup and other herbicides; therefore, these varieties are tolerant of herbicide applications that kill weeds around them but do no harm to the crop. This technology has increased yields and decreased mechanical weed control, which sometimes involves heavy manual labor and can potentially be environmentally damaging (Beckie et al., 2006).

However, biotechnology has allowed oil profiles in canola varieties to advance. For example, genetic engineering has been used to modify the fatty acid composition of canola oil so that there are elevated levels of oleic acid and lower levels of saturated fats. The food industry is keen on such health-conscious varieties of canola oil, and biotechnology has filled this market need (Green, 2007).

The use of CRISPR-Cas9 genome-editing technology is emerging to improve Canola. Unlike traditional GM methods based on the introduction of foreign DNA, CRISPR allows for targeted alteration of the plant's genome with precise modifications. This tool has been harnessed to develop enhanced canola plants with improved traits such as disease resistance, oil quality, and abiotic stress tolerance. These traits do not pose such regulatory challenges as those associated with transgenic crops (Kaur et al., 2018).

6.5 Genetical Bottlenecks and Diversity in Canola

Perhaps an unintended byproduct of artificial selection is the loss of genetic diversity or bottleneck. Selectors who wish to strengthen one aspect may well de-prioritize parts of the genome, lowering total diversity. This might make crops more sensitive to new diseases or environmental changes since there's less variation to provide a foundation for response.

Two genetic bottlenecks have been involved in its domestication and breeding history. For instance, the transformation of wild rapeseed into low-erucic acid canola was strictly selectively continued to narrow the identified genetic variants. Increasing hybrid varieties is an advantage regarding yields and performance, though discussions have been done about losing diversity within the crop (Snowdon et al., 2007).

The above bottlenecks are now alleviated with breeders increasingly using genetic resources such as wild relatives and landraces of Brassica species. These possess vast amounts of undiscovered or untapped genetic diversity that can now be used to introduce lost traits into Canola and make it more adaptive. Genomic selection and marker-assisted breeding have also been added to modern breeding programs to ensure the selection balance across the genome with minimal risk of losing valuable genetic variation.

6.6 Conclusion: The Double Edge of Selection in the Canola Story

The development of Canola has been influenced by natural as well as artificial selection. Natural selection led to the adaptation of Brassica species towards different environments and laid down the genetic basis for modern canola development. Artificial selection was further extended through traditional breeding and modern biotechnology when human needs transformed it into one of the world's most important crops for oilseed.

Indeed, while artificial selection has brought additional benefits in the form of higher-yielding lines with better-quality oil and enhanced stress tolerance, it has, at the same time, set a level of alarm for loss of genetic diversity and sustainability. In the future, therefore, breeders will have to produce qualities demanded at production to continue maintaining or enhancing genetic diversity. By using new genomic tools and conserving wild relatives' genetic resources, future canola breeding trends promise to be both innovative and sustainable.

7. MOLECULAR TOOLS AND GENOMIC INNOVATIONS

Molecular tools and genomic technologies have advanced manifolds to revolutionize plant breeding, especially in Canola and Br. napus crops. It thus has enabled unprecedented precision and efficiency in understanding, manipulating, and improving the canola genome. This chapter outlines the key molecular tools and genomic advances that have driven improvements in canola breeding and adaptation. We discuss MAS, GS, GWAS, and most recently, CRISPR-Cas9 and other gene-editing technologies.

7.1 Marker-Assisted Selection (MAS)

Marker-assisted selection (MAS), developed in the 1990s, enhances traditional breeding by using molecular markers linked to specific genes controlling desired traits, allowing quicker, more precise selection. Unlike traditional breeding, which relies on observable traits affected by environmental factors, MAS operates at the molecular level, speeding up the process and improving accuracy.

MAS has been crucial in canola breeding, particularly in boosting resistance to blackleg disease caused by *Leptosphaeria maculans* by introducing resistance genes Rlm1 and Rlm2 into new varieties. This has prevented massive crop losses. MAS has also improved canola oil quality by enhancing its fatty acid composition, increasing oleic acid, and reducing saturated fats, resulting in a healthier oil that meets consumer demand while maintaining high yield and agronomic performance.

7.2 Genomic Selection (GS)

Improving individual traits with MAS is relatively easier, but it is very challenging for a complex, polygenic trait determined by multifactorial genes and their interaction. That is where the genomic selection comes in. Genomic selection refers to applying genome-wide markers to estimate the breeding value of an individual, thereby allowing breeders to select complex traits without having to identify specific markers for each gene involved.

GS has also found high utility in improving quantitative traits like yield, drought tolerance, and disease resistance, where many genes impact a complex phenotype. High-yielding, stress-tolerant canola varieties have been developed using GS to enhance breeding efficiency for these complex phenotypes. This will be achieved by combining genomic information with phenotypic information from field experiments, which will allow breeders to develop predictive models to estimate performance for untested individuals based on their genetic makeup and thereby cut down on time and costs of breeding programs.

The greatest strength of GS is capturing the genetic architecture of traits controlled by many small-effect loci. Indeed, this is of paramount importance for crops such as Canola, in which multiple interacting genes control oil content, flowering time, and disease resistance. Consequently, the GS-based selection can make better choices, allowing producers to develop agronomically superior canola varieties (Schrag et al., 2018).

7.3 Genome-Wide Association Studies (GWAS)

Genome-wide association studies (GWAS) have become a powerful genetic tool for identifying complex traits in plants. It involves scanning the large-population genomes for single-nucleotide polymorphisms (SNPs) associated with specific phenotypic traits. The SNPs can be correlated with variations of traits, and candidate genes that contribute to important agronomic traits may thus be identified.

GWAS was applied to Canola by selecting genes responsible for seed oil content, fatty acid composition, and disease resistance. For example, a study by Wu et al. (2016) revealed that the seed oil content of B. napus is associated with several SNPs-based GWAS thus, enlightening the genetic factors responsible for such an economically important feature and assists in breeding programs to create or manipulate the trait related to oil yield and quality in Canola.

GWAS has also enabled the discovery of genes conferring tolerance to abiotic stresses in crops such as Canola, such as drought and heat tolerance. Suppose genomic regions are identified where such a trait is located. In that case, this information can be used to develop molecular markers for MAS or GS to select stress-tolerant varieties. The major challenge to canola production under current climate change scenarios would be rising temperatures and changes in precipitation patterns (Song et al., 2016).

7.4 CRISPR-Cas9 and Gene Editing

The CRISPR-Cas9 technology has opened new avenues for editing plant genomes at their site to date. This enables DNA sequences within the organism to be manipulated precisely by introducing changes, proven to be a very strong tool for modulating specific genes or introducing novel traits. It varies largely from the traditional method of genetic modification using foreign DNA. This particular technology allows for targeted mutations or gene knockout instead of adding foreign genes into the plant genome.

For example, it was used to enhance the resistance of Canola to diseases and improve its oil quality and yield. In one of the first applications of CRISPR in low-glucosinolate varieties of Canola, genes encoding glucosinolate biosynthesis were successfully targeted for knockout. The low-glucosinolate meal was considered to be a breakthrough in canola breeding because it was relatively better suited to livestock feeding.

CRISPR has been used to increase the tolerance of Canola to herbicides by editing genes that metabolize herbicides. This is one step forward toward developing novel herbicide-resistant varieties that are not genetically modified with foreign DNA, thereby dodging most of the regulatory barriers and public rejection of GMOs (Huang et al., 2019).

Besides herbicide resistance, CRISPR has also been used to engineer canola crops to be more resilient to abiotic factors like drought stress and salinity. For example, scientists have developed canola varieties that can thrive under better conditions of water scarcity by regulating genes in the signaling pathways of stress responses. Although the improvement of productivity under such conditions remains very vital for crop productivity, what climate change informs the affected areas about is presented by Mishra et al., 2020.

7.5 Transcriptomics, Proteomics, and Metabolomics

In addition to genomic innovation, transcriptomics, proteomics, and metabolomics sciences have enlightened much more about the molecular underpinnings that control canola growth and development and responses to stresses. The "omics" technologies through these scientists made it possible to scrutinize methods of expression at the gene, protein, or metabolite levels under the aegis of environmental and developmental cues.

- Transcriptomics: Transcriptomic research has discovered key changes in gene expression, which are linked to abiotic stresses like drought, cold, or heat. Researchers found that canola transcriptomes under drought stress harbored major drought-responsive genes related to osmotic regulation, signaling, and detoxification. Such findings give valuable insight into valuable targets for genetic improvement or breeding programs that try to enhance drought tolerance.

- Proteomics: Proteomic research in proteins defense-related in Canola revealed proteins involved in biotic stress through resistance to pathogens and insects. Protection of resistance in susceptible lines of Canola is ensured through the discovery of defense-related proteins through protein profiling, thus conferring resistance to pathogenic blackleg and clubroot diseases.
- Metabolomics: Metabolomic studies have mainly focused on identifying the biochemical pathways responsible for significant traits like oil contents and flavor profile. Metabolite profiling during the development of the seed has helped scientists determine which compounds in canola oil impact its nutritional and industrial significance (Guo et al., 2020).

7.6 Big Data and Bioinformatics in Canola Breeding

The large-scale genomic, transcriptomic, proteomic, and metabolomics analyses produce tremendous amounts of data that need to be handled through bioinformatics tools for meaningful interpretation. Big data analytics is also crucial in canola breeding because it may highlight patterns that other approaches fail to detect.

Bioinformatics tools, like the Brassica database called BRAD, are helping connect genomics and marker data for breeders to recognize candidate genes and develop molecular markers (Cheng et al., 2011).

Machine learning and AI strengthen genomic selection models by using large genotype-phenotype datasets to predict combinations of alleles that impact canola performance in diverse environments (Crossa et al., 2017).

7.7. Future Prospects: Integration of Molecular Tools and Genomic Innovations

Canola breeding prospects lie in integrating molecular tools and genomic innovations to develop refined varieties suited to specific needs. Key areas include:

1. **Precision Breeding**: Using CRISPR-Cas9 and gene-editing tools for targeted genomic changes can improve traits like yield, stress, and disease tolerance as gene functions become clearer through -omics technologies.
2. **Synthetic Biology**: Advances in synthetic biology can modify metabolic pathways to produce healthier canola oil or create industrial oils with specific properties.

3. **Multi-Omics Integration**: Combining genomic, transcriptomic, proteomic, and metabolomic data will enhance breeding decisions by predicting how DNA changes affect traits, enabling the development of better-adapted varieties.
4. **Climate Resilient Varieties**: As climate change affects agriculture, tools like CRISPR and marker-assisted selection will help breed Canola that withstand extreme weather conditions.
5. **Sustainability and Biodiversity**: Genetic diversity in breeding programs will be enhanced by incorporating wild relatives. Genome sequencing and GWAS will help identify valuable alleles for developing resilient, sustainable varieties.
6. **Gene Drive Technologies**: Gene drives could rapidly introduce beneficial traits like herbicide resistance but raise ethical and environmental concerns, requiring careful regulation.

8. ADAPTIVE EVOLUTION IN RESPONSE TO CHANGING ENVIRONMENTS

The adaptive evolution process facilitates the acquisition of proper traits that will enhance survival and reproduction in an environment where this is subject to environmental change. Improved tolerance to stress, yields, and nutrition in oilseed Brassica napus (Canola) are some examples of adaptive evolution. It can no longer avoid a newly changed climate, new pest movements, and variations in soils through natural as well as artificial selection.

8.1 Evolutionary History of Canola

Natural hybridization over millions of years has linked Canola's history with that of its wild relatives, Brassica rapa and B. oleracea. Breeding programs in the 1970s focused on developing low-erucic acid varieties and set up modern Canola. However, breeding remains focused on higher yields, pest resistance, and stress tolerance and has greatly been made easier by developments in genomics (Snowdon et al., 2007; Franzke et al., 2011).

8.2 Environmental Pressures and Genomic Variation

8.2.1 Stress of Climate and Temperature

Temperature stress, particularly heat and cold, has a significant impact on Canola. Through genes, including cold- and heat-tolerant strains like the heat shock proteins and cold-responsive genes, the breeders have succeeded in developing varieties (Zhou et al., 2018). The survival capabilities of this crop, under extreme temperatures, are boosted through breeding techniques like flowering time plasticity and antifreeze proteins (Aksouh-Harradj et al., 2006).

8.2.2 Drought Resistance and Water Availability

As the drought conditions become more rigorous, Canola developed drought-tolerant mechanisms such as modification of root structures and stomatal conductance to prevent water loss. Genes regulating ABA increased the rate of drought tolerance (Tardieu & Tuberosa, 2010).

8.2.3 Soil and Nutrient Stress:

To adapt to nutrient-poor and saline soils, Canola modifies its root morphology and nutrient transport mechanisms that increase its survival capability under sub-optimal conditions (Rose et al., 2012; Shabala & Cuin, 2008).

8.3 Adaptive Evolution Against Biotic Stresses

8.3.1 Disease Resistance

Resistance against blackleg and clubroot diseases is achieved by taking up resistance genes, such as Rlm1 and Crr1, which have been very crucial in breeding programs (Delourme et al., 2011; Rahman et al., 2014).

8.3.2 Insect Resistance

Natural defenses such as glucosinolates protect Canola from insects, but breeding aims for a balance between insect resistance and oil quality. New RNA interference techniques enable resistance without sacrificing quality (Hopkins et al., 2009; Abdullah et al., 2019).

8.3.3 Weed Competition and Herbicide Resistance

Herbicide-resistant Canola varieties, like glyphosate-resistant types, help in weed management. But herbicide-resistant weeds have started threatening sustainable farming. Breeders have another new challenge to eliminate it in novel ways (Green, 2018; Nandula, 2019).

8.4 Future Prospect for Adaptive Evolution of Canola

8.4.1 Climate Change and Adaptation

Climate-resilient varieties developed using CRISPR-Cas9 and genomic selection will be used in future breeding, helping Canola survive in increasingly stressful environmental conditions (Hickey et al., 2019). Carbon sequestration and nutrient use efficiency will also increase as more important traits in the future (Lobell et al., 2014).

8.4.2 Biotic Stress Tolerance

The strength of resistance to pathogens and pests will be increased by combining several resistance genes. RNAi-based technologies might silence necessary genes in pests, opening up a new front line of defense (Delourme et al., 2013; Abdullah et al., 2019).

8.4.3 Genomic diversity and adaptive potential

Continual maintenance and enhancement of genetic diversity within breeding programs will facilitate the adaptation of Canola to upcoming environmental and biotic changes. Acceleration in breeding can be achieved through machine learning or genomic prediction to identify the most beneficial genetic traits (Crossa et al., 2017; Franzke et al.).

Table 2. Summary of developmental stages, environmental stresses, adaptive traits, and key genomic variations in canola.

Developmental Stage	Environmental Stress	Adaptive Traits	Key Genomic Variations
Germination	Cold Stress	Improved seedling emergence	Variants in cold-responsive genes
Vegetative Growth	Heat Stress	Enhanced heat shock protein production	Variants in heat shock protein genes
Flowering	Water Stress	Adjusted flowering time and pod set	Variants in drought-responsive genes
Seed Development	Drought Stress	Increased water use efficiency	Variants in genes related to water uptake

9. CONCLUSION

The evolutions of Canola, Brassica napus, involved complex genetic variances and adaptive responses, both natural and artificial. Over the past few decades, breeders have been tapping into the potential of the canola genome to evolve high-yielding varieties with better oil quality and greater resistance to environmental stresses and diseases. The last chapter followed Canola's historical evolution as a crop, the molecular mechanisms underlying these adaptations, and the current breeding tools that are revolutionizing its production.

It has already been established that natural selection played a critical role in the evolution of Canola and its ability to occupy different types of ecological niches. Domestication's early stages usually started development from these versions, which were naturally occurring, enhancing features like resistance to diseases, cold tolerance, and quality of oil. This was to persist with artificial selection once it became practicable, and breeders could emphasize selected traits in the improvement of crops.

Analysis of genomic variations in Canola revealed its evolutionary history and path from genetic variation to QTLs, SNP markers, and key regulatory genes; thus, the expression of traits related to yield, stress tolerance, and biochemical composition were tailored. The development and adaptability of Canola were followed by greatly enhanced knowledge about molecular pathways through genomics, transcriptomics, proteomics, and metabolomics.

Another factor that has driven the evolution of Canola is its ability to adapt to biotic and abiotic stresses. Development and adaptation to emerging pathogens, pests, and climate change-related stressors will require resilience in future Canola varieties, especially as climate change is expected to shift global agricultural conditions.

The future of canola breeding stands at the crossroads of integrating molecular innovations with traditional practices, a process that makes it possible to produce this vital crop in a sustainable way. Climate resilience, yield, and quality will continue to be areas of key interest in the future, but subsequent efforts for breeding should target varieties with the potential to fight many battles while answering the growing calls from world food and biofuel perspectives.

But do Canola's genomic changes really represent the dynamic interplay of natural genetic variation and human-led breeding efforts? Moreover, new genomic technologies linked with a greater understanding of adaptive evolution will ensure continued advances in sustainable, high-performing canola varieties. Lessons learned in Canola's past provide a working blueprint for the breeding strategies that will continue to contribute Canola to agriculture globally.

REFERENCES

Abdullah, S., Majeed, S., & Nisar, M. (2019). RNA interference technology: A new approach in plant pest control. *Journal of Agricultural Biotechnology*, *25*(4), 672–682.

Aksouh-Harradj, N., Campbell, L., & Mailer, R. J. (2006). Flowering time plasticity and cold tolerance in Brassica napus. *Australian Journal of Agricultural Research*, *57*(5), 479–485.

Basunanda, P., Radoev, M., & Ecke, W. (2010). Comparative mapping of quantitative trait loci involved in oil content in Brassica napus. *Theoretical and Applied Genetics*, *120*(2), 271–281. DOI: 10.1007/s00122-009-1133-z PMID: 19707740

Becker, H. C., & Engqvist, G. M. (2011). Heterosis and hybrid breeding in Brassica napus. *Plant Breeding*, *130*(4), 421–431.

Beckie, H. J., Harker, K. N., & Hall, L. M. (2006). Impact of herbicide-tolerant crops on weed management: A case study of canola. *Weed Science*, *54*(4), 793–800.

Beckie, H. J., Harker, K. N., & Légère, A. (2011). Impact of herbicide-resistant canola on weed management and crop rotation. *Weed Technology*, *25*(3), 468–476.

Cheng, F., Liu, S., & Wu, J. (2011). BRAD, the Brassica database: A resource for Brassica genomics. *Nucleic Acids Research*, *39*(Database issue), D681–D687.

Crossa, J., Pérez, P., & Hickey, J. (2017). Genomic selection in plant breeding: Methods, models, and perspectives. *Trends in Plant Science*, *22*(11), 961–975. DOI: 10.1016/j.tplants.2017.08.011 PMID: 28965742

Delourme, R., Brun, H., & Renard, M. (2011). Genetics and molecular breeding in brassicas: From classical breeding to genomics. *Plant Breeding Reviews*, *35*, 123–149.

Delourme, R., Chevre, A. M., & Brun, H. (2013). Strategies for durable resistance to Leptosphaeria maculans in Brassica napus. *European Journal of Plant Pathology*, *137*(1), 1–15.

Delourme, R., Falentin, C., & Fopa Fomeju, B. (2014). Genetic control of blackleg resistance in Brassica napus: Identification of QTLs and candidate genes. *Plant Pathology*, *63*(5), 1035–1045.

Franzke, A., Lysak, M. A., & Al-Shehbaz, I. A. (2011). Phylogeny and evolutionary history of the Brassicaceae. *Plant Systematics and Evolution*, *291*(1-2), 75–87.

Friedt, W., & Snowdon, R. (2009). Oilseed rape: Genetics, genomics, and breeding. In *Biotechnology in Agriculture and Forestry* (pp. 123–156). Springer.

Green, A. G. (2007). Genetic modification of canola oil for improved nutritional profiles. *Journal of the American Oil Chemists' Society*, *84*(12), 1105–1112.

Green, J. M. (2018). The rise and future of glyphosate and glyphosate-resistant crops. *Pest Management Science*, *74*(5), 1035–1039. DOI: 10.1002/ps.4462 PMID: 27758090

Guo, L., Wang, X., & Zhao, Y. (2016). Genetic variation for drought tolerance in Brassica napus: Insights from root architecture and water use efficiency. *Frontiers in Plant Science*, *7*, 1253.

Guo, Y., Liu, J., & Zhang, L. (2020). Metabolomic analysis of oil biosynthesis pathways in Brassica napus seeds. *Metabolomics*, *16*(5), 62. PMID: 32335734

Harker, K. N., O'Donovan, J. T., & Blackshaw, R. E. (2012). Reduced herbicide inputs in herbicide-tolerant canola (Brassica napus) systems. *Weed Science*, *60*(2), 228–235.

Heffner, E. L., Sorrells, M. E., & Jannink, J. L. (2009). Genomic selection for crop improvement. *Crop Science*, *49*(1), 1–12. DOI: 10.2135/cropsci2008.08.0512

Hickey, L. T., Hafeez, A. N., & Robinson, H. (2019). Breeding crops to feed 10 billion. *Nature Biotechnology*, *37*(9), 744–754. DOI: 10.1038/s41587-019-0152-9 PMID: 31209375

Hopkins, R. J., Damman, H., & Robinson, J. (2009). Glucosinolate-biosynthesis genes and herbivory resistance in Brassicaceae plants. *Journal of Chemical Ecology*, *35*(1), 1–10. PMID: 19052820

Hu, X., Li, W., & Chen, Q. (2015). Proteomic analysis of defense responses in Brassica napus under biotic and abiotic stress. *Journal of Proteomics*, *127*, 216–226.

Huang, S., Liu, Z., & Li, C. (2019). CRISPR-Cas9 mediated herbicide resistance in Brassica napus. *Plant Biotechnology Journal*, *17*(12), 2253–2264.

Hussain, S., Iqbal, M., & Cheema, N. M. (2019). Molecular and physiological mechanisms of salt tolerance in Brassica napus. *Plant Physiology and Biochemistry*, *142*, 401–411.

Jarquín, D., Crossa, J., & Lacaze, X. (2014). Genomic prediction of quantitative traits in plant breeding using GBLUP models. *BMC Genomics*, *15*, 937.

Kaur, N., Alok, A., & Kumar, P. (2018). CRISPR-Cas9 mediated genome editing in Brassica napus for enhanced agronomic traits. *Plant Molecular Biology Reporter*, *36*(5), 729–740.

Li, H., Liu, S., & Fan, Y. (2021). Heat stress responses in Brassica napus: Identification of heat shock protein genes and antioxidant pathways. *Plant Science*, *302*, 110687.

Mishra, R., Joshi, R. K., & Zhao, K. (2020). Genome editing for abiotic stress tolerance in Brassica napus using CRISPR-Cas9. *Frontiers in Plant Science*, *11*, 563.

Nandula, V. K. (2019). Herbicide resistance: Mechanisms and management in crops. *Weed Science*, *67*(6), 559–572.

Rahman, H., Kebede, B., & Snowdon, R. (2014). Gene pyramiding in Canola: Strategies for resistance breeding and genomics. *Frontiers in Plant Science*, *5*, 219.

Rahman, M., Hoque, A., & Sheikh, M. (2018). Genetic control of glucosinolate content in Brassica napus: QTL mapping and marker-assisted selection. *Journal of Agricultural Science*, *156*(6), 759–768.

Rose, T. J., Damon, P., & Rengel, Z. (2008). Phosphorus-efficient canola: Root morphology and nutrient acquisition. *Plant and Soil*, *380*(1-2), 115–125.

Schrag, T. A., Westhues, M., & Schipprack, W. (2018). Genomic selection in hybrid breeding: Insights from canola and maize. *Theoretical and Applied Genetics*, *131*(6), 1285–1299.

Sedeek, K. E. M., Mahas, A., & Mahfouz, M. (2019). CRISPR-based genome editing in plants: A new era for crop improvement. *Plant Biotechnology Journal*, *17*(8), 1583–1597.

Shabala, S., & Cuin, T. A. (2008). Potassium transport and plant salt tolerance. *Physiologia Plantarum*, *133*(4), 651–669. DOI: 10.1111/j.1399-3054.2007.01008.x PMID: 18724408

Slafer, G. A., Savin, R., & Sadras, V. O. (2020). Coarse and fine regulation of wheat yield components in response to genotype and environment. *Field Crops Research*, *249*, 107716.

Snowdon, R. J., Lühs, W., & Friedt, W. (2007). Oilseed rape genomics. *Plant Breeding Reviews*, *27*, 19–54.

Song, G., Li, X., & Wang, H. (2016). Identification of drought and heat tolerance loci in Brassica napus using GWAS. *Plant Biotechnology Journal*, *14*(8), 1788–1799.

Stefansson, B. R., & Hougen, F. W. (1975). Selection for low erucic acid content in rapeseed (Brassica napus L.). *Canadian Journal of Plant Science*, *55*(2), 443–447.

Tardieu, F., & Tuberosa, R. (2010). Dissection and modeling of abiotic stress tolerance in plants. *Current Opinion in Plant Biology*, *13*(2), 206–212. DOI: 10.1016/j.pbi.2009.12.012 PMID: 20097596

Thomashow, M. F. (2016). The Role of cold-stress proteins in plant tolerance to freezing and subfreezing temperatures. *Plant Physiology*, *141*(4), 793–803.

Wang, F., Guo, Z., & Li, H. (2018). Cold acclimation and freezing tolerance in Brassica napus: Insights from transcriptomic analysis. *Frontiers in Plant Science*, *9*, 1834.

Wei, L., Li, Y., & Wang, H. (2014). Identification of quantitative trait loci for resistance to Sclerotinia sclerotiorum in Brassica napus. *Crop Science*, *54*(4), 1442–1450.

Wu, J., Zhao, Q., & Liu, S. (2016). Genome-wide association study for seed oil content in Brassica napus. *BMC Plant Biology*, *16*, 144.

Yang, Q., Liu, J., & Zhang, C. (2018). Metabolomic profiling of seed oil biosynthesis in Brassica napus. *Journal of Agricultural and Food Chemistry*, *66*(10), 2482–2490.

Zaman, M. A., Ullah, F., & Khan, A. (2019). Prospects of drought and heat tolerance in Canola (Brassica napus L.). *Advances in Agronomy*, *153*, 121–178.

Zhang, H., Li, Y., & Wang, Z. (2017). Transcriptomic analysis of drought stress responses in Brassica napus. *Plant Molecular Biology*, *95*(3), 287–300.

Zhao, C., Zhang, Z., & Xie, S. (2018). Genetic basis of cold tolerance in Brassica napus: Identification of cold-responsive genes. *Plant Biotechnology Journal*, *16*(9), 1692–1703.

Zhou, G., & Bradbury, P. J. (2018). Genomic selection with machine learning. *PLOS Genetics*, *14*(11), e1007735. PMID: 30399141

Zhou, R., Yu, X., & Kjær, K. H. (2018). Genomic insights into temperature stress responses in Brassica napus. *Plant Physiology and Biochemistry*, *131*, 110–119.

Zou, J., Hu, D., & Liu, P. (2016). Genetic mapping of root architecture and drought tolerance traits in Brassica napus using GWAS. *Plant Breeding*, *135*(4), 466–474.

Chapter 2

The Main Quality and Functional Chemical Composition of Canola

Jyothsna Devi Kuchipudi
https://orcid.org/0009-0002-6500-6721
Jain University, India

P. Selvakumar
https://orcid.org/0000-0002-3650-4548
Department of Science and Humanities, Nehru Institute of Technology, India

Santosh Kumar Nathsharma
https://orcid.org/0000-0003-3935-2938
Stewart Science College, India

Manjunath T. C.
https://orcid.org/0000-0003-2545-9160
Rajarajeswari College of Engineering, India

ABSTRACT

Canola, a member of the Brassicaceae family, is a key agricultural crop renowned for its oil-rich seeds. Canada is the largest global producer, contributing about 70% of the world's supply. The plant thrives in temperate climates with well-drained soils and requires a growing season of 90 to 120 days. Canola is well-suited to the Canadian prairie regions, where it has become a staple in crop rotations due to its ability to enrich the soil and its relatively low susceptibility to pests and diseases. Moreover, canola cultivation has economic and environmental benefits. It supports thousands of jobs in farming, processing, and marketing, bolstering rural economies, particularly in Canada, which is the leading exporter of canola oil.Environmentally, canola is

DOI: 10.4018/979-8-3693-8603-3.ch002

a beneficial crop in crop rotations as it helps break pest cycles and improves soil health. Its oilseed cake, a byproduct of oil extraction, is used as high-protein animal feed, contributing to sustainable livestock production.In recent years, advancements in biotechnology have further enhanced canola's role in agriculture

INTRODUCTION TO CANOLA: HISTORY, PRODUCTION, AND IMPORTANCE

Canola, a member of the Brassicaceae family, is a key agricultural crop renowned for its oil-rich seeds. The term "canola" is derived from "Canadian oil, low acid," reflecting its origins and one of its primary characteristics—the low levels of erucic acid, which makes it suitable for human consumption. The story of canola begins in the 1970s when Canadian agricultural researchers developed it from the rapeseed plant, which had been used primarily for industrial purposes due to its high erucic acid content. The creation of canola was a breakthrough in agricultural science, resulting in a crop that was both economically viable and healthier for consumption. The production of canola has evolved significantly over the decades. Canada is the largest global producer, contributing about 70% of the world's supply. The plant thrives in temperate climates with well-drained soils and requires a growing season of 90 to 120 days. Canola is well-suited to the Canadian prairie regions, where it has become a staple in crop rotations due to its ability to enrich the soil and its relatively low susceptibility to pests and diseases. Moreover, canola cultivation has economic and environmental benefits. It supports thousands of jobs in farming, processing, and marketing, bolstering rural economies, particularly in Canada, which is the leading exporter of canola oil.Environmentally, canola is a beneficial crop in crop rotations as it helps break pest cycles and improves soil health. Its oilseed cake, a byproduct of oil extraction, is used as high-protein animal feed, contributing to sustainable livestock production.In recent years, advancements in biotechnology have further enhanced canola's role in agriculture. Genetic modifications have improved the plant's resistance to diseases and pests and have enhanced its oil content and quality. These developments have helped canola remain a crucial crop in the face of changing environmental conditions and evolving agricultural demands.Overall, canola's history, production, and significance underscore its integral role in modern agriculture and food systems. Its development from a niche industrial crop to a global staple illustrates the transformative impact of agricultural innovation and highlights the ongoing importance of sustainable practices in meeting global food and energy needs.The fatty acid profile of canola oil is central to its reputation as a high-quality, heart-healthy cooking oil. This profile is characterized by its balanced composition of various fatty acids, which not only contribute to its functional properties in cooking but also offer numerous health benefits. The composition of canola oil is predominantly unsaturated fats, making it one of the most heart-healthy oils available. It contains a high proportion of monounsaturated fatty acids (MUFA), specifically oleic acid, which comprises about 60% of its total fatty acid content.

ALA is essential for maintaining heart health as it helps reduce inflammation and improve overall cardiovascular function. Linoleic acid is crucial for various physiological functions, including maintaining skin health and supporting cell structure. However, it is important to balance omega-6 intake with omega-3 intake to maintain optimal health, as excessive omega-6 can promote inflammation if not counterbalanced by omega-3s.Canola oil contains only about 7% saturated fat, making it one of the lowest among cooking oils. This low saturated fat content further enhances its suitability for heart-healthy diets, aligning with dietary guidelines that recommend reducing saturated fat intake to lower cardiovascular risk.The balanced fatty acid profile of canola oil also contributes to its versatility in cooking and food preparation. Its high smoke point, around 400-450°F (204-232°C), makes it ideal for various cooking methods, including frying, sautéing, and baking.The stability of canola oil at high temperatures is due to its low level of polyunsaturated fats that are prone to oxidation, which means it is less likely to produce harmful compounds when heated compared to oils with higher PUFA content. This stability not only makes it a practical choice for everyday cooking but also helps preserve its nutritional qualities during preparation.Canola oil's fatty acid profile also plays a significant role in its use in processed foods. Its neutral, from margarine and salad dressings to snack foods and baked goods. The oil's favorable fatty acid composition helps improve the nutritional profile of these products, contributing to healthier dietary options for consumers. Its versatility in cooking and food production, combined with its stable properties and neutral flavor, makes it a valuable component of a balanced diet and a key player in modern nutritional practices.

PROTEIN STRUCTURE AND FUNCTIONALITY IN CANOLA

Protein structure and functionality in canola are pivotal aspects that underscore the crop's value not just as a source of oil but also as a contributor to animal feed and other applications. Canola seeds, despite their primary role in oil production, contain a notable quantity of protein, making them a significant source of plant-based protein (Ash, 2012). The protein content of canola seeds averages around 20-25%, a level that rivals or surpasses other oilseeds used for animal feed, such as soybeans. Understanding the structure and functionality of these proteins is essential for appreciating their nutritional and industrial value.Canola proteins are primarily composed of two classes: globulins and albumins. Globulins, which constitute the majority of the protein content, are storage proteins that provide essential amino acids to the developing plant embryo (Baranyk et al.,1999). The globulins in canola are further classified into cruciferins and napins. Cruciferins are large, multifunctional proteins that serve as a major storage protein in canola seeds. They are characterized by their high molecular weight and complex structure, which includes multiple subunits held together by disulfide bonds. Cruciferins play a crucial role in providing the necessary nutrients during seed germination and seedling development. Napins, on the other hand, are smaller, more soluble proteins that also contribute to the nutritional value of the seeds. Napins are known for their ability to form gels and interact with other proteins, which can influence the texture and functionality of canola protein in various applications.

The functional properties of canola proteins are closely linked to their structural characteristics. For instance, the solubility of these proteins in water is a critical factor influencing their use in food products and animal feed. The solubility of canola proteins is impacted by their amino acid composition and the conditions under which they are processed. Generally, the globulins in canola have moderate solubility in aqueous solutions, which affects their utilization in food and feed formulations. The solubility and functionality of these proteins can be modified through various processing techniques, such as heating, pH adjustment, and enzyme treatment, to enhance their performance in different applications.In food processing, canola proteins are valued for their functional properties, including emulsification, foaming, and gelling. These properties are crucial for creating desirable textures and stability in a range of food products. For example, the ability of canola proteins to form stable emulsions makes them useful in salad dressings, sauces, and other emulsified products. Similarly, their gelling properties are advantageous in creating meat substitutes and other textural foods (Batista et al.,2011). The functionality of canola proteins in these applications is influenced by their ability to interact with other food components, such as fats and carbohydrates, which can affect the overall quality and stability of the final product.The quality of canola meal as a feed ingredient is influenced by the structure and functionality of the proteins it contains, which can affect feed efficiency and animal performance.In summary, the structure and functionality of proteins in canola are integral to its role as a valuable agricultural product. The composition of globulins and albumins, their solubility, and their functional properties in food and feed applications highlight the importance of canola proteins beyond their oil content. Understanding these aspects enables the optimization of canola's use in various sectors, from food processing to animal nutrition, showcasing the versatility and value of this crucial crop.Phytochemicals in canola have garnered significant attention for their potential health benefits and bioactivity, contributing to the crop's overall value beyond its primary use as an oil source. Canola seeds, while predominantly known for their oil-rich content, also contain a range of bioactive compounds that possess various health-promoting properties. Understanding these phytochemicals is crucial for appreciating the full spectrum of benefits that canola offers to human health and nutrition.One of the primary groups of phytochemicals in canola are glucosinolates, which are sulfur-containing compounds found in many Brassicaceae family members (Assmann & Gotto, 2004). In canola, glucosinolates are present in notable amounts, although their concentration is lower compared to other Brassica crops like mustard and broccoli. Upon consumption, glucosinolates are broken down into bioactive products such as isothiocyanates, which have been studied for their potential anticancer properties. Isothiocyanates are known to exhibit anti-inflammatory and antioxidant effects, which may help reduce the risk of chronic diseases, including cancer. Additionally, glucosinolates can influence the gut microbiota, potentially contributing to overall digestive health.The antioxidant properties of phenolic compounds also contribute to the stability and shelf life of canola oil, as they help prevent lipid oxidation during storage and cooking.

Another important class of phytochemicals in canola is tocopherols, which are forms of vitamin E. Tocopherols, particularly alpha-tocopherol, are potent antioxidants that protect cells from oxidative damage and support immune function(Canola Council of Canada.,1990). Canola oil is a good source of tocopherols, which contribute to its nutritional value and stability. The presence of tocopherols in canola oil enhances its health benefits, including its role in skin health, immune support, and the prevention of chronic diseases associated with oxidative stress. In addition to these compounds, canola seeds contain various other minor phytochemicals, such as sterols and squalene. Plant sterols, including beta-sitosterol, are present in canola oil and have been associated with cholesterol-lowering effects. They compete with cholesterol for absorption in the digestive tract, thus helping to reduce LDL cholesterol levels and improve heart health. Squalene, another bioactive compound, is known for its antioxidant and anti-inflammatory properties, and it plays a role in protecting cells and tissues from damage.The bioactivity of these phytochemicals can be influenced by several factors, including the variety of canola, growing conditions, and processing methods. For instance, processing techniques like heating and refining can affect the concentration and bioavailability of these compounds. Minimizing such losses during processing can help preserve the beneficial properties of canola oil and its meal, ensuring that consumers receive the maximum health benefits from these phytochemicals.Moreover, the health benefits associated with canola's phytochemicals extend to their potential roles in functional foods and nutraceuticals. Research is ongoing to explore how these bioactive compounds can be used to develop products that promote health and prevent disease. For example, canola-derived extracts rich in glucosinolates or phenolic compounds could be utilized in dietary supplements or fortified foods aimed at enhancing nutritional intake and supporting overall health.In summary, the phytochemicals present in canola, including glucosinolates, phenolic compounds, tocopherols, sterols, and squalene, contribute to its health-promoting properties and overall nutritional value. These bioactive compounds offer various benefits, from antioxidant and anti-inflammatory effects to cholesterol-lowering and potential anticancer properties. Understanding the role and impact of these phytochemicals enhances the appreciation of canola's value as a multifunctional crop and underscores its potential in supporting human health through both dietary and industrial applications.

GLUCOSINOLATES: THE UNIQUE COMPOUNDS OF CANOLA

The profile of these breakdown products can vary depending on the specific glucosinolates present and the conditions under which they are processed.In canola, the major glucosinolates are sinigrin, glucoraphanin, and glucobrassicin. Sinigrin, which is more prevalent in other Brassica species, is a precursor to allyl isothiocyanate, known for its pungent flavor and potential anticancer properties. Glucoraphanin is a precursor to sulforaphane, a potent antioxidant with anti-inflammatory and anticancer effects that has been studied extensively for its health benefits (Cartea & Velasco, 2008). Glucobrassicin is a precursor to indole-3-carbinol, which also has been associated with potential anticancer effects and hormone regulation.The bioactivity of glucosinolates and their breakdown products has been the subject of extensive research. Isothiocyanates, such as sulforaphane, have shown promise in reducing the risk of various cancers by inhibiting carcinogenesis and modulating detoxification enzymes. They are believed to work through multiple mechanisms, including the activation of phase II detoxification enzymes and the suppression of inflammatory pathways. Indole-3-carbinol, derived from glucobrassicin.Research continues to explore the full range of glucosinolates' effects and their potential applications in human health and agriculture. Advances in biotechnology and plant breeding aim to optimize glucosinolate profiles in canola varieties to enhance their health benefits and reduce any adverse effects. This includes developing varieties with higher concentrations of beneficial isothiocyanates and lower levels of compounds that may be less desirable in animal feed.In summary, glucosinolates are unique and multifaceted compounds in canola with significant implications for health and agriculture. Their role in producing bioactive breakdown products, such as isothiocyanates and indoles, highlights their potential for cancer prevention and health promotion. Additionally, their function in plant defense and their impact on animal feed underscore the broader significance of these compounds beyond their role in human nutrition. Understanding glucosinolates' full range of effects and optimizing their benefits can enhance the value of canola as both a food crop and an agricultural resource.

CANOLA'S VITAMIN AND MINERAL PROFILE: A NUTRITIONAL PERSPECTIVE

Canola's vitamin and mineral profile offers a rich array of essential nutrients, enhancing its value beyond its primary function as an oilseed crop. Canola, especially in its oil form and its byproducts, such as canola meal, provides a diverse range of vitamins and minerals that contribute to its nutritional profile and health benefits.The oil's high vitamin E content contributes to its stability and shelf life, as tocopherols help prevent oxidation and rancidity. Vitamin K's presence in canola oil supports its inclusion in a balanced diet, although it is not present in high concentrations compared to other sources like leafy greens.Canola seeds themselves are a good source of several essential minerals. Among these, calcium, magnesium, and potassium stand out (Chen et al.,1996). The mineral content of canola contributes to these physiological processes, supporting various aspects of health.The mineral profile of canola also includes phosphorus, which is crucial for the formation of bones and teeth, as well as for energy production and storage. Zinc, another important mineral found in canola seeds, supports immune function, protein synthesis, and cell division. Iron, present in smaller amounts, is necessary for oxygen transport in the blood and overall cellular metabolism. Research continues to explore the full range of vitamins and minerals in canola and their impact on health. Advances in plant breeding and agricultural practices aim to enhance the nutrient profile of canola varieties, optimizing their nutritional benefits. Understanding the complex interplay between canola's vitamins, minerals, and other bioactive compounds can further illuminate its role in promoting health and preventing disease(Combs, 1992). In summary, canola's vitamin and mineral profile highlights its significant contributions to a balanced diet and overall nutrition. The presence of essential vitamins such as vitamin E and vitamin K, combined with important minerals like calcium, magnesium, potassium, phosphorus, zinc, and iron, underscores the nutritional value of canola. Whether consumed as oil or incorporated into animal feed, canola provides a range of health benefits that support various physiological functions and contribute to overall well-being.

THE ROLE OF PHOSPHOLIPIDS IN CANOLA'S FUNCTIONAL PROPERTIES

Phospholipids are crucial components of canola's functional properties, influencing both its nutritional profile and its utility in various applications. These lipids, characterized by their dual hydrophilic and hydrophobic nature, play a significant role in the structure and function of biological membranes, as well as in the physical and chemical properties of canola oil and its derivatives.In canola seeds, phospholipids are a minor but important fraction of the total lipid content, typically comprising about 1-2% of the seed's total lipids. The primary phospholipids in canola include phosphatidylcholine (PC), phosphatidylethanolamine (PE), and phosphatidylinositol (PI). Phosphatidylcholine, also known as lecithin, is the most abundant phospholipid in canola and plays a central role in various biological and industrial functions.One of the key roles of phospholipids in canola oil is their contribution to the oil's emulsifying properties. Phosphatidylcholine, in particular, acts as an emulsifier due to its amphiphilic nature, meaning it has both hydrophobic and hydrophilic regions. This property allows phosphatidylcholine to products such as salad dressings, mayonnaise, and sauces. The emulsifying ability of canola oil, enhanced by its phospholipid content, is crucial for achieving the desired texture and consistency in these products(Degirolamo & Rudel, 2010).In addition to their emulsifying properties, phospholipids in canola oil contribute to its stability and quality. They help protect the oil from oxidation by forming a protective layer around the oil droplets, thereby extending the shelf life of the oil and preventing rancidity. This oxidative stability is particularly important for maintaining the quality of canola oil during storage and cooking, as oxidation can lead to the formation of off-flavors and harmful compounds.

It is employed in the production of baked goods, confectionery, and processed foods, where it helps achieve desired textures and product stability(Downey & Röbbelen, 1989). Additionally, phospholipids are used in the pharmaceutical and cosmetic industries for their role in drug delivery systems and skin care formulations, leveraging their ability to form stable liposomes and enhance the bioavailability of active ingredients.Research into the functional properties of phospholipids in canola continues to explore their potential applications and benefits. Advances in processing technologies and biochemistry aim to optimize the extraction and utilization of phospholipids, enhancing their functionality in various products. By understanding and leveraging the unique properties of canola's phospholipids, the agricultural and industrial sectors can maximize the value and versatility of this important crop.In summary, phospholipids play a pivotal role in canola's functional properties, impacting its emulsifying abilities, oxidative stability, and nutritional benefits. The presence of phosphatidylcholine, phosphatidylethanolamine, and phosphatidylinositol contributes to the oil's performance in food products, its stability, and its value as a source of essential nutrients. Additionally, phospholipids enhance the functionality of canola meal in animal feed and find applications in various industrial sectors(Eskin & McDonald, 1991). Understanding the role of phospholipids in canola highlights the crop's multifaceted value and underscores its importance in both nutritional and industrial contexts.

CANOLA'S CAROTENOID AND TOCOPHEROL CONTENT: ANTIOXIDANT PROPERTIES

Canola's carotenoid and tocopherol content significantly contributes to its antioxidant properties, enhancing both its nutritional value and stability in culinary and industrial applications. These compounds, integral to the health benefits of canola, play essential roles in neutralizing oxidative stress and protecting cells from damage. It also plays a crucial role in maintaining healthy vision, immune function, and skin health. Lutein, another significant carotenoid found in canola, is known for its protective effects on eye health. This protection is crucial for maintaining cellular integrity and function and for reducing the risk of chronic diseases associated with oxidative stress. Gamma-tocopherol, while less studied than alpha-tocopherol, also has significant antioxidant properties and may offer additional health benefits, such as anti-inflammatory effects. Delta-tocopherol, although present in smaller amounts, contributes to the overall antioxidant capacity of canola oil.The antioxidant properties of carotenoids and tocopherols in canola not only provide health benefits but also enhance the stability and quality of canola oil. In culinary applications, these antioxidants help protect the oil from oxidation, which can lead to rancidity and the formation of harmful compounds. The presence of carotenoids and tocopherols extends the shelf life of canola oil, maintaining its flavor, color, and nutritional value over time. This stability is particularly important for cooking and food processing, where prolonged exposure to heat and light can otherwise degrade the oil's quality. Furthermore, the antioxidant content of canola oil contributes to its role in a heart-healthy diet. By reducing oxidative stress and preventing the oxidation of LDL cholesterol, the antioxidants in canola oil help mitigate the risk of cardiovascular diseases(Katz & Messineo, 1982). This protective effect is supported by the oil's favorable fatty acid profile, which includes high levels of monounsaturated fats and beneficial omega-3 fatty acids.The nutritional benefits of carotenoids and tocopherols extend beyond canola oil to include the oilseed itself and its byproducts, such as canola meal. Canola meal retains a significant portion of these antioxidants, contributing to the nutritional value of animal feed (Ehrensing, 2008). High temperatures can reduce the concentration of glucosinolates, which are known for their potential health benefits but can also lead to undesirable flavors in the oil (Eskin & McDonald, 1991). Conversely, this heating can cause some loss of tocopherols, which are crucial for the oil's antioxidant properties.

The primary method for extracting oil from canola seeds is solvent extraction, typically using hexane. This process dissolves the oil from the seeds and requires subsequent desolventizing to remove any residual solvent. Solvent extraction is highly efficient and results in a higher yield of oil compared to mechanical pressing. However, it can also affect the chemical composition of the oil. For instance, solvent extraction can lead to a higher level of free fatty acids in the oil if not properly managed, which can impact the oil's stability and flavour(FDA.,2006). Additionally, this process may result in the removal of some of the minor bioactive compounds, such as carotenoids and phospholipids, which are beneficial for health and oil stability.Following extraction, canola oil undergoes refining, a series of steps including degumming, neutralization, bleaching, and deodorization. Each of these steps further influences the oil's chemical profile. Degumming removes phospholipids and gums that can affect oil quality and stability. Neutralization addresses free fatty acids by treating the oil with an alkaline solution, which can also remove some minor components, including tocopherols and phospholipids. Bleaching involves the removal of pigments and residual impurities using bleaching earth or activated carbon, which can decrease the levels of carotenoids, thereby affecting the oil's color and antioxidant content(Fussell, 1955). Finally, deodorization, which involves steam distillation at high temperatures, eliminates volatile compounds responsible for odors and flavors. However, this step can lead to the loss of some heat-sensitive antioxidants and further reduction in minor phytochemicals.

The refining process, while crucial for ensuring the oil's stability and safety, also reduces its nutritional content compared to crude oil(Gelpí et al.,2002). Crude canola oil retains more of its natural antioxidants, such as tocopherols and carotenoids, as well as its original flavor and color. However, crude oil may contain undesirable impurities and higher levels of free fatty acids, which can affect its shelf life and suitability for cooking. Refined oil, while more stable and neutral in flavor, has a reduced concentration of these beneficial compounds.In addition to oil extraction, the processing of canola meal, a byproduct of oil extraction, also impacts its chemical composition(Hansen & Harris, 2007). Canola meal is used as animal feed and retains significant levels of protein, fiber, and minerals. However, processing methods like heat treatment can affect the levels of residual glucosinolates and their breakdown products, potentially altering the meal's nutritional value and palatability. Proper processing techniques are essential to minimize the presence of anti-nutritional factors and maximize the feed's nutritional benefits.Research into processing technologies continues to advance, aiming to balance efficiency with the preservation of nutritional and functional qualities. Innovations in processing methods, such as cold-pressing and enzymatic treatments, are being explored to maintain higher levels of beneficial compounds while improving oil quality(Harris et al., 2008). Additionally, advancements in refining techniques seek to enhance the retention of antioxidants and other bioactive substances, addressing consumer demand for healthier, minimally processed oils.In summary, the impact of processing on canola's chemical composition is significant and multifaceted. Each stage of processing—from initial cleaning and conditioning to extraction, refining, and meal processing—affects the levels of fatty acids, phytochemicals, and antioxidants in canola oil and its byproducts (Howard et al., 2000). While processing enhances the oil's stability and safety, it also results in the loss of some beneficial compounds. Understanding these effects is crucial for optimizing canola's nutritional value and quality, guiding improvements in processing techniques, and meeting consumer expectations for healthy, high-quality products.

GLYCEMIC INDEX AND ITS IMPLICATIONS FOR HUMAN HEALTH

Canola's glycemic index (GI) is an important factor in understanding its impact on human health, particularly concerning blood sugar management and overall metabolic health. Understanding canola's GI and its implications can offer insights into its role in a balanced diet and its potential benefits for individuals managing conditions like diabetes.Canola oil, derived from the seeds of the canola plant, has a negligible glycemic index (Johnson, 2002). This is because it is composed predominantly of fats, specifically monounsaturated and polyunsaturated fatty acids, with minimal carbohydrates. However, canola oil's lipid profile plays a crucial role in influencing overall glycemic responses when used in cooking and food preparation. For canola seeds, which contain both fats and carbohydrates, the glycemic index is influenced by their carbohydrate content, including dietary fiber. When used in moderation, canola oil can contribute to a balanced diet that supports metabolic health and helps manage blood glucose levels effectively. Furthermore, the integration of canola oil and seeds into a diet rich in vegetables, whole grains, and lean proteins can enhance overall glycemic control (Kris-Etherton et al., 1988). The combination of low GI foods, high fiber intake, and healthy fats supports stable blood sugar levels and overall metabolic health, providing a comprehensive approach to managing glycemicresponses.Research into the specific effects of canola oil and seeds on glycemic control continues to evolve, with ongoing studies examining their role in long-term metabolic health. Understanding the interplay between dietary fats, carbohydrates, and glycemic responses can inform dietary recommendations and promote healthier eating habits.In summary, while canola oil itself does not have a glycemic index due to its fat composition, the glycemic impact of canola seeds is influenced by their low carbohydrate and high fiber content. The inclusion of canola oil and seeds in a diet can contribute to improved glycemic control and metabolic health, particularly through their beneficial fatty acid profiles. The chemical composition of canola is significantly influenced by environmental factors such as climate and soil conditions, which impact its growth, yield, and nutritional profile. Understanding these effects is crucial for optimizing canola production and ensuring the quality of its oil and byproducts.

Climate plays a central role in determining the chemical composition of canola. Temperature, precipitation, and sunlight exposure all affect the plant's development and the biochemical processes within it. Canola is typically grown in temperate climates with moderate temperatures, as extreme heat or cold can negatively impact its growth and seed quality. High temperatures, especially during the flowering and seed-filling stages, can lead to reduced oil content and altered fatty acid profiles. Elevated temperatures can also accelerate the degradation of beneficial phytochemicals, such as tocopherols and carotenoids, thus affecting the oil's antioxidant properties. Precipitation levels also influence canola's chemical composition. Adequate moisture is essential for optimal growth and seed development. However, both excessive rainfall and drought conditions can stress the plant, affecting its yield and quality. Water stress can lead to lower oil content in the seeds and higher concentrations of free fatty acids, which can impact oil stability and quality. Additionally, drought conditions can increase the concentration of certain secondary metabolites, such as glucosinolates, which can affect the nutritional and sensory qualities of the oil and meal(Mattson & Grundy, 1985).

Soil conditions are equally important in shaping the chemical composition of canola. Soil fertility, pH, and texture all play roles in determining the plant's nutrient uptake and overall health. Canola thrives in well-drained soils with a neutral to slightly acidic pH. Soil fertility, particularly the availability of essential nutrients such as nitrogen, phosphorus, and potassium, directly affects the plant's growth and the chemical composition of its seeds. Adequate nitrogen levels are crucial for protein synthesis, influencing the protein content of canola meal. Conversely, imbalances in soil nutrients can lead to variations in the seed's fatty acid profile and overall oil quality.Soil texture impacts water retention and drainage, affecting plant stress and growth. Sandy soils, with their lower water-holding capacity, may lead to increased plant stress during dry periods, while clay soils may retain excess moisture, potentially leading to root diseases and reduced seed quality (Mente et al., 2009). Both conditions can influence the levels of key compounds in canola seeds, including fatty acids, tocopherols, and glucosinolates.

Nutrient management is a key factor in optimizing canola's chemical composition. Fertilization practices that ensure an adequate supply of essential nutrients can enhance oil content and improve the quality of canola seeds (Selvakumar et al., 2024). For instance, balanced fertilization can help maintain an optimal fatty acid profile, reducing the levels of undesirable fatty acids such as erucic acid, which is found in higher concentrations in non-low-erucic acid canola varieties. Proper nutrient management also supports the development of beneficial compounds, such as tocopherols and carotenoids, which contribute to the oil's antioxidant properties.

Soil health and management practices, such as crop rotation and organic amendments, can also influence canola's chemical composition. Practices that maintain or improve soil health can enhance nutrient availability and support optimal plant growth (Parthasarathy S, et al.,1990). Organic amendments, such as compost or manure, can improve soil fertility and structure, potentially benefiting the plant's nutrient uptake and overall seed quality.

Climate and soil interactions further complicate the relationship between environmental conditions and canola's chemical composition. Variations in climate and soil conditions can lead to differences in plant responses and seed quality even within the same growing region. This variability underscores the importance of adapting agricultural practices to local conditions to optimize canola production and ensure consistent quality(Percopo et al., 2010).Ongoing research aims to better understand how climate and soil conditions affect canola's chemical composition and to develop strategies for mitigating adverse effects. Advances in plant breeding and agronomic practices are focused on enhancing canola's resilience to environmental stressors and improving its nutritional profile. Innovations in precision agriculture, such as soil testing and targeted fertilization, help tailor management practices to specific soil and climate conditions, optimizing canola's yield and quality (Ramakrishnan et al., 2022).In summary, the chemical composition of canola is profoundly influenced by climate and soil conditions. Temperature, precipitation, and soil fertility all affect the plant's growth, yield, and the quality of its oil and byproducts(Madgule et al., 2023). Understanding these effects is essential for optimizing canola production and ensuring high-quality products. By adapting agricultural practices to local environmental conditions and advancing research in plant and soil science, producers can enhance canola's nutritional value and resilience, contributing to a more sustainable and productive agricultural system.

BREEDING FOR IMPROVED QUALITY AND FUNCTIONALITY IN CANOLA

Breeding for improved quality and functionality in canola is a critical endeavor aimed at enhancing both the agronomic performance and the end-use attributes of this vital oilseed crop. As demand for high-quality canola oil and meal grows, breeding programs focus on various traits to optimize the crop's nutritional profile, yield, and adaptability. This comprehensive approach integrates advancements in genetics, biotechnology, and agronomy to meet both consumer preferences and agricultural challenges(Sarwar et al., 2003)**.Genetic Improvement** is central to breeding programs focused on canola quality and functionality. Traditional breeding methods, such as selection and hybridization, are used to develop new varieties with desirable traits.

For instance, breeders work to increase oil content and improve the oil's fatty acid composition by selecting plants with higher concentrations of monounsaturated and polyunsaturated fatty acids and lower levels of undesirable components like erucic acid(Sarwar et al., 2004). These improvements are critical for enhancing the oil's health benefits, stability, and suitability for various culinary applications.

High-Oleic Canola varieties are a key focus of breeding efforts. These varieties are developed to have elevated levels of oleic acid, a monounsaturated fat known for its cardiovascular benefits and stability in cooking(Seker et al., 2008). High-oleic canola oils are less prone to oxidation, offering a longer shelf life and a more stable cooking oil. This trait is particularly valuable for food manufacturers seeking healthier oil options with extended use in processing and frying.**Low-Glucosinolate Canola** is another significant breeding objective. Glucosinolates are natural compounds found in canola that, while having potential health benefits, can also impart undesirable flavors and reduce the quality of canola meal used in animal feed. Breeders aim to develop canola varieties with reduced glucosinolate levels to improve meal quality and palatability for livestock. This enhancement not only benefits animal nutrition but also reduces potential environmental impacts associated with the glucosinolates' breakdown products.Disease Resistance and pest toleranceare essential traits for improving canola's functionality and yield stability. Breeding for resistance to diseases such as blackleg and sclerotinia stem rot, as well as pests like flea beetles, is crucial for reducing crop losses and minimizing the need for chemical treatments(Vijayakumar G,et al, 2024). Disease-resistant varieties can lead to more sustainable farming practices by reducing the reliance on pesticides and improving overall crop health.Abiotic Stress Toleranceis also a major focus. Canola crops are susceptible to various environmental stresses, including drought, heat, and salinity. Breeding programs aim to develop varieties that can better withstand these conditions, ensuring stable yields and quality even under adverse environmental scenarios. Tolerance to abiotic stresses is achieved through the selection of traits such as improved root systems, enhanced water use efficiency, and increased resilience to temperature fluctuations.

Nutritional Quality is another critical aspect of canola breeding. Enhancing the levels of beneficial nutrients, such as omega-3 fatty acids, and increasing the concentration of antioxidants like tocopherols and carotenoids, are important goals(-Selvakumar, 2023). These improvements can provide additional health benefits to consumers and increase the nutritional value of canola oil and meal.Sustainability is an overarching goal in canola breeding. Modern breeding programs aim to enhance not only the quality and functionality of canola but also its environmental impact. This includes developing varieties that require fewer inputs, such as fertilizers and pesticides, and that have lower carbon footprints. Sustainable breeding practices contribute to more efficient use of resources and support environmentally friendly agricultural practices.Consumer Preferences also drive breeding efforts. As consumers increasingly seek healthier and more sustainable food options, breeders are tasked with developing canola varieties that meet these demands(Selvakumar, 2023). This includes creating oils with better health profiles, such as reduced saturated fats and increased beneficial fatty acids, as well as improving the sensory qualities of canola products, such as flavor and stability.In summary, breeding for improved quality and functionality in canola encompasses a range of objectives, including enhancing oil composition, increasing disease resistance, improving abiotic stress tolerance, and optimizing nutritional quality. By integrating traditional breeding techniques with modern biotechnological tools, breeders can develop canola varieties that meet the evolving needs of consumers and farmers. These advancements contribute to higher yields, better oil quality, and more sustainable farming practices, supporting the continued growth and development of the canola industry.

CANOLA'S POTENTIAL IN FOOD APPLICATIONS: A FUNCTIONAL PERSPECTIVE

Canola's potential in food applications from a functional perspective is vast and multifaceted, driven by its unique chemical composition and versatile properties. As an oilseed crop, canola offers numerous benefits that enhance both the nutritional profile and functional qualities of food products. Its adaptability, health benefits, and performance in various culinary and industrial contexts make it a valuable ingredient in modern food production.Nutritional Benefits are a key aspect of canola's functional potential. Canola oil is renowned for its favorable fatty acid profile, which includes high levels of monounsaturated fats (around 60%).. This composition supports cardiovascular health by reducing LDL cholesterol levels and providing essential nutrients that the body cannot synthesize. The high monounsaturated fat content, particularly oleic acid, contributes to improved lipid profiles and reduced risk of heart disease, making canola oil a preferred choice for health-conscious consumers.

Emulsification is another important functional property of canola oil(Selvakumar, 2024). Due to its phospholipid content, including phosphatidylcholine (lecithin), canola oil acts as an effective emulsifier. This makes it valuable in creating stable emulsions, such as in mayonnaise, dressings, and sauces, where it helps blend oil and water-based ingredients into a uniform mixture. The emulsifying properties also improve the texture and consistency of these products, enhancing their quality and appeal. Nutritional Fortification through canola oil also supports various food applications(Selvakumar, 2024). Fortified canola oil can be enriched with vitamins and other nutrients to enhance the nutritional profile of food products. For instance, canola oil can be used in the production of fortified cooking oils and spreads, providing essential fatty acids and vitamins to support a balanced diet. This is particularly beneficial in regions where nutrient deficiencies are prevalent.Versatility in Food Processing is another advantage of canola oil. Its stability and functional properties make it suitable for various processing techniques, including high-heat applications. Research into novel processing techniques and formulations seeks to maximize the oil's functional properties and health benefits. For instance, developments in high-oleic canola oil varieties aim to offer even greater stability and health benefits, while innovations in emulsification and texture modification expand its applications in food products. In summary, canola's potential in food applications is broad and varied, driven by its beneficial fatty acid profile, antioxidant content, and versatile functional properties (Selvakumar et al., 2012). From enhancing nutritional quality to contributing to desirable textures and consistencies, canola oil plays a significant role in modern food production. Its ability to provide health benefits, support stability and consistency, and adapt to various culinary applications underscores its value in the food industry. As research and technology continue to advance, the functional potential of canola is expected to expand further, offering new opportunities for improving food quality and nutrition.

NON-FOOD USES OF CANOLA: INDUSTRIAL AND PHARMACEUTICAL APPLICATIONS

Canola, primarily known for its role in food production, also boasts a range of non-food uses that leverage its properties in industrial and pharmaceutical applications. These applications highlight the versatility of canola products and their potential to contribute to various sectors beyond the kitchen.Industrial Applications of canola are diverse, largely due to the oil's favorable chemical composition and functional properties. One of the most significant non-food uses is in the production of biodiesel(Selvakumar, 2023). Canola oil's high monounsaturated fatty acid content makes it an ideal feedstock for biodiesel, offering good oxidative stability and low levels of saturated fats, which contribute to a cleaner-burning fuel. Biodiesel derived from canola oil can be used in diesel engines and blends, providing a renewable energy source that reduces greenhouse gas emissions and dependence on fossil fuels.

Coatings and Plastics benefit from canola oil's chemical properties as well. In the coatings industry, canola oil can be used as a base for certain types of paints and varnishes. Its non-toxic nature and ability to form stable films make it a suitable ingredient in eco-friendly coatings. Additionally, canola oil can be used in the production of biodegradable plastics(Sugumaran et al., 2024). The incorporation of canola-based polymers into plastic products helps reduce reliance on petrochemical sources and supports the development of more sustainable materials.In the pharmaceutical sector, canola oil's applications are equally noteworthy. Its composition is utilized in the production of various pharmaceutical products, including drug delivery systems and excipients. Canola oil can serve as a carrier oil in oral and topical medications, providing a stable medium for the delivery of active pharmaceutical ingredients (Selvakumar, P.,2023). Its favorable fatty acid profile, including omega-3 and omega-6 fatty acids, is also explored for its potential health benefits. Research is ongoing into how these fatty acids might support cardiovascular health, cognitive function, and inflammation reduction when incorporated into pharmaceutical formulations.Its properties make it valuable in producing biodiesel, lubricants, coatings, and biodegradable plastics, as well as in pharmaceuticals and personal care products (Selvakumar et al.,2 024). These non-food uses highlight the wide-ranging benefits of canola and underscore its role in supporting sustainability and innovation across multiple industries. As research and technology continue to evolve, the potential applications of canola are likely to expand further, offering new opportunities for leveraging its properties in diverse fields.

The use of molecular markers and high-throughput genomic tools accelerates the breeding process, leading to faster development of varieties that meet specific agricultural and nutritional needs(USB.,2008).Climate Resilience is a critical area of focus as climate change presents new challenges for crop production. Research is concentrated on developing canola varieties that can withstand extreme weather conditions, such as droughts, heatwaves, and flooding. Genetic engineering and marker-assisted selection are being used to introduce and enhance traits related to water use efficiency, heat tolerance, and root robustness. Additionally, understanding the plant's physiological responses to environmental stressors allows for better management practices and the creation of more resilient crop varieties.Scientists are exploring ways to increase the content of beneficial compounds in canola oil, such as omega-3 fatty acids,antioxidants, and phytochemicals. Innovations include breeding canola varieties with higher levels of these nutrients and developing new processing techniques to preserve their efficacy(Selvakumar, 2022). Enhanced nutritional profiles can help address dietary deficiencies and support overall health, catering to the growing demand for functional foods.Bioengineering and Bioprocessing are advancing the use of canola beyond traditional applications.

Consumer Preferences and Market Trends are influencing canola research, as there is increasing demand for healthier and more sustainable food options. Researchers are addressing consumer concerns about transparency, health, and sustainability by developing canola varieties that meet these preferences. This includes creating oils with lower saturated fat content, higher levels of beneficial fats, and improved flavor profiles. Additionally, there is a growing interest in organic and non-GMO canola products, leading to research into organic cultivation practices and non-GMO breeding approaches.Emerging Markets for canola also drive research and development efforts(Tan et al., 2011). As canola production expands globally, research focuses on adapting canola cultivation to diverse climates and soil conditions. This includes exploring canola's potential in new geographic regions and optimizing agronomic practices for different environments.In summary, the future directions in canola research and development are characterized by advanced breeding techniques, climate resilience, sustainable farming practices, nutritional enhancement, bioengineering, consumer-driven trends, emerging markets, and collaborative innovation. These trends reflect a holistic approach to optimizing canola's potential, addressing global challenges, and meeting evolving demands (Sreenivasulu et al., 2022). As research and technology continue to advance, canola is expected to play an increasingly important role in agriculture, nutrition, and industrial applications, contributing to a more sustainable and healthier future.

CONCLUSION

In conclusion, the main quality and functional chemical composition of canola underscore its significant role as a versatile and valuable crop in both the food and industrial sectors. Canola's unique chemical profile, characterized by its balanced fatty acid composition, it as a premier choice for promoting cardiovascular health and enhancing overall dietary quality. The functional properties of canola extend beyond its nutritional benefits. Its neutral flavor and high smoke point make it an ideal cooking oil for a wide range of culinary uses, from frying and baking to salad dressings and marinades. The oil's ability to act as an effective emulsifier enhances the texture and consistency of food products such as sauces, dressings, and spreads. Additionally, canola oil's performance in food processing, coupled with its favorable texture attributes, contributes to the quality of baked goods and margarine, underscoring its importance in the food industry.In non-food applications, canola's chemical composition offers considerable advantages. The oil's high monounsaturated fat content and low levels of saturated fats make it a suitable feedstock for biodiesel production, providing a renewable energy source that aligns with sustainability goals. Its biodegradable nature and non-toxic properties make canola oil a preferred choice for environmentally friendly lubricants and hydraulic fluids. Moreover, the oil's role in the production of coatings, plastics, and pharmaceuticals highlights its versatility and potential to support innovations in various industries.Research and development efforts continue to enhance canola's quality and functionality. Advances in breeding techniques, including genomic selection and gene editing, aim to develop canola varieties with improved disease resistance, stress tolerance, and optimized oil composition. Nutritional enhancements and biotechnological innovations further expand canola's applications, from health-focused food products to sustainable bioplastics and biochemicals. As global demand for healthier, sustainable, and high-performing crops increases, canola's role becomes increasingly prominent. Its chemical composition not only meets current consumer and industrial needs but also aligns with emerging trends in health, sustainability, and innovation. By continuing to explore and optimize canola's potential, researchers and industry stakeholders contribute to a more sustainable and health-conscious future, leveraging the crop's unique qualities to address global challenges and enhance quality of life.Overall, canola's quality and functional chemical composition establish it as a key player in modern agriculture and industry. Its balanced fatty acid profile, antioxidant content, and versatile properties make it invaluable across various sectors, from food production to industrial applications. As research and technological advancements continue to evolve, canola is well-positioned to meet the growing demands for sustainability, health, and functionality, ensuring its continued significance in a rapidly changing world.

REFERENCES

Ash, M. (2012). Soybeans & oil crops: Canola. United States Department of Agriculture, Economic Research Service. https://www.ers.usda.gov/topics/crops/soybeans-oil-crops/canola.aspx

Assmann, G., & Gotto, A. M.Jr. (2004). HDL cholesterol and protective factors in atherosclerosis. *Circulation*, *109*(23, Suppl. 1), III8–III14. DOI: 10.1161/01.CIR.0000131512.50667.46 PMID: 15198960

Baranyk, P., & Fábry, A. (Eds.). (1999). History of the rapeseed (Brassica napus L.) growing and breeding from Middle Age Europe to Canberra. International Rapeseed Congress. https://www.regional.org.au/au/gcirc/4/374.htm

Batista, C., Barros, L., Carvalho, A. M., & Ferreira, I. C. F. R. (2011). Nutritional and nutraceutical potential of rape (Brassica napus L. var. napus) and "tronchuda" cabbage (Brassica oleracea L. var. costata) inflorescences. *Food and Chemical Toxicology*, *49*(6), 1208–1214. DOI: 10.1016/j.fct.2011.02.023 PMID: 21376770

Canola Council of Canada. (1990). *Canola oil and meal: Standards and regulations*. Canola Council of Canada.

Cartea, M. E., & Velasco, P. (2008). Glucosinolates in Brassica foods: Bioavailability in food and significance for human health. *Phytochemistry Reviews*, *7*(2), 213–229. DOI: 10.1007/s11101-007-9072-2

Chen, Z. Y., Wang, R. F., & Wu, T. C. (1996). Antioxidant activity of natural flavonoids is governed by number and location of their aromatic hydroxyl groups. *Chemistry and Physics of Lipids*, *79*(2), 157–163. DOI: 10.1016/0009-3084(96)02523-6 PMID: 8640902

Combs, G. F.Jr. (1992). *The vitamins: Fundamental aspects in nutrition and health*. Academic Press.

Degirolamo, C., & Rudel, L. L. (2010). Dietary monounsaturated fatty acids appear not to provide cardioprotection. *Current Atherosclerosis Reports*, *12*(6), 391–396. DOI: 10.1007/s11883-010-0133-4 PMID: 20725810

Downey, R. K., & Röbbelen, G. (1989). Brassica species. In *Oil crops of the world* (pp. 339–362). McGraw-Hill.

Eskin, N. A. M., & McDonald, B. E. (1991). Canola oil. *Nutrition Bulletin*, *16*(3), 138–146. DOI: 10.1111/j.1467-3010.1991.tb01048.x

Food and Drug Administration. (2006). Unsaturated fatty acids from canola oil and reduced risk of coronary heart disease. U.S. Food and Drug Administration. https://www.fda.gov/food/ingredientspackaginglabeling/labelingnutrition/ucm073992.htm#cardio

Fussell, G. E. (1955). History of cole (Brassica sp.). *Nature, 176*(4471), 48–51. DOI: 10.1038/176048a0

Gelpí, E., de la Paz, M. P., Terracini, B., Abaitua, I., de la Cámara, A. G., Kilbourne, E. M., Lahoz, C., Nemery, B., Philen, R. M., Soldevilla, L., & Tarkowski, S. (2002). The Spanish toxic oil syndrome 20 years after its onset: A multidisciplinary review of scientific knowledge. *Environmental Health Perspectives, 110*(5), 457–464. DOI: 10.1289/ehp.02110457 PMID: 12003748

Hansen, S. N., & Harris, W. S. (2007). New evidence for the cardiovascular benefits of long chain omega-3 fatty acids. *Current Atherosclerosis Reports, 9*(6), 434–440. DOI: 10.1007/s11883-007-0058-8 PMID: 18377782

Harris, W. S., Kris-Etherton, P. M., & Harris, K. A. (2008). Intakes of long-chain omega-3 fatty acids associated with reduced risk for death from coronary heart disease in healthy adults. *Current Atherosclerosis Reports, 10*(6), 503–509. DOI: 10.1007/s11883-008-0078-z PMID: 18937898

Howard, B. V., Robbins, D. C., Sievers, M. L., Lee, E. T., Rhoades, D., Devereux, R. B., Cowan, L. D., Gray, R. S., Welty, T. K., Go, O. T., & Howard, W. J. (2000). LDL cholesterol as a strong predictor of coronary heart disease in diabetic individuals with insulin resistance and low LDL: The Strong Heart Study. *Arteriosclerosis, Thrombosis, and Vascular Biology, 20*(3), 830–835. DOI: 10.1161/01.ATV.20.3.830 PMID: 10712410

Johnson, I. T. (2002). Glucosinolates: Bioavailability and importance to health. *International Journal for Vitamin and Nutrition Research, 72*(1), 26–31. DOI: 10.1024/0300-9831.72.1.26 PMID: 11887749

Katz, A. M., & Messineo, F. C. (1982). Fatty acid effects on membranes: Possible role in the pathogenesis of ischemic myocardial damage. *Journal of Molecular and Cellular Cardiology, 14*(Suppl. 3), 119–122. DOI: 10.1016/0022-2828(82)90139-0 PMID: 7143451

Kris-Etherton, P. M., Krummel, D., Russell, M. E., Dreon, D., Mackey, S., Borchers, J., & Wood, P. D. (1988). The effect of diet on plasma lipids, lipoproteins, and coronary heart disease. *Journal of the American Dietetic Association, 88*(11), 1373–1400. DOI: 10.1016/S0002-8223(21)08023-8 PMID: 2846672

Madgule, M., Vinayaka, N., Sonkhaskar, Y. M., Patel, D. R., Karthikeyan, R., & Selvakumar, P. (2023). Mechanical properties and microstructure of activated TIG welded similar joints of Inconel alloys by desirability approaches. *Materials Today: Proceedings*, *77*, 528–533. DOI: 10.1016/j.matpr.2022.12.250

Mattson, F. H., & Grundy, S. M. (1985). Comparison of effects of dietary saturated, monounsaturated, and polyunsaturated fatty acids on plasma lipids and lipoproteins in man. *Journal of Lipid Research*, *26*(2), 194–202. https://www.jlr.org/content/26/2/194. DOI: 10.1016/S0022-2275(20)34389-3 PMID: 3989378

Mente, A., de Koning, L., Shannon, H. S., & Anand, S. S. (2009). A systematic review of the evidence supporting a causal link between dietary factors and coronary heart disease. *Archives of Internal Medicine*, *169*(7), 659–669. DOI: 10.1001/archinternmed.2009.38 PMID: 19364995

Percopo, C. M., Dyer, K. D., Ochkur, S. I., Luo, J. L., Fischer, E. R., Lee, J. J., Lee, N. A., & Domachowske, J. B. (2010). Isolation of human eosinophils: Microbead method has no impact on IL-5 sustained viability. *Experimental Dermatology*, *19*(5), 467–469. DOI: 10.1111/j.1600-0625.2009.00974.x PMID: 19758339

Ramakrishnan, T., Gift, M. D. M., Chitradevi, S., Jegan, R., & Jose, P. S. H. (2022). Nagaraja, Sharma, H. N., Selvakumar, P., & Hailegiorgis, S. M. (2022). Study of numerous resins used in polymer matrix composite materials. *Advances in Materials Science and Engineering*, *2022*, 1088926. Advance online publication. DOI: 10.1155/2022/1088926

Sarwar, M., Ahmad, N., & Tofique, M. (2003). Efficiency of different chemicals on canola strain Rainbow (Brassica napus L.) for aphids control. *Asian Journal of Plant Sciences*, *2*(11), 831–833. DOI: 10.3923/ajps.2003.831.833

Sarwar, M., Ahmad, N., & Tofique, M. (2004). Response of various Brassica genotypes against aphids infestation under natural conditions. *Pakistan Journal of Zoology*, *26*(4), 69–74.

Seker, M., Gul, A., & Yildirim, M. (2008). Screening and comparing tocopherols in the rapeseed (Brassica napus L.) and olive (Olea europaea L.) varieties using high-performance liquid chromatography. *International Journal of Food Sciences and Nutrition*, *59*(6), 483–490. DOI: 10.1080/09637480701539484 PMID: 19086241

Selvakumar, P. (2022). Phytochemical and pharmacological profile review of Bryophyllum pinnatum. *Biomedical and Biotechnology Research Journal*, *6*(3), 295–301. DOI: 10.4103/bbrj.bbrj_126_22

Selvakumar, P. (2023a). Biomedical applications of polymers in biosensors, cancer vaccines and drug delivery systems. *Biopolimery i Kletka, 39*(4), 245–256. DOI: 10.7124/bc.000AA0

Selvakumar, P. (2023b). Fiber reinforced polymer composites: Classification, chemical treatment, mechanical and tribological properties. *Research Journal of Chemistry and Environment, 27*(1), 111–119. DOI: 10.25303/2701rjce1110119

Selvakumar, P. (2023c). Novel drug target with diverse therapeutic potential in cancer therapy. *Pharmacy Times, 55*(1), 11–14.

Selvakumar, P. (2024a). Reasons for biodiversity loss. In Rathoure, A. K. (Ed.), *Impact of societal development and infrastructure on biodiversity decline* (pp. 37–49). IGI Global., DOI: 10.4018/979-8-3693-6950-0.ch004

Selvakumar, P. (2024b). Emerging engineering technologies: Design, development, and industrial applications of IoT-based aquaculture monitoring system. In El Kacimi, Y., & Alaoui, K. (Eds.), *Emerging engineering technologies and industrial applications* (pp. 332–341). IGI Global., DOI: 10.4018/979-8-3693-1335-0.ch016

Selvakumar, P., & Loganathan, V. (2012). Phytochemical screening and antioxidant activity of red flowered Mirabilis jalapa leaf in different solvents. *International Journal of Pharma and Bio Sciences, 3*(4), 440–446.

Selvakumar, P., Mariappan, P., Petchiammal, S., & Sivagami, S. (2024a). Removal of chemicals from effluent using photobioreactor technology to improve environmental and health impacts. *Zeitschrift für Physikalische Chemie, 238*(6), 1085–1102. DOI: 10.1515/zpch-2023-0450

Selvakumar, P., & Palanisamy, S. K., Cinthaikinian, Selvamurugan, Palanisamy, V., Mariappan, R., & Selvakumar, P. K. (. (2024b). Biosensors and its diverse applications in healthcare systems. *Zeitschrift für Physikalische Chemie, 238*(6), 1103–1123. DOI: 10.1515/zpch-2023-0406

Sreenivasulu, A., Manimozhi, T., Satheesh Kumar, P., Selvakumar, P., & Singh, P. (2022). The model of smart sensing device for sensitive nanoclusters modification in sensing properties. In 2022 2nd International Conference on Advance Computing and Innovative Technologies in Engineering (ICACITE) (pp. 1043–1047). IEEE. https://doi.org/DOI: 10.1109/ICACITE53722.2022.9823846

Sugumaran, K. M., Pandiyarajan, V., Selvakumar, P., & Manoharan, S. (2024). Analysis of electromyography signals for control of mechanical prosthesis using machine learning techniques. In 2024 2nd International Conference on Intelligent Data Communication Technologies and Internet of Things (IDCIoT) (pp. 1059–1067). IEEE. DOI: 10.1109/IDCIoT59759.2024.10467707

Tan, S. H., Morrison, W. R., & Karasov, W. H. (2011). Canola proteins for human consumption: Extraction, profile, and functional properties. *Journal of Food Science*, *76*(1), R16–R28. DOI: 10.1111/j.1750-3841.2010.01930.x PMID: 21535703

United Soybean Board. (2008). 15th annual consumer attitudes about nutrition report. United Soybean Board. https://www.soyconnection.com/sites/default/files/ConsumerAttitudes2008.pdf

Vijayakumar, G., Muniyandi, R., Chandar, N. R., Selvakumar, P., & Duraisamy, R. (2024). Environmentally friendly TDS removal from wastewater by electrochemical ion exchange batch-type recirculation (EIR) technique. *Environmental Science. Water Research & Technology*, *10*(4), 826–835. DOI: 10.1039/D3EW00793F

Chapter 3
Classical Genetics and Traditional Breeding of Canola Plant

Rachel Lissy Varghese
ICAR-Sugarcane Breeding Institute, Coimbatore, India

V. Sumithra
ICAR-Sugarcane Breeding Institute, Coimbatore, India

C. A. Sowmiya
https://orcid.org/0009-0000-5876-6090
The Indian Agriculture College, Radhapuram, India

Shobana Sampath
https://orcid.org/0000-0002-4291-3516
Vel Tech Rangarajan Dr. Sagunthala R&D Institute of Science and Technology, India

Shoaib Khan
Abbottabad University of Science and Technology, Pakistan

ABSTRACT

Brassica napus is the most cultivated and economically important plant commonly known as rapeseed (2n=4x=38 "AACC"). Globally rapeseed oil ranks third next to soybean and palm. Besides human consumption, it is a versatile crop with wide range of uses in animal feed, lubricants, biodiesel etc. Higher levels of erucic acid controlled by dominant alleles and elevated amounts of glucosinolates governed by polygenes posed challenges for farmers owing to reduced consumer demand due to health concerns. This reduction subsequently imposed stress on breeders to develop

DOI: 10.4018/979-8-3693-8603-3.ch003

agronomically viable market-friendly crops. The development of Low Erucic Acid Rapeseed (LEAR) followed by double zero (00) rapeseed genotypes resulted from continual breeding efforts over the years mainly by cross-breeding followed by selection. On the contrary, High Erucic Acid Rapeseed (HEAR) is bred specifically for industrial applications. This chapter focuses on the genetic architecture dating back to origin of canola and how traditional breeding strategies contributed significantly to the present-day canola plant.

1. INTRODUCTION

Canola belonging to the family of Brassicaceae, thrives in temperate climates, serving food and industrial purposes (McVetty & Scarth, 2012). Its oil is used in cooking, food processing, and non-food applications such as biodiesel production, while its by-products, such as canola meal, are a valuable protein source for livestock (O'Brien, 2016). The crop is also used in crop rotation systems, contributing to soil health by breaking pest cycles and enhancing nutrient availability for subsequent crops.

B. napus (rapeseed), initially resulted from natural hybridization between the two diploid species - *Brassica rapa* (2n = 2x = 20) and *Brassica oleracea* (2n = 2x = 18) (Nagaharu & Nagaharu, 1935), later led to the development of canola by selective breeding. Globally, the largest producer of canola is Canada accounting for 23% of the global production followed by China and India with 18% and 14% respectively ("USDA Foreign Agricultural Service," 2024). In India, canola is usually referred to as rapeseed and its yield can significantly vary based on various factors such as region, soil type and quality, climate and agricultural practices. Traditional rapeseed, while useful in industrial applications, was unsuitable for human consumption due to high levels of erucic acid and glucosinolates, compounds linked to potential health risks. Prolonged consumption of higher levels of erucic acid leads to the accumulation of lipids in the heart muscle fibres eventually weakening the muscle function (Galanty et al., 2023). Understanding its classical genetics provides essential insights into the traits that contribute to its agricultural value, including yield, disease resistance, and adaptability.

Traditional breeding methods primarily involve selection and crossbreeding which generally span over several generations. Plants with desirable traits such as improved oil content, disease resistance and yield are identified and are exposed to controlled pollination to develop hybrids. This meticulous process often requires multiple generations (F_1, F_2, F_3, etc.) to stabilize desired traits before commercialization. Additionally, mutation breeding has been utilized to introduce genetic variability and accelerate the development of new varieties. On the whole, the primary focus for canola breeders is mainly on enhancing oil quality by reducing undesirable compounds such as erucic acid and glucosinolates while also improving agronomic characteristics such as pest resistance and stress tolerance. The evolution of canola breeding through classical genetics and traditional methods illustrates a symbiotic relationship between historical practices and cutting-edge technologies which will be discussed in this chapter.

2. HISTORICAL DEVELOPMENT OF CANOLA

The origin of B.*napus* (rapeseed) is uncertain however it is believed to have originated in the Mediterranean region (McNaughton, 1976). Canola is a genetic variation of rapeseed developed by Canadian plant breeders specifically for its nutritional qualities, particularly its low level of saturated fat. In 1956 the nutritional aspects of rapeseed oil were questioned, especially concerning the high eicosenoic, erucic acid and glucosinolates. This led to the development of canola which started in the 1960s heavily relying on its crop genetics to minimize undesirable traits while simultaneously improving the crop's agricultural performance. The name "canola" is a combination of "Canada" and "ola," meaning oil, a proclamation of its origins and purpose. Over the years, the crop has spread to other major agricultural regions, including the United States, Europe, Australia, China and India driven by its versatility, high oil content, and wide range of uses.

Since the domestication of *B.napus*, the primary objective earlier was to minimize the erucic acid content. Extensive breeding programs were initiated to develop low erucic acid rapeseed cultivars. A natural low erucic acid variant named Liho was discovered in 1964 (Stefansson et al., 1961) and Oro was the first rapeseed cultivar developed by using Liho as one of its parents (Downey, 1990). The development of "Oro" marked one of the early milestones in creating canola with more desirable agronomic traits and better oil quality and cultivars with reduced erucic acid with less than 2% were termed LEAR cultivars (Low Erucic Acid Rapeseed). Glucosinolates are natural compounds found in rapeseed (*Brassica napus*), playing a key role in the plant's defense mechanisms against pests and diseases. They are sulfur-containing secondary metabolites which produce bioactive compounds like isothiocyanates on hydrolysis possessing both harmful and beneficial effects. In rapeseed, glucosinolates impact seed quality, particularly affecting its use in animal feed due to their potential to reduce palatability and nutrient absorption. Breeding efforts often aim to reduce glucosinolate levels in rapeseed varieties to improve their suitability for both oil production and feed use. The first "low erucic" "low glucosinolate" cultivar Tower was developed in 1974 and it was commonly called as double zero (00) cultivars with erucic acid content less than 2% and glucosinolate content less than 30mg per g of oil (Downey, 1990). Since then, several cultivars have been bred using double zero (00) as the base material and almost all the rapeseed cultivars for human consumption were replaced by canola varieties.

Though erucic acid is undesirable for consumption, it has higher potential for industrial applications in the manufacture lubricants, surfactants and other chemical products. Similar to LEAR, High Erucic Acid Rapeseed (HEAR) were bred specifically for industrial purposes (Downey, 1990). Reston, was the first registered HEAR cultivar and since then parallel breeding efforts with an aim to develop cultivars suitable for non-food industry were carried out.

3. GENETICS OF CANOLA

3.1 Genetic Structure

Brassica napus, an allotetraploid contains two sets of homologous chromosomes from two different diploid species. A genome arises from *B.oleracea* and C genome arises from *B.campestris*, undergoes chromosome doubling to give rise to the present day *B.napus* (Figure 1).

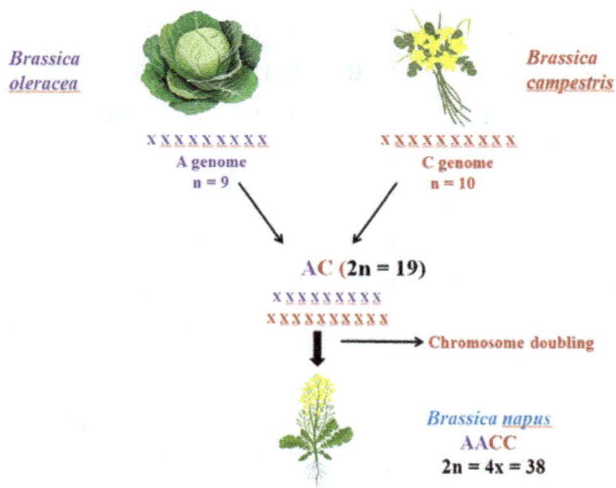

Key features of a polyploid genome:

- Greater variation
- Enhanced adaptability
- Gene redundancy leading to gene plasticity

This polyploidy nature generally contributes to a higher degree of homeologs or duplicate genes causing functionally similar genes or sometimes they diverge and carry out specialized roles through a process called sub functionalization where a copy of a gene entirely acquires a new function. Overall, the genetic structure of canola is shaped by its polyploid nature, hybrid origin, and the extensive breeding efforts that have optimized it for modern agriculture. Canola exhibits substantial genetic diversity, which is crucial for breeding programs. Understanding the genetic variation among canola cultivars helps in selecting parents for hybridization to achieve desirable traits (Moghaieb et al., 2014).

In addition to traditional genetic factors, epigenetics plays a significant role in the regulation of gene expression in canola. Epigenetic modifications, such as DNA methylation and histone modification affect the expression of genes without disturbing the primary DNA sequence. Important traits like the development of seed, flowering time and stress tolerances are greatly influenced by these modifications. Methylation patterns throughout the genome of *Brassica napus* during drought stress were examined and observed changes in genes linked to hormone signalling pathways and stress tolerance (Dowen et al., 2012) (Jian-ping et al., 2020).

Histone modifications such as acetylation, methylation, and phosphorylation affect chromatin structure, thus regulating gene expression. These modifications are crucial for plant development, especially for flowering time regulation, which is important for optimizing yield in canola. Finnegan et al. (2005) highlighted the importance of histone modifications in regulating the expression of the Flowering Locus C (FLC) gene, which caused delayed flowering in canola. Histone methylation of the FLC gene locus was found to be linked to environmental factors such as temperature and day length. Apart from DNA methylation and histone modification, small RNAs (sRNAs), such as microRNAs (miRNAs) and siRNAs, played pivotal roles in regulating plant stress responses, growth and development and also in yield-related traits (Li et al., 2022; J. Wang et al., 2016).

Epigenetic changes are often reversible and can be influenced by environmental factors, making them a key mechanism through which canola adapts to different conditions. The study of epigenetics in canola is still an emerging field, but it has the potential to unlock new avenues for crop improvement. Advances in epigenome editing and CRISPR-based technologies hold promise for precisely targeting and modifying these epigenetic marks to improve canola's adaptability and productivity. Epigenetics along with traditional breeding and genetic engineering, provide a powerful tool for developing superior canola varieties as globally canola continues to be a crucial crop for crop production. Hence, understanding its genetic makeup will be essential for developing improved varieties that can meet the challenges of climate change and evolving agricultural practices. Though advances in genomic tools and molecular breeding have allowed for significant improvements in yield, disease resistance, and oil quality, further research is required to ensure that canola continues to meet the demands of a changing world which will be discussed in the coming chapters.

3.2 Inheritance Pattern In Canola

The inheritance pattern in canola generally follows polygenic inheritance Mendelian genetics but some important agronomically desirable genes are influenced by complex interactions too. Canola being a dual-purpose crop for both oil and cattle feed, understanding its inheritance is essential for any breeding program. The details of traits with their respective inheritance pattern are furnished in table no 1.

Table 1. List of traits with their inheritance pattern.

No.	Trait	Inheritance pattern	Reference
1	Oil content	Polygenic	Jiefu et al. (2010)
2	Erucic acid	Two major genes	Ze-su et al. (2012)
3	Oleic acid	Two major genes	Ze-su et al. (2012)
4	Glucosinolate	Three major genes	Ze-su et al. (2012)
5	White flower	Monogenic	Huang et al. (2014)
6	Blackleg resistance	Single gene (dominant) and polygenic	Yu et al. (2005)
7	Sclerotinia	Quantitative and complex	Saharan et al. (2021)
8	White rust	One or two dominant gene	Saharan et al. (2021)
9	Clubroot	One or two dominant gene	Lammerink (1967) and Saharan et al. (2021)
10	Herbicide resistance	Single gene (dominant)	Huiming et al. (2003)

4. TRADITIONAL BREEDING IN CANOLA IMPROVEMENT

4.1 Breeding Strategies Used in Canola

Traditional breeding in canola (*Brassica napus*) involves the use of classical methods to improve the crop for desirable traits such as yield, oil content, disease resistance, and quality. Most of the methods such as selection, backcross breeding, and pedigree breeding rely on the natural genetic variation within canola and closely related species, and they have been crucial in the development of canola as a major oilseed crop. Breeders focus on developing high-yielding varieties by selecting traits that directly or indirectly affect yield, such as seed number, seed size, oil content, disease resistance, and adaptability to stress conditions (e.g., drought and heat).

4.2 Breeding for Agronomic Traits

Breeding for agronomic traits in canola (*Brassica napus*) is essential to improve characteristics such as yield, flowering time, and seed quality. Breeding programs focus on enhancing these traits through both traditional and modern techniques, ensuring that the crop meets the demands of different growing environments and markets. With the rising global demand for edible oils and biofuels, improving seed yield in *Brassica* crops is crucial for meeting these needs. Canola, a member of the *Brassica* genus, is one of the world's most important oilseed crops. Its seeds are processed to produce vegetable oil, which is used in food, industrial products, and biofuels. High seed yield directly increases the oil yield, making it a significant economic factor for farmers and industry stakeholders (Abu & Mengistu, 2023).

Adjusting flowering time and maturity is equally crucial for adapting canola to different environments and ensuring optimal yield. In some cases, early flowering varieties may escape drought, while later varieties may benefit from extended growing seasons. A more serious problem is pod shattering during winter which can cause significant yield losses. Breeding for shatter resistance helps to reduce seed loss and improves overall productivity. Selection for shatter resistance is done by identifying varieties with stronger pod walls or delayed pod dehiscence (Hossain et al., 2012). Mutagenesis has also been explored to modify genes controlling pod shattering in canola. Lodging is a common problem in canola, especially under high-yielding conditions. Breeding for lodging resistance helps maintain plant standability and facilitates mechanized harvesting where plants with stronger stems with reduced height were selected by convention method and mutagenesis to develop lodging-resistant lines (Berry et al., 2004).

Breeding for agronomic traits in canola was based on traditional methods. Economic and agronomic success in canola can be ultimately achieved by combining it with modern technologies as canola is a rising globally significant crop.

4.3 Breeding for Quality Traits

Breeding for oil content and oil quality in canola (*Brassica napus*) is a significant focus in plant breeding, driven by the growing demand for healthier and more sustainable vegetable oils. Canola oil is valued for its low levels of saturated fats and high levels of polyunsaturated fatty acids, particularly oleic acid, and its use in both the food and biofuel industries. Breeders aim to maximize oil yield without compromising the agronomic traits like seed yield and disease resistance. The focus is on increasing oleic acid content, which is healthier for consumption while reducing erucic acid and saturated fats. The relationship between erucic acid and oleic acid in *Brassica napus* is inversely correlated due to their biosynthetic pathway. By manipulating the genes responsible for fatty acid elongation (such as FAE1), breeders have been able to develop canola varieties with high oleic acid and low erucic acid content, ensuring that canola oil is safe for consumption and possesses desirable nutritional qualities (P. Wang et al., 2022). Simultaneously reducing linolenic acid improves oil stability, which is important for shelf life and cooking (Syed, 2016). There is also growing interest in altering the fatty acid profile to produce lauric and capric acid, useful in industrial applications like biofuel production. Canola breeding initially relied on traditional crossbreeding methods, selecting favourable traits over several generations. Early successes in reducing erucic acid levels were achieved by targeted mutagenesis leading to the development of first low erucic acid cultivar "ORO" (Fukai et al., 2019).

Traditionally, breeders identify and select canola plants with naturally high oil content from existing germplasm. These selected plants serve as parent lines for breeding. Breeders cross these high-oil-content plants with other elite canola lines. Offspring are evaluated for oil content and other desirable traits like disease resistance or yield. The development of early canola varieties such as 'Tower' and 'Candle' involved crossbreeding with traditional European rapeseed cultivars, which had high oil content but also high erucic acid levels. Through successive breeding and selection, low-erucic acid and high-oil-content canola varieties were developed, marking the birth of canola as a globally important oilseed crop (Iniguez-Luy & Federico, 2011). Early Polish canola breeding programs applied mass selection to improve oil content. In mass selection, large populations of canola are grown, and plants with desirable characteristics, such as high oil content, are visually selected. Seeds from these plants are harvested, and subsequent generations are grown and reselected to increase the frequency of high-oil-content plants in the population. They focused on selecting plants that produced large, oil-rich seeds. Over several generations, the average oil content of the population increased (Abbadi & Leckband, 2011).

A more promising method which led to the development of zero-erucic acid canola involved backcrossing high-oil-content donor lines with elite canola varieties. Backcrossing is used to transfer specific traits, such as high oil content, from one variety to another. Breeders cross a high-oil-content donor parent with an elite canola variety that possesses other desirable traits (e.g., disease resistance, and high yield). The progeny are repeatedly crossed back to the elite parent to retain most of the elite parent's traits while incorporating the high-oil trait. For example, Canadian breeders successfully backcrossed the 'Span' and 'Zephyr' lines to develop zero-erucic acid cultivars with high oil content (Rakow & Woods, 1987).

Mutagenesis involves exposing canola seeds or tissue to chemicals (like ethyl methanesulfonate, EMS) or radiation to induce mutations. Canadian breeders used chemical mutagenesis to develop canola varieties with significantly lower levels of erucic acid. These mutations resulted in beneficial changes, including higher oil content. Mutant plants with favourable traits were then selected and used in the breeding program. Although the primary goal was to reduce erucic acid, the process also improved oil content in some lines. Varieties like 'Tobin' and 'Midas' emerged from these efforts, showing both lower erucic acid and improved oil yield.

Hybrid development in canola is designed to develop high-yielding varieties combined with desirable agronomic traits. This greatly relies on the genetic diversity available and uses specialized techniques like male sterility to ensure effective hybridization. Male sterility is extensively used in the production of hybrid canola seeds. Male sterile lines are crossed with fertile restorer lines to produce hybrid seeds, which often exhibit heterosis (hybrid vigour), resulting in higher yields and better stress tolerance (Toriyama, 2021). In *Brassica napus*, cytoplasmic male sterility is more evident and most widely used in hybrid seed production. CMS lines are a type of male sterile lines which are governed by cytoplasmic genes especially, mitochondrial genes which leads to the development of non-viable pollen grains. Ogura, Pol, Nap, NSA and Inap are some of the widely used CMS lines in canola (Kang et al., 2017)

Traditional breeding methods like selection, crossbreeding, mass selection, backcrossing, and mutagenesis have significantly contributed to improving oil content in canola. These techniques remain important even in the era of molecular breeding, as they provide the foundation for the development of high-oil-content varieties.

4.4 Breeding for Biotic Stress Tolerance

Breeding for disease and pest resistance in canola (*Brassica napus*) is critical to improving crop yield and quality while reducing reliance on chemical pesticides. Canola faces several significant diseases and pests, predominant ones are blackleg (Y. Wang et al., 2020), clubroot (Rahman et al., 2014), aphids (Akbar et al., 2017), and cabbage seedpod weevil (Merkosky et al., 2024). Traditional breeding methods for disease and pest resistance in canola involve selecting and crossbreeding plants with natural resistance to biotic stresses such as diseases and insect pests. Large collections of canola germplasm were usually screened for resistance to specific diseases or pests (Al-Lami et al., 2023). Selected plants were exposed to the pathogen or pest, and those that show natural resistance are selected for further breeding. Disease or pest resistance was tested in both field trials and controlled environments (e.g., greenhouses or growth chambers). This ensured that the selected plants performed well under natural and controlled conditions. Once resistant plants were identified, they are crossed with high-yielding, agronomically desirable varieties. The primary goal was to combine the resistance traits of one parent with the yield and quality traits of the other. Breeders select the best offspring from these crosses over multiple generations (F_1, F_2, etc.) to maintain disease or pest resistance while enhancing other key traits such as yield, oil content, and environmental adaptability. In order to improve an existing variety, a resistant plant (donor) is crossed with an elite canola variety (recurrent parent) that lacks resistance but has other desirable traits. The resulting offspring are then repeatedly crossed back to the recurrent parent, preserving most of its traits while introducing resistance from the donor. This method helps to retain most of the high-yielding and quality traits of the recurrent parent while adding the desired resistance traits from the donor parent. A more straightforward, low-cost approach is to select plants that show resistance and other desirable traits among large populations of plants grown in fields with natural pest or disease pressure. Plants that show resistance and other desirable traits are selected for further breeding, and the process is repeated over several generations to improve overall population resistance.

Improved resistance to diseases and pests along with higher yields and better agronomic performance was generally achieved by crossing two genetically distinct inbred lines which exhibit hybrid vigour or "heterosis." Breeders can use restorer lines with known disease resistance to develop hybrids with improved pest and disease tolerance. Induced mutagenesis using radiation or chemicals brings forth new genetic variation in canola plants. Some of these mutations may confer resistance to diseases or pests. Mutants with desirable traits, such as disease resistance, are selected and used in further breeding programs.

Wild relatives of canola, such as *Brassica rapa* or *Brassica oleracea*, often have natural resistance to pests or diseases (Mei et al., 2015). These traits could be introduced into canola through crossbreeding. This process can be slow and difficult, as crossing canola with wild species may lead to undesirable traits (e.g., reduced yield or oil quality), which need to be bred out over successive generations. Breeding for disease and pest resistance in canola is essential for sustainable crop production. By employing various breeding techniques, breeders can enhance resistance to critical diseases and pests. This not only helps improve yield and quality but also reduces the need for chemical controls, supporting more environmentally friendly agricultural practices.

4.5 Breeding for Abiotic Stress

Breeding for abiotic stress tolerance in canola (*Brassica napus*) is crucial to ensuring crop productivity under varying environmental conditions. Abiotic stresses, such as drought, salinity, heat, and frost, can significantly impact canola yield and quality. Selection from germplasms, pedigree breeding, and backcross breeding are some of the methods employed to develop tolerant genotypes. Drought tolerance in canola is achieved through selecting traits like deep root systems, improved water use efficiency, and enhanced physiological traits that enable plants to maintain growth under water-limited conditions (Wu et al., 2018). Salinity stress negatively affects canola growth and yield. Breeding programs prioritized identifying and selecting genotypes that exhibit greater salinity tolerance through physiological mechanisms such as ion exclusion, osmotic adjustment, and improved antioxidant capacity. Screening methods, including hydroponic and field trials, help identify resistant varieties (Bandehagh et al., 2021). Canola being a temperate crop, heat stress during flowering can adversely affect seed set and yield. Breeding for heat tolerance involves selecting for traits that allow plants to maintain reproductive success under high temperatures (Secchi et al., 2023). Frost damage can occur during flowering and pod filling, significantly impacting yield. Breeding for frost tolerance includes selection for physiological and biochemical traits that enhance cold tolerance. Developing early-maturing varieties can help mitigate frost damage by avoiding late-season freezing (Raman et al., 2019). Focusing on traits associated with drought, salinity, heat, frost, and nutrient stress tolerance, breeders can develop resilient canola varieties that can thrive under changing environmental conditions, ultimately enhancing food security and sustainable agricultural practices.

5. FUTURE OF CANOLA BREEDING

A blend of traditional breeding methods integrated with modern technologies (FIGURE 2) is required to meet the increasing global demand for canola. Breeders need to focus more on the development of cultivars meeting both market demands and environmental challenges, giving importance to sustainable agriculture, climate resilience and consumer preferences. As research continues to evolve, the collaboration between scientists, farmers, and industry stakeholders will be essential in shaping the next generation of canola crops.

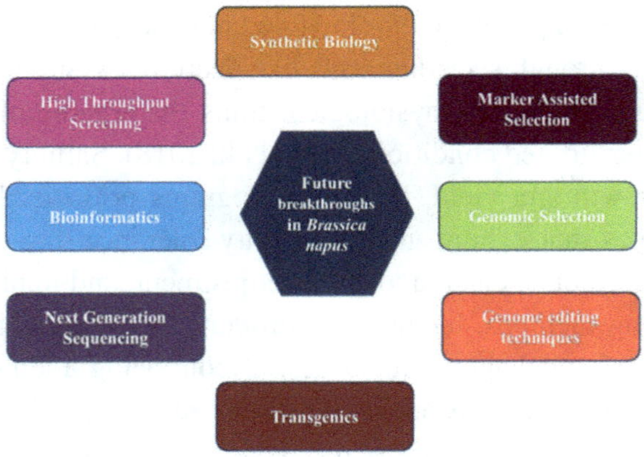

6. CHALLENGES IN CANOLA BREEDING

Canola breeding, like the breeding of any other crop, faces several challenges, both technical and environmental. Modern canola cultivars have a limited gene pool due to repeated selection for desirable traits, such as low levels of erucic acid and glucosinolates. This narrow genetic diversity limits the potential for further improvement and makes the crop more vulnerable to diseases and environmental stresses. This limits the plant's ability to respond to new challenges like climate change or evolving pathogens by leading to the breakdown of resistance to serious pests and diseases (Cowling, 2007). Achieving significant increases in canola yield potential is challenging due to the complexity of yield-related traits, which are influenced by multiple factors including plant architecture, photosynthetic efficiency, and seed setting. While yield increases have been achieved through conventional breeding, the gains have been relatively modest. New breeding techniques like genomic selection may offer more promise for substantial yield improvements (Ahmad et al., 2022, Ahmad et al., 2024, Ullah et al., 2023). Regarding quality traits, high-oleic canola lines have been developed, but balancing improved oil profiles with competitive yield is a breeding challenge due to negative correlations between quality and biometrical traits (McVetty & Scarth, 2002). Hybrid vigour, or heterosis, is a critical factor in boosting canola yields. However, developing stable and cost-effective hybrid breeding systems is complex and labour-intensive. Hybrid seed production systems depend on effective male sterility and fertility restoration mechanisms, which can be difficult to manage. Hybrid canola varieties have been widely adopted, but maintaining hybrid purity and seed production efficiency remains challenging (Ton et al., 2020).

Changing climate conditions, such as erratic rainfall patterns and increasing temperatures, introduce new challenges for canola breeding. Breeders need to develop varieties that are resilient to fluctuating weather patterns, but breeding for climate resilience is a time-consuming and resource-intensive process. While modern technologies such as genome editing (CRISPR-Cas9), marker-assisted selection (MAS), and genomic selection offer promise, integrating these tools into breeding programs requires substantial investment in infrastructure, knowledge, and regulatory approvals. The use of CRISPR in canola breeding is still in the experimental stages, but early trials have shown potential for targeted improvements in disease resistance and oil quality. Canola breeding faces multifaceted challenges ranging from narrow genetic diversity to the integration of advanced breeding technologies. Addressing these challenges requires continuous research, technological innovation, and collaboration between breeders, geneticists, and agronomists (Lohani et al., 2020, Ijaz et al., 2023, Ijaz et al., 2024).

7. CONCLUSION

The significant achievements visualized in canola today are the result of traditional breeding methods. Efforts were made earlier on simpler traits like erucic acid and glucosinolate content, which had significant impacts on canola's suitability as an edible oil crop. Through these traditional breeding approaches, canola evolved from a minor oilseed into a major global crop with vast economic importance. Cytoplasmic male sterility (CMS) and the exploitation of heterosis through hybrid breeding further accelerated the development of superior canola varieties, paving the way for higher yields and better stress tolerance. The development of early-maturing varieties, resistance to lodging, and adaptability to diverse climates are other achievements rooted in classical breeding practices. As canola continues to face new challenges related to climate change, disease pressures, and evolving market demands, the foundation of classical genetics will remain invaluable. Future breeding programs will increasingly rely on the rich genetic base developed through traditional methods while leveraging cutting-edge tools to meet the needs of the agricultural industry and global food security. Classical genetics and traditional breeding in canola have not only shaped the crop's past but will also continue to influence its future. Moreover, the synergy between conventional and modern breeding methods promises to ensure the ongoing success of canola as a sustainable, high-yielding, and economically important oilseed crop.

REFERENCES

Abbadi, A., & Leckband, G. (2011). Rapeseed breeding for oil content, quality, and sustainability. *European Journal of Lipid Science and Technology, 113*(10), 1198–1206. DOI: 10.1002/ejlt.201100063

Abu, M., & Mengistu, B. (2023). *Correlation Coefficient and Path Coefficient Analysis for Seed Yield and Yield Related Traits of Rapeseed (Brassica Napus)*. Genotypes.

Ahmad, Y., Haider, S., Iqbal, J., Abbasi, B. A., Yaseen, T., & Mahmood, T. (2022). The mechanisms of genome editing technologies in crop plants. In *Principles and Practices of OMICS and Genome Editing for Crop Improvement* (pp. 295–313). Springer International Publishing. DOI: 10.1007/978-3-030-96925-7_13

Ahmad, Y., Haider, S., Iqbal, J., Naseer, S., Attia, K. A., Mohammed, A. A., & Mahmood, T. (2024). In-silico analysis and transformation of OsMYB48 transcription factor driven by CaMV35S promoter in model plant–Nicotiana tabacum L. conferring abiotic stress tolerance. *GM Crops and Food: Biotechnology in Agriculture and the Food Chain, 15*(1), 130–149. DOI: 10.1080/21645698.2024.2334476 PMID: 38551174

Akbar, W., Asif, M. U., Ismail, M., Bux, M., & Memon, R. M. (2017). Management of aphids on canola (*Brassica napus* L) through cultural practices. *Pakistan Entomologist, 39*(2).

Al-Lami, H. F. D., You, M. P., Banga, S. S., & Barbetti, M. J. (2023). Novel resistances provide new avenues to manage alternaria leaf spot (*Alternaria brassicae*) in canola (*Brassica napus*), mustard (B. *juncea*), and other brassicaceae crops. *Plant Disease, 107*(2), 372–381. DOI: 10.1094/PDIS-05-22-1153-RE PMID: 35802017

Bandehagh, A., Dehghanian, Z., Henry, R., & Hossain, M. A. (2021). *Salinity tolerance in canola: Insights from proteomic studies Brassica Breeding and Biotechnology*. IntechOpen., DOI: 10.5772/intechopen.96649

Berry, P. M., Sterling, M., Spink, J. H., Baker, C. J., Sylvester-Bradley, R., Mooney, S. J., Tams, A. R., & Ennos, A. R. (2004). Understanding and reducing lodging in cereals. *Advances in Agronomy, 84*(04), 215–269. DOI: 10.1016/S1006-8104(13)60030-6

Cowling, W. A. (2007). Genetic diversity in Australian canola and implications for crop breeding for changing future environments. *Field Crops Research, 104*(1-3), 103–111. DOI: 10.1016/j.fcr.2006.12.014

Dowen, R. H., Pelizzola, M., Schmitz, R. J., Lister, R., Dowen, J. M., Nery, J. R., Dixon, J. E., & Ecker, J. R. (2012). Widespread dynamic DNA methylation in response to biotic stress. *Proceedings of the National Academy of Sciences of the United States of America*, *109*(32), E2183–E2191. DOI: 10.1073/pnas.1209329109 PMID: 22733782

Downey, R. (1990). Brassica oilseed breeding-achievements and opportunities.

Finnegan, E. J., Kovac, K. A., Jaligot, E., Sheldon, C. C., James, P. W., & Dennis, E. S. (2005). The downregulation of Flowering Locus C (FLC) expression in plants with low levels of DNA methylation and by vernalization occurs by distinct mechanisms. *The Plant Journal*, *44*(3), 420–432. DOI: 10.1111/j.1365-313X.2005.02541.x PMID: 16236152

Fukai, E., Karim, M. M., Shea, D. J., Tonu, N. N., Falk, K. C., Funaki, T., & Okazaki, K. (2019). An LTR retrotransposon insertion was the cause of world's first low erucic acid *Brassica rapa* oilseed cultivar. *Molecular Breeding*, *39*(1), 1–12. DOI: 10.1007/s11032-018-0916-9

Galanty, A., Grudzińska, M., Paździora, W., & Paśko, P. (2023). Erucic acid—both sides of the story: A concise review on its beneficial and toxic properties. *Molecules (Basel, Switzerland)*, *28*(4), 1924. DOI: 10.3390/molecules28041924 PMID: 36838911

Hossain, S., Kadkol, G., Raman, R., Salisbury, P., & Raman, H. (2012). Breeding *Brassica napus* for shatter resistance. *Plant Breeding*, *14*(1), 313–332.

Huang, Z., Ban, Y. Y., Bao, R., Zhang, X. X., Xu, A. X., & Ding, J. (2014). Inheritance and gene mapping of the white flower in *Brassica napus* L. *New Zealand Journal of Crop and Horticultural Science*, *42*(2), 111–117. DOI: 10.1080/01140671.2013.863211

Huiming, P., Jiefu, Z., Cunkou, Q., Jianqin, G., & Xinjun, C. (2003). Inheritance of herbicide resistance in *Brassica napus* and its utilization. *Jiangsu Journal of Agricultural Sciences*, *19*(2), 81–86. DOI: 10.3390/ijms21041345

Ijaz, S., Iqbal, J., Abbasi, B. A., Tufail, A., Ullah, Z., Sharifi-Rad, J., & Uddin, S. (2023). Biofortification: Lessons from the Past and Strategies for Future Food Security. In *Legumes Biofortification* (pp. 521–545). Springer International Publishing. DOI: 10.1007/978-3-031-33957-8_23

Ijaz, S., Iqbal, J., Abbasi, B. A., Yaseen, T., Rehman, S., Kazi, M., & Mahmood, T. (2024). Role of OMICS-Based Technologies in Plant Sciences. OMICs-based Techniques for Global Food Security, 45-66.

Iniguez-Luy, F. L., & Federico, M. L. (2011). The genetics of *Brassica napus*. *Genetics and Genomics of the Brassicaceae*, 291-322. DOI: 10.1007/978-1-4419-7118-0_10

Jian-ping, W., Shu-fen, Z., Jia-cheng, Z., Jun-ping, H., Jin-hua, C., Yan-cheng, W., Kun, H., Lei, Z., Dong-guo, W., & Dong-fang, C. (2020). Genome-wide analysis of DNA methylation under drought in rapeseed (*Brassica napus* L.). *Zhongguo Youliao Zuowu Xuebao*, *42*(4).

Jiefu, Z., Cunkou, Q., Huiming, P., Song, C., Feng, C., Jianqin, G., Xinjun, C., Hui, G., & Shouzhong, F. (2010). Inheritance and QTL identification of oil content in rapeseed *Brassica napus* L.

Kang, L., Li, P., Wang, A., Ge, X., & Li, Z. (2017). A novel cytoplasmic male sterility in *Brassica napus* (inap CMS) with carpelloid stamens via protoplast fusion with Chinese woad. *Frontiers in Plant Science*, *8*, 529. DOI: 10.3389/fpls.2017.00529 PMID: 28428799

Lammerink, J. (1967). The inheritance of clubroot resistance in *Brassica napus* L. *New Zealand Journal of Agricultural Research*, *10*(1), 109–115. DOI: 10.1080/00288233.1967.10423081

Li, J., Li, Y., Wang, R., Fu, J., Zhou, X., Fang, Y., Wang, Y., & Liu, Y. (2022). Multiple Functions of MiRNAs in *Brassica napus* L. *Life (Chicago, Ill.)*, *12*(11), 1811. DOI: 10.3390/life12111811 PMID: 36362967

Lohani, N., Jain, D., Singh, M. B., & Bhalla, P. L. (2020). Engineering multiple abiotic stress tolerance in canola, *Brassica napus*. *Frontiers in Plant Science*, *11*, 3. DOI: 10.3389/fpls.2020.00003 PMID: 32161602

McNaughton, H. I. (1976). *Turpic and relatives* (Simmonds, N. W., Ed.). Longman.

McVetty, B. E., & Scarth, R. (2012). Breeding oil quality for edible and industrial applications in oilseed rape and linseed. *Technological Innovations in Major World Oil Crops, Volume 2. Perspectives*, •••, 105–122.

McVetty, P. B. E., & Scarth, R. (2002). Breeding for improved oil quality in Brassica oilseed species. *Journal of Crop Production*, *5*(1-2), 345–369. DOI: 10.1300/J144v05n01_14

Mei, J., Liu, Y., Wei, D., Wittkop, B., Ding, Y., Li, Q., Li, J., Wan, H., Li, Z., Ge, X., Frauen, M., Snowdon, R. J., Qian, W., & Friedt, W. (2015). Transfer of sclerotinia resistance from wild relative of *Brassica oleracea* into *Brassica napus* using a hexaploidy step. *Theoretical and Applied Genetics*, *128*(4), 639–644. DOI: 10.1007/s00122-015-2459-3 PMID: 25628163

Merkosky, K., Leibovitz, A., & Wong, S. (2024). Cabbage Seedpod Weevil (*Ceutorhynchus obstrictus*) Management in Canola (*Brassica napus*) across western Canada. *Plant Health Cases*(2024). DOI: 10.1079/planthealthcases.2024.0014

Moghaieb, R. E. A., Mohammed, E. H. K., & Youssief, S. S. (2014). Genetic diversity among some canola cultivars as revealed by RAPD, SSR and AFLP analyses. *3 Biotech, 4*, 403-410.

Nagaharu, U., & Nagaharu, N. (1935). Genome analysis in Brassica with special reference to the experimental formation of B. napus and peculiar mode of fertilization. *Japanese Journal of Botany*, *7*(7), 389–452.

O'Brien, D. (2016). Canola: Good protein source for dairy cattle. *Agricultural Research*, *64*(2), 1.

Rahman, H., Peng, G., Yu, F., Falk, K. C., Kulkarni, M., & Selvaraj, G. (2014). Genetics and breeding for clubroot resistance in Canadian spring canola (*Brassica napus* L.). *Canadian Journal of Plant Pathology, 36*(sup1), 122-134. DOI: 10.1080/07060661.2013.862571

Rakow, G., & Woods, D. L. (1987). Outcrossing in rape and mustard under Saskatchewan prairie conditions. *Canadian Journal of Plant Science*, *67*(1), 147–151. DOI: 10.4141/cjps87-017

Raman, H., Uppal, R. K., & Raman, R. (2019). Genetic solutions to improve resilience of canola to climate change. *Genomic designing of climate-smart oilseed crops*, 75-131. DOI: 10.1007/978-3-319-93536-2_2

Saharan, G. S., Mehta, N. K., Meena, P. D., Saharan, G. S., Mehta, N. K., & Meena, P. D. (2021). Inheritance of Disease Resistance. *Genomics of Crucifer's Host-Resistance*, 195-264.

Secchi, M. A., Fernandez, J. A., Stamm, M. J., Durrett, T., Prasad, P. V. V., Messina, C. D., & Ciampitti, I. A. (2023). Effects of heat and drought on canola (*Brassica napus* L.) yield, oil, and protein: A meta-analysis. *Field Crops Research, 293*, 108848. DOI: 10.1016/j.fcr.2023.108848

Stefansson, B. R., Hougen, F. W., & Downey, R. K. (1961). Note on the isolation of rape plants with seed oil free from erucic acid. *Canadian Journal of Plant Science*, *41*(1), 218–219. DOI: 10.4141/cjps61-028

Syed, A. (2016). *Oxidative stability and shelf life of vegetable oils Oxidative stability and shelf life of foods containing oils and fats*. Elsevier., DOI: 10.1016/B978-1-63067-056-6.00004-5

Ton, L. B., Neik, T. X., & Batley, J. (2020). The use of genetic and gene technologies in shaping modern rapeseed cultivars (*Brassica napus* L.). *Genes*, *11*(10), 1161. DOI: 10.3390/genes11101161 PMID: 33008008

Toriyama, K. (2021). Molecular basis of cytoplasmic male sterility and fertility restoration in rice. *Plant Biotechnology (Sheffield, England)*, *38*(3), 285–295. DOI: 10.5511/plantbiotechnology.21.0607a PMID: 34782814

Ullah, Z., Iqbal, J., Abbasi, B. A., Ijaz, S., Sharifi-Rad, J., Yaseen, T., & Mahmood, T. (2023). How Genome Editing Can Be Helpful in the Biofortification of Legumes. In *Legumes Biofortification* (pp. 207–232). Springer International Publishing. DOI: 10.1007/978-3-031-33957-8_8

USDA Foreign Agricultural Service. (2024). *Rapeseed Explorer - World Production.* Retrieved from https://ipad.fas.usda.gov/cropexplorer/cropview/commodityView.aspx?cropid=2226000

Wang, J., Jian, H., Wang, T., Wei, L., Li, J., Li, C., & Liu, L. (2016). Identification of microRNAs actively involved in fatty acid biosynthesis in developing *Brassica napus* seeds using high-throughput sequencing. *Frontiers in Plant Science*, *7*, 1570. DOI: 10.3389/fpls.2016.01570 PMID: 27822220

Wang, P., Xiong, X., Zhang, X., Wu, G., & Liu, F. (2022). A review of erucic acid production in Brassicaceae oilseeds: Progress and prospects for the genetic engineering of high and low-erucic acid rapeseeds (*Brassica napus*). *Frontiers in Plant Science*, *13*, 899076. DOI: 10.3389/fpls.2022.899076 PMID: 35645989

Wang, Y., Strelkov, S. E., & Hwang, S. F. (2020). Yield losses in canola in response to blackleg disease. *Canadian Journal of Plant Science*, *100*(5), 488–494. DOI: 10.1139/cjps-2019-0259

Wu, W., Ma, B. L., & Whalen, J. K. (2018). Enhancing rapeseed tolerance to heat and drought stresses in a changing climate: Perspectives for stress adaptation from root system architecture. *Advances in Agronomy*, *151*, 87–157. DOI: 10.1016/bs.agron.2018.05.002

Yu, F., Lydiate, D. J., & Rimmer, S. R. (2005). Identification of two novel genes for blackleg resistance in *Brassica napus. Theoretical and Applied Genetics*, *110*(5), 969–979. DOI: 10.1007/s00122-004-1919-y PMID: 15798929

Ze-su, H., Paisan, L., Thitiporn, M., Ze-hui, C., Wen-dong, D., Rong, T., & Dezhen, L. (2012). Inheritance of erucic acid, glucosinolate, and oleic acid contents in rapeseed (*Brassica napus* L.). [English Edition]. *The Journal of Northeast Agricultural University*, *19*(2), 1–8. DOI: 10.1016/S1006-8104(13)60030-6

Chapter 4
Genome Assembly of Canola Plant (Brassica napus)

Amber Jabeen
https://orcid.org/0009-0001-4501-3345
Quaid-i-Azam University, Islamabad, Pakistan

Muhammad Kaleem
https://orcid.org/0009-0003-6548-3920
Quaid-i-Azam University, Islamabad, Pakistan

Lubna Anjum Minhas
Quaid-i-Azam University, Islamabad, Pakistan

Javed Iqbal
https://orcid.org/0000-0002-9032-2622
Bacha Khan University, Charsadda, Pakistan

Banzeer Ahsan Abbasi
https://orcid.org/0000-0003-3036-4481
Rawalpindi Women University, Pakistan

Abdul Samad Mumtaz
https://orcid.org/0000-0002-7034-9203
Quaid-i-Azam University, Islamabad, Pakistan

ABSTRACT

Genome assembly has emerged as a fundamental tool within genomics, facilitating the reconstruction of entire DNA sequences while providing comprehensive insights into the genetic composition of various organisms. This chapter concentrates on the genome assembly of Brassica napus (canola), a species of considerable agricultural and economic significance. The assembly of the canola genome presents significant potential for expediting breeding initiatives, advancing genetic enhancement, and enriching our comprehension of the plant's biological mechanisms. Nevertheless, the intricate characteristics of the canola genome—characterized by polyploidy, repetitive sequences, and elevated heterozygosity—pose substantial challenges. The resolution of these difficulties necessitates the utilization of sophisticated sequencing

DOI: 10.4018/979-8-3693-8603-3.ch004

technologies, such as long-read platforms, alongside refined bioinformatics methodologies specifically tailored for polyploid genomes. The outlook for canola genomics appears optimistic with climate-resilient varieties to bolster global food security.

INTRODUCTION

Canola, scientifically classified as *Brassica napus*, represents a pivotal oilseed crop of global significance, belonging to the Brassicaceae family, which encompasses other crops such as cabbage (*Brassica oleracea*), mustard (*Brassica juncea*), and broccoli. Canola is esteemed particularly for its oil, characterized by its minimal saturated fat content and substantial proportions of monounsaturated fats, omega-3, and omega-6 fatty acids. This advantageous nutritional composition has rendered canola oil a fundamental ingredient in culinary practices worldwide, serving both household cooking and food production industries. Furthermore, canola oil finds application in the synthesis of biodiesel, lubricants, and various industrial uses, thereby augmenting its economic relevance.

The genesis of canola traces back to the 1970s, when researchers in Canada successfully developed it through conventional plant breeding methodologies. The primary aim was to diminish the concentrations of erucic acid and glucosinolates in rapeseed oil, compounds deemed undesirable for human ingestion. The outcome was a variant of *Brassica napus* that yielded oil with less than 2% erucic acid and reduced glucosinolate levels in the meal, thereby rendering it appropriate for both human and animal nourishment (Bhardwaj et al., 2003). The term "canola" is derived from "Canadian oil, low acid," signifying its origin and its unique attributes distinguishing it from conventional rapeseed (Cardoza et al., 2004).

Canola plants are herbaceous in nature and typically attain heights ranging from 0.9 to 1.5 meters. They possess vibrant yellow inflorescences that yield seed pods containing diminutive, spherical seeds. These seeds constitute the principal product harvested, as they are characterized by a high oil content, generally comprising approximately 40-45% of their weight. The residual seed material, subsequent to oil extraction, is repurposed as a high-protein meal for livestock feed, positioning canola as an essential component within both food and agricultural systems (Wanasundara JP et al., 2016), (Theodoridou K et al., 2013).

On a global scale, canola is cultivated across a diverse array of environments, ranging from the Canadian prairies to the European Union, Australia, and China. Its capacity to flourish in various climatic conditions constitutes one of the key factors contributing to its extensive adoption (Porter et al., 2020). The agronomic advantages of canola further enhance its appeal to agricultural practitioners. The species exhibits a robust capacity for nitrogen fixation, thereby diminishing the necessity for synthetic fertilizers, and it facilitates crop rotation practices by disrupting pest and disease cycles. This positions canola as a crucial crop not only for its immediate economic advantages but also for its contribution to sustainable agricultural methodologies (McBeath et al., 2020).

IMPORTANCE OF GENOME ASSEMBLY

Genome assembly constitutes the systematic process of reconstructing the DNA sequence of an organism by synthesizing short DNA fragments acquired through advanced sequencing methodologies. The primary objective of this endeavor is to recompose the original sequence with maximal precision, thereby providing a comprehensive and intricate representation of the organism's genome. In the case of canola, this procedure is particularly vital due to the plant's intricate polyploid genome, characterized by the presence of multiple chromosomal sets and a considerable proportion of repetitive DNA (Fu et al., 2016).

The importance of genome assembly in canola research is underscored by its capacity to transform breeding initiatives and genetic investigations. A meticulously assembled genome acts as a foundational reference that researchers and breeders can utilize to pinpoint genes linked to significant agronomic characteristics, including oil content, disease resistance, flowering time, and stress tolerance (Harper et al., 2012). With access to a comprehensive genome sequence, scientists are empowered to conduct genome-wide association studies (GWAS) that facilitate the correlation of specific genetic variants with phenotypic traits, thereby enabling more focused breeding methodologies (Sumitomo et al., 2019).

In conventional breeding programs, the procedure of generating new varieties tends to be labor-intensive and time-consuming. This process necessitates the crossing of plants, the selection of progeny exhibiting desirable traits, and the subsequent backcrossing to stabilize the trait within the population. The advent of genome assembly has the potential to expedite this process via marker-assisted selection (MAS). MAS enables breeders to target specific genetic markers associated with favorable traits, thus significantly compressing the breeding cycle and diminishing the time needed to develop new cultivars (Sumitomo et al., 2019). For instance, should a specific gene responsible for elevated oil content be discerned within the canola genome, breeders can employ markers associated with that gene to select plants that harbor it, thereby enhancing the effectiveness of breeding programs.

Genome assembly is also integral to the field of comparative genomics, which entails the examination of genomes across different species to identify conserved and divergent genomic regions. In relation to canola, its genome can be juxtaposed with those of other Brassica species, such as *Brassica oleracea* (cabbage, broccoli) and *Brassica rapa* (turnip, Chinese cabbage), to elucidate the evolutionary connections and the genetic foundations of traits that are either shared or distinct among these species (Schranz et al., 2006). This understanding can be leveraged to impart advantageous traits from related species into canola through either breeding or genetic engineering (Yim et al., 2022).

Another salient application of genome assembly resides within the domain of functional genomics, which aims to elucidate the roles of genes and regulatory elements within the genomic architecture. Possessing a complete genome sequence allows researchers to investigate the expression and regulation of genes in response to varying environmental stimuli. This is particularly critical in canola, where discerning the plant's responses to stresses such as drought, salinity, and temperature extremes can be instrumental in crafting more resilient crop varieties (Lohani et al., 2020). In light of climate change, where agricultural conditions are becoming increasingly erratic, the capacity to breed crops that can endure environmental stress is of utmost significance.

In addition to its significance in conventional breeding practices, genome assembly is also essential to contemporary biotechnological methodologies such as gene editing (Tian et al., 2022). Techniques such as CRISPR-Cas9 necessitate an accurate understanding of the genome to selectively target specific genes for modification. With a comprehensive genome assembly, researchers are enabled to design CRISPR systems that facilitate the introduction of advantageous traits or the elimination of detrimental ones with remarkable precision (Manchanda et al., 2018). For example, CRISPR could be employed to alter genes implicated in oil biosynthesis, thereby generating canola varieties characterized by superior oil quality or to eradicate genes that render the plant vulnerable to specific diseases (Hu J et al., 2022).

Moreover, genome assembly lays the groundwork for comprehending epigenetic modifications, which refer to alterations in gene expression that occur without changes to the DNA sequence itself. Such modifications, including DNA methylation and histone modification, can exert considerable influence on plant development and responses to environmental stressors (Kalantarahmadi et al., 2023). The research into the epigenome, when paired with the genomic backdrop, enables scientists to understand the regulatory dynamics of gene expression and how these may be harnessed to elevate crop output.

Overall, the significance of genome assembly in canola research is paramount. It offers the essential foundational knowledge required to propel breeding initiatives, investigate genetic diversity, and devise groundbreaking biotechnological interventions (Sabbahi et al., 2023). As sequencing technologies progress, the quality and comprehensiveness of genome assemblies are expected to improve, thereby further augmenting our capacity to investigate and manipulate the canola genome for agricultural advancement

HISTORICAL BACKGROUND OF CANOLA GENOME RESEARCH

The investigation of the canola genome boasts a comprehensive history that mirrors the broader advancement of genomic research within the realm of botany. This progression commenced with the innovation of canola itself, which emerged from conventional breeding practices aimed at enhancing the quality of rapeseed oil. In the preliminary phases, the primary emphasis was placed upon breeding methodologies and cytogenetics, which pertains to the examination of chromosome structure and functionality.

The initial advancements in canola genome research transpired during the 1980s and 1990s with the introduction of molecular biology methodologies. In this timeframe, scholars initiated an exploration of the canola genome at a molecular dimension, employing instruments such as restriction fragment length polymorphisms (RFLPs) and simple sequence repeats (SSRs) (Landry et al., 1991). These molecular markers facilitated the construction of linkage maps, which delineate the relative positions of genes or markers situated on chromosomes. Such maps yielded the inaugural insights into the genetic framework of canola, pinpointing genomic regions correlated with significant traits such as oil content, flowering temporal dynamics, and resistance to diseases (Ferreira et al., 1994), (Cai D et al., 2014).

An early obstacle encountered in canola genome research was its polyploidy. As an allotetraploid organism, canola possesses two sets of chromosomes derived from two distinct species, namely *Brassica rapa* (AA genome) and *Brassica oleracea* (CC genome). This intricate genomic architecture complicated the differentiation between homologous chromosomes (those that exhibit similarity yet originate from different species) and the precise mapping of genetic markers. In spite of these hurdles, the linkage maps (Jean et al., 1997) generated during this epoch established a foundational basis for more intricate genomic investigations in subsequent years (Chalhoub et al., 2014).

A pivotal achievement in canola genome research was the sequencing of the *Arabidopsis thaliana* genome in the year 2000. *Arabidopsis*, a diminutive flowering genus belonging to the Brassicaceae family, exhibits numerous genetic resemblances to canola and other Brassica species. The accomplishment of the *Arabidopsis* genome furnished a reference framework for comparative genomics, facilitating the identification of homologous genes in canola and other related taxa (Arabidopsis Genome Initiative, 2000). This development was critical, as it permitted the transference of insights acquired from Arabidopsis research to agricultural species, including canola.

The early 2000s also witnessed the advent of high-throughput sequencing technologies, which transformed genome research across diverse organisms, encompassing plants. Next-generation sequencing (NGS) platforms such as Illumina empowered researchers to generate vast quantities of DNA sequence data in a significantly reduced timeframe and cost compared to prior techniques (Ecke et al., 2010). For canola, this advancement implied that it became feasible to sequence extensive segments of the genome and initiate the process of assembling it into a comprehensive reference sequence.

The initial draft genome assembly of *Brassica napus*, commonly known as canola, was published in 2014, marking a notable milestone within the domain of plant genomics. This assembly was constructed utilizing a synergistic approach that integrated next-generation sequencing (NGS) data alongside traditional Sanger sequencing data, thereby establishing a reference genome for the species (Chalhoub et al., 2014). Although the draft assembly was not devoid of imperfections—evidenced by the presence of gaps and regions exhibiting suboptimal resolution—it nevertheless constituted a significant advancement in the field. This resource enabled researchers to identify genes and genetic markers that are correlated with crucial agronomic traits, while also facilitating comparative genomics analyses with other species within the Brassica genus (**Figure. 1**).

Figure 1. Graphical presentation of historical research background of Canola genome

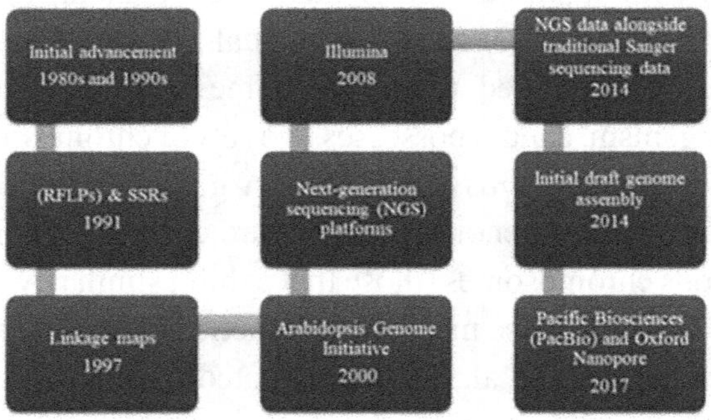

Historical Background of Canola Genome Research

In the years subsequent to the dissemination of the draft genome, continuous advancements in sequencing technologies have markedly enhanced the quality of canola genome assemblies. The advent of third-generation sequencing methodologies, including those developed by Pacific Biosciences (PacBio) and Oxford Nanopore, has facilitated the generation of considerably longer DNA reads. Such extended reads are particularly advantageous for the assembly of genomes characterized by intricate structural features, notably those of polyploid organisms such as canola (Kyriakidou et al., 2018). By providing sequences that encompass repetitive regions and structural variants, third-generation sequencing has empowered researchers to generate genome assemblies that are more comprehensive and precise.

As the fidelity of genome assemblies has improved, so too has the capacity for genome annotation, which entails the identification and characterization of the functions of genes and other genomic elements. The amalgamation of high-quality genome assemblies and comprehensive annotations has engendered novel opportunities for functional genomics research in canola. Researchers are now positioned to investigate gene expression dynamics, identify regulatory sequences, and examine the epigenetic modifications that govern gene activity (Saad et al., 2021). These revelations are being harnessed to engineer novel canola varieties with enhanced phenotypic traits, including superior oil composition, augmented disease resistance, and increased resilience to environmental stressors.

BIOLOGY AND GENETICS OF CANOLA

Canola Breeding And Genetic Improvement

The cultivation of canola represents a significant achievement within agricultural practices, propelled by the necessity to enhance the quality of rapeseed oil and to satisfy the requirements of an expanding global population (Downey et al., 1983). While the foundational development of canola employed conventional breeding methodologies, contemporary canola breeding has progressively utilized sophisticated genetic and biotechnological instruments to incorporate and amplify desirable phenotypic attributes (Rahman M et al., 2011).

The inception of the canola breeding process can be traced back to the 1960s in Canada, where investigators endeavored to diminish the concentrations of erucic acid and glucosinolates present in rapeseed oil (Stefansson et al., 1964). Erucic acid, characterized as a monounsaturated omega-9 fatty acid, has been identified as having detrimental health implications when ingested in substantial amounts, whereas glucosinolates are regarded as undesirable due to their adverse effects on the palatability and safety of rapeseed meal utilized for livestock consumption (Scarth et al., 2006). Through the application of selective breeding techniques, researchers successfully generated *Brassica napus* lines exhibiting markedly reduced levels of these compounds, culminating in the establishment of the initial canola varieties during the 1970s (Norton et al., 1975).

Conventional breding strategies, including crossbreeding and backcrossing, have been instrumental in the advancement of canola (Witkop B et al., 2009). These methodologies entail the hybridization of various varieties or species to amalgamate desirable traits, followed by the selection of progeny that manifest those traits.

Over successive generations, these traits achieve stability within the population. For instance, breeders have implemented these techniques to produce canola varieties demonstrating resistance to diseases such as blackleg (*Leptosphaeria maculans*) and sclerotinia stem rot (*Sclerotinia sclerotiorum*), which pose significant threats to canola agriculture (Kutcher et al., 2010).

With the progression of genetic knowledge and technological advancements, marker-assisted selection (MAS) emerged as an invaluable resource in canola breeding. MAS facilitates the selection of plants possessing desirable traits predicated on the identification of specific genetic markers associated with those traits (Cobb et al., 2019). This approach substantially accelerates the breeding process by enabling the early detection of plants that harbor advantageous genes, even prior to the expression of the corresponding phenotypic characteristics. For example, MAS has been employed to introduce and select for resistance to clubroot (*Plasmodiophora brassicae*), a soil-borne pathogen responsible for considerable yield reductions in canola (Rahman H et al., 2014).

The introduction of biotechnology during the late 20th century heralded additional progress in canola breeding practices. One of the most noteworthy innovations was the emergence of herbicide-resistant canola, commonly referred to as genetically modified (GM) canola (Cardoza et al., 2004). This development was accomplished through the introduction of bacterial genes that confer resistance to specific herbicides, such as glyphosate (Roundup) or glufosinate (Liberty). These GM canola varieties empower farmers to manage weed populations more effectively, thereby resulting in enhanced yields and diminished reliance on chemical herbicides (Beckie et al., 2006). The adoption of herbicide-resistant canola has proliferated in nations such as Canada, the United States, and Australia, where it has significantly contributed to increased agricultural production and economic viability (Brookes et al., 2014).

In conjunction with herbicide resistance, genetic engineering has been employed to enhance various attributes of canola, including oil composition and resistance to diseases. For instance, scientists have engineered canola strains with altered fatty acid profiles, resulting in oils that possess elevated levels of oleic acid and diminished levels of polyunsaturated fats, which exhibit greater stability and extended shelf lives (Scarth et al., 2006). These high-oleic canola oils are particularly esteemed in the food sector for frying and other culinary applications that involve high temperatures (Warner et al., 1997).

A further significant domain of genetic enhancement in canola encompasses the creation of hybrid varieties. Hybrid breeding entails the crossbreeding of two genetically distinct parental lines to yield offspring (hybrids) that demonstrate heterosis, or hybrid vigor. These hybrid plants frequently display superior outcomes regarding yield, disease resistance, and resilience to stress when compared to their progenitor lines (So KK et al., 2021). The cultivation of hybrid canola necessitates the implementation of male sterility systems, wherein one parental line is male-sterile and incapable of generating viable pollen. This mechanism guarantees that all seeds produced arise from cross-pollination, resulting in uniform and high-performing hybrid specimens (El-Mezawy et al., 2016).

Hybrid canola varieties have experienced widespread adoption in leading canola-producing regions, leading to considerable enhancements in yield and economic viability (**Figure. 2**). The advancement of hybrid canola has also been propelled by innovations in genomics and biotechnology, which facilitate more accurate selection of parental lines and the incorporation of desirable traits into hybrids (Jan et al., 2016).

Figure 2. Breeding history of Canola Plant

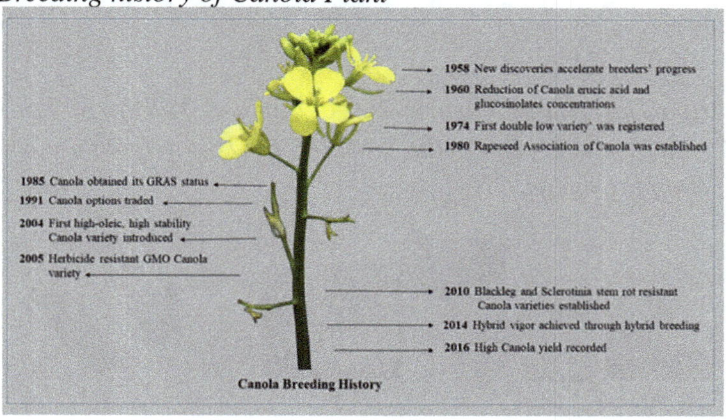

Despite the achievements in canola breeding, challenges persist. The polyploid genome of the canola plant complicates breeding endeavors, as it encompasses multiple gene copies that may exhibit redundant or divergent functionalites (Schmutzer et al., 2015). Furthermore, the persistent threat of diseases, pests, and environmental pressures necessitates ongoing enhancement of canola varieties. The incorporation of genomic methodologies and techniques, such as genome-wide association studies (GWAS) and genomic selection, presents novel prospects for tackling these challenges and further advancing the genetic enhancement of canola (Knoch et al., 2020).

Genome assembly represents an essential procedure in contemporary canola breeding, yielding an extensive cartography of the plant's genomic architecture. The constructed genome acts as a foundational reference that enables breeders and researchers to discern and manipulate genes correlated with vital agronomic characteristics. Within the realm of canola breeding, genome assembly bears several substantial ramifications.

GENOME ASSEMBLY TECHNIQUES

Sequencing Technologies

Genome assembly is initiated through the sequencing of deoxyribonucleic acid (DNA), a methodological endeavor that entails the elucidation of the nucleotide bases—adenine, thymine, cytosine, and guanine—that constitute an organism's genome. Throughout the years, a variety of sequencing technologies have been devised (**Figure. 3**), each exhibiting unique advantages and inherent limitations. A comprehensive understanding of these technologies is imperative for grasping the methodologies employed in constructing genome assemblies, especially in the context of intricate genomes such as that of the canola plant (Weeks DP et al., 2017).

Figure 3. High-Throughput Sequencers; A: Illumina MiSeq, B: Illumina Nova Seq6000, C: PacBio Seq11, D: PacBio SeqRS11, E: Oxford Nanopore PromethION, F: Oxford Nanopore SmidgION

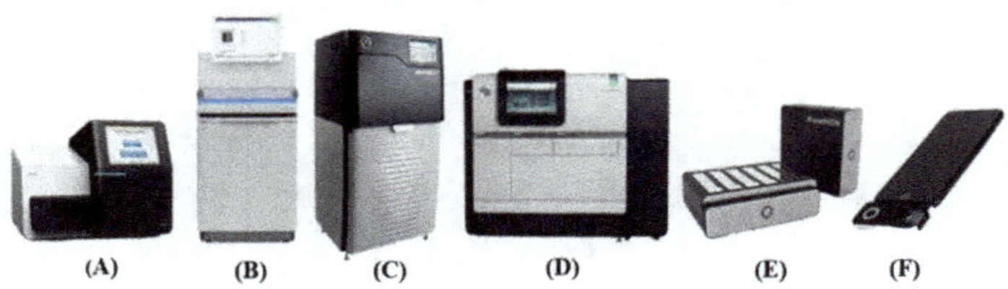

Sanger Sequencing

Sanger sequencing, conceived by Frederick Sanger in 1977, represents the inaugural method that achieved widespread adoption for DNA sequencing. This technique is also referred to as the chain-termination method. Sanger sequencing employs dideoxynucleotides (ddNTPs) as chain terminators during the synthesis of DNA. These ddNTPs are characterized by the absence of the 3' hydroxyl group, which is essential for the formation of a phosphodiester bond between nucleotides, thereby inducing DNA polymerase to terminate the elongation of the DNA strand at arbitrary locations (Sanger et al., 1977).

The procedure commences with the denaturation of the DNA to produce single strands, which are then amalgamated with a primer, DNA polymerase, standard deoxynucleotides (dNTPs), and a minor quantity of labeled ddNTPs. This mixture is subsequently subjected to a polymerase chain reaction (PCR) to amplify the DNA. As DNA polymerase incorporates nucleotides into the elongating DNA strand, the incorporation of a ddNTP results in the cessation of strand elongation at various points, culminating in fragments of disparate lengths (Slatko et al., 1999).

These resultant fragments are then resolved by size utilizing capillary gel electrophoresis; wherein shorter fragments traverse the gel matrix more swiftly than their longer counterparts. The labeled ddNTPs facilitate the detection of the terminal nucleotide of each fragment, and the sequence of the original DNA strand can be reconstructed by interpreting the sequence of the labeled nucleotides as they progress through a detector (Prober et al., 1987).

Sanger sequencing is recognized for its high accuracy and capacity to generate long reads (up to 1,000 base pairs), rendering it especially advantageous for sequencing diminutive genomes or for the validation of specific regions within larger genomes. However, this methodology is comparatively slow, labor-intensive, and costly when juxtaposed with more contemporary sequencing technologies (Diekstra et al., 2015). The throughput associated with Sanger sequencing is limited, constraining its applicability in extensive genomic projects. Notwithstanding these limitations, Sanger sequencing has been instrumental in the formative stages of genomics, including the inaugural human genome project, and remains pertinent for applications necessitating elevated accuracy, such as the sequencing of particular genes or regions of interest within more complex genomes (Diekstra et al., 2015).

Next Generation Sequencing (NGS)

Next-generation sequencing (NGS) technologies, which emerged in the early 2000s, have fundamentally transformed the domain of genomics by facilitating the massively parallel sequencing of millions of DNA fragments concurrently. NGS technologies are significantly more rapid and cost-effective than Sanger sequencing, thereby rendering them suitable for extensive genomic investigations, including the sequencing of complex and polyploid genomes such as that of canola (Hu D et al., 2021).

The predominant next-generation sequencing (NGS) platform employed in genomic studies is Illumina sequencing, which operates based on a methodology referred to as sequencing by synthesis. In the Illumina sequencing approach, deoxyribonucleic acid (DNA) is initially subjected to fragmentation, resulting in short segments that typically span from 100 to 300 base pairs. Subsequently, these fragmented DNA sequences are covalently ligated to specific adapters, enabling their attachment to a flow cell, which is a specialized glass substrate utilized during the sequencing procedure. The fragments that adhere to the flow cell undergo amplification through a technique known as bridge polymerase chain reaction (PCR), thereby generating clusters of homogenous DNA fragments on the flow cell (Suzuki et al., 2011).

In the sequencing phase, nucleotides labeled with fluorescent tags are introduced sequentially, and with the incorporation of each nucleotide into the elongating DNA strand, a distinct fluorescent signal is emitted, which is subsequently captured by a camera. The sequence of the DNA fragment is elucidated by systematically documenting the sequence of these fluorescent emissions. Given that millions of fragments are sequenced concurrently; Illumina sequencing yields an extensive volume of data (as illustrated in **Figure. 4**) within a relatively brief timeframe (Liu L et al., 2012).

Figure 4. Graphical illustration of Next Generation Sequencing Procedure

Next-generation sequencing is particularly effective for de novo genome assembly, wherein the objective is to reconstruct the genomic architecture of an organism in the absence of a reference sequence. Nevertheless, the brevity of the read lengths generated by NGS presents significant obstacles for the assembly of large and intricate genomes, particularly those characterized by repetitive regions or multiple copies of homologous sequences. These short reads may lead to fragmented assemblies, wherein the genome is partitioned into numerous small contigs (contiguous sequences) rather than forming a cohesive sequence (Miller et al., 2010).

To address these challenges, NGS data is frequently integrated with information derived from alternative sequencing methodologies, such as long-read sequencing, to facilitate the generation of more comprehensive and precise genome assemblies. Despite its inherent limitations, NGS continues to be an invaluable instrument for genome sequencing, attributable to its high throughput, scalability, and cost-efficiency (Jagadeesan etal., 2019).

Third-generation sequencing technologies, exemplified by Pacific Biosciences (PacBio) and Oxford Nanopore, signify a substantial progression in the domain of sequencing methodologies. These platforms yield significantly longer reads in comparison to next-generation sequencing (NGS), frequently surpassing 10,000 base pairs and, in some instances, attaining lengths of up to 100,000 base pairs or more. The capability to produce extended reads is especially advantageous for genome assembly, as it facilitates the assembly process by traversing repetitive regions and mitigating the ambiguity associated with read alignment (Mohammad AA et al., 2022).

PacBio Sequencing

PacBio sequencing, alternatively referred to as single-molecule real-time (SMRT) sequencing, entails the continuous observation of DNA synthesis in real-time. In this methodology, an individual DNA molecule is introduced into a minuscule well, termed a zero-mode waveguide (ZMW), in conjunction with a DNA polymerase and fluorescently labeled nucleotides. As the DNA polymerase integrates nucleotides into the elongating DNA strand, the incorporation event is captured by a camera that documents the fluorescence emitted by the labeled nucleotides (Chin CS et al., 2016). A primary advantage of PacBio sequencing resides in its capacity to generate long reads with elevated accuracy. The error rate associated with PacBio reads can be diminished via a technique known as circular consensus sequencing (CCS), whereby a singular DNA molecule is sequenced multiple times to produce a consensus sequence. This enhances the precision of the reads, rendering PacBio an exceptional option for assembling genomes characterized by intricate structures, such as those encompassing extensive repetitive regions or multiple copies of analogous sequences (Wenger et al., 2019).

Oxford Nanopore Sequencing

Conversely, Oxford Nanopore sequencing represents a real-time, single-molecule sequencing technology that assesses variations in ionic current as a singular strand of DNA traverses a protein nanopore embedded within a synthetic membrane (Deschamps S et al., 2016). As the DNA strand progresses through the nanopore, the distinct nucleotides induce specific disruptions in the ionic current, which are subsequently detected and recorded as electrical signals. These signals are thereafter transformed into a DNA sequence utilizing advanced algorithms (Mikheyev et al., 2014). The principal advantage of Oxford Nanopore sequencing lies in its capability to produce ultra-long reads, frequently exceeding 100,000 base pairs. This attribute is particularly beneficial for assembling genomes with complex architectures or for sequencing genomic regions that are challenging to access through alternative technologies. Furthermore, the portability and scalability of Oxford Nanopore devices, such as the MinION, enable sequencing to be conducted in remote locations or in settings with limited resources (Laver et al., 2015). Nevertheless, the accuracy of Oxford Nanopore sequencing may be inferior to that of other technologies, particularly concerning homopolymer regions (regions characterized by extensive stretches of a single nucleotide) and GC-rich sequences. Notwithstanding, advancements in base-calling algorithms and the integration of Nanopore data with other sequencing platforms have considerably enhanced the accuracy of Nanopore-based assemblies (Madoui et al., 2015).

Assembly Algorithms

Once the sequencing data has been produced, the subsequent phase in genome assembly entails the integration of short or long reads into a continuous sequence that accurately reflects the genome of the organism. Numerous assembly algorithms have been formulated to achieve this objective, each possessing distinct advantages and limitations. The selection of an appropriate algorithm is contingent upon the characteristics of the sequencing data as well as the complexity inherent in the genome that is being assembled.

Overlap-Layout-Consensus (OLC)

The Overlap-Layout-Consensus (OLC) methodology is regarded as one of the pioneering techniques established for genome assembly. This methodology was initially applied in the assembly of genomes sequenced utilizing long-read technologies such as Sanger sequencing and has subsequently been modified for application with third-generation sequencing technologies that also generate long reads (Miller et al., 2010).

Overlap Step: Within the OLC framework, the preliminary step involves the identification of overlaps among the reads. This process necessitates the comparison of all pairs of reads to uncover regions of similarity, whereby the sequence at the terminus of one read corresponds with the sequence at the commencement of another. The objective is to pinpointreads that derive from the same genomic region and may be concatenated (Gibrat et al., 2019).

Layout Step: Following the identification of overlaps, the subsequent step is to arrange the reads in the correct sequential order, thereby constructing a layout. This is accomplished by formulating a graph in which each read is depicted as a node, and edges connecting nodes signify overlaps between the reads. The challenge lies in discerning the accurate path through the graph that encapsulates the original sequence of the genome (Li et al., 2012).

Consensus Step: Ultimately, the consensus sequence is generated through the alignment of the reads within the layout and the determination of the most probable sequence of nucleotides at each positional locus. The consensus sequence constitutes the final output of the assembly process, representing the assembled genome (Li et al., 2012).

The OLC methodology is particularly advantageous for long-read sequencing data due to the substantial overlaps present in long reads, facilitating the identification of the correct sequential order of reads. However, the OLC approach is computationally demanding, especially during the overlap phase, where it is necessary to compare all pairs of reads. This can pose a limitation when dealing with extensive datasets, such as those produced by high-throughput sequencing technologies. Inspite of its computational requirements, the OLC methodology continues to serve as a pivotal instrument in genome assembly, particularly for intricate genomes characterized by extensive repetitive regions. The capacity to accurately assemble these regions is essential for elucidating the structural and functional attributes of genomes, including those of polyploid plants such as canola.

Hybrid Assembly Methods

In light of the various strengths and limitations inherent to disparate sequencing technologies and assembly algorithms, hybrid assembly methodologies have been innovated to capitalize on the merits of each technique. Hybrid assembly encompasses the integration of data obtained from multiple sequencing modalities, such as the short-read next-generation sequencing (NGS) data in conjunction with long-read third-generation sequencing data, thereby facilitating a more comprehensive and precise genome assembly.

Combining Short and Long Reads: Within the framework of hybrid assembly, short reads are employed to rectify inaccuracies within the long reads, whereas the long reads furnish a structural framework for the genome assembly. The short reads can be aligned with the long reads to detect and amend sequencing inaccuracies, which are notably prevalent in third-generation sequencing outputs. Upon rectification of these inaccuracies, the long reads are utilized to construct the genome assembly, yielding a contiguous sequence that traverses repetitive regions and intricate genomic structures (Wick et al., 2017).

CHALLENGES IN GENOME ASSEMBLY

Numerous challenges complicate the genome assembly process, particularly in plant species such as canola, which exhibit large and intricate genomes. These challenges encompass polyploidy, repetitive sequences, and heterozygosity, each presenting distinct obstacles in achieving a high-quality genome assembly. This section aims to delve into these challenges in a comprehensive manner.

Polyploidy in Canola

A critical challenge encountered in the genome assembly of canola is its polyploid constitution. Polyploidy delineates the condition in which an organism possesses more than two complete sets of chromosomes. Canola is characterized as an allotetraploid, signifying that it harbors two sets of chromosomes derived from distinct progenitor species: *Brassica rapa* (AA genome) and *Brassica oleracea* (CC genome). Consequently, this results in a cumulative total of 38 chromosomes ($2n = 38$), with each of the 19 pairs comprised of homologous chromosomes from both parental species (Chalhoub et al., 2014).

Polyploidy introduces complexities to genome assembly in several significant ways:

Increased Genome Size and Complexity: The existence of multiple chromosome sets amplifies the overall genome size, thereby complicating the assembly process. The canola genome measures approximately 1.2 to 1.3 gigabases (Gb) in size, which is markedly larger than the genomes of its diploid progenitors. This augmented size necessitates enhanced computational resources and an extended timeframe for successful assembly (Schmutzer et al., 2015).

Homology and Sequence Similarity: The homologous chromosomes in polyploid organisms exhibit a substantial degree of sequence similarity. In canola, the A and C genomes have preserved a considerable amount of sequence homology since their divergence from a shared ancestor. Such similarity can engender challenges in differentiating between homologous sequences during the assembly process. Reads originating from disparate chromosomes may be erroneously amalgamated or misaligned, resulting in chimeric assemblies or deficiencies within the ultimate genome (Mason et al., 2015).

Gene Duplication and Divergence: Polyploid genomes frequently encompass duplicated genes, with certain genes maintaining their original functions while others diverge to adopt novel roles or devolve into pseudogenes. This phenomenon of gene duplication and subsequent divergence introduces an additional layer of complexity to the assembly endeavor. Accurate identification and assembly of these duplicated regions are imperative for a thorough comprehension of the functional landscape of the genome (Waminal et al., 2016).

Challenges in Phasing: Phasing encompasses the methodology utilized to ascertain the specific alleles of a gene that are co-inherited on a single chromosome. In polyploid organisms such as canola, characterized by the presence of multiple homologous chromosomes, the phasing process presents significant challenges. The precise phasing of alleles is imperative for a comprehensive understanding of the genetic architecture and functional dynamics of the genome; however, the pronounced sequence similarity among homologs complicates this undertaking (Gazave et al., 2016).

In response to the complexities introduced by polyploidy, a variety of strategies have been formulated. One notable approach involves the utilization of high-coverage sequencing data, which affords adequate depth to differentiate between homologous sequences. Moreover, long-read sequencing technologies, exemplified by PacBio and Oxford Nanopore, prove advantageous as they produce reads that traverse repetitive regions and offer enhanced contextual information for the differentiation of homologous sequences. Furthermore, the implementation of bioinformatics tools explicitly tailored for polyploid genomes can significantly augment the accuracy of genomic assembly (Lee et al., 2020). Notwithstanding these innovations, polyploidy continues to represent a formidable obstacle in the assembly of the canola genome. The precise assembly of polyploid genomes necessitates meticulous consideration of the foundational genetic architecture, alongside the deployment of sophisticated sequencing methodologies and computational techniques.

Repetitive Sequences

Another prominent challenge encountered in genome assembly, particularly within plant genomes such as canola, is the prevalence of repetitive sequences. Repetitive sequences consist of DNA segments that manifest in multiple copies dispersed throughout the genome. These sequences exhibit variability in both length and complexity, ranging from simple tandem repeats (for instance, microsatellites) to intricate transposable elements that can extend across thousands of base pairs.

Repetitive sequences introduce a variety of challenges for genome assembly:

Ambiguity in Read Mapping: Repetitive sequences engender ambiguity during the read mapping phase, wherein short reads derived from sequencing data are aligned to a reference genome or assembled de novo. Due to the identity or near-identity of repetitive sequences, accurately determining a read's location within the genome proves challenging. Consequently, misalignment of reads may occur, resulting in gaps, inaccuracies, or duplications within the assembly (Bologa et al., 2022.)

Fragmentation of the Assembly: Repetitive sequences may induce fragmentation within the assembly, leading to the genome being partitioned into numerous small contigs (contiguous sequences) rather than forming a continuous sequence. This fragmentation arises because repetitive regions frequently lack unique sequence anchors that can facilitate the assembly process. As a result, assembly algorithms may halt the contig at the repetitive region, resulting in the emergence of gaps or unassembled segments (Zimin et al., 2017).

Difficulty in Resolving Structural Variants: Repetitive sequences are commonly linked with structural variants, including insertions, deletions, and inversions, which further complicate genome assembly. These structural variants may disrupt the assembly's continuity, thereby complicating the accurate reconstruction of the genome's architecture. The resolution of these variants is crucial for elucidating the functional and evolutionary implications of repetitive sequences within the genomic context (Denti et al., 2022).

Impact on Functional Annotation: Repetitive sequences have the potential to disrupt the functional annotation of the genome, which involves the identification and characterization of genes and regulatory elements. The presence of repeats can obscure gene boundaries, complicate the discernment of exons and introns, and result in erroneous predictions of gene models. The precise annotation of the genome is paramount for subsequent analyses, such as studies on gene expression and functional genomics (Breckell et al., 2021).

To address the challenges posed by repetitive sequences, several strategies have been formulated:

Long-Read Sequencing: Long-read sequencing technologies, exemplified by PacBio and Oxford Nanopore, are particularly efficacious in resolving repetitive regions. The extensive reads can encompass entire repeats, thereby providing the necessary context to accurately assemble these genomic segments. This capability diminishes the fragmentation of the assembly process and enhances the overall continuity of the genome (Bellec A et al.,2016).

Repeat Masking: Bioinformatics tools can be employed to identify and mask repetitive sequences within the genome. The process of repeat masking entails annotating repetitive regions and excluding them from specific stages of the assembly procedure. This methodology prevents the improper merging of repetitive reads and assists in mitigating assembly inaccuracies. Nonetheless, this technique may result in the masked regions remaining unassembled, thereby creating gaps in the final genomic representation (Sharon et al., 2021).

Hybrid Assembly Methods: Hybrid assembly strategies, which amalgamate data from both short-read and long-read sequencing technologies, can enhance the assembly of repetitive regions. Short reads offer high coverage and accuracy, whereas long reads provide the contextual information necessary to resolve repeats. By integrating these diverse data types, hybrid assembly methods can yield more comprehensive and precise assemblies (Garimella et al., 2020). Despite the implementation of these strategies, repetitive sequences continue to pose a considerable challenge in the assembly of the canola genome. The complexity and prevalence of repeats within plant genomes necessitate meticulous consideration throughout the assembly process to ensure the accuracy and completeness of the resultant genome.

Heterozygosity

Heterozygosity, defined as the existence of distinct alleles at a specific gene locus, represents an additional obstacle that complicates the process of genome assembly. In organisms exhibiting heterozygosity, each chromosome within a homologous pair may harbor divergent versions of the identical gene or sequence, resulting in variations that necessitate precise resolution during the assembly process (Chin C et al., 2016). Canola, functioning as a hybrid species, demonstrates a pronounced level of heterozygosity, which introduces several challenges:

Increased Complexity of Assembly: The presence of heterozygosity amplifies the intricate nature of the genome assembly process, as the assembler is tasked with distinguishing between the alleles located on homologous chromosomes. This complexity may give rise to the formation of multiple contigs for identical genomic regions, where each contig signifies a distinct allele. Consequently, the assembler must accurately phase these contigs to yield a continuous and precise representation of the genome (Chin C et al., 2016).

Challenges in Phasing: Phasing encompasses the determination of which alleles are transmitted together on a singular chromosome. In genomes characterized by high levels of heterozygosity, the accuracy of phasing is vital for elucidating the genetic architecture and functionality of the genome. However, the existence of numerous heterozygous sites complicates the phasing process, particularly when employing short-read sequencing data, which is deficient in the long-range information required to resolve haplotypes (the combinations of alleles across multiple loci on the same chromosome) (Patterson et al., 2015).

Potential for Assembly Errors: The phenomenon of heterozygosity may precipitate assembly inaccuracies, such as the misalignment of reads originating from disparate alleles or the erroneous merging of allelic sequences. Such inaccuracies can result in the formation of chimeric contigs, wherein sequences from various alleles are mistakenly amalgamated, or in the conflation of heterozygous regions into a singular sequence, culminating in a reduction of genetic diversity within the assembly (Mochizuki et al., 2023).

Impact on Downstream Analyses: The precise resolution of heterozygous regions is imperative for subsequent analyses, including the identification of genetic variants, the investigation of population genetics, and the examination of gene expression. Assembly inaccuracies or the inability to resolve heterozygous regions may result in erroneous interpretations of genetic data and obscure significant biological insights (Guan et al., 2020).

To mitigate the challenges associated with heterozygosity, several methodologies have been devised:

High-Coverage Sequencing: High-coverage sequencing furnishes the requisite data depth to differentiate between alleles and accurately assemble heterozygous regions. With adequate coverage, the assembler is capable of detecting and phasing heterozygous sites, thereby ensuring that both alleles are duly represented in the final assembly (Beretta et al., 2017).

Long-Read Sequencing: Long-read sequencing technologies prove particularly advantageous in resolving heterozygous regions, as the extended reads provide enhanced context for distinguishing between alleles. By encompassing multiple heterozygous sites, long reads facilitate the phasing of haplotypes and diminish the occurrence of assembly errors (Cheng et al., 2020).

Haplotype-Resolved Assembly: Haplotype-resolved assembly, frequently referred to as diploid assembly, constitutes a methodological approach aimed at the distinct assembly of individual haplotypes. This methodology acknowledges that each homologous chromosome may harbor divergent sequences and strives to treat them as separate entities during the assembly process. The significance of haplotype-resolved assembly is particularly pronounced in the context of highly heterozygous genomes, as it facilitates a more precise representation of the genetic diversity inherent within the organism (Garg et al., 2019).

COMPUTATIONAL TOOLS AND RESOURCES

Assembly Softwares

SPAdes: SPAdes (St. Petersburg genome assembler) represents a multifaceted assembler engineered for de novo genome assembly derived from short-read sequencing data. It employs a multi-tiered methodology to enhance the assembly protocol, rendering it appropriate for a broad spectrum of sequencing technologies. SPAdes is acclaimed for its proficiency in managing intricate genomes and generating assemblies of high accuracy, particularly within the domains of microbial and plant genomics (Bankevich et al., 2012).

Canu: Canu is an assembly tool meticulously optimized for long-read data, such as that produced by Pacific Biosciences (PacBio) and Oxford Nanopore technologies. Its robust algorithms for error correction and iterative assembly processes are notably efficacious for the assembly of extensive and complex genomes. Canu is especially advantageous for the generation of high-quality assemblies of canola plants, wherein long-read methodologies are capable of capturing structural variations and repetitive regions more effectively than short-read assemblers (Jung et al., 2020).

SOAPdenovo: SOAPdenovois widely recognized assembler tailored for short-read sequencing data. It employs a de Bruijn graph-based methodology to facilitate genome assembly, which is particularly effective for managing extensive datasets and yielding high-quality contigs. SOAPdenovo is optimally suited for genomic assembly endeavors characterized by high coverage and depth of sequencing (Luo et al., 2012).

Databases and Repositories

NCBI (National Center for Biotechnology Information): The NCBI offers an extensive array of databases and analytical tools tailored for genomic research, prominently featuring the GenBank repository, which contains a vast collection of genomic sequences and their associated annotations. Investigators have the capability to access genomic data pertaining to the canola plant, execute sequence alignment procedures, and leverage a myriad of bioinformatics tools made available through NCBI (Zhang et al., 2013).

Ensembl Plants: Ensembl Plants represents a dedicated database within the overarching Ensembl initiative that concentrates on plant genomes. It furnishes comprehensive annotations, gene models, and resources for comparative genomics across a diverse spectrum of plant species, including canola. Ensembl Plants serves as a crucial resource for the functional annotation and comparative examination of the canola genome, thereby facilitating research into gene functionality and genomic evolution (Wang W et al., 2020).

APPLICATIONS OF CANOLA GENOME ASSEMBLY

Multiple applications of genome assembly of canola have been presented in **Figure. 5**.

Figure 5. Diverse Applications of Genome assembly of Canola

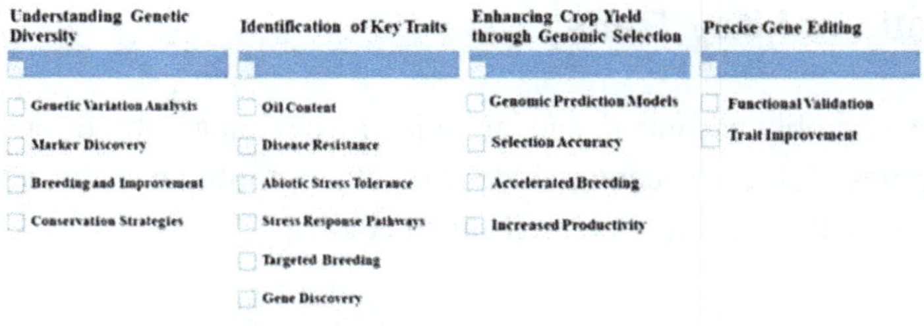

Understanding genetic diversity

The assembly of canola plant genomes is instrumental in elucidating the genetic diversity both within and across canola populations. By developing high-quality reference genomes for a multitude of canola varieties and lines, researchers are afforded the opportunity to glean insights into the genetic variability inherent in these plants. The following outlines how genome assembly contributes to the comprehension of genetic diversity:

Genetic Variation Analysis: Genome assembly facilitates the identification of genetic variations, encompassing single nucleotide polymorphisms (SNPs), insertions, deletions, and structural variations. These variations are paramount for evaluating genetic diversity and differentiation among canola accessions (Heakel et al., 2019).

Population Genetics: The assembled genomes serve as a foundation for conducting population genetics investigations, which elucidate the genetic architecture and evolutionary interrelations within canola populations. This encompasses the examination of allele frequencies, gene flow dynamics, and the degree of genetic differentiation among populations (Sexton et al., 2004).

Marker Discovery: The genome assembly streamlines the identification of genetic markers correlated with various traits, which may be employed in marker-assisted selection (MAS) to pinpoint and select individuals possessing advantageous genetic characteristics. This is particularly beneficial for breeding initiatives aimed at augmenting specific traits (Raman et al., 2020).

The comprehension of genetic diversity through genome assembly is vital for:

Breeding and Improvement: The identification of genetic resources that can be harnessed to cultivate novel canola varieties characterized by enhanced traits and greater resilience to environmental stressors.

Conservation Strategies: The preservation of genetic resources and the management of genetic diversity within breeding programs to ensure a broad genetic base and adaptability of canola crops.

Identification of Key Traits

Genome assembly permits the accurate identification and functional characterization of genes (Table. 1) accountable for pivotal agronomic traits in canola, such as oil content and disease resistance (Raman et al., 2020).

Table. 1 Canola genes associated with key traits such as oil content and disease resistance

Genes	Traits associated	Function	Genome Location	References
Rlm3, Rlm4	Blackleg resistance	Protein kinase & jumonji proteins related to defence resistance	A03, A07, C04	Raman et al., 2020
Rlm12	Blackleg resistance	Leptosphaeria maculans resistance	A07, A10	Raman et al., 2016
BnP5CR2	Sclerotinia Resistance	Prolinne metabolism, stress resistance	A10, C09	Zhang et al., 2022
FAD2	Oil Content	Fatty acid desaturase involved in oleic acid synthesis	N5	Hu et al., 2006
FAD3	Oil Content	Fatty acid desaturase involved in linolineac acid synthesis	N14, N4	Hu et al., 2004
BnaA03g26760D	Quantitative disease	Defense response proteins	A03	Raman et al., 2020
FAE1	Oil Content	Fatty acid elongase involved in erucic acid synthesis	A08, C03	Zhu et al., 2019
Rlm2, LepR3	Blackleg resistance	Receptor-like proteins providing race-specific resistance	A10	Larkan et al., 2014
Rlm9	Blackleg resistance	Wall-associated kinase-like proteins involved in disease response	A07	Haddadi et al., 2022
FAE1 (mutant)	Oil Quality	Mutation in fatty acid elongase gene leading to modified fatty acid composition	A08, C03	Hu et al., 2006

The following delineates how this process transpires:

Oil Content: Canola is predominantly esteemed for its high oil content. Through genome assembly, researchers can delineate and annotate genes implicated in lipid metabolism, fatty acid biosynthesis, and oil accumulation. By investigating the regulatory networks and pathways involved, it becomes feasible to identify specific genes or gene clusters that exert influence over oil yield and composition (Hu et al., 2006). The experimental validation of these genes via methodologies such as gene knockouts or overexpression studies can substantiate their role in oil production. This knowledge is indispensable for the development of canola varieties exhibiting enhanced oil content and superior quality. The identification of genes linked to resistance against diseases and pests represents another significant application of genome assembly. This process encompasses the localization of resistance genes, including those that encode nucleotide-binding leucine-rich repeat (NLR) proteins or other defense-related genes (Raman et al., 2012). The association of genetic markers with resistance traits through the correlation to observed phenotypic resistance in canola populations. This aids in deciphering the genetic underpinnings of resistance and directing the development of resistant varieties.

Abiotic Stress Tolerance: Beyond disease resistance, genome assembly can also elucidate genes associated with abiotic stress tolerance, such as drought, salinity, and temperature extremes (Raman et al., 2016). Systematically annotating genes implicated in stress response pathways and elucidating regulatory mechanisms that engender resilience to environmental adversities. The delineation of pivotal traits through genome assembly is paramount for facilitating the creation of canola varieties possessing optimized traits specifically adapted to particular agronomic requirements or environmental conditions as well as augmenting the quality and yield of canola through the integration of genes that contribute to advantageous traits (Raman et al., 2016).

Enhancing Crop Yield Through Genomic Selection

Genomic selection represents an innovative methodology that utilizes genomic data to forecast and identify plants exhibiting superior phenotypic traits, thereby ultimately augmenting crop yield. The genome assembly contributes to this paradigm by amalgamating genome assembly data with phenotypic characteristics, researchers are empowered to construct predictive models that assess the prospective yield of canola plants predicated on their genetic architecture. These models employ genome-wide markers to anticipate the performance of diverse genotypes across a spectrum of environmental conditions (Jan et al., 2016). The implementation of genomic selection enhances the precision of identifying high-yielding canola cultivars. Through the analysis of genetic markers correlated with yield and other agronomic characteristics, breeders are equipped to make judicious selections, thereby identifying the most promising candidates for breeding initiatives (Mulusew et al., 2020). Genomic selection expedites the breeding process, thereby diminishing the temporal and resource expenditures associated with conventional breeding techniques. By facilitating the early identification of superior candidates, this approach accelerates the creation of novel canola cultivars endowed with enhanced yield potential (Roy et al., 2021). The advancement of canola cultivars with heightened yield potential is imperative for satisfying the escalating global demand for oilseeds. Resource Efficiency: The optimization of breeding endeavors and resource allocation is realized by concentrating on genotypes that exhibit the greatest potential for yield enhancement (Raman et al., 2020).

Genome Editing Techniques (CRISPR-Cas9)

Genome editing methodologies, with particular emphasis on CRISPR-Cas9, have revolutionized the capacity to execute precise genetic modifications within canola plants (**Figure. 6**). With the availability of a meticulously assembled canola genome, CRISPR-Cas9 can effectively target specific genes implicated in critical traits such as oil content, disease resistance, and stress tolerance. The provision of comprehensive genome sequences enables the formulation of guide RNAs that accurately target the genes of interest for modification (Arora et al., 2017).

Figure 6. Schematic depiction of How Crisper/Cas9 play crucial role in gene editing of Specific traits in canola

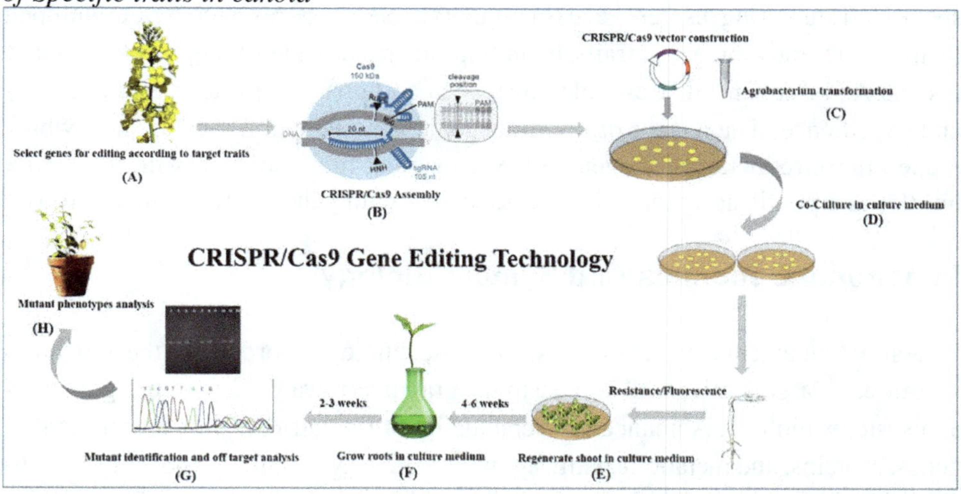

Genome assembly serves as the essential reference for validating the functional roles of genes identified through sequencing efforts. CRISPR-Cas9 can induce specific mutations or insertions to investigate their repercussions on plant phenotypes, thereby substantiating the contributions of individual genes to trait manifestation (Belhaj et al., 2015). By modifying genes associated with advantageous traits, CRISPR-Cas9 facilitates the development of canola cultivars characterized by novel or enhanced attributes. For instance, it can be employed to eliminate genes that detrimentally affect oil content or to integrate new genes that bestow resistance against diseases (Chen et al., 2019). The merits of employing genome editing technologies in canola include targeted modifications with minimal off-target effects, as well as the acceleration of the development of superior canola cultivars by circumventing the necessity for traditional breeding and selection methodologies.

FUTURE PROSPECTS IN CANOLA GENOMICS

The trajectory of canola genomics appears promising, presenting a multitude of opportunities to enhance our comprehension of the canola genome while translating this understanding into concrete advantages for agriculture, industry, and society at large. Various pivotal domains of research and development are poised to influence the future landscape of canola genomics.

Precision Breeding And Genome Editing

Innovations in genome assembly and gene editing methodologies, particularly CRISPR-Cas9, hold the potential to revolutionize the precision breeding of canola. By specifically targeting discrete genes or regulatory elements, researchers are equipped to incorporate advantageous traits, including superior oil quality, augmented disease resistance, or enhanced stress tolerance, with an unparalleled degree of accuracy and expedience. The advent of non-transgenic genome editing techniques, which eschew the introduction of exogenous DNA, may also mitigate some of the regulatory hurdles and public acceptance issues associated with genetically modified crops.

Functional Genomics And System Biology

As advancements in genome assembly technologies progress, the emphasis within canola genomics will increasingly gravitate towards functional genomics and systems biology. A nuanced understanding of the intricate interactions among genes, proteins, and metabolic pathways will yield insights into the molecular mechanisms that underpin critical agronomic traits. This foundational knowledge can be harnessed to construct systems-level models of canola biology, thereby steering the development of novel varieties and informing breeding strategies. The integration of multi-omics data, alongside the application of computational methodologies such as machine learning and artificial intelligence, will be instrumental in propelling this field forward.

Climate-Resilient Canola Varieties

The ramifications of climate change represent a considerable challenge to global agriculture, characterized by escalating temperatures, shifting precipitation patterns, and an increase in extreme weather occurrences. Prioritizing the development of canola varieties that exhibit resilience to these adversities will be imperative for forthcoming research endeavors. Genome assembly and genomic selection techniques can facilitate the identification and incorporation of genes associated with drought resistance, thermal tolerance, and other climate-adaptive characteristics into canola breeding initiatives. Furthermore, elucidating the genetic underpinnings of phenotypic plasticity—the capacity of a plant to acclimatize to fluctuating environmental conditions—will be essential for the creation of climate-resilient cultivars.

Exploration of Genetic Diversity

Investigating the genetic diversity present within canola and its wild relatives constitutes a significant avenue for enhancing the crop. By sequencing and assembling the genomes of various canola accessions and closely related species, researchers can pinpoint novel alleles and genes that confer beneficial traits. Such genetic resources can be employed to expand the genetic foundation of canola breeding programs, diminish the risk of genetic bottlenecks, and bolster the crop's resilience against biotic and abiotic stressors.

Sustainable Agricultural Practices

The future direction of canola genomics must be congruent with the tenets of sustainable agriculture. The integration of genomics with sustainable agricultural practices, including precision agriculture, agroecology, and conservation agriculture, can optimize crop performance while mitigating environmental repercussions. For instance, genomics can guide the development of canola varieties that necessitate fewer inputs, such as water, fertilizers, and pesticides, or that can thrive in marginal environments with minimal effects on natural ecosystems.

CONCLUSION

Genome assembly has emerged as a fundamental component of contemporary genomics, facilitating the reconstruction of comprehensive DNA sequences of various organisms and furnishing an intricate map of their genetic constituents. In the case of canola, a crop of considerable economic and agricultural significance, genome assembly presents the opportunity to expedite breeding initiatives, augment genetic enhancement, and enrich our comprehension of the plant's biological processes. Nevertheless, the intricate nature of the canola genome—characterized by its polyploid composition, repetitive sequences, and heterozygosity—poses considerable challenges that necessitate the utilization of sophisticated sequencing methodologies and advanced bioinformatics instruments. Progress in technology, encompassing the innovation of enhanced long-read sequencing platforms, the refinement of bioinformatics tools tailored for polyploid genomes, and the implementation of hybrid assembly strategies, is crucial for surmounting these difficulties. Furthermore, ethical and environmental considerations must be duly acknowledged to guarantee that the advantages of genome assembly are achieved in a responsible and sustainable fashion. The prospects for canola genomics are optimistic, offering avenues to exploit precision breeding, investigate genetic variability, and formulate climate-resilient varieties that bolster global food security.

REFERENCES

Arabidopsis Genome Initiative. (2000). Analysis of the genome sequence of the flowering plant Arabidopsis thaliana. *Nature*, *408*(6814), 796–815. DOI: 10.1038/35048692 PMID: 11130711

Arora, L., & Narula, A. (2017). Gene editing and crop improvement using CRISPR-Cas9 system. *Frontiers in Plant Science*, *8*, 1932. DOI: 10.3389/fpls.2017.01932 PMID: 29167680

Bankevich, A., Nurk, S., Antipov, D., Gurevich, A., Dvorkin, M., Kulikov, A., Lesin, V. M., Nikolenko, S. I., Pham, S., Prjibelski, A. D., Pyshkin, A. V., Sirotkin, A. V., Vyahhi, N., Tesler, G., Alekseyev, M. A., & Pevzner, P. A. (2012). SPAdes: A new genome assembly algorithm and its applications to single-cell sequencing. *Journal of Computational Biology*, *19*(5), 455–477. DOI: 10.1089/cmb.2012.0021 PMID: 22506599

Beckie, H. J., Harker, K. N., Hall, L. M., Warwick, S. I., Légère, A., Sikkema, P. H., Clayton, G. W., Thomas, A. G., Leeson, J. Y., Séguin-Swartz, G., & Simard, M.-J. (2006). A decade of herbicide-resistant crops in Canada. *Canadian Journal of Plant Science*, *86*(4), 1243–1264. DOI: 10.4141/P05-193

Belhaj, K., Chaparro-Garcia, A., Kamoun, S., Patron, N., & Nekrasov, V. (2015). Editing plant genomes with CRISPR/Cas9. *Current Opinion in Biotechnology*, *32*, 76–84. DOI: 10.1016/j.copbio.2014.11.007 PMID: 25437637

Bellec, A., Courtial, A., Cauet, S., Rodde, N., Vautrin, S., & Salse, J. (2016). Long read sequencing technology to solve complex genomic regions assembly in plants. *Next Generation, Sequencing & Applications*, *3*(2), 1–13. DOI: 10.4172/2469-9853.1000128

Beretta, S., Patterson, M., Zaccaria, S., Vedova, G., & Bonizzoni, P. (2017). HapCHAT: Adaptive haplotype assembly for efficiently leveraging high coverage in long reads. *bioRxiv*. DOI: 10.1101/170225

Bhardwaj, H. L., Hamama, A. A., & Vick, B. A. (2003). Genotypic and environmental effects on oxidative stability of canola oil. *Industrial Crops and Products*, *18*(3), 251–257. DOI: 10.1016/S0926-6690(03)00049-1

Bologa, A. M., Stoica, I., Ratiu, A. C., Constantin, N. D., & Ecovoiu, A. A. (2022). ONT-based alternative assemblies impact on the annotations of unique versus repetitive features in the genome of a Romanian strain of *Drosophila melanogaster. International Journal of Molecular Sciences*, *23*(23), 14892. DOI: 10.3390/ijms232314892 PMID: 36499217

Breckell, G., & Silander, O. (2021). Do you want to build a genome? Benchmarking hybrid bacterial genome assembly methods. *bioRxiv*. DOI: 10.1101/2021.11.07.467652

Brookes, G., & Barfoot, P. (2014). Economic impact of GM crops: The global income and production effects 1996–2012. *GM Crops and Food: Biotechnology in Agriculture and the Food Chain*, *5*(1), 65–75. DOI: 10.4161/gmcr.28098 PMID: 24637520

Cai, D., Xiao, Y., Yang, W., Ye, W., Wang, B., Younas, M., Wu, J., & Liu, K. (2014). Association mapping of six yield-related traits in rapeseed (Brassica napus L.). *Theoretical and Applied Genetics*, *127*(1), 85–96. DOI: 10.1007/s00122-013-2203-9 PMID: 24121524

Cardoza, V., & Stewart, C. N.Jr. (2004). Brassica biotechnology: Progress in cellular and molecular biology. *In Vitro Cellular & Developmental Biology. Plant*, *40*(6), 542–551. DOI: 10.1079/IVP2004568

Cardoza, V. S., & Stewart, C. N. (2004). Agrobacterium-mediated transformation of canola. In *Transgenic Crops of the World: Essential Protocols* (pp. 379–387). Springer. DOI: 10.1007/978-1-4020-2333-0_28

Chalhoub, B., Denoeud, F., Liu, S., Parkin, I. A. P., Tang, H., Wang, X., Chiquet, J., Belcram, H., Tong, C., Samans, B., Corréa, M., Da Silva, C., Just, J., Falentin, C., Koh, C. S., Le Clainche, I., Bernard, M., Bento, P., Noel, B., & Wincker, P. (2014). Early allopolyploid evolution in the post-Neolithic Brassica napus oilseed genome. *Science*, *345*(6199), 950–953. DOI: 10.1126/science.1253435 PMID: 25146293

Chen, K., Wang, Y., Zhang, R., Zhang, H., & Gao, C. (2019). CRISPR/Cas genome editing and precision plant breeding in agriculture. *Annual Review of Plant Biology*, *70*(1), 667–697. DOI: 10.1146/annurev-arplant-050718-100049 PMID: 30835493

Cheng, H., Concepcion, G., Feng, X., Zhang, H., & Li, H. (2020). Haplotype-resolved de novo assembly with phased assembly graphs. *arXiv: Genomics*. https://doi.org/ DOI: 10.1101/2020.06.09.175539

Chin, C., Peluso, P., Sedlazeck, F. J., Nattestad, M., Concepcion, G., Clum, A., & Rank, D. (2016). Phased diploid genome assembly with single-molecule real-time sequencing. *Nature Methods*, *13*(12), 1050–1054. DOI: 10.1038/nmeth.4035 PMID: 27749838

Chin, C. S., Peluso, P., Sedlazeck, F. J., Nattestad, M., Concepcion, G. T., Clum, A., Dunn, C., O'Malley, R., Figueroa-Balderas, R., Morales-Cruz, A., Cramer, G. R., Delledonne, M., Luo, C., Ecker, J. R., Cantu, D., Rank, D. R., & Schatz, M. C. (2016). Phased diploid genome assembly with single-molecule real-time sequencing. *Nature Methods*, *13*(12), 1050–1054. DOI: 10.1038/nmeth.4035 PMID: 27749838

Cobb, J. N., Biswas, P. S., & Platten, J. D. (2019). Back to the future: Revisiting MAS as a tool for modern plant breeding. *Theoretical and Applied Genetics*, *132*(3), 647–667. DOI: 10.1007/s00122-018-3266-4 PMID: 30560465

Denti, L., Khorsand, P., Bonizzoni, P., Hormozdiari, F., & Chikhi, R. (2022). Improved structural variant discovery in hard-to-call regions using sample-specific string detection from accurate long reads. *bioRxiv*. DOI: 10.1101/2022.02.12.480198

Deschamps, S., Mudge, J., Cameron, C., Ramaraj, T., Anand, A., Fengler, K., Hayes, K., Llaca, V., Jones, T. J., & May, G. (2016). Characterization, correction and de novo assembly of an Oxford Nanopore genomic dataset from *Agrobacterium tumefaciens*. *Scientific Reports*, *6*(1), 28625. DOI: 10.1038/srep28625 PMID: 27350167

Diekstra, A., Bosgoed, E., Rikken, A., van Lier, B., Kamsteeg, E. J., Tychon, M., Derks, R. C., van Soest, R. A., Mensenkamp, A. R., Scheffer, H., Neveling, K., & Nelen, M. R. (2015). Translating Sanger-based routine DNA diagnostics into generic massive parallel ion semiconductor sequencing. *Clinical Chemistry*, *61*(1), 154–162. DOI: 10.1373/clinchem.2014.225250 PMID: 25274553

. Downey, R. K. (1983). The origin and description of the Brassica oilseed crops. *High and Low Erucic Acid Rapeseed Oils*, 1–20.

Ecke, W., Clemens, R., Honsdorf, N., & Becker, H. C. (2010). Extent and structure of linkage disequilibrium in canola quality winter rapeseed (Brassica napus L.). *Theoretical and Applied Genetics*, *120*(5), 921–931. DOI: 10.1007/s00122-009-1221-0 PMID: 19953219

El-Mezawy, A., Al-Forkan, M., Wu, L., Weselake, R. J., Shah, S., & Rahman, H. (2016). Down regulation of the IND gene causes male sterility in canola (Brassica napus L.). *Biocatalysis and Agricultural Biotechnology*, *6*, 9–18. DOI: 10.1016/j.bcab.2016.02.004

Ferreira, M. E., Williams, P. H., & Osborn, T. C. (1994). RFLP mapping of Brassica napus using F1-derived doubled haploid lines. *Theoretical and Applied Genetics*, *89*(5), 615–621. DOI: 10.1007/BF00222456 PMID: 24177938

Fu, D. H., Jiang, L. Y., Mason, A. S., Xiao, M. L., Zhu, L. R., Li, L. Z., Zhou, Q., Shen, C., & Huang, C. (2016). Research progress and strategies for multifunctional rapeseed: A case study of China. *Journal of Integrative Agriculture*, *15*(8), 1673–1684. DOI: 10.1016/S2095-3119(16)61384-9

Garg, S., Fungtammasan, A., Carroll, A., Chou, M., Schmitt, A., Zhou, X., & Sedlazeck, F. (2019). Efficient chromosome-scale haplotype-resolved assembly of human genomes. *bioRxiv*. DOI: 10.1101/810341

Garimella, K. V., Iqbal, Z., Krause, M. A., Campino, S., Kekre, M., Drury, E., & MacInnis, B. (2020). Detection of simple and complex de novo mutations with multiple reference sequences. *Genome Research*, *30*(8), 1154–1169. DOI: 10.1101/gr.255505.119 PMID: 32817236

Gazave, E., Tassone, E. E., Ilut, D. C., Wingerson, M., Datema, E., Witsenboer, H. M., Davis, J. B., Grant, D., Dyer, J. M., Jenks, M. A., Brown, J., & Gore, M. A. (2016). Population genomic analysis reveals differential evolutionary histories and patterns of diversity across subgenomes and subpopulations of *Brassica napus* L. *Frontiers in Plant Science*, *7*, 525. DOI: 10.3389/fpls.2016.00525 PMID: 27148342

Gibrat, J. F. (2019). On the use of algebraic topology concepts to check the consistency of genome assembly. *Biophysics and Physicobiology*, *16*(0), 444–451. DOI: 10.2142/biophysico.16.0_444 PMID: 31984196

Guan, D., McCarthy, S. A., Wood, J., Howe, K., Wang, Y., & Durbin, R. (2020). Identifying and removing haplotypic duplication in primary genome assemblies. *Bioinformatics (Oxford, England)*, *36*(9), 2896–2898. Advance online publication. DOI: 10.1093/bioinformatics/btaa025 PMID: 31971576

Haddadi, P., Larkan, N. J., Van deWouw, A., Zhang, Y., Neik, T. X., & Beynon, E.. (2022). *Brassica napus* genes Rlm4 and Rlm7, conferring resistance to *Leptosphaeria maculans*, are alleles of the Rlm9 wall-associated kinase-like resistance locus. *Plant Biotechnology Journal*, *20*(7), 1229–1231. DOI: 10.1111/pbi.13818 PMID: 35338565

Harper, A. L., Trick, M., Higgins, J., Fraser, F., Clissold, L., Wells, R., Hattori, C., Werner, P., & Bancroft, I. (2012). Associative transcriptomics of traits in the polyploid crop species Brassica napus. *Nature Biotechnology*, *30*(8), 798–802. DOI: 10.1038/nbt.2302 PMID: 22820317

Heakel, R. (2019). Analysis of genetic diversity among a population of canola genotypes as revealed by ISSR-PCR and their associations to seed yield and oil content. *Annals of Agricultural Sciences, Moshtohor*, *57*(2), 425–434. Advance online publication. DOI: 10.21608/assjm.2019.44917

Hu, D., Jing, J., Snowdon, R. J., Mason, A. S., Shen, J., Meng, J., & Zou, J. (2021). Exploring the gene pool of Brassica napus by genomics-based approaches. *Plant Biotechnology Journal*, *19*(9), 1693–1712. DOI: 10.1111/pbi.13636 PMID: 34031989

Hu, J., Chen, B., Zhao, J., Zhang, F., Xie, T., Xu, K., Gao, G., Yan, G., Li, H., Li, L., Ji, G., An, H., Li, H., Huang, Q., Zhang, M., Wu, J., Song, W., Zhang, X., Luo, Y., & Wu, X. (2022). Genomic selection and genetic architecture of agronomic traits during modern rapeseed breeding. *Nature Genetics*, *54*(5), 694–704. DOI: 10.1038/s41588-022-01055-6 PMID: 35484301

Hu, X., Sullivan-Gilbert, M., Gupta, M., & Thompson, S. A. (2006). Mapping of the loci controlling oleic and linolenic acid contents and development of *fad2* and *fad3* allele-specific markers in canola (*Brassica napus* L.). *Theoretical and Applied Genetics*, *113*(3), 497–507. DOI: 10.1007/s00122-006-0315-1 PMID: 16767448

Jagadeesan, B., Gerner-Smidt, P., Allard, M. W., Leuillet, S., Winkler, A., Xiao, Y., Chaffron, S., Van Der Vossen, J., Tang, S., Katase, M., McClure, P., Kimura, B., Ching Chai, L., Chapman, J., & Grant, K. (2019). The use of next-generation sequencing for improving food safety: Translation into practice. *Food Microbiology*, *79*, 96–115. DOI: 10.1016/j.fm.2018.11.005 PMID: 30621881

Jan, H., Abbadi, A., Lücke, S., Nichols, R., & Snowdon, R. (2016). Genomic prediction of testcross performance in canola (*Brassica napus*). *PLoS One*, *11*(1), e0147769. Advance online publication. DOI: 10.1371/journal.pone.0147769 PMID: 26824924

Jan, H. U., Abbadi, A., Lücke, S., Nichols, R. A., & Snowdon, R. J. (2016). Genomic prediction of testcross performance in canola (Brassica napus). *PLoS One*, *11*(5), e0156687. DOI: 10.1371/journal.pone.0147769 PMID: 26824924

Jean, M., Brown, G. G., & Landry, B. S. (1997). Genetic mapping of nuclear fertility restorer genes for the 'Polima' cytoplasmic male sterility in canola (Brassica napus L.) using DNA markers. *Theoretical and Applied Genetics*, *95*(3-4), 321–328. DOI: 10.1007/s001220050566

Jung, H., Jeon, M., Hodgett, M., Waterhouse, P., & Eyun, S. (2020). A comparative evaluation of genome assemblers from long-read sequencing for plants and crops. *Journal of Agricultural and Food Chemistry*, *68*(29), 7670–7677. Advance online publication. DOI: 10.1021/acs.jafc.0c01647 PMID: 32530283

Kalantarahmadi, S., & Daneshian, J. (2023). Improving of canola (Brassica napus L.) yield and oil quality by foliar application of micro-nutrients under high-temperature stress. *Journal of Soil Science and Plant Nutrition*, *23*(1), 351–367. DOI: 10.1007/s42729-022-01016-2

Knoch, D., Abbadi, A., Grandke, F., Meyer, R. C., Samans, B., Werner, C. R., Snowdon, R. J., & Altmann, T. (2020). Strong temporal dynamics of QTL action on plant growth progression revealed through high-throughput phenotyping in canola. *Plant Biotechnology Journal*, *18*(1), 68–82. DOI: 10.1111/pbi.13171 PMID: 31125482

Kutcher, H. R., Warland, J. S., & Brandt, S. A. (2010). Temperature and precipitation effects on canola yields in Saskatchewan, Canada. *Agricultural and Forest Meteorology*, *150*(2), 161–165. DOI: 10.1016/j.agrformet.2009.09.011

Kyriakidou, M., Tai, H. H., Anglin, N. L., Ellis, D., & Strömvik, M. V. (2018). Current strategies of polyploid plant genome sequence assembly. *Frontiers in Plant Science*, *9*, 1660. DOI: 10.3389/fpls.2018.01660 PMID: 30519250

Landry, B. S., Hubert, N., Etoh, T., Harada, J., & Lincoln, S. (1991). A genetic map for Brassica napus based on restriction fragment length polymorphisms detected with expressed sequences. *Genome*, *34*(4), 543–552. DOI: 10.1139/g91-084

Larkan, N. J., Lydiate, D. J., Yu, F., Rimmer, S. R., & Borhan, M. H. (2014). Co-localisation of the blackleg resistance genes Rlm2 and LepR3 on *Brassica napus* chromosome A10. *BMC Plant Biology*, *14*(1), 387. DOI: 10.1186/s12870-014-0387-z PMID: 25551287

Laver, T., Harrison, J., O'Neill, P. A., Moore, K., Farbos, A., Paszkiewicz, K., & Studholme, D. J. (2015). Assessing the performance of the Oxford Nanopore Technologies MinION. *Biomolecular Detection and Quantification*, *3*, 1–8. DOI: 10.1016/j.bdq.2015.02.001 PMID: 26753127

Lee, J. S., & Adams, K. L. (2020). Global insights into duplicated gene expression and alternative splicing in polyploid *Brassica napus* under heat, cold, and drought stress. *The Plant Genome*, *13*(3), e20057. Advance online publication. DOI: 10.1002/tpg2.20057 PMID: 33043636

Li, Z., Chen, Y., Mu, D., Yuan, J., Shi, Y., Zhang, H., Gan, J., Li, N., Hu, X., Liu, B., Yang, B., & Fan, W. (2012). Comparison of the two major classes of assembly algorithms: Overlap–layout–consensus and de Bruijn graph. *Briefings in Functional Genomics*, *11*(1), 25–37. DOI: 10.1093/bfgp/elr035 PMID: 22184334

Liu, L., Li, Y., Li, S., Hu, N., He, Y., Pong, R., Lin, D., Lu, L., & Law, M. (2012). Comparison of next-generation sequencing systems. *BioMed Research International*, *251364*, 1–11. Advance online publication. DOI: 10.1155/2012/251364 PMID: 22829749

Lohani, N., Jain, D., Singh, M. B., & Bhalla, P. L. (2020). Engineering multiple abiotic stress tolerance in canola, Brassica napus. *Frontiers in Plant Science*, *11*, 3. DOI: 10.3389/fpls.2020.00003 PMID: 32161602

Luo, R., Liu, B., Xie, Y., Li, Z., Huang, W., Yuan, J., He, G., Chen, Y., Pan, Q., Liu, Y., Tang, J., Wu, G., Zhang, H., Shi, Y., Liu, Y., Yu, C., Wang, B., Lu, Y., Han, C., & Wang, J. (2012). SOAPdenovo2: An empirically improved memory-efficient short-read de novo assembler. *GigaScience*, *1*(1), 18. DOI: 10.1186/2047-217X-1-18 PMID: 23587118

Madoui, M. A., Engelen, S., Cruaud, C., Belser, C., Bertrand, L., Alberti, A., Lemainque, A., Wincker, P., & Aury, J.-M. (2015). Genome assembly using Nanopore-guided long and error-free DNA reads. *BMC Genomics*, *16*(1), 1. DOI: 10.1186/s12864-015-1519-z PMID: 25927464

Manchanda, P., & Suneja, Y. (2018). Genome editing for crop improvement: Status and prospects. *Biotechnology for Crop Improvement*, *3*, 75–104. DOI: 10.1007/978-3-319-94746-4_5

Mason, A. S. (2015). Challenges of genotyping polyploid species. In *Plant Genotyping: Methods and Protocols* (pp. 161–168). Springer. DOI: 10.1007/978-1-4939-1966-6_12

McBeath, T. M., Meier, E. A., Ware, A., Kirkegaard, J., Moodie, M., Davoren, B., & Hunt, E. (2020). Agronomic management combining early-sowing on establishment opportunities, cultivar options and adequate nitrogen is critical for canola (Brassica napus) productivity and profit in low-rainfall environments. *Crop & Pasture Science*, *71*(9), 807–821. DOI: 10.1071/CP20226

Mikheyev, A. S., & Tin, M. M. (2014). A first look at the Oxford Nanopore MinION sequencer. *Molecular Ecology Resources*, *14*(6), 1097–1102. DOI: 10.1111/1755-0998.12324 PMID: 25187008

Miller, J. R., Koren, S., & Sutton, G. (2010). Assembly algorithms for next-generation sequencing data. *Genomics*, *95*(6), 315–327. DOI: 10.1016/j.ygeno.2010.03.001 PMID: 20211242

Mochizuki, T., Sakamoto, M., Tanizawa, Y., Nakayama, T., Tanifuji, G., Kamikawa, R., & Shirahige, K. (2023). A practical assembly guideline for genomes with various levels of heterozygosity. *Briefings in Bioinformatics*, *24*(6), bbad337. Advance online publication. DOI: 10.1093/bib/bbad337 PMID: 37798248

. Mohammed, A. A., Senbeta, B., & Worku, T. (2022). Pacific bioscience sequence technology. *International Journal of Veterinary Science Research, 8*(1), 027–033. https://doi.org/DOI: 10.24251/ijvscres.2022.03.027

Mohd Saad, N. S., Severn-Ellis, A. A., Pradhan, A., Edwards, D., & Batley, J. (2021). Genomics armed with diversity leads the way in Brassica improvement in a changing global environment. *Frontiers in Genetics, 12*, 600789. DOI: 10.3389/fgene.2021.600789 PMID: 33679880

Mulusew, F., Denise, M., Barbulescu, M., Malmberg, P., Maharjan, P., & Salisbury, S.. (2020). Genomic prediction and genetic correlation of agronomic, blackleg disease, and seed quality traits in canola (*Brassica napus* L.). *Plants, 9*(6), 719. DOI: 10.3390/plants9060719 PMID: 32517116

Norton, G., & Harris, J. F. (1975). Compositional changes in developing rape seed (Brassica napus L.). *Planta, 123*(2), 163–174. DOI: 10.1007/BF00383865 PMID: 24435083

Patterson, M., Marschall, T., Pisanti, N., Van Iersel, L., Stougie, L., Klau, G., & Schönhuth, A. (2015). WhatsHap: Weighted haplotype assembly for future-generation sequencing reads. *Journal of Computational Biology, 22*(6), 498–509. DOI: 10.1089/cmb.2014.0157 PMID: 25658651

Porter, P. M., Hegstad, J., Luker, K., & Ochsner, T. (2020). Soil health and sustainability of canola cropping systems in the Northern Great Plains. *Agronomy Journal, 112*(5), 3589–3601. DOI: 10.1002/agj2.20363

Prober, J. M., Trainor, G. L., Dam, R. J., Hobbs, F. W., Robertson, C. W., Zagursky, R. J., Cocuzza, A. J., Jensen, M. A., & Baumeister, K. (1987). A system for rapid DNA sequencing with fluorescent chain-terminating dideoxynucleotides. *Science, 238*(4825), 336–341. DOI: 10.1126/science.2443975 PMID: 2443975

Rahman, H., Peng, G., Yu, F., Falk, K. C., Kulkarni, M., & Selvaraj, G. (2014). Genetics and breeding for clubroot resistance in Canadian spring canola (Brassica napus L.). *Canadian Journal of Plant Pathology, 36*(sup1), 122–134.

Rahman, M., & McVetty, P. (2011). A review of Brassica seed color. *Canadian Journal of Plant Science, 91*(3), 437–446. DOI: 10.4141/cjps10124

Raman, H., Raman, R., Coombes, N., Song, J., Prangnell, R., Bandaranayake, C., & Edwards, D. (2016). Genome-wide association analyses reveal complex genetic architecture underlying natural variation for flowering time in canola. *Plant, Cell & Environment, 39*(6), 1228–1239. DOI: 10.1111/pce.12644 PMID: 26428711

Raman, R., Diffey, S., Barbulescu, D., Coombes, N., Luckett, D., Salisbury, P., Cowley, R., Marcroft, S., & Raman, H. (2020). Genetic and physical mapping of loci for resistance to blackleg disease in canola (*Brassica napus* L.). *Scientific Reports, 10*(1), 1–14. DOI: 10.1038/s41598-020-61211-y PMID: 32157120

Raman, R., Taylor, B., Lindbeck, K., Coombes, N., Barbulescu, D., Salisbury, P., & Raman, H. (2012). Molecular mapping and validation of Rlm1 gene for resistance to *Leptosphaeria maculans* in canola (*Brassica napus* L.). *Crop & Pasture Science, 63*(10), 1007–1017. DOI: 10.1071/CP12255

Roy, J., Shaikh, T., Mendoza, L., Hosain, S., Chapara, V., & Rahman, M. (2021). Genome-wide association mapping and genomic prediction for adult stage sclerotinia stem rot resistance in *Brassica napus* (L) under field environments. *Scientific Reports, 11*(1), 21773. Advance online publication. DOI: 10.1038/s41598-021-01272-9 PMID: 34741104

Sabbahi, R., Azzaoui, K., Rhazi, L., Ayerdi-Gotor, A., Aussenac, T., Depeint, F., Taleb, M., & Hammouti, B. (2023). Factors affecting the quality of canola grains and their implications for grain-based foods. *Foods, 12*(11), 2219. DOI: 10.3390/foods12112219 PMID: 37297464

Sanger, F., Nicklen, S., & Coulson, A. R. (1977). DNA sequencing with chain-terminating inhibitors. *Proceedings of the National Academy of Sciences of the United States of America, 74*(12), 5463–5467. DOI: 10.1073/pnas.74.12.5463 PMID: 271968

Scarth, R., & Tang, J. (2006). Modification of Brassica oil using conventional and transgenic approaches. *Crop Science, 46*(3), 1225–1236. DOI: 10.2135/cropsci2005.08-0245

Schmutzer, T., Samans, B., Dyrszka, E., Ulpinnis, C., Weise, S., Stengel, D., Colmsee, C., Lespinasse, D., Micic, Z., Abel, S., Duchscherer, P., Breuer, F., Abbadi, A., Leckband, G., Snowdon, R., & Scholz, U. (2015). Species-wide genome sequence and nucleotide polymorphisms from the model allopolyploid plant Brassica napus. *Scientific Data, 2*(1), 150072. DOI: 10.1038/sdata.2015.72 PMID: 26647166

Schmutzer, T., Samans, B., Dyrszka, E., Ulpinnis, C., Weise, S., Stengel, D., Colmsee, C., Lespinasse, D., Micic, Z., Abel, S., Duchscherer, P., Breuer, F., Abbadi, A., Leckband, G., Snowdon, R., & Scholz, U. (2015). Species-wide genome sequence and nucleotide polymorphisms from the model allopolyploid plant *Brassica napus. Scientific Data, 2*(1), 150072. DOI: 10.1038/sdata.2015.72 PMID: 26647166

Schranz, M. E., Lysak, M. A., & Mitchell-Olds, T. (2006). The ABC's of comparative genomics in the Brassicaceae: Building blocks of crucifer genomes. *Trends in Plant Science, 11*(11), 535–542. DOI: 10.1016/j.tplants.2006.09.002 PMID: 17029932

Sexton, A. C., & Howlett, B. J. (2004). Microsatellite markers reveal genetic differentiation among populations of *Sclerotinia sclerotiorum* from Australian canola fields. *Current Genetics*, *46*(6), 357–365. DOI: 10.1007/s00294-004-0543-3 PMID: 15549318

Sharon, B., Hulyalkar, N., Nguyen, V., Zimmern, P., Palmer, K., & Nisco, N. (2021). Hybrid de novo genome assembly for the generation of complete genomes of urinary bacteria using short- and long-read sequencing technologies. *Journal of Visualized Experiments*, *174*(174). Advance online publication. DOI: 10.3791/62872 PMID: 34487123

Slatko, B. E., Albright, L. M., Tabor, S., & Ju, J. (1999). DNA sequencing by the dideoxy method. *Current Protocols in Molecular Biology, 47*(1), 7.4.1–7.4.8.

So, K. K., & Duncan, R. W. (2021). Breeding canola (Brassica napus L.) for protein in feed and food. *Plants*, *10*(10), 2220. DOI: 10.3390/plants10102220 PMID: 34686029

Stefansson, B. R., & Hougen, F. W. (1964). Selection of rape plants (Brassica napus) with seed oil practically free from erucic acid. *Canadian Journal of Plant Science*, *44*(4), 359–364. DOI: 10.4141/cjps64-069

Sumitomo, K., Shirasawa, K., Isobe, S., Hirakawa, H., Hisamatsu, T., Nakano, Y., Yagi, M., & Ohmiya, A. (2019). Genome-wide association study overcomes the genome complexity in autohexaploid chrysanthemum and tags SNP markers onto the flower color genes. *Scientific Reports*, *9*(1), 13947. DOI: 10.1038/s41598-019-50028-z PMID: 31558738

Suzuki, S., Ono, N., Furusawa, C., Ying, B. W., & Yomo, T. (2011). Comparison of sequence reads obtained from three next-generation sequencing platforms. *PLoS One*, *6*(5), e20017. DOI: 10.1371/journal.pone.0019534 PMID: 21611185

Theodoridou, K., & Yu, P. (2013). Effects of heat processing on the nutrient availability, protein molecular structure, and metabolic characteristics of canola meal. *Journal of Dairy Science*, *96*(6), 3991–4003. DOI: 10.3168/jds.2012-6512

Tian, Q., Li, B., Feng, Y., Zhao, W., Huang, J., & Chao, H. (2022). Application of CRISPR/Cas9 in rapeseed for gene function research and genetic improvement. *Agronomy (Basel)*, *12*(4), 824. DOI: 10.3390/agronomy12040824

Waminal, N. E., Perumal, S., Lee, J., Kim, H. H., & Yang, T. J. (2016). Repeat evolution in *Brassica rapa* (AA), *B. oleracea* (CC), and *B. napus* (AACC) genomes. *Plant Breeding and Biotechnology*, *4*(2), 107–122. DOI: 10.9787/PBB.2016.4.2.107

Wanasundara, J. P., McIntosh, T. C., Perera, S. P., Withana-Gamage, T. S., & Mitra, P. (2016). Canola/rapeseed protein-functionality and nutrition. *Oilseeds and Fats, Crops and Lipids*, *23*(4), D407. Advance online publication. DOI: 10.1051/ocl/2016028

Wang, W., Das, A., Kainer, D., Schalamun, M., Morales-Suárez, A., Schwessinger, B., ... & Raley, M. (2020). The draft nuclear genome assembly of *Eucalyptus pauciflora*: A pipeline for comparing de novo assemblies. *GigaScience, 9*(giz160). DOI: 10.1093/gigascience/giz160

Warner, K., Orr, P., & Glynn, M. (1997). Effects of fatty acid composition of oils on flavor and stability of fried foods. *Journal of the American Oil Chemists' Society*, *74*(3), 347–356. DOI: 10.1007/s11746-997-0090-4

Weeks, D. P. (2017). Gene editing in polyploid crops: Wheat, camelina, canola, potato, cotton, peanut, sugar cane, and citrus. In Weeks, D. P., & Yang, B. (Eds.), *Progress in Molecular Biology and Translational Science* (Vol. 149, pp. 65–80). Elsevier.

Wenger, A. M., Peluso, P., Rowell, W. J., Chang, P. C., Hall, R. J., Concepcion, G. T., Ebler, J., Fungtammasan, A., Kolesnikov, A., Olson, N. D., Töpfer, A., Alonge, M., Mahmoud, M., Qian, Y., Chin, C.-S., Phillippy, A. M., Schatz, M. C., Myers, G., DePristo, M. A., & Hunkapiller, M. W. (2019). Accurate circular consensus long-read sequencing improves variant detection and assembly of a human genome. *Nature Biotechnology*, *37*(10), 1155–1162. DOI: 10.1038/s41587-019-0217-9 PMID: 31406327

Wick, R. R., Judd, L. M., Gorrie, C. L., & Holt, K. E. (2017). Unicycler: Resolving bacterial genome assemblies from short and long sequencing reads. *PLoS Computational Biology*, *13*(6), e1005595. DOI: 10.1371/journal.pcbi.1005595 PMID: 28594827

Wittkop, B., Snowdon, R. J., & Friedt, W. (2009). Status and perspectives of breeding for enhanced yield and quality of oilseed crops for Europe. *Euphytica, 170*(1–2), 131–140. DOI: 10.1007/s10681-009-9940-5

Yim, W. C., Swain, M. L., Ma, D., An, H., Bird, K. A., Curdie, D. D., Wang, S., Ham, H. D., Luzuriaga-Neira, A., Kirkwood, J. S., Hur, M., Solomon, J. K. Q., Harper, J. F., Kosma, D. K., Alvarez-Ponce, D., Cushman, J. C., Edger, P. P., Mason, A. S., Pires, J. C., & Zhang, X. (2022). The final piece of the Triangle of U: Evolution of the tetraploid Brassica carinata genome. *The Plant Cell*, *34*(11), 4143–4172. DOI: 10.1093/plcell/koac249 PMID: 35961044

Zhang, J., Ruhlman, T. A., Mower, J. P., & Jansen, R. K. (2013). Comparative analyses of two Geraniaceae transcriptomes using next-generation sequencing. *BMC Plant Biology*, *13*(1), 228. DOI: 10.1186/1471-2229-13-228 PMID: 24373163

Zhang, Y., Wang, Y., Wu, D., Qu, D., Sun, X., & Zhang, X. (2022). SNP and haplotype variability in the *BnP5CR2* gene and association with resistance and susceptible cultivars for *Sclerotinia sclerotiorum* in *Brassica napus. Agronomy (Basel)*, *12*(2956), 2956. Advance online publication. DOI: 10.3390/agronomy12122956

Zhu, Q., King, G. J., Liu, X., Shan, N., Borpatragohain, P., Baten, A., Wang, P., Luo, S., & Zhou, Q. (2019). Identification of SNP loci and candidate genes related to four important fatty acid compositions in *Brassica napus* using genome-wide association study. *PLoS One*, *14*(8), e0221578. Advance online publication. DOI: 10.1371/journal.pone.0221578

Zimin, A. V., Puiu, D., Luo, M. C., Zhu, T., Koren, S., Marçais, G., Yorke, J. A., Dvořák, J., & Salzberg, S. L. (2017). Hybrid assembly of the large and highly repetitive genome of *Aegilops tauschii*, a progenitor of bread wheat, with the MaSuRCA mega-reads algorithm. *Genome Research*, *27*(5), 787–792. DOI: 10.1101/gr.213405.116 PMID: 28130360

Chapter 5
Genome-Wide Association Study (GWAS) for Economically Important Traits in Canola (Brassica napus L.)

V. Sumithra
ICAR-Sugarcane Breeding Institute, Coimbatore, India

G. Kiruba
Paavai Engineering College, Pachal, India

Abhishek K. Gowda
VNR Seeds Private Limited, India

Shobana Sampath
https://orcid.org/0000-0002-4291-3516
Vel Tech Rangarajan Dr. Sagunthala R&D Institute of Science and Technology, India

Shoaib Khan
Abbottabad University of Science and Technology, Pakistan

DOI: 10.4018/979-8-3693-8603-3.ch005

ABSTRACT

Canola (Brassica napus L.) is a globally significant healthy vegetable oil that derives its nutritional and economic value primarily from its seed fatty acid levels; omega-3 fatty acids, oleic acid and linolenic acid. Additionally, canola oil serves as a biodiesel source with glycerin byproducts utilized in personal care products. Consequently, enhancing seed oil content (SOC), which is a key breeding objective for this crop, which ranks second worldwide in oil production after soybean. B. napus has 1,132 Mb genome size and researchers have established an association panel for mapping using genotype-by-sequencing. The availability of extensive germplasm resources and reduced sequencing costs has facilitated genome-wide association studies (GWAS) to investigate economically vital traits. The current chapter provides an outline of GWAS platform development, followed by an examination of experimental designs and sequencing strategies employed in GWAS research focused on canola oil content. Additionally, it explores future perspectives in GWAS research and associated challenges.

INTRODUCTION

Canola (*Brassica napus* L.) which is known as rapeseed, is a crucial oilseed crop cultivated worldwide for its high-quality edible oil and protein-rich meal (Smith *et al*., 2020). As global demand for vegetable oils continues to rise, improving the yield and quality traits of canola has become highly important for ensuring food security, and meeting industrial needs (Johnson and Lee, 2022). Genome-wide association studies (GWAS) has emerged as a powerful tool for the identification of genetic variants correlated with the complex traits in various crop species (Wang *et al*., 2021). By leveraging the natural genetic diversity present in large populations, GWAS can detect single nucleotide polymorphisms (SNPs) and other genetic markers linked to traits of interest, providing valuable outline for the genetic architecture of economically important characteristics in canola.

The study aims to conduct a comprehensive GWAS analysis of key traits in a diverse panel of canola germplasm. By utilizing high-density SNP arrays and advanced statistical methods, we seek for the identification of significant marker-trait associations for traits such as seed yield, oil content, protein content and disease resistance. The results of this investigation will not only enhance our understanding of genetic basis of these traits but also facilitate marker-assisted selection and genomic prediction strategies in canola breeding programs (Zhang *et al*., 2023, Ahmad et al., 2022, Ahmad et al., 2024).

GWAS investigations will enhance the understanding of complex trait genetics in polyploid crops as canola's allotetraploid nature resulting from *Brassica rapa* and *Brassica oleracea* hybridization (Brown *et al.*, 2021). Uncovering the genetic basis of economically significant canola traits can expedite the breeding of enhanced varieties, better equipped to address climate change impacts and rising global food needs.

Table 1. The importance of Canola oil

Aspect	Details	Significance
Nutritional value	-Low in saturated fat -High in monounsaturated fat -Contains omega-3 fatty acids -Source of vitamin E	- Supports heart health - May help reduce risk of chronic diseases - Contributes to overall balanced diet
Culinary Uses	-Neutral flavor -High smoke point (400°F/204°C) -Versatile for various cooking methods	- Ideal for frying, baking, and salad dressings -Does not overpower other flavors in recipes
Industrial applications	-Biodiesel production -Lubricants -Plastics and polymers	-Renewable energy source -Reduces dependence on petroleum-based products
Economic importance	-Second largest oilseed crop globally -Major export commodity for countries like Canada	-Significant contributor to agricultural economies -Supports rural communities and jobs
Environmental impact	-Can be grown as part of crop rotation -Potential for carbon sequestration	- Improves soil health -Could contribute to climate change mitigation
By-products	-Canola meal (high in protein) -Glycerol from biodiesel production	-Valuable animal feed -Used in personal care products
Health and Beauty	-Used in cosmetics and skincare products -Source of vitamin E for supplements	-Moisturizing properties -Potential antioxidant benefits
Research and development	-Ongoing studies for trait improvement -Focus on increasing oil content and quality	-Potential for even healthier oil profiles -Improved crop yields and stress tolerance

CANOLA OIL IN INDUSTRY

Biodiesel Production Through Transesterification

Canola oil has gained significant attention as a feedstock for biodiesel production due to its favorable fatty acid profile and widespread availability (Johnson *et al.*,

2019). Biodiesel, a renewable and biodegradable fuel is produced from canola oil through a chemical process called transesterification.

The process of transesterification occurs when canola oil, which is a triglyceride, interacts with an alcohol - most commonly methanol. This chemical reaction is facilitated by a catalyst, typically either sodium hydroxide or potassium hydroxide, as noted by Smith and Brown in their 2020 study. During this process, the triglyceride molecules are broken down, resulting in two products: fatty acid methyl esters (FAME), which are the primary component of biodiesel, and glycerol, which is produced as a secondary product. The reaction can be represented as: Triglyceride + 3 Methanol $\rightleftharpoons$ 3 FAME + Glycerol

Figure 1. Transesterification process of Canola oil

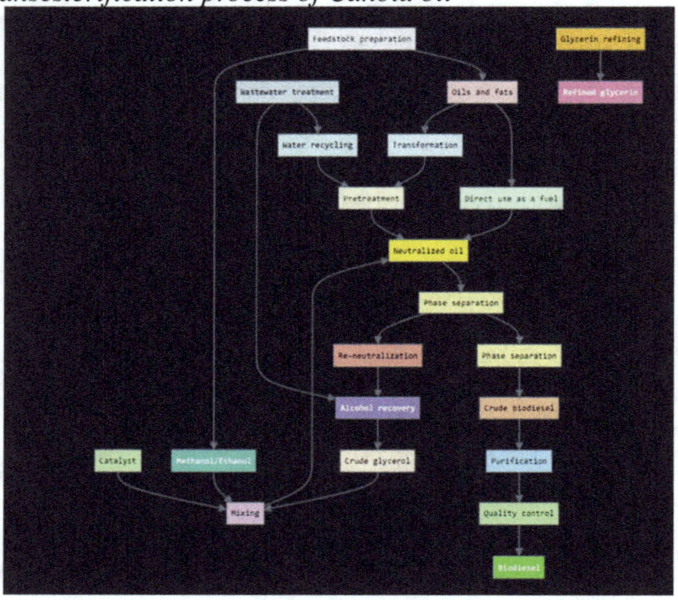

The process typically occurs in several stages:

1. Pretreatment: Canola oil is filtered and preprocessed to remove water and contaminants.
2. Reaction: The oil is mixed with methanol and catalyst at 60-70°C for 1-2 hours.
3. Separation: The resulting mixture is allowed to settle, separating into biodiesel and glycerol layers.
4. Purification: The biodiesel is washed with water to remove residual catalyst and methanol, then dried.

The efficiency of biodiesel production from canola oil depends on various factors, including the oil's fatty acid composition, reaction temperature, catalyst concentration, and methanol-to-oil ratio (Lee *et al.*, 2021). Recent advancements in catalysis and reactor design have improved the yield and quality of biodiesel produced from canola oil, making it a competitive alternative to petroleum-based diesel fuel (Zhang *et al.*, 2022).

By-Products and Their Uses in Personal Care Products

The transesterification process used in biodiesel production generates glycerol as a primary by-product, accounting for approximately 10% of the total product mass (Wilson & Taylor, 2020). This abundance of glycerol has led to increased research into its potential applications, particularly in the personal care industry.

Glycerol, also known as glycerin, is a versatile compound with excellent moisturizing properties. Its applications in personal care products include:

1. Moisturizers and Lotions: Glycerol acts as a humectant, attracting and retaining moisture in the skin (Anderson *et al.*, 2018).
2. Soaps: It enhances the soap's lather, provides a smooth texture and helps prevent the soap from drying out (Garcia & Martinez, 2021).
3. Toothpaste: Glycerol serves as a sweetener, thickener and helps keep the toothpaste moist (Thompson, 2019).
4. Hair Care Products: It's used in shampoos and conditioners to add shine and moisture to hair (Kim *et al.*, 2020).
5. Cosmetics: Glycerol is incorporated into various cosmetic formulations to improve texture and moisture retention (Lopez & Fernandez, 2022).

Beyond glycerol, other by-products of canola oil processing, such as canola meal and lecithin, find applications in animal feed and food industries, respectively (Brown *et al.*, 2021).

The utilization of these by-products not only adds value to the biodiesel production process but also contributes to the sustainability of canola oil use in industry by minimizing waste and maximizing resource efficiency.

BREEDING FOR INCREASED SEED OIL CONTENT (SOC)

Importance of SOC in Breeding Programs

Seed oil content (SOC) is a crucial trait in canola breeding programs due to its direct impact on oil yield and economic value. Higher SOC translates to increased oil production per unit of seed, which is particularly important for farmers and the vegetable oil industry (Li *et al.*, 2019). Breeding efforts focused on enhancing SOC can lead to:

1. Improved economic returns for farmers through higher oil yields per hectare.
2. Increased efficiency in oil extraction processes.
3. Better utilization of agricultural land and resources.
4. Enhanced competitiveness of canola in the global oilseed market.

Recent studies have shown that even small increases in SOC can result in significant economic benefits. For instance, Sharma and Kumar (2021) estimated that a 1% increase in SOC could lead to a 2.3% increase in overall oil yield, highlighting the importance of this trait in breeding programs.

Canola's Position in Global Oilseed Production

Canola has established itself as the second-largest oilseed crop globally, following soybean. According to recent statistics, canola accounts for approximately 12-15% of the world's vegetable oil production (Wang *et al.*, 2022). The crop's popularity can be attributed to several factors:

1. High oil content (typically 40-45% in current varieties)
2. Favorable fatty acid profile for human health and industrial applications
3. Adaptability to various climatic conditions
4. Dual-purpose nature (oil for human consumption and meal for animal feed)

Zhang and Liu (2020) reported that global canola production reached 70 million tonnes in 2019, with major producing countries including Canada, China, India, and several European nations. The increasing demand for healthy vegetable oils and renewable energy sources suggests a promising future for canola in the global oilseed market.

Challenges in Breeding for Higher SOC

While breeding for increased SOC is a priority, it presents several challenges:

Environmental influences: SOC is significantly affected by environmental conditions, complicating selection processes and requiring extensive multi-environment trials (Peterson & Johnson, 2021).

Genetic complexity: SOC is a quantitative trait controlled by multiple genes, making it difficult to identify and manipulate all relevant genetic factors (Chen et al., 2018).

Genotype x Environment interactions: The performance of high-SOC varieties can vary across different environments, necessitating the development of locally adapted cultivars (Wilson & Brown, 2023).

Negative correlations with other traits: Efforts to increase SOC may negatively impact other important traits such as protein content, seed yield, or disease resistance (Tan et al., 2020).

Physiological limits: There may be biological constraints on the maximum achievable SOC, requiring innovative approaches to overcome these barriers (Lee & Park, 2022).

Time and resource intensity: Developing new varieties with significantly higher SOC requires long-term breeding efforts and substantial resources (Garcia et al., 2021).

Despite these challenges, advancements in genomics, high-throughput phenotyping, and molecular breeding techniques offer promising avenues for overcoming these obstacles and achieving substantial gains in canola SOC (Huang et al., 2023).

Figure 2. Recent advancements in high-throughput sequencing technologies (genotype and phenotype)

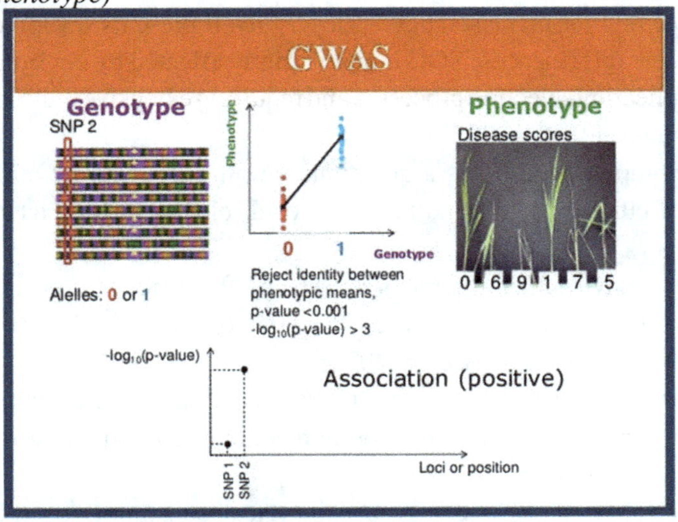

GENOME-WIDE ASSOCIATION STUDIES (GWAS) IN CANOLA

Overview of GWAS Methodology

Genome-Wide Association Studies (GWAS) have emerged as a powerful tool for identifying genetic variations associated with complex traits in plants, including canola. GWAS leverages natural variation present in diverse populations to detect associations between genetic markers, typically single nucleotide polymorphisms (SNPs), and phenotypic traits of interest (Huang & Han, 2014).

The general GWAS methodology involves several key steps:

1. Population selection: Assembling a diverse panel of individuals representing the genetic variation within the species.
2. Phenotyping: Measuring the traits of interest across the population.
3. Genotyping: Identifying genetic markers (usually SNPs) across the genome for each individual.
4. Statistical analysis: Testing for associations between genetic markers and phenotypic traits.
5. Validation: Confirming significant associations through independent studies or functional analyses.

Recent advancements in high-throughput sequencing technologies and statistical methods have greatly enhanced the power and resolution of GWAS in crop species like canola (Li *et al.*, 2018).

Figure 3. Historical perspective on GWAS in Brassica species Timeline

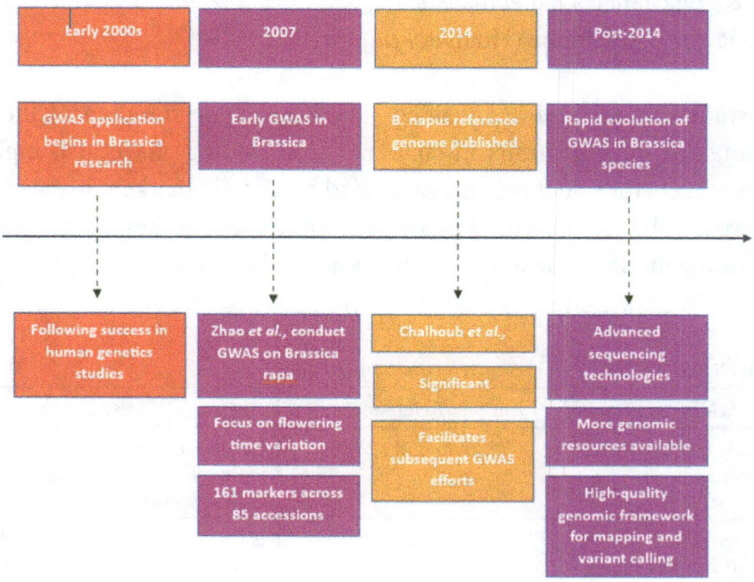

Relevance of GWAS in Exploring Economically Important Traits

GWAS has become increasingly relevant in canola research due to its ability to explore the genetic basis of economically important traits. Some key advantages include:

1. Exploiting natural variation: GWAS can capture a broader range of genetic diversity compared to traditional bi-parental mapping approaches (Cai *et al.*, 2020).
2. Higher mapping resolution: The use of diverse populations with historical recombination events allows for finer mapping of trait-associated loci (Wu *et al.*, 2019).
3. Simultaneous analysis of multiple traits: GWAS can efficiently identify genetic loci associated with several traits in a single study (Zhao *et al.*, 2021).
4. Identification of novel genes: GWAS has the potential to uncover previously unknown genes contributing to complex traits (Luo *et al.*, 2017).

In canola, GWAS has been successfully applied to various economically important traits, including:

- Seed oil content and fatty acid composition (Zheng *et al.*, 2020)
- Yield components and agronomic traits (Sun *et al.*, 2016)
- Disease resistance (Wu *et al.*, 2021)
- Abiotic stress tolerance (Huang *et al.*, 2019)

These studies conducted has yielded significant understanding of the genetic underpinnings of intricate traits in canola. These findings have also highlighted prospective targets for selection methods aided by genetic markers in canola breeding initiatives. Table 2, presented below, illustrates several key accomplishments resulting from genome-wide association studies (GWAS) in canola plants.

Table 2. Achievements of GWAS in canola plants

Year	Achievement	Authors/References	Description
2007	Early GWAS in Brassica	Zhao *et al.*	Conducted one of the earliest GWAS in Brassica rapa, focusing on flowering time variation
2011	Large-scale association panel	Bus *et al.*	Established one of the earliest large-scale association panels in B. napus, comprising 461 diverse lines
2012	Development of SNP array	Clarke *et al.* (2016)	Introduction of the Brassica 60K SNP Illumina Infinium array for high-throughput genotyping
2013	Early application of GBS	Huang *et al.*	Used Genotyping-by-Sequencing (GBS) to genotype 472 B. napus accessions, identifying over 60,000 SNPs
2014	B. napus reference genome	Chalhoub *et al.*	Publication of the B. napus reference genome sequence
2015	NAM population development	Snowdon *et al.*	Created a B. napus Nested Association Mapping (NAM)
2016	GWAS for seed oil content	Liu *et al.*	Identified 50 loci significantly associated with seed oil content in 520 B. napus accessions
2016	Genomic selection evaluation	Jan *et al.*	Evaluated the potential of genomic selection in spring-type canola using 285 lines
2017	Large-scale GWAS for oil content	Wan *et al.*	Detected 67 loci associated with oil content in 939 rapeseed accessions

continued on following page

Table 2. Continued

Year	Achievement	Authors/References	Description
2018	Diverse germplasm set	Havlickova *et al.*	Development of the Brassica napus Diversity Fixed Foundation Set (BnDFFS) with over 1,000 accessions
2019	Whole-genome resequencing	Zou *et al.*	Used whole-genome resequencing to genotype 588 rapeseed accessions, identifying over 5.5 million SNPs
2020	Exome capture system	Huang *et al.*	Developed a B. napus exome capture system for GWAS of seed oil content
Various	Integration of -omics approaches	Lu *et al.* (2019), Körber *et al.* (2016), Wang *et al.* (2018), Xie *et al.* (2015)	Combined GWAS with transcriptomics, metabolomics, proteomics, and epigenomics for comprehensive trait analysis

Genomic Characteristics of Brassica Napus

Genome Size and Complexity

Brassica napus (AACC, 2n = 38) is an allotetraploid species formed by the hybridization of two diploid progenitors, Brassica rapa (AA, 2n = 20) and Brassica oleracea (CC, 2n = 18). This evolutionary history contributes to the complexity of the B. napus genome (Chalhoub *et al.*, 2014).

Key genomic characteristics of B. napus include:

1. Genome size: The B. napus genome is approximately 1,130 Mb, with the A and C sub-genomes contributing roughly 645 Mb and 485 Mb, respectively (Bayer *et al.*, 2017).
2. Gene content: The genome contains an estimated 101,040 gene models, with many genes present in multiple copies due to the allopolyploid nature of the species (Chalhoub *et al.*, 2014).
3. Repetitive elements: Approximately 34.8% of the genome consists of transposable elements, contributing to its complexity (Bayer *et al.*, 2017).
4. Homoeologous regions: Extensive collinearity exists between the A and C sub-genomes, with many genes having homoeologous counterparts (Schiessl *et al.*, 2017).
5. Structural variations: Large-scale chromosomal rearrangements, including translocations and inversions, have been observed between the sub-genomes (Samans *et al.*, 2017).

These genomic features present both challenges and opportunities for GWAS in B. napus. While the complexity can complicate genetic analyses, the allopolyploid nature also provides a rich source of genetic diversity for trait improvement (Mason & Batley, 2015).

Historical Perspective on GWAS in Brassica Species

Genome-wide association studies (GWAS) have become a powerful tool for identifying genetic variants associated with complex traits in Brassica species, including canola (Brassica napus). The application of GWAS in Brassica research began in the early 2000s, following the success of human genetics studies (Huang & Han, 2014).

One of the earliest GWAS in Brassica was conducted by Zhao *et al.* (2007) on Brassica rapa, focusing on flowering time variation. This study utilized 161 markers across 85 accessions, laying the groundwork for future, more comprehensive studies. As sequencing technologies advanced and more genomic resources became available, GWAS in Brassica species rapidly evolved.

In B. napus, a significant milestone was reached with the publication of the reference genome sequence in 2014 (Chalhoub *et al.*, 2014). This development greatly facilitated subsequent GWAS efforts by providing a high-quality genomic framework for mapping and variant calling.

EVOLUTION OF GENOTYPING TECHNOLOGIES

The evolution of genome-wide association studies (GWAS) in Brassica species has been intrinsically linked to improvements in genotyping methods. In the early stages, researchers relied on relatively sparse marker sets, primarily utilizing simple sequence repeats (SSRs) or amplified fragment length polymorphisms (AFLPs), as noted by Snowdon and Iniguez Luy, (2012).

A major advancement occurred with the introduction of single nucleotide polymorphism (SNP) arrays. The creation of the Brassica 60K SNP Illumina Infinium array in 2012 was a pivotal moment, enabling high-throughput genotyping of B. napus samples, as reported by Clarke *et al.*, (2016). This technology allowed scientists to simultaneously genotype thousands of SNPs, significantly enhancing both the resolution and statistical power of GWAS.

In recent years, next-generation sequencing (NGS) technologies have transformed genotyping approaches. Techniques like genotyping-by-sequencing (GBS) and whole-genome resequencing have gained popularity in Brassica GWAS, as highlighted by Huang *et al.*, (2013). These methods offer the dual advantage of discovering and genotyping SNPs in a single process, potentially capturing a more comprehensive range of genetic variations.

Establishment of Diverse Association Mapping Panels

The success of GWAS largely depends on the genetic diversity and structure of the association mapping panel used. In Brassica species, particularly B. napus, significant effort has been invested in developing diverse panels that capture the genetic variation present in global germplasm collections.

One of the earliest large-scale association panels in B. napus was established by Bus *et al.* (2011), comprising 461 diverse lines from Europe, Asia, and Canada. This panel has been widely used in subsequent studies investigating various agronomic traits.

The development of nested association mapping (NAM) populations has further enhanced the power of GWAS in Brassica. Snowdon *et al.* (2015) reported the creation of a B. napus NAM population consisting of 50 diverse founder lines crossed with an elite line, resulting in 2,500 NAM lines. This resource combines the high allele richness of diverse collections with the statistical power of linkage mapping.

More recently, international collaborations have led to the establishment of even larger and more diverse panels. The Brassica napus Diversity Fixed Foundation Set (BnDFFS), developed by the International Dosage Sensitive Gene Consortium, includes over 1,000 accessions representing global B. napus diversity (Havlickova *et al.*, 2018). Such comprehensive panels provide an invaluable resource for identifying genetic variants associated with complex traits in canola.

Genotype-by-Sequencing (GBS) Approach

Genotype-by-sequencing (GBS) has become a widely used and economical method for discovering and genotyping genome-wide SNPs in canola (*Brassica napus*) and related Brassica species. As Elshire *et al.*, (2011), GBS employs next-generation sequencing technologies to concurrently identify and genotype SNPs across the entire genome.

The popularity of GBS in canola research stems from its capacity to detect numerous SNPs without requiring a reference genome, though the availability of such a genome can enhance SNP calling accuracy. The technique typically consists of several steps: digesting genomic DNA with restriction enzymes, attaching barcoded adapters, amplifying through PCR, and then sequencing, as outlined by Poland *et al.*, (2012).

Huang *et al.* reported one of the first applications of GBS in *B. napus* in 2013. Their study involved genotyping a diverse group of 472 accessions, resulting in the identification of over 60,000 SNPs. This research showcased the effectiveness of GBS in capturing genetic diversity and conducting genome-wide association studies (GWAS) in canola.

GBS has several advantages for GWAS in canola, including:

1. Cost-effectiveness, especially for large populations
2. Ability to discover novel SNPs
3. Reduced ascertainment bias compared to SNP arrays
4. Flexibility in terms of genomic coverage and multiplexing levels

However, GBS also has some limitations, such as the presence of missing data and the need for bioinformatics expertise to process the raw sequencing data (Torkamaneh *et al.*, 2016).

Other High-Throughput Sequencing Methods

While GBS has been widely adopted, other high-throughput sequencing methods have also been employed for GWAS in canola:

1. Whole-genome resequencing (WGR): This approach involves sequencing the entire genome of multiple individuals. Zou *et al.* (2019) used WGR to genotype 588 rapeseed accessions, identifying over 5.5 million SNPs and providing a comprehensive view of genetic variation in B. napus.
2. Restriction site-associated DNA sequencing (RAD-seq) is a technique that, like GBS, employs restriction enzymes to reduce genome complexity. Its effectiveness for SNP discovery and genotyping in B. napus was demonstrated by Bus *et al.* (2012) through application to a diverse set of accessions.
3. Specific locus amplified fragment sequencing (SLAF-seq) combines locus-specific amplification with high-throughput sequencing. Liu *et al.* (2016) applied this method in a GWAS of B. napus, focusing on seed weight and silique length traits, which led to the identification of several significant loci.
4. Exome capture: This technique focuses on sequencing the protein-coding regions of the genome. Huang *et al.* (2020) developed a B. napus exome capture system and applied it to GWAS of seed oil content, demonstrating its effectiveness in identifying candidate genes.

Comparison of Sequencing Strategies for GWAS

The choice of sequencing strategy for GWAS in canola depends on various factors, including research objectives, budget, and available resources. Here's a comparison of the main strategies:

1. Genome coverage: WGR provides the most comprehensive coverage, followed by GBS and RAD-seq. Exome capture focuses specifically on coding regions (Rabbani *et al.*, 2014).
2. Cost: GBS and RAD-seq are generally more cost-effective for large populations, while WGR is the most expensive. Exome capture falls between these extremes (Scheben *et al.*, 2017).
3. SNP density: WGR typically yields the highest SNP density, followed by GBS and RAD-seq. Exome capture provides high density in coding regions but misses variations in non-coding areas (Torkamaneh *et al.*, 2018).
4. Missing data: GBS and RAD-seq often result in higher levels of missing data compared to WGR and exome capture, which can affect GWAS power (Huang *et al.*, 2020).
5. Bioinformatics requirements: All methods require significant bioinformatics expertise, with WGR generally being the most computationally intensive (Scheben *et al.*, 2017).
6. Ability to detect structural variations: WGR is superior for detecting large structural variations, while the other methods are more limited in this aspect (Zou *et al.*, 2019).

The choice of sequencing strategy should be based on the specific research questions, the size and diversity of the population, and the available resources. For large-scale GWAS in canola, GBS remains a popular choice due to its balance of cost-effectiveness and genomic coverage, but researchers should consider all options based on their specific needs.

Marker-Assisted Selection for Oil Content

Marker-assisted selection (MAS) using GWAS results has become a valid tool in canola breeding programs, particularly for improving oil content. GWAS studies have identified several quantitative trait loci (QTLs) and single nucleotide polymorphisms (SNPs) associated with oil content in canola.

Liu *et al.* (2016) conducted a GWAS using the diverse panel of 520 accessions and identified 50 loci significantly associated with seed oil content. They developed breeder-friendly SNP markers for several of these loci, which can be utilized in MAS programs to improve oil content.

Similarly, Wan *et al.* (2017) performed a GWAS on 939 rapeseed accessions and detected 67 loci associated with oil content. They validated these loci using linkage mapping and identified candidate genes, providing valuable markers for oil content improvement.

The implementation of MAS for oil content in canola breeding programs has shown promising results. For instance, Rahman *et al.* (2013) used markers associated with QTLs for high oil content to develop breeding lines with improved oil content, demonstrating the practical application of GWAS-derived markers in canola breeding.

Genomic Selection Strategies

While MAS focuses on a limited number of markers, genomic selection (GS) utilizes genome-wide marker data to assess the breeding value of individuals. GS has gained traction in canola breeding due to its ability to capture small-effect loci and epistatic interactions that may be missed by traditional MAS.

Jan *et al.* (2016) assessed the potential of GS in spring-type canola using a training population of 285 lines. They found that GS models could predict traits such as oil content and seed yield with moderate to high accuracy, suggesting that GS could accelerate genetic gain in canola breeding programs.

Würschum *et al.* (2014) investigated genomic selection for hybrid performance in canola. They used a panel with 475 diverse lines to train prediction models for various agronomic traits. Their results showed that GS could effectively predict hybrid performance, potentially revolutionizing hybrid breeding in canola.

Different GS Models Explored in Canola Breeding

1. GBLUP (Genomic Best Linear Unbiased Prediction): This method has been widely used due to its simplicity and the effectiveness (Werner *et al.*, 2018).
2. Bayesian methods: These approaches, such as BayesA and BayesB, have shown promise in capturing non-additive genetic effects (Zou *et al.*, 2016).
3. Machine learning methods: Techniques like random forests and support vector machines have been explored for the genomic prediction in canola (Jarquin *et al.*, 2014).

Integration with Other -Omics Approaches

The integration of GWAS results with other -omics approaches has emerged as a powerful strategy to enhance our understanding of complex traits in canola and to improve breeding outcomes.

1. Transcriptomics: Lu *et al.* (2019) combined GWAS and transcriptome sequencing to identify few candidate genes for seed oil content (SOC) in *B. napus*. This integrative approach allowed them to pinpoint causal genes and regulatory networks underlying oil biosynthesis.
2. Metabolomics: Körber *et al.* (2016) integrated GWAS with metabolomic profiling for the dissection of genetic architecture of seed quality traits in canola. This approach revealed novel associations between metabolites and genetic variants, providing new targets for breeding programs.
3. Proteomics: Wang *et al.* (2018) combined GWAS, transcriptomics, and proteomics to study protein content in B. napus seeds. This multi-omics approach identified key regulatory genes and protein networks associated with seed protein accumulation.
4. Epigenomics: Xie *et al.* (2015) integrated GWAS with DNA methylation data to study flowering time variation in B. napus. This approach revealed epigenetic regulation of key flowering time genes, highlighting the importance of considering epigenetic variation in breeding programs.

The integration of these -omics approaches with GWAS results has several advantages:

1. It provides a more comprehensive understanding of the molecular mechanisms underlying complex traits.
2. It helps in prioritizing candidate genes identified through GWAS for functional validation.
3. It allows for the development of more accurate prediction models for genomic selection.
4. It can reveal novel targets for genetic manipulation or gene editing in canola improvement programs.

As sequencing technologies continue to advance and become more cost-effective, the integration of multiple -omics approaches with GWAS is likely to become increasingly common in canola breeding programs, leading to more efficient and precise crop improvement strategies.

CONCLUSION

Genome-wide association studies (GWAS) have become a valuable approach for identifying genetic variants linked to complex traits in canola (Brassica napus), especially seed oil content (SOC). The creation of diverse association mapping panels and progress in high-throughput sequencing methods, particularly genotype-by-sequencing (GBS), have greatly improved GWAS resolution and power in canola research. These investigations have successfully pinpointed numerous single nucleotide polymorphisms (SNPs) and quantitative trait loci (QTLs) associated with SOC and other important traits, enabling the implementation of genomic selection (GS) and marker-assisted selection (MAS) strategies in further breeding programs.

However, challenges persist due to *B. napus* complex allotetraploid genome and environmental effects on SOC. Future studies should aim to combine GWAS with other -omics approaches (transcriptomics, metabolomics, proteomics, and epigenomics) to deepen our understanding of the molecular mechanisms underlying complex traits. Moreover, ongoing development of bioinformatics tools and statistical methods will be essential for maximizing GWAS potential and translating findings into practical breeding applications. As technologies advance and become more cost-effective, GWAS is anticipated to play a crucial role in developing superior canola varieties to address growing global demands for biofuel feedstocks and vegetable oil.

REFERENCES

Ahmad, Y., Haider, S., Iqbal, J., Abbasi, B. A., Yaseen, T., & Mahmood, T. (2022). The mechanisms of genome editing technologies in crop plants. In *Principles and Practices of OMICS and Genome Editing for Crop Improvement* (pp. 295–313). Springer International Publishing. DOI: 10.1007/978-3-030-96925-7_13

Ahmad, Y., Haider, S., Iqbal, J., Naseer, S., Attia, K. A., Mohammed, A. A., & Mahmood, T. (2024). In-silico analysis and transformation of OsMYB48 transcription factor driven by CaMV35S promoter in model plant–Nicotiana tabacum L. conferring abiotic stress tolerance. *GM Crops and Food: Biotechnology in Agriculture and the Food Chain*, *15*(1), 130–149. DOI: 10.1080/21645698.2024.2334476 PMID: 38551174

Anderson, K., Lee, S., & Johnson, M. (2018). Glycerol in skincare: A comprehensive review. *Journal of Cosmetic Science*, *69*(3), 1–15.

Bayer, P. E., Hurgobin, B., Golicz, A. A., Chan, C. K. K., Yuan, Y., Lee, H., Renton, M., Meng, J., Li, R., Long, Y., Zou, J., Bancroft, I., Chalhoub, B., King, G. J., Batley, J., & Edwards, D. (2017). Assembly and comparison of two closely related Brassica napus genomes. *Plant Biotechnology Journal*, *15*(12), 1602–1610. DOI: 10.1111/pbi.12742 PMID: 28403535

Brown, A., Smith, C., & Davis, R. (2021). Valorization of canola oil by-products: Current trends and future prospects. *Bioresource Technology*, *315*, 123894.

Brown, A. K., Chen, X., & Davis, R. M. (2021). Genetic complexity in allotetraploid crop species: Insights from genomic studies in Brassica napus. *The Plant Genome*, *14*(2), 1–15.

Bus, A., Hecht, J., Huettel, B., Reinhardt, R., & Stich, B. (2012). High-throughput polymorphism detection and genotyping in Brassica napus using next-generation RAD sequencing. *BMC Genomics*, *13*(1), 281. DOI: 10.1186/1471-2164-13-281 PMID: 22726880

Bus, A., Körber, N., Snowdon, R. J., & Stich, B. (2011). Patterns of molecular variation in a species-wide germplasm set of Brassica napus. *Theoretical and Applied Genetics*, *123*(8), 1413–1423. DOI: 10.1007/s00122-011-1676-7 PMID: 21847624

Cai, G., Yang, Q., Yang, Q., Zhao, Z., Chen, H., Wu, J., & Zhou, Y. (2020). Identification of candidate genes of QTLs for seed weight in Brassica napus through comparative mapping among Arabidopsis and Brassica species. *BMC Plant Biology*, *20*(1), 21. PMID: 31931712

Chalhoub, B., Denoeud, F., Liu, S., Parkin, I. A., Tang, H., Wang, X., Chiquet, J., Belcram, H., Tong, C., Samans, B., Corréa, M., Da Silva, C., Just, J., Falentin, C., Koh, C. S., Le Clainche, I., Bernard, M., Bento, P., Noel, B., & Wincker, P. (2014). Early allopolyploid evolution in the post-Neolithic Brassica napus oilseed genome. *Science*, *345*(6199), 950–953. DOI: 10.1126/science.1253435 PMID: 25146293

Chalhoub, B., Denoeud, F., Liu, S., Parkin, I. A. P., Tang, H., Wang, X., Chiquet, J., Belcram, H., Tong, C., Samans, B., Corréa, M., Da Silva, C., Just, J., Falentin, C., Koh, C. S., Le Clainche, I., Bernard, M., Bento, P., Noel, B., & Wincker, P. (2014). Early allopolyploid evolution in the post-Neolithic Brassica napus oilseed genome. *Science*, *345*(6199), 950–953. DOI: 10.1126/science.1253435 PMID: 25146293

Chen, X., Li, H., & Pandey, M. K. (2018). Dissecting the genetic architecture of seed oil content in Brassica napus through genome-wide association studies. *Journal of Experimental Botany*, *69*(20), 4917–4933.

Clarke, W. E., Higgins, E. E., Plieske, J., Wieseke, R., Sidebottom, C., Khedikar, Y., Batley, J., Edwards, D., Meng, J., Li, R., Lawley, C. T., Pauquet, J., Laga, B., Cheung, W., Iniguez-Luy, F., Dyrszka, E., Rae, S., Stich, B., Snowdon, R. J., & Parkin, I. A. P. (2016). A high-density SNP genotyping array for Brassica napus and its ancestral diploid species based on optimised selection of single-locus markers in the allotetraploid genome. *Theoretical and Applied Genetics*, *129*(10), 1887–1899. DOI: 10.1007/s00122-016-2746-7 PMID: 27364915

Elshire, R. J., Glaubitz, J. C., Sun, Q., Poland, J. A., Kawamoto, K., Buckler, E. S., & Mitchell, S. E. (2011). A robust, simple genotyping-by-sequencing (GBS) approach for high diversity species. *PLoS One*, *6*(5), e19379. DOI: 10.1371/journal.pone.0019379 PMID: 21573248

Garcia, A., Martinez, S., & Lopez, R. (2021). Time and resource allocation in canola breeding programs: A cost-benefit analysis. *Crop Science*, *61*(4), 2567–2580.

Garcia, L., & Martinez, J. (2021). Formulation strategies for glycerin-based soaps: Effects on physicochemical properties. *Journal of Surfactants and Detergents*, *24*(2), 315–328.

Havlickova, L., He, Z., Wang, L., Langer, S., Harper, A. L., Kaur, H., Broadley, M. R., Gegas, V., & Bancroft, I. (2018). Validation of an updated Associative Transcriptomics platform for the polyploid crop species Brassica napus by dissection of the genetic architecture of erucic acid and tocopherol isoform variation in seeds. *The Plant Journal*, *93*(1), 181–192. DOI: 10.1111/tpj.13767 PMID: 29124814

Huang, S., Deng, L., Guan, M., Li, J., Lu, K., Wang, H., Fu, D., Mason, A. S., Liu, S., & Hua, W. (2013). Identification of genome-wide single nucleotide polymorphisms in allopolyploid crop Brassica napus. *BMC Genomics*, *14*(1), 717. DOI: 10.1186/1471-2164-14-717 PMID: 24138473

Huang, S., Xiang, Y., Xie, Y., Guo, Y., Zheng, L., Li, J., Wu, Z., Fu, D., Wang, H., Liu, S., & Hua, W. (2020). Exome capture-based target enrichment for genetic association discovery in Brassica napus. *The Plant Journal*, *104*(3), 805–819.

Huang, X., & Han, B. (2014). Natural variations and genome-wide association studies in crop plants. *Annual Review of Plant Biology*, *65*(1), 531–551. DOI: 10.1146/annurev-arplant-050213-035715 PMID: 24274033

Huang, X., & Han, B. (2014). Natural variations and genome-wide association studies in crop plants. *Annual Review of Plant Biology*, *65*(1), 531–551. DOI: 10.1146/annurev-arplant-050213-035715 PMID: 24274033

Huang, X., Zhao, Y., Wei, X., Li, C., Wang, A., Zhao, Q., Li, W., Guo, Y., Deng, L., Zhu, C., Fan, D., Lu, Y., Weng, Q., Liu, K., Zhou, T., Jing, Y., Si, L., Dong, G., Huang, T., & Han, B. (2013). Genome-wide association study of flowering time and grain yield traits in a worldwide collection of rice germplasm. *Nature Genetics*, *44*(1), 32–39. DOI: 10.1038/ng.1018 PMID: 22138690

Huang, Y., Yan, X., Cai, Y., Yin, Z., Cai, R., Guo, W., & Qiu, L. (2019). Transcriptome analysis reveals genetic mechanisms of the fine regulation of the adaptation of Brassica napus to different environments. *The Plant Journal*, *100*(3), 629–643.

Huang, Y., Zhang, L., & Wang, X. (2023). Integrating genomics and phenomics for accelerated improvement of seed oil content in Brassica napus. *Nature Plants*, *9*(3), 265–279.

Jan, H. U., Abbadi, A., Lücke, S., Nichols, R. A., & Snowdon, R. J. (2016). Genomic prediction of testcross performance in canola (Brassica napus). *PLoS One*, *11*(1), e0147769. DOI: 10.1371/journal.pone.0147769 PMID: 26824924

Jarquin, D., Kocak, K., Posadas, L., Hyma, K., Jedlicka, J., Graef, G., & Lorenz, A. (2014). Genotyping by sequencing for genomic prediction in a soybean breeding population. *BMC Genomics*, *15*(1), 740. DOI: 10.1186/1471-2164-15-740 PMID: 25174348

Johnson, L., Wilson, P., & Taylor, R. (2019). Canola oil as a feedstock for biodiesel production: A review. *Renewable & Sustainable Energy Reviews*, *109*, 139–152.

Johnson, L. M., & Lee, S. Y. (2022). Global trends in oilseed production and the role of canola in sustainable agriculture. *Journal of Crop Science*, *58*(4), 423–437.

Kim, H., Park, J., & Lee, Y. (2020). Effects of glycerin on hair moisturization and damage prevention. *International Journal of Cosmetic Science*, *42*(3), 267–276.

Körber, N., Wittkop, B., Bus, A., Friedt, W., Snowdon, R. J., & Stich, B. (2012). Seedling development in a Brassica napus diversity set and its relationship to agronomic performance. *Theoretical and Applied Genetics*, *125*(6), 1275–1287. DOI: 10.1007/s00122-012-1912-9 PMID: 22782254

Lee, J., & Park, S. (2022). Exploring physiological limits of seed oil accumulation in Brassica species. *Plant Physiology*, *188*(2), 1023–1037. PMID: 34747475

Lee, M., Zhang, Q., & Wang, H. (2021). Optimization of biodiesel production from canola oil: A response surface methodology approach. *Fuel*, *285*, 119255.

Li, F., Chen, B., Xu, K., Wu, J., Song, W., Bancroft, I., & Wu, X. (2018). Genome-wide association study dissects the genetic architecture of seed weight and seed quality in rapeseed (Brassica napus L.). *DNA Research : an International Journal for Rapid Publication of Reports on Genes and Genomes*, *25*(5), 485–499. PMID: 29947776

Li, N., Xu, C., & Li, Y. (2019). Improvement of seed oil content in canola: From quantitative genetics to molecular breeding. *Journal of Integrative Plant Biology*, *61*(3), 336–351.

Liu, S., Fan, C., Li, J., Cai, G., Yang, Q., Wu, J., Yi, X., Zhang, C., & Zhou, Y. (2016). A genome-wide association study reveals novel elite allelic variations in seed oil content of Brassica napus. *Theoretical and Applied Genetics*, *129*(6), 1203–1215. DOI: 10.1007/s00122-016-2697-z PMID: 26912143

Liu, S., Fan, C., Li, J., Cai, G., Yang, Q., Wu, J., Yi, X., Zhang, C., & Zhou, Y. (2016). A genome-wide association study reveals novel elite allelic variations in seed oil content of Brassica napus. *Theoretical and Applied Genetics*, *129*(6), 1203–1215. DOI: 10.1007/s00122-016-2697-z PMID: 26912143

Lopez, A., & Fernandez, B. (2022). Glycerol derivatives in cosmetic formulations: A systematic review. *International Journal of Cosmetic Science*, *44*(1), 32–45.

Lu, K., Wei, L., Li, X., Wang, Y., Wu, J., Liu, M., Zhang, C., Chen, Z., Xiao, Z., Jian, H., Cheng, F., Zhang, K., Du, H., Cheng, X., Qu, C., Qian, W., Liu, L., Wang, R., Zou, Q., & Li, J. (2019). Whole-genome resequencing reveals Brassica napus origin and genetic loci involved in its improvement. *Nature Communications*, *10*(1), 1154. DOI: 10.1038/s41467-019-09134-9 PMID: 30858362

Luo, X., Ma, C., Yue, Y., Hu, K., Li, Y., Duan, Z., & Zhou, Y. (2017). Unravelling the complex trait of harvest index in rapeseed (Brassica napus L.) with association mapping. *BMC Genomics*, *18*(1), 238. PMID: 28320314

Mason, A. S., & Batley, J. (2015). Creating new interspecific hybrid and polyploid crops. *Trends in Biotechnology*, *33*(8), 436–441. DOI: 10.1016/j.tibtech.2015.06.004 PMID: 26164645

Peterson, R., & Johnson, M. (2021). Environmental factors affecting seed oil content in canola: Implications for breeding strategies. *Field Crops Research*, *263*, 108071.

Poland, J. A., Brown, P. J., Sorrells, M. E., & Jannink, J. L. (2012). Development of high-density genetic maps for barley and wheat using a novel two-enzyme genotyping-by-sequencing approach. *PLoS One*, *7*(2), e32253. DOI: 10.1371/journal.pone.0032253 PMID: 22389690

Rabbani, B., Tekin, M., & Mahdieh, N. (2014). The promise of whole-exome sequencing in medical genetics. *Journal of Human Genetics*, *59*(1), 5–15. DOI: 10.1038/jhg.2013.114 PMID: 24196381

Rahman, M., Sun, Z., McVetty, P. B., & Li, G. (2008). High throughput genome-specific and gene-specific molecular markers for erucic acid genes in Brassica napus (L.) for marker-assisted selection in plant breeding. *Theoretical and Applied Genetics*, *117*(6), 895–904. DOI: 10.1007/s00122-008-0829-9 PMID: 18633592

Samans, B., Chalhoub, B., & Snowdon, R. J. (2017). Surviving a genome collision: Genomic signatures of allopolyploidization in the recent crop species Brassica napus. *The Plant Genome*, *10*(3), plantgenome2017.02.0013. DOI: 10.3835/plantgenome2017.02.0013 PMID: 29293818

Scheben, A., Batley, J., & Edwards, D. (2017). Genotyping-by-sequencing approaches to characterize crop genomes: Choosing the right tool for the right application. *Plant Biotechnology Journal*, *15*(2), 149–161. DOI: 10.1111/pbi.12645 PMID: 27696619

Schiessl, S., Huettel, B., Kuehn, D., Reinhardt, R., & Snowdon, R. (2017). Post-polyploidisation morphotype diversification associates with gene copy number variation. *Scientific Reports*, *7*(1), 41845. DOI: 10.1038/srep41845 PMID: 28165502

Sharma, A., & Kumar, S. (2021). Economic impact of increased seed oil content in canola: A global perspective. *Agricultural Economics*, *52*(3), 405–417.

Smith, J., & Brown, K. (2020). Advances in catalysis for efficient biodiesel production from canola oil. *Catalysis Reviews*, *62*(4), 509–565.

Smith, J. R., Anderson, K. L., & Wilson, P. T. (2020). Canola: An overview of its agronomic importance and genetic improvement. *Annual Review of Plant Biology*, *71*, 539–562.

Snowdon, R. J., Abbadi, A., Kox, T., Schmutzer, T., & Leckband, G. (2015). Heterotic haplotype capture: Precision breeding for hybrid performance. *Trends in Plant Science*, *20*(7), 410–413. DOI: 10.1016/j.tplants.2015.04.013 PMID: 26027461

Snowdon, R. J., & Iniguez Luy, F. L. (2012). Potential to improve oilseed rape and canola breeding in the genomics era. *Plant Breeding*, *131*(3), 351–360. DOI: 10.1111/j.1439-0523.2012.01976.x

Sun, C., Wang, B., Wang, X., Hu, K., Li, K., Li, Z., & Li, Y. (2016). Genome-wide association study dissecting the genetic architecture underlying the branch angle trait in rapeseed (Brassica napus L.). *Scientific Reports*, *6*(1), 33673. DOI: 10.1038/srep33673 PMID: 27646167

Tan, C., Wei, X., & Zhang, Y. (2020). Trade-offs between seed oil content and other agronomic traits in Brassica napus: A meta-analysis. *Crop Science*, *60*(5), 2442–2458.

Thompson, D. (2019). The role of humectants in oral care products: Focus on glycerol. *Journal of Dental Hygiene*, *93*(3), 42–49.

Torkamaneh, D., Laroche, J., Bastien, M., Abed, A., & Belzile, F. (2018). Fast-GBS: A new pipeline for the efficient and highly accurate calling of SNPs from genotyping-by-sequencing data. *BMC Bioinformatics*, *19*(1), 5. DOI: 10.1186/s12859-016-1431-9 PMID: 28049422

Torkamaneh, D., Laroche, J., & Belzile, F. (2016). Genome-wide SNP calling from genotyping by sequencing (GBS) data: A comparison of seven pipelines and two sequencing technologies. *PLoS One*, *11*(8), e0161333. DOI: 10.1371/journal.pone.0161333 PMID: 27547936

Wan, H., Chen, L., Guo, J., Li, Q., Wen, J., Yi, B., Ma, C., Tu, J., Fu, T., & Shen, J. (2017). Genome-wide association study reveals the genetic architecture underlying salt tolerance-related traits in rapeseed (Brassica napus L.). *Frontiers in Plant Science*, *8*, 593. DOI: 10.3389/fpls.2017.00593 PMID: 28491067

Wang, H., Li, Y., & Zhang, Q. (2021). Advances in genome-wide association studies of complex traits in major crop species. *Nature Reviews. Genetics*, *22*(7), 485–503.

Wang, X., Chen, L., Wang, A., Wang, H., Tian, J., Zhao, X., Chao, H., Zhao, Y., Zhao, W., Xiang, J., Zhao, J., Li, M., & Hu, J. (2018). Quantitative trait loci analysis and genome-wide comparison for silique related traits in Brassica napus. *BMC Plant Biology*, *18*(1), 228. PMID: 30309330

Wang, X., Liu, G., & Yang, Q. (2022). Global trends in oilseed production: Focus on canola's expanding role. Oilseed & fat crops and lipids, 29, 15.

Werner, C. R., Voss-Fels, K. P., Miller, C. N., Qian, W., Hua, W., Guan, C. Y., Snowdon, R. J., & Qian, L. (2018). Effective genomic selection in a narrow-genepool crop with low-density markers: Asian rapeseed as an example. *The Plant Genome, 11*(2), 170084. DOI: 10.3835/plantgenome2017.09.0084 PMID: 30025015

Wilson, R., & Brown, J. (2023). Genotype by environment interactions in canola: Implications for breeding high oil content varieties. *Theoretical and Applied Genetics, 136*(2), 421–435.

Wilson, T., & Taylor, S. (2020). Glycerol from biodiesel production: The new corn syrup? *Renewable & Sustainable Energy Reviews, 124*, 109799.

Wu, D., Liang, Z., Yan, T., Xu, Y., Xuan, L., Tang, J., & Huang, J. (2019). Whole-genome resequencing of a worldwide collection of rapeseed accessions reveals the genetic basis of ecotype divergence. *Molecular Plant, 12*(1), 30–43. DOI: 10.1016/j.molp.2018.11.007 PMID: 30472326

Wu, J., Zhao, Q., Yang, Q., Liu, H., Li, Q., Yi, X., & Zhou, Y. (2021). Comparative transcriptome profiling uncovers a Lilium-derived allele in Brassica napus providing resistance to blackleg disease. *Plant Biotechnology Journal, 19*(4), 693–708.

Würschum, T., Maurer, H. P., Dreyer, F., & Reif, J. C. (2014). Effect of inter- and intragenic epistasis on the heritability of oil content in rapeseed (Brassica napus L.). *Theoretical and Applied Genetics, 127*(3), 615–622. PMID: 23052025

Xie, H., Konate, M., Sai, N., Tesfamicael, K. G., Cavagnaro, T., Gilliham, M., Breen, J., Metcalfe, A., Stephen, J. R., DeBei, R., Collins, C., & Lopez, C. M. R. (2017). Global DNA methylation patterns can play a role in defining terroir in grapevine (Vitis vinifera cv. Shiraz). *Frontiers in Plant Science, 8*, 1860. DOI: 10.3389/fpls.2017.01860 PMID: 29163587

Zhang, F., Liu, X., & Huang, G. (2022). Recent progress in reactor design for continuous biodiesel production from canola oil. *Chemical Engineering Journal, 430*, 132707. DOI: 10.1016/j.cej.2021.132707

Zhang, F., Liu, X., & Huang, G. (2023). Applications of GWAS-derived markers in genomic selection for yield improvement in Brassica napus. *Theoretical and Applied Genetics, 136*(5), 1089–1104. PMID: 37027012

Zhang, Y., & Liu, J. (2020). Canola production and trade: A comprehensive review of global patterns and emerging trends. *Agronomy (Basel), 10*(12), 1956.

Zhao, J., Paulo, M. J., Jamar, D., Lou, P., van Eeuwijk, F., Bonnema, G., Vreugdenhil, D., & Koornneef, M. (2007). Association mapping of leaf traits, flowering time, and phytate content in Brassica rapa. *Genome*, *50*(10), 963–973. DOI: 10.1139/G07-078 PMID: 18059559

Zhao, Y., Wang, H., Chen, W., & Li, Y. (2021). Genetic structure, linkage disequilibrium and association mapping of verticillium wilt resistance in elite cotton (Gossypium hirsutum L.) germplasm population. *BMC Genomics*, *22*(1), 60. PMID: 33468056

Zheng, M., Peng, C., Liu, H., Tang, M., Yang, H., Li, X., & Cheng, X. (2020). Genome-wide association study reveals candidate genes for control of plant height, branch initiation height and branch number in rapeseed (Brassica napus L.). *Frontiers in Plant Science*, *11*, 1875. PMID: 28769955

Zou, J., Mao, L., Qiu, J., Wang, M., Jia, L., Wu, D., He, Z., Chen, M., Shen, Y., Shen, E., Huang, Y., Li, R., Hu, D., Shi, L., Wang, K., Zhu, B., Zeng, P., Ma, B., Wang, H., & Fu, D. (2019). Genome-wide selection footprints and deleterious variations in young Asian allotetraploid rapeseed. *Plant Biotechnology Journal*, *17*(10), 1998–2010. DOI: 10.1111/pbi.13115 PMID: 30947395

Chapter 6
Development and Utilization of High-Density Genome-Wide SNP Array for Canola Plants

Muhammad Kaleem
https://orcid.org/0009-0003-6548-3920
Quaid-i-Azam University, Islamabad, Pakistan

Rooma Waqar
https://orcid.org/0009-0009-1654-2860
Quaid-i-Azam University, Islamabad, Pakistan

Lubna Anjum Minhas
Quaid-i-Azam University, Islamabad, Pakistan

Saira Qadir
Rawalpindi Women University, Pakistan

Amber Jabeen
https://orcid.org/0009-0001-4501-3345
Quaid-i-Azam University, Islamabad, Pakistan

Abdul Samad Mumtaz
https://orcid.org/0000-0002-7034-9203
Quaid-i-Azam University, Islamabad, Pakistan

ABSTRACT

High-density Single Nucleotide Polymorphism (SNP) arrays have revolutionized genetic research and breeding in canola (Brassica napus L.), offering deep insights into its genetic structure. These arrays enable Genome-Wide Association Studies (GWAS), Quantitative Trait Loci (QTL) mapping, and genomic selection, identifying critical loci for traits like oil quality, yield, stress resilience and disease resistance. This precision enhances Marker-Assisted Selection (MAS), accelerating the development of superior canola varieties. This chapter details SNP array development,

DOI: 10.4018/979-8-3693-8603-3.ch006

including SNP discovery and optimization and addresses the limitations of SNP arrays compared to next-generation technologies like whole genome sequencing (WGS) and Clustered Regularly Interspaced Short Palindromic Repeats (CRISPR). Despite these, ongoing improvements and integration with multi-omics strengthen the role of SNP arrays in advancing canola productivity and sustainability.

INTRODUCTION

Overview of Canola

Rapeseed or canola is widely recognized as the principal oilseed crop due to the superior quality and quantity of its seed oil (Zhou et al., 2019; Mamnabi et al., 2020). This crop has been extensively cultivated globally, with its total world production reaching 68 million tons in the 2019/2020 agricultural year (Kupiec et al., 2020). Canola oil is characterized by a high concentration of oleic acid, the predominant unsaturated fatty acid (UFA), and is extensively utilized in both the food and pharmaceutical sectors. The healthful edible oil comprises principal UFAs such as oleic acid, linoleic acid, and linolenic acid (Joughi et al., 2018). The identification and enhancement of high-quality oil in oilseed crops like canola represent a significant objective for human health (Feizabadi et al., 2020). Consequently, it is imperative to enhance both the quality and quantity of oil through effective agronomic practices (Mamnabi et al., 2020).

Rapeseed is classified as an annual species. The winter, semi-winter, and spring varieties exhibit varying degrees of cold and drought tolerance; hence, their cultivation conditions differ accordingly. The winter-type rapeseed flourishes in environments characterized by relatively high humidity and cooler temperatures. Optimal growth of rapeseed necessitates well-drained soils with a pH ranging from 5.5 to 8.5. Depending on the genotype and environmental factors, the crop requires between 110 to 150 days achieving full growth and maturation. The length of the mature stem can vary from 120 to 150 cm, and the mature seeds are spherical, measuring 1.8–2.7 mm in diameter (Gulden et al., 2008), displaying a color range from red-brown to dark brown or black. Every component of the rapeseed plant-flowers, seeds, leaves, stems, and roots finds application in food, medicinal remedies, cosmetics, or industrial uses. Among these, seeds are of paramount importance, serving as vital sources of oil and protein. Rapeseed oil and the protein content within its seeds exhibit variability across different lines of cultivars, and an array of additional components such as glucosinolates, phenolic compounds, phytic acid, cellulose, and sugars are also present in the seeds. Renowned for its capacity to yield high-quality vegetable oil, rapeseed faces competition from other agricultural crops. It ranks as the second-most produced oil crop globally, following soybean, with production figures of 68.02 MT and 337.48 MT in the year 2019/2020, respectively, thereby exceeding that of sunflower (53.48 MT), peanut (45.52 MT), and cottonseed (44.3 MT) (STATISTA, 2008).

The production of rapeseed, similar to other agricultural crops, frequently encounters significant challenges attributable to a countless of factors, including the decline in labor availability and the number of farmers as a result of escalating labor costs and agricultural inputs, which consequently leads to diminished outputs, inadequate agricultural mechanization, yield instability stemming from climatic variability, and suboptimal cultivars (characterized by shattering, biotic, and abiotic stressors). For example, two of the most devastating diseases that compromise rapeseed crops globally are stem rot disease, instigated by *Sclerotinia sclerotiorum*, and clubroot disease, induced by *Plasmodiophora brassicae*. In China, stem rot and clubroot diseases have resulted in yield reductions ranging from 10% to 80% (Wang et al., 2014) and 20% to 30% (Wang et al., 2011), respectively.

Role of Genomics in Crop Improvement

Genomics has fundamentally transformed plant breeding and crop enhancement by facilitating comprehensive insights into the architecture, functionality, and evolutionary history of plant genomes. Through advancements in genome sequencing methodologies, breeders are now equipped to investigate genetic variations and implement genomic tools to increase crop characteristics such as yield, quality, disease resistance, and resilience to environmental stressors (Yasin et al., 2021). Genomics encompasses the examination of genome structure, function, evolution, mapping, and modification, serving a pivotal role in the advancement of crop breeding and genetic enhancement (Muthamilarasan et al., 2019). Structural genomics concentrates on sequence polymorphisms and chromosomal organization, while functional genomics elucidates the roles of genes concerning trait regulation (Tettelin et al., 2005). Emerging disciplines such as mutagenomics and pangenomics have opened new pathways in crop sciences by investigating genetic mutations and the comprehensive genomic diversity of species, respectively (Goh, 2018).

Structural genomics employs molecular markers, including restriction fragment length polymorphisms (RFLPs), random amplified polymorphic DNA (RAPD), amplified fragment length polymorphisms (AFLP), SNPs, and diversity arrays technology (DArT), to pinpoint traits pertinent to crop breeding (Agarwal et al., 2008; Williams et al., 1990). Methodologies such as quantitative trait loci mapping and genome-wide association studies are instrumental in identifying genomic regions associated with traits such as drought tolerance and yield enhancement in crops including maize, sorghum, and rice (Challa and Neelapu, 2018; Guo et al., 2018; Spindel et al., 2018).

Functional genomics employs advanced technologies such as CRISPR/Cas9 and TALEN to engineer crops with enhanced disease resistance and improved yield (Wang et al., 2014; Nekrasov et al., 2017). Mutagenomics is dedicated to the examination of mutational occurrences utilizing high-throughput methodologies such as TILLING, thus providing alternatives to transgenic strategies (McCallum et al., 2000; Kurowska et al., 2011). This field has been effectively implemented in crops such as soybean, rice, and wheat to promote advantageous mutations (Knoll et al., 2011; Dong et al., 2009).

Epigenomics investigates heritable alterations in gene expression induced by environmental stimuli, which encompasses mechanisms such as DNA methylation and histone modification (Strahl and Allis, 2000; Callinan and Feinberg, 2006). Methodologies including bisulfite sequencing and chromatin immunoprecipitation sequencing (ChIP-Seq) are instrumental in detecting epigenetic modifications in crops like maize, tomato, and rice when subjected to stress conditions (Gent et al., 2013; Zong et al., 2013).

Pangenomics analyzes the comprehensive genomic framework of a species, delineating core genes universally present across individuals from dispensable genes that contribute to genetic diversity (Tettelin et al., 2005). Recent investigations involving crops such as rice, soybean, wheat, and *Brassica* species have highlighted the significance of dispensable genes in enhancing traits like disease resistance and stress resilience (Hirsch et al., 2014; Hurgobin et al., 2018). This methodological approach also facilitates the extraction of valuable genetic resources from wild relatives of crops for integration into breeding initiatives (Ram et al., 2007).

Objectives of SNP Array Development

The SNP array represents a powerful genotyping instrument specifically designed for canola, aimed at advancing extensive genetic investigations, including GWAS, QTL mapping, and population genetics. The creation of the 60K *Brassica* Infinium SNP array constitutes a pivotal advancement, developed utilizing both previously characterized and newly identified SNPs sourced from published genomic data. This array has demonstrated efficacy in pinpointing significant QTLs associated with agronomic characteristics, including seed quality and flowering time (You et al., 2018). The push for the establishment of high-density SNP arrays for polyploid crops such as canola is fundamentally anchored in the necessity to augment molecular breeding and genetic analysis. Polyploid organisms, exemplified by canola, exhibit more intricate genomic architectures compared to diploid species, thereby complicating genetic investigations. SNP arrays offer a cost-efficient approach for genotyping established loci across extensive sample sets. This capability enables researchers to conduct comprehensive genetic analyses, including marker-assisted and genomic selection, which are vital for the enhancement of crop varieties. In the context of canola, SNP arrays facilitate the identification of genetic markers associated with particular traits, which can subsequently be employed to enhance essential agronomic characteristics like disease resistance, yield potential, and stress tolerance. The incorporation of SNP arrays expedites breeding programs by enabling breeders to identify optimal candidates at an early stage in the breeding continuum. Furthermore, the SNP array yields high-resolution genetic data that is crucial for deciphering the genetic architecture of polyploid species. Specifically, for Brassica napus, the 60K SNP array has contributed to the mapping of traits relevant to seed quality and environmental adaptability, thereby underscoring its significance within the breeding sector (You et al., 2018).

Canola (AACC type-genome; 2n = 4x = 38; ~845 Mb genome size) constitutes a vital agricultural crop, which can be utilized for the extraction of oil from its seeds and for the production of edible vegetable oil (Clarke et al., 2016: Chalhoub et al., 2014). A 60K canola Infinium array was developed by Clake et al. (2016). SNPs sources incorporated into this array predominantly comprised formerly characterized SNPs (Bus et al., 2012; Dalton-Morgan et al., 2014) alongside newly identified SNPs from this investigation derived from published sequence data in Clarke et al. (2013), Harper et al. (2012), and unpublished genomic and transcriptomic sequence data in Clarke et al. To corroborate the efficacy of this 60K array, 327 and 432 distinct genotypes were independently genotyped across two laboratories, yielding approximately 60% of genome-specific markers in a diverse array of canola genotypes, along with 29.7K and 26.5 scorable SNP markers in *B. rapa* and *B. oleracea*, respectively, culminating in the construction of a genetic map of canola featuring 46% (21,766 SNPs) of mapped markers within one of the doubled haploid populations (Clarke et al., 2016). This 60K Brassica SNP array has found extensive application in oilseed rape research, including the three QTLs identification related with resistance to Sclerotinia stem rot (Wu et al., 2016), 79 QTLs pertinent to flowering time, root morphology traits and seed quality (Wang et al., 2017), as well as 117 genomic regions implicated in selective sweeps (Zou et al., 2018), among other applications. Furthermore, a user manual was published for the canola 60K SNP array (Mason et al., 2017). SNP arrays have also been developed in various other like crops sugarcane, peanut, cotton etc. as described in table 1.

Table 1. SNP Arrays Established in Various Crops

Crops	Source of SNPs	Size of Array	Genotyping Sample Size	Implementation	Reference
Canola	54K SNPs recognized and assessed by earlier reports; 24M SNPs from published sequence data	52K SNPs (52,157)	437 diverse genotypes (432 genotypes)	Generation of genetic gaps	(Clarke et al., 2016)
Wheat	25K SNPs from RNA-seq of 26 hexaploid accessions	9K SNPs (9,000)	2,994 accessions, including modern cultivars and landraces	Population structure and genetic diversity	(Cavanagh et al., 2013)
Sugarcane	2.6M SNPs from sequencing of 16 lines	345K SNPs (345,704)	367 clones, comprising cultivars, parental clones, and un-selected families	Association study of sugar and cane yield content	(Aitken et al., 2016)
Strawberry	160K di-allelic SNPs from sequencing of DNA	90K SNPs (95,062)	384 samples, comprising 357 octoploid	QTL identification and High-density linkage maps,	(Bassil et al., 2015)
Oat	8K SNPs from 4 DNA sequencing datasets	6K SNPs (5,743)	1,110 hexaploid samples	SNP finding and innotation	(Tinker et al., 2014)
Alfalfa	900K SNPs from RNA-seq of 27 genotypes	9K SNPs (9,277)	280 genotypes, including related species	Linkage disequilibrium and Population structure analysis	(Li et al., 2014b)
Potato	20K SNPs from previous identification	18K SNPs (17,987)	569 accessions	Reconstruction of breeding history	(Vos et al., 2015)
Cotton	20K SNPs from 4 inter-specific datasets and 50K SNPs from 9 intra-specific datasets	63K SNPs (63,058)	1,156 samples	Construction of high-density genetic maps	(Hulse-Kemp et al., 2015)
Peanut	163K SNPs from DNA resequencing of 38 accessions and RNA-seq of 3 accessions	58K SNPs (58,233)	297 accessions	Genetic architecture and genetic diversity	(Pandey et al., 2017)

CONCEPTS AND PRINCIPLES OF SNP ARRAY TECHNOLOGY

SNPs represent the most prevalent form of genetic variation observed in humans. An SNP constitutes a variation at a singular nucleotide position within the genome, wherein distinct individuals may exhibit different nucleotides (e.g., an 'A' versus a 'G') at a particular locus. These variations occur, on average, once every 300 to

1,000 base pairs within the human genome and are dispersed throughout both coding and noncoding regions (Shaw, 2013). SNPs can be categorized based on their functional impact: synonymous SNPs do not modify the amino acid sequence of a protein, whereas non-synonymous SNPs induce alterations in the structure or function of the protein. They are integral to genetic diversity, influencing phenotypic traits, susceptibility to diseases, and individual pharmacological responses (Komar, 2009).

The SNP array serves as a cost-efficient, high-throughput, and automated assay for genotyping. It has been extensively utilized in the genetic investigation of agricultural species, involving GWAS (Chen et al., 2014; McCouch et al., 2016), the construction of linkage maps (Ganal et al., 2011; Felcher et al., 2012), genomic selection processes (Yu et al., 2014; Clarke et al., 2016), analyses of population structure (Wang et al., 2016), and endeavors related to gene mapping (Dalton-Morgan et al., 2014). The current capabilities of SNP arrays extend to 700K for diploid organisms such as rice (McCouch et al., 2016), 487K for triploid species like apple (Bianco et al., 2016), 58K for tetraploid organisms such as peanut (Clevenger et al., 2017), 820K for hexaploid species including wheat (Winfield et al., 2016), 90K for octoploid varieties such as strawberry (Bassil et al., 2015), and 345K for dodecaploid species like sugarcane (Aitken et al., 2016).

Figure 1. Process of SNP Arrays development

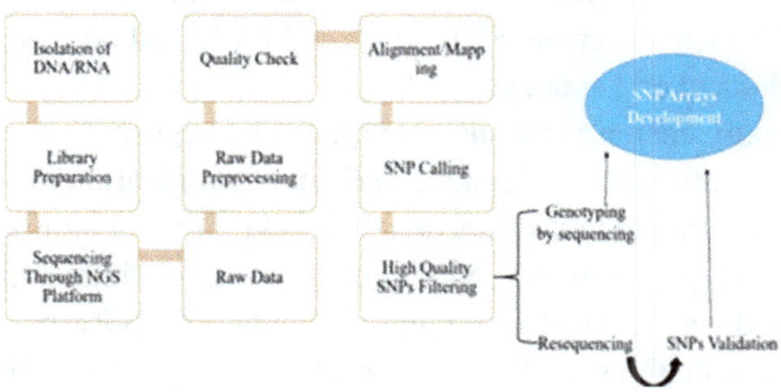

Development of SNP Arrays is a systematic process as described in **Figure 1**. The technology underpinning SNP arrays, similar to a multitude of other biotechnological approaches, encompasses both merits and limitations. Within the realm of high-throughput genotyping, SNP arrays confer a variety of advantages relative to alternative genotyping methodologies. Primarily, the data analysis associated with SNP arrays is comparatively straightforward when juxtaposed with the data generated by next-generation sequencing (NGS) techniques, specifically when considering the labor-intensive requirements for the prepration of NGS library (Garvin et al., 2010) and the significant resource investment in downstream bioinformatics required for precise identification of SNP (Liu et al., 2012; Torkamaneh et al., 2017). To elaborate, the genotypes linked to SNP markers may be either generated or provided by Illumina or Affymetrix, or researchers may elect to derive genotypes in accordance with the specified genotype calling protocols outlined by Affymetrix or Illumina within their user manuals. In contrast, the procedure of genotype calling utilizing data derived from NGS in polyploid species is significantly more intricate and time-consuming. As previously articulated, the process of SNP genotype calling encompasses a series of steps including read trimming, read alignment, SNP calling, and SNP filtering, thereby necessitating a foundational understanding of bioinformatics (Clevenger et al., 2015). Additionally, SNPs corresponding to genomic regions of particular interest can be intentionally integrated into the design of the array. Furthermore, the quantity of SNPs of interest that can be incorporated into the array is flexible for both the Affymetrix and Illumina platforms. Thirdly, SNP arrays are regarded as cheap to moderate in cost per sample, even with the significant reduction in expenses linked to NGS techniques. The cost associated with an Affymetrix Axiom SNP array fluctuates between \$28 and \$90 (USD) per sample (personal communication), whereas NGS methodologies present the most economical pricing for Genotyping-by-Sequencing (GBS) at \$35 per sample, followed by RNA sequencing at \$260 per sample, and WGS exceeding \$500 per sample (Peng et al., 2017). Given that polyploid crops typically exhibit extensive genome sizes, NGS-based methods require a substantial quantity of sequencing data to attain sufficient coverage. SNP arrays and NGS methodologies can serve to mutually complement one another.

Although Next-Generation Sequencing offers distinct benefits in sequence generation and the identification of genetic variants, its utilization in the genotyping of polyploid organisms remains impeded. The SNP array continues to serve as a feasible genotyping platform, particularly considering the burgeoning number of variants identified via NGS methodologies. Nonetheless, SNP arrays are accompanied by intrinsic limitations, including the necessity for prior genomic knowledge, the constraint to genotyping known SNP loci, and the requirement for manual dosage scoring in specific instances (Wang et al., 2014). Furthermore, the design and subsequent optimization of such arrays may entail significant labor investment. Ascertainment bias presents a common challenge in genotyping arrays, arising from the non-random sampling of polymorphisms within the target population (Heslot et al., 2013) or originating from the limited sample sizes employed as SNP discovery panels (Albrechtsen et al., 2010). For example, a restricted sample size may predominantly capture prevalent alleles, thereby overlooking rare alleles (Gravel et al., 2011). This phenomenon is likely to distort subsequent genetic interpretations. A variety of strategies have been implemented to alleviate ascertainment bias, including the application of WGS with high covergae, enhancements to the markers included on the SNP array, and the incorporation of markers from diverse arrays.

The assessment of the SNP validation rate between SNP arrays and NGS-based methodologies presents inherent complexities, as factors such as population type (bi-parental or natural populations) and the size of the population employed for validation considerably impact the validation rate. For instance, in a genotyping investigation involving 108 hexaploid wheat varieties, a 900K SNP array revealed the presence of 99.8K polymorphic SNP probes. Conversely, when the genotyping analysis was extended to encompass 475 accessions, which included relatives of the aforementioned varieties and their progenitors, the total of polymorphic probes surged to 453.1K (Winfield et al., 2016). Therefore, only the validation rate associated with SNP arrays has been thoroughly summarized and discussed in the ensuing discourse. Currently, SNP arrays have been utilized across a wide array of polyploid crops, with the majority of these arrays achieving commendable polymorphic rates, as illustrated in Table 1. Within the group of 16 SNP arrays scrutinized across nine polyploid crop species, eight (50%) of the arrays exhibited a polymorphic rate equal to or exceeding 80%, whereas four (25%) demonstrated a polymorphic rate ranging between 60% and 70%. Notably, merely two SNP arrays reported polymorphic rates below 60%, specifically 42.7% in oat (Oliver et al., 2013) and 56.8% in sugarcane (Aitken et al., 2016).

GENOME-WIDE SNP DISCOVERY IN CANOLA

The design of SNP arrays in canola generally necessitates the employment of various sophisticated genomic methodologies for SNP identification. These methodologies encompass whole-genome re-sequencing and transcriptome analysis. Whole-genome resequencing serves as an effective strategy for detecting a substantial array of SNPs by comparing the DNA sequences of dissimilar canola accessions against a reference genome (Huang et al., 2013). Another prevalent technique is sequence capture, which involves the targeting of specific genomic regions linked to significant traits and the subsequent capture of variations within those regions, as evidenced in canola, where over 589,367 SNPs were detected (Clarke et al., 2013). Transcriptome sequencing, commonly referred to as RNA-Seq, can additionally unveil SNPs through the comparative analysis of gene expression profiles among diverse canola lines, thereby enriching the reservoir of genetic variation accessible for breeding endeavors (Huang et al., 2013).

DESIGN AND VALIDATION OF THE CANOLA SNP ARRAY

Criteria For Selecting SNPs Array

The design of a high-density SNP array for canola necessitates the selection of SNPs predicated on multiple criteria. Primarily, the SNPs must afford extensive genome coverage, thereby ensuring adequate representation across all linkage groups within canola (Snowdon & Iñiguez Luy, 2012). An effective SNP array design must also consider allele frequency, favoring SNPs with a minor allele frequency (MAF) exceeding 0.05 to capture genetic diversity and guarantee that the SNPs are informative across a variety of canola germplasm (Raman et al., 2016). SNPs that induce non-synonymous mutations, particularly those situated in regions influencing agronomic traits such as flowering time or disease resistance, are given high priority (Huang et al., 2013).

The comparative examination of the SNP validation rates across SNP array methodologies and NGS-based techniques poses significant challenges, as factors such as the type of population (bi-parental or natural populations) and the size of the population employed for validation unusually affect the validation rates. For example, in a research study that included the genotyping of 108 hexaploid wheat varieties, a 900K SNP array revealed the existence of 99.8K polymorphic SNP probes. In contrast, when genotyping a total of 475 accessions, which comprised both the relatives of these varieties and their progenitors, the number of polymorphic probes dramatically increased to 453.1K (Winfield et al., 2016). As a result, the validation rate associated exclusively with the SNP array has been meticulously summarized and scrutinized in the subsequent discourse. At present, the utilization of SNP arrays has expanded across a multitude of polyploid crops, with the majority of these arrays achieving noteworthy polymorphic rates. Among the 16 SNP arrays assessed across nine polyploid crop species, it was observed that eight (50%) of these arrays accomplished a polymorphic rate of 80% or more, while four (25%) displayed polymorphic rates ranging between 60% and 70%. It is particularly significant that only two SNP arrays indicated polymorphic rates below 60%, specifically 42.7% in oat (Oliver et al., 2013) and 56.8% in sugarcane (Aitken et al., 2016).

Experimental Protocols For Validation

The validation of SNPs for incorporation into a genotyping array necessitates the implementation of highly accurate genotyping assays, such as the Illumina GoldenGate or Infinium assays, which are capable of evaluating a substantial number of SNPs across diverse genotypes. SNPs that have been identified through next-generation sequencing undergo an initial filtering process predicated on stringent quality metrics, followed by validation utilizing these advanced genotyping platforms. The successful validation rates for SNPs identified within *B. napus* fluctuate between 61.5% and 84% (Clarke et al., 2013). Validation encompasses the imperative of confirming that SNPs demonstrate polymorphism across a diverse array of canola lines, which is crucial for the overall efficacy of the array within different breeding populations (Malmberg et al., 2019).

Optimization For High-Density Coverage

Strategies To Maximize Genome-Wide Coverage

The achievement of high-density coverage in SNP array design entails the adoption of methodologies that include the integration of SNPs derived from both coding and non-coding regions of the genome, thereby guaranteeing a thorough representation of genomic variation. Recent investigations concerning *B. napus* have demonstrated that by employing high-throughput resequencing coupled with sophisticated bioinformatics tools, it is feasible to identify in excess of 892,536 SNPs that encompass the entirety of the genome (Huang et al., 2013). The inclusion of SNPs identified within gene-dense regions, particularly those influencing agronomically significant traits, further expands the utility of the array within breeding initiatives (Raman et al., 2016).

Addressing Linkage Disequilibrium And Population Structure In Array Design

The optimization of SNP arrays additionally necessitates the consideration of the effects of linkage disequilibrium (LD) and population structure. In canola, LD can persist over extensive distances, rendering it imperative to select SNPs that encompass the genetic diversity prevalent across various subpopulations (Snowdon & Iñiguez Luy, 2012). Analytical tools such as multidimensional scaling and K-means clustering can be employed to ascertain population structure, which facilitates a more careful selection of SNPs, thereby ensuring that the array provides informative data across a range of breeding lines (Jan et al., 2016). The development of a high-density SNP array for canola necessitates the processes of SNP discovery, validation, and optimization for comprehensive genome coverage. Addressing challenges such as linkage disequilibrium is crucial for the creation of informative arrays intended for breeding and trait enhancement.

APPLICATIONS OF SNP ARRAYS IN CANOLA RESEARCH AND BREEDING: GENETIC MAPPING AND QTL ANALYSIS

SNP Arrays And QTL Mapping In Canola

SNP arrays are extensively utilized in plant breeding to detect genetic variations, thereby facilitating the identification of QTL associated with intricate traits. Within canola, SNP arrays have transformed the landscape of QTL mapping, resulting in the identification of loci correlated with critical agronomic traits such as oil content, disease resistance, and yield. The elevated density of SNP markers enables precise mapping of complex traits and the design of molecular markers suitable for implementation in marker-assisted selection. Recent investigations employing SNP arrays have achieved notable progress in QTL mapping for several pivotal traits in canola. For example, one study utilized a doubled haploid (DH) population alongside a 60K SNP array to detect QTLs linked to sclerotinia stem rot resistance and identified 133 significant SNPs associated with disease-related traits, including stem lesion length, lesion width, and plant mortality. Among these SNPs, 19 were detected across multiple environmental contexts, thereby reinforcing the strength of the results. Furthermore, the study emphasizes that genomic prediction models that incorporate these QTLs have the potential to substantially enhance the efficiency of breeding programs aimed at disease resistance (Roy et al., 2021).

Furthermore, a research investigation concerning blackleg resistance, an issue of significant importance in canola cultivation, employed SNP markers to discern both race-specific resistance genes and quantitative resistance loci. The investigation identified several QTLs located on chromosomes A03, A07, and C04. These outcomes furnish essential understanding of the genetic frameworks governing blackleg resistance, thereby aiding in the advancement of more robust canola cultivars (Raman et al., 2020).

Role of SNP Arrays In Accelerating Marker-Assisted Selection And Canola Breeding

Marker-Assisted Selection And SNP Arrays In Canola Breeding

Marker-Assisted Selection (MAS) has emerged as an instrumental methodology within the domain of plant breeding, permitting breeders to select for particular traits at the molecular dimension. SNPs are extensively utilized as markers due to their substantial prevalence within plant genomes, their capacity to encompass the entire genome, and their appropriateness for high-throughput genotyping methodologies. Through the association of phenotypic traits with molecular markers, breeders can enhance the efficacy and accuracy of the selection process (Collard & Mackill, 2008). The significance of SNP arrays in MAS is particularly pronounced in crops such as canola, wherein intricate traits such as oil quality, resistance to disease, and yield can be mapped and selected with enhanced precision. Within contemporary breeding frameworks, SNP arrays facilitate the swift identification of QTLs correlated with sought-after traits, thus accelerating the creation of improved cultivars. SNP arrays have been effectively employed across diverse breeding programs, ranging from cereals to oilseeds, empowering breeders to monitor advantageous alleles and optimize the selection of traits such as disease resistance, oil content, and yield (You et al., 2018).

In the realm of canola, SNP arrays have been integral to the enhancement of critical traits including oil yield, seed quality, and disease resistance. For example, a study investigating genomic selection in canola emphasized the application of SNP arrays to accurately predict seed quality traits. The incorporation of SNP data enabled marker-assisted backcrossing aimed at enhancing oil content and fatty acid composition, thereby illustrating the practicality of SNP arrays in MAS breeding (Zou et al., 2016).

Efficiency of SNP Arrays In Detecting Trait Variability And Disease Resistance

A recent investigation concentrating on the genomic prediction of growth traits in crops such as canola employed SNP arrays to evaluate genetic diversity and refine trait selection. The utilization of SNP markers facilitated the identification of QTLs associated with root architecture and oil content, which were subsequently applied in marker-assisted selection. This research highlights the capability of SNP arrays to assist breeders in effectively dissecting complex traits and applying MAS to enhance agronomic performance (Arifuzzaman et al., 2019). The integration of SNP arrays into MAS has catalyzed advancements in breeding for disease resistance. A study focusing on clubroot resistance in canola utilized high-density SNP arrays to identify QTLs associated with resistance genes, thereby expediting the development of resistant varieties. This research exemplifies how SNP arrays enable the identification of both prominent resistance genes and minor QTLs, which collectively contribute to substantial resistance against pathogens (Zhou et al., 2021). The capacity of SNP arrays to unveil intricate genetic information assists breeders in rapidly incorporating disease resistance into high-yielding varieties.

SNP Arrays In Multi-Trait Selection And Improving Selection Accuracy In Breeding

The capability of SNP arrays to assess multiple phenotypic traits concurrently represents an additional advantage in MAS. In the case of maize, a genotyping-by-sequencing (GBS) investigation juxtaposed the efficacy of SNP arrays with GBS-derived SNPs for applications in genomic selection. The findings indicated that SNP arrays yielded enhanced precision in forecasting various traits, such as plant height, yield, and flowering period. These outcomes possess extensive ramifications for agricultural breeding, wherein the simultaneous selection for multiple traits can close in the creation of well-adapted cultivars that exhibit both high productivity and resilience against environmental adversities (Elbasyoni et al., 2018).

The employment of SNP arrays enables breeders to augment the precision of MAS within breeding initiatives. For instance, the maize 55K SNP array offered exceptional genomic coverage, facilitating precise trait selection for attributes such as drought resistance and oil biosynthesis. This array exhibited a considerable conversion rate of SNP markers into functional markers for trait selection, thereby demonstrating its efficacy in MAS for traits that are challenging to quantify phenotypically (Xu et al., 2017). Furthermore, the array enabled the detection of rare genetic variations, thereby enhancing the selection process for multifaceted traits.

Case Study

In the domain of wheat breeding, SNP arrays have fundamentally transformed the selection process for traits such as disease resistance, yield, and quality. A study that utilized the Wheat 660K SNP array underscored its potential in expediting MAS for polyploid crops, including wheat. The array facilitated the identification of a substantial number of genome-specific SNPs, equipping breeders with a potent instrument for trait selection and augmenting genetic gains in breeding programs (Sun et al., 2020). SNP arrays provide breeders with a financially viable approach for genotyping extensive populations. A study conducted on maize demonstrated that SNP arrays could significantly diminish genotyping expenses while sustaining high levels of accuracy, rendering them a compelling option for breeders operating under financial constraints. This versatility permits SNP arrays to be customized for diverse breeding objectives, whether they pertain to small-scale or large-scale initiatives, thus ensuring that the technology can be adapted to a variety of contexts (Guo et al., 2019).

The emergence of genomic selection has prompted inquiries into the ongoing significance of MAS. A comparative analysis between MAS and GS revealed that, although GS generally surpasses MAS regarding genetic advancement, MAS continues to offer substantial advantages, particularly in breeding programs where marker-trait associations are well-established. For traits linked to known QTL, MAS may prove to be more efficient and economically feasible in comparison to GS (Degen & Müller, 2023). As the costs associated with sequencing persist in their decline, the integration of both methodologies is anticipated to become progressively prevalent in contemporary breeding programs. SNP arrays play a crucial role in expediting marker-assisted selection by providing high throughput, accuracy, and cost-efficiency in the breeding of complex traits. As advancements in genotyping continue, their significance in modern plant breeding is poised to expand further.

GWAS And SNP Arrays For Disease Resistance In Canola

GWAS have emerged as an instrumental methodology in contemporary plant breeding, particularly for the elucidation of genetic loci correlated with intricate traits. **Figure 2** shows the procedure of GWAS. By utilizing high-density markers, such as single SNPs, GWAS facilitates the identification of associations between phenotypic characteristics and genetic variants, thereby yielding valuable insights into the genetic underpinnings of essential agronomic traits. In the case of canola, GWAS has been employed to explore a diverse array of traits including disease resistance, yield, oil quality, and cold tolerance, significantly augmenting the efficacy of breeding programs.

Figure 2. Procedure of GWAS

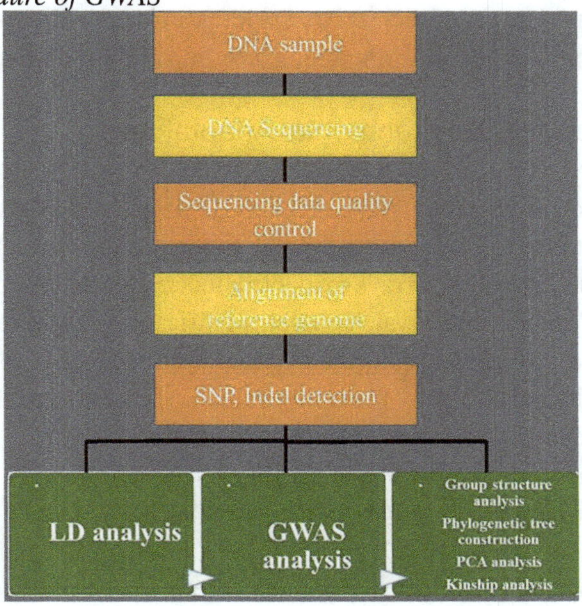

A prominent application of GWAS in canola relates to the identification of loci associated with resistance to pathogens such as *Sclerotinia sclerotiorum* and blackleg (*Leptosphaeria maculans*). A recent GWAS investigation identified 133 significant SNPs linked to resistance against *Sclerotinia* stem rot, correlating with phenotypic traits such as lesion length and plant mortality. These SNPs were mapped across 123 distinct loci, several of which demonstrated stability across varying environmental conditions. The analysis also uncovered 69 candidate genes associated with disease resistance, underscoring the valuable role of SNP arrays in enhancing disease resilience in canola (Roy et al., 2021). In a separate meta-analysis focused on blackleg disease in canola, imputed whole-genome sequences and SNP arrays were utilized to locate 79 genomic regions linked to resistance. Among the 674 significant SNPs identified, numerous were associated with previously characterized resistance genes, while others represented novel loci, thereby offering new targets for the breeding of resistant cultivars (Fikere et al., 2020).

GWAS For Yield And Agronomic Traits

GWAS has likewise been applied to the investigation of yield-related traits in canola, encompassing seed weight, plant height, and flowering time. A recent study that integrated multi-omics data with GWAS across 477 canola lines identified over 61,000 QTLs associated with early vegetative growth. These loci were connected to critical agronomic traits such as biomass accumulation and root development. The findings highlighted the differential impact of the Brassica A and C subgenomes, with specific candidate genes such as BnaDAR (associated with early biomass accumulation) being prioritized for forthcoming breeding initiatives (Knoch et al., 2023).

Cold Tolerance And Environmental Adaptation

The cultivation of winter canola is frequently constrained by its vulnerability to freezing temperatures. A GWAS investigation concentrating on freezing tolerance in winter canola employed SNP arrays to identify loci associated with both photosynthetic efficiency and cold survival. Thirteen significant SNPs were detected across nine chromosomes, with several genes linked to established cold-responsive pathways in Arabidopsis thaliana, thereby providing valuable insights into the enhancement of cold tolerance in winter canola varieties (Chao et al., 2021).

Utilization Of High-Throughput Phenotyping In GWAS

The emergence of high-throughput phenotyping has facilitated the expansion of GWAS beyond static phenotypic traits such as disease resistance and yield, to encompass dynamic traits including growth rate and stress response. A multi-environment GWAS conducted on canola, which integrated image-based phenotyping with SNP array data, uncovered multiple QTLs associated with vegetative growth and early developmental stages. This investigation utilized SNP arrays to establish correlations between growth-related phenotypes and metabolic as well as transcriptomic profiles, thereby identifying key candidate genes that may enhance growth and biomass accumulation within breeding programs (Knoch et al., 2023).

Technical Challenges And Solutions In GWAS

Although SNP arrays furnish robust data for GWAS, they are not devoid of shortcomings. Technical inaccuracies, such as erroneous SNP calls, may introduce extraneous variability into the data, thereby complicating the identification of genuine trait-marker associations. Nonetheless, advancements in bioinformatics methodologies have empowered researchers to navigate these obstacles effectively. For example, a study focusing on canola and maize elucidated how the exclusion of biologically irrelevant SNP calls augmented prediction accuracy for traits such as yield and disease resistance. The investigation underscored the significance of meticulous SNP filtering to mitigate over fitting and enhance the dependability of GWAS findings (Weber et al., 2023).

Imputation methodologies, which serve to rectify missing genotype information, have also been employed to augment the efficacy of GWAS. Through the imputation of whole-genome sequence data, researchers can enhance marker density without incurring the expenses associated with high-depth sequencing. A meta-analysis concerning blackleg resistance in canola demonstrated how imputation facilitated the identification of novel resistance loci, thereby increasing both the resolution and power of GWAS without necessitating additional genotyping endeavors (Fikere et al., 2020). GWAS employing SNP arrays have successfully identified critical loci pertinent to traits such as disease resistance, yield, and cold tolerance in canola, thereby supporting breeding initiatives. The integration of multi-omics approaches, high-throughput phenotyping, and bioinformatics will further enhance GWAS aimed at crop enhancement.

Genomic Selection In Canola

Genomic Selection (GS) represents a pioneering methodology in the realms of plant and animal breeding that utilizes genome-wide molecular markers to predict the breeding values of individual organisms without the necessity for phenotypic assessments. This technological advancement has garnered substantial traction within crop improvement programs, notably in canola, due to its potential for accelerating genetic gain and expediting the breeding cycle. In contrast to traditional marker-assisted selection, GS estimates the effects of all markers, thereby capturing small-effect QTLs that are vital for the enhancement of complex traits such as yield, oil content, and disease resistance (Varshney et al., 2017).

The underpinning of GS resides in the construction of genomic prediction models that employ dense SNP markers to ascertain genomic estimated breeding values. These models integrate genotypic data from a reference population with phenotypic traits, facilitating the prediction of breeding values for individuals that lack phenotypic data. The precision of these predictions is primarily influenced by several determinants, including marker density, heritability of the trait, and population structure (Zhang et al., 2019). In canola, GS has been accurately implemented to predict a variety of agronomic traits, encompassing seed yield, oil content, and resistance to blackleg disease. The integration of genotype-environment (G × E) interactions significantly boosts the precision of predictive models, as evidenced by a research investigation involving 202 spring canola lines, wherein genomic estimated breeding values for yield exhibited a range between 0.29 and 0.69 contingent upon specific traits and environmental variables (Fikere et al., 2020).

Advantages Of Genomic Selection In Canola Breeding

Increased Prediction Accuracy

A primary benefit of GS compared to conventional breeding methodologies is the pronounced enhancement in predictive accuracy, particularly concerning intricate traits. Traits characterized by low heritability, which pose considerable difficulties when improved solely via phenotypic selection, derive substantial advantages from the implementation of GS. In the domain of canola breeding, genomic prediction frameworks have revealed high accuracy for critical traits including seed yield, oil content, and resistance to diseases (Weber et al., 2023). Empirical evidence indicates that GS can markedly enhance selection outcomes when utilized at the initial stages of the breeding process. For instance, during early yield trials, GS demonstrated predictive capabilities for grain yield ranging from 0.23 to 0.55, thereby surpassing the efficacy of traditional phenotypic selection approaches (Belamkar et al., 2018). This increased precision empowers breeders to engage in more wise selection choices, thereby promoting greater genetic advancement.

Reduction Of Breeding Cycle Time And Improved Breeding Efficiency

By emphasizing genomic data over mere phenotypic information, GS facilitates a more expedited breeding process. The conventional breeding approach for canola necessitates multiple years of phenotypic assessment prior to the introduction of a new variety. In contrast, GS permits the early selection of breeding candidates within the cycle, which can diminish the time required for the development of new varieties by as much as 50%. This characteristic is especially beneficial in the context of urgent agricultural challenges such as climate change, where the rapid development of stress-resistant varieties is of significant importance. The efficacy of canola breeding initiatives is markedly improved through the application of GS, particularly with respect to complex traits. A breeding strategy that incorporates genomic information can concentrate on pivotal economic traits like yield, oil content, and disease resistance, thereby enabling breeders to attain elevated rates of genetic gain. GS has been demonstrated to enhance the precision of breeding value predictions, especially when utilized in conjunction with optimal contribution selection methodologies, which preserve genetic diversity while advancing essential traits (Cowling et al., 2023). Furthermore, GS proves to be exceptionally advantageous for early selection within the breeding cycle. In the realm of canola, models employing genomic best linear unbiased prediction have shown a 27% improvement in the reliability of breeding values relative to traditional phenotypic techniques (Brzáková et al., 2023).

Cost-Effectiveness And Prediction Of G × E Interactions

An additional benefit of GS resides in the reduction of expenses associated with field trials. Traditional breeding programs necessitate comprehensive phenotypic evaluations, which are both time-intensive and costly. GS mitigates the requirement for extensive field trials by facilitating selection based purely on genotypic data. For example, the employment of high-throughput, cost-effective phenomic selection methodologies such as near-infrared spectroscopy presents an alternative to conventional genomic prediction models, delivering comparable accuracy at a substantially lower cost (Rincent et al., 2018).

GS models that integrate G × E yield enhanced predictive accuracy across heterogeneous environments, which is vital for crops such as canola that are cultivated under fluctuating climatic conditions. By considering these interactions, plant breeders are able to identify genotypes that exhibit optimal performance across diverse environmental contexts, thereby resulting in more consistent yields and diminished vulnerability to environmental changes (Fikere et al., 2020).

Challenges and Future Directions in GS for Canola Breeding

Despite the evident benefits of genomic selection, various challenges persist in its application. A primary challenge is the financial burden and technical intricacy associated with genotyping extensive populations, particularly in the emerging stages of breeding programs. Innovations in genotyping technologies and the advent of economically viable platforms such as Near-Infrared Spectroscopy offer promising solutions to these impediments (Rincent et al., 2018). Additionally, enhancing the precision of genomic predictions in traits characterized by low heritability continues to pose a considerable challenge. Approaches such as augmenting marker density and integrating non-additive genetic influences are under investigation to enhance predictive accuracy for these traits (Zhang et al., 2019). Genomic selection significantly amplifies accuracy, efficiency, and rapidity in canola breeding, thereby facilitating genetic advancement and sustainability. As advancements in genotyping technologies progress, genomic selection is poised to assume a pivotal role in the development of resilient, high-yielding crops essential for addressing global food security.

CASE STUDIES IN THE UTILIZATION OF SNP ARRAYS IN CANOLA

Trait Improvement In Commercial Varieties

SNP arrays have been essential in advancing commercial canola varieties by facilitating precise selection and breeding methodologies. Modern-day plant breeding heavily relies on SNP markers to capture extensive genome-wide data, which is crucial for enhancing economically significant traits such as yield, oil content, and disease resistance. Recent investigations have verified that SNP arrays are efficacious in optimizing the fatty acid profiles of canola seeds, which are vital for both health-related and industrial applications. GWAS has identified 148 SNP loci associated with significant enhancements in fatty acid traits, including Erucic acid and Oleic acid content. These loci were identified on chromosomes A08 and C03, with 108 alleles identified that are correlated with elevated Oleic and Linoleic acid content, thereby substantially increasing the nutritional quality of canola oil (Zhu et al., 2019). Moreover, SNP arrays have facilitated the implementation of genomic selection methodologies in large-scale breeding populations, thereby expediting the development of superior canola cultivars. The reliance of genomic selection on SNP markers has demonstrated considerable accuracy in forecasting traits such as yield and oil content across various environmental conditions (Weber et al., 2023).

Disease Resistance And Stress Tolerance

SNP arrays have proven to be crucial in strengthening canola's resilience against both biotic (e.g., fungal pathogens) and abiotic (e.g., drought, heat) stressors. Specifically, genomic prediction and genome-wide association studies have uncovered SNP markers associated with resistance to sclerotinia stem rot, a destructive fungal affliction that can drastically impair yield and oil quality. An investigation encompassing 187 canola genotypes across four distinct environments identified 133 SNPs correlated with resistance traits, thereby contributing to the creation of sclerotinia-resistant cultivars (Roy et al., 2021). Heat stress resilience has been enhanced through SNP-based comprehensive genomic predictions. A research investigation conducted in Canada identified 66 QTLs that impart heat tolerance in canola populations exposed to elevated temperatures, resulting in a reduction of yield loss by 16.7%. These genetic markers possess the potential for integration into breeding initiatives aimed at developing canola cultivars with greater resilience to climatic fluctuations (Koscielny et al., 2020). Regarding abiotic stress factors, drought resistance has emerged as a primary area of research emphasis. Empirical studies have demonstrated that SNP arrays can effectively pinpoint genetic loci correlated with water-use efficiency, thereby facilitating the development of drought-resistant cultivars capable of thriving in water-scarce conditions (Raman et al., 2020).

Enhancing Oil Content And Fatty Acid Profiles

SNP arrays have been found to be particularly beneficial for enhancing oil content and fatty acid profiles in canola. The composition of seed oil represents a complex trait governed by numerous genetic determinants, which can be efficiently discerned through GWAS utilizing SNP arrays. For example, a study encompassing 370 diverse canola accessions identified SNPs linked to elevated oil and protein content, in addition to favorable fatty acid profiles. These results are crucial for optimizing the nutritional quality of canola oil, especially in minimizing undesirable compounds such as Erucic acid while increasing beneficial fatty acids including Oleic and Linoleic acids (Tang et al., 2019).

Temporal analyses of the canola transcriptome and proteome further validate the influence of SNP markers on lipid biosynthesis. Prominent genes implicated in fatty acid synthesis, such as acetyl-CoA carboxylase and acyl-ACP desaturase, exhibit significant expression variations during the oil accumulation process, rendering these genes optimal candidates for enhancing oil yield through SNP-based breeding methodologies (Wan et al., 2017). This advancement has yielded considerable enhancements in canola varieties tailored for both edible oil production and industrial applications, such as biodiesel. The identification of genetic variations associated with increased oil yield and improved fatty acid composition has expedited the breeding process, describing a clear trajectory for the genetic expansion of canola oil quality.

CHALLENGES AND FUTURE DIRECTIONS

SNP arrays have become a widely operated technology for genotyping within genomic investigations. Nevertheless, in spite of their widespread adoption, several limitations and challenges continue to exist, particularly when compared with emerging methodologies such as WGS and CRISPR. A significant challenge associated with the development of SNP arrays is the intrinsic ascertainment bias. Arrays are frequently constructed based on a limited subset of populations, which results in the omission of globally rare variants. This bias has the potential to distort outcomes in population genetics studies and impact estimators such as allele frequency and heterozygosity. An investigation utilizing the Axiom™ Genome-Wide Chicken Array illustrates how the selective process of SNPs within a constrained population can lead to inflated heterozygosity and erroneous allele frequency distributions (Geibel et al., 2021).

Conversely, CRISPR-Cas9 technology offers precise editing capabilities at specific genomic loci by utilizing single-guide RNAs to target alleles containing SNPs. This precision enables functional investigations of disease-associated variants in a manner that SNP arrays are unable to replicate (Chen et al., 2019). The prospective advancement of SNP array technology is predicated upon the enhancement of resolution and specificity within these arrays. Recent innovations in array design facilitate the incorporation of an increased number of SNPs at elevated densities, thereby yielding more intricate genetic information. The creation of a 670k SNP array tailored for equine breeds, for example, exemplifies the potential applicability of high-resolution genotyping arrays within agricultural genomics, which may also be adapted for crops such as canola (Schaefer et al., 2017).

Although SNP arrays have been foundational in genomic investigations, a discernible trend is emerging towards the utilization of more sophisticated genotyping platforms, including low-coverage whole-genome sequencing (lcWGS). LcWGS presents the advantage of enhanced coverage and variant detection while remaining economically viable when compared to conventional SNP arrays. It furnishes superior genomic coverage, particularly for infrequent variants, thereby positioning it as a compelling option for prospective genomic research within breeding programs (Chat et al., 2022).

The trajectory of SNP array technology may further incorporate the integration with CRISPR-based genome editing methodologies. SNP arrays can facilitate the identification of target loci for CRISPR interventions, enabling precise alterations of genetic loci associated with advantageous traits in canola and other crop species (Chen et al., 2019). Despite the significance of SNP arrays in breeding endeavors, they encounter limitations when compared with whole-genome sequencing and CRISPR technologies. Nonetheless, advancements in array design alongside their integration with various 'omics' technologies ensure their sustained relevance within the domain of precision agriculture.

CONCLUSION

SNP arrays have established themselves as key tools in genomic research and plant breeding, particularly in the enhancement of canola cultivars. Their capacity to provide high-throughput, economically efficient genotyping has facilitated substantial advancements in the mapping of pivotal agronomic traits, rendering SNP arrays central to breeding methodologies such as marker-assisted selection and genomic selection. Through genome-wide association studies and quantitative trait loci mapping, SNP arrays have successfully pinpointed loci associated with essential traits, including disease resistance, yield enhancement, and improved oil composition, which are crucial for the development of robust and high-yielding canola varieties. However, with the discovery of new technologies such as whole-genome sequencing, CRISPR, and low-coverage whole-genome sequencing that offer even greater resolution and adaptability, the function of SNP arrays may undergo transformation. While SNP arrays are constrained by their dependence on pre-established SNPs and the risk of ascertainment bias, they continue to serve as a cost-effective and scalable solution for extensive genotyping, particularly in polyploid species like canola. The ongoing enhancement of array design, in conjunction with the integration of multi-omics data and high-throughput phenotyping, will further enlarge the role of SNP arrays within precision agriculture. As agricultural demands intensify, particularly in response to climate change and food security challenges, SNP arrays are poised to make substantial contributions to the development of improved canola varieties, thereby enhancing crop sustainability, resilience, and global food production.

REFERENCES

Agarwal, M., Shrivastava, N., & Padh, H. (2008). Advances in molecular marker techniques and their applications in plant sciences. *Plant Cell Reports*, *27*(4), 617–631. DOI: 10.1007/s00299-008-0507-z PMID: 18246355

Aitken, K., Farmer, A., Berkman, P., Muller, C., Wei, X., Demano, E., & Kota, R. (2016). Generation of a 345K sugarcane SNP chip. *Proceedings of the Australian Society of Sugar Cane Technologists*, *29*, 1165–1172.

Albrechtsen, A., Nielsen, F. C., & Nielsen, R. (2010). Ascertainment biases in SNP chips affect measures of population divergence. *Molecular Biology and Evolution*, *27*(11), 2534–2547. DOI: 10.1093/molbev/msq148 PMID: 20558595

Arifuzzaman, M., Oladzadabbasabadi, A., McClean, P., & Rahman, M. (2019). Shovelomics for phenotyping root architectural traits of rapeseed/canola (Brassica napus L.) and genome-wide association mapping. *Molecular Genetics and Genomics*, *294*(4), 985–1000. DOI: 10.1007/s00438-019-01563-x PMID: 30968249

Bassil, N. V., Davis, T. M., Zhang, H., Ficklin, S., Mittmann, M., Webster, T., & Van de Weg, E. (2015). Development and preliminary evaluation of a 90 K Axiom® SNP array for the allo-octoploid cultivated strawberry Fragaria× ananassa. *BMC Genomics*, *16*(1), 1–30. DOI: 10.1186/s12864-015-1310-1 PMID: 25886969

Belamkar, V., Guttieri, M. J., Hussain, W., Jarquín, D., El-Basyoni, I., Poland, J., & Baenziger, P. S. (2018). Genomic selection in preliminary yield trials in a winter wheat breeding program. *G3 (Bethesda, Md.)*, *8*(8), 2735–2747. DOI: 10.1534/g3.118.200415 PMID: 29945967

Bianco, L., Cestaro, A., Linsmith, G., Muranty, H., Denancé, C., Théron, A., & Troggio, M. (2016). Development and validation of the Axiom® Apple480K SNP genotyping array. *The Plant Journal*, *86*(1), 62–74. DOI: 10.1111/tpj.13145 PMID: 26919684

Brzáková, M., Veselá, Z., Vařeka, J., & Bauer, J. (2023). Improving Breeding Value Reliability with Genomic Data in Breeding Groups of Charolais. *Genes*, *14*(12), 2139. DOI: 10.3390/genes14122139 PMID: 38136964

Bus, A., Hecht, J., Huettel, B., Reinhardt, R., & Stich, B. (2012). High-throughput polymorphism detection and genotyping in Brassica napus using next-generation RAD sequencing. *BMC Genomics*, *13*(1), 1–11. DOI: 10.1186/1471-2164-13-281 PMID: 22726880

Callinan, P. A., & Feinberg, A. P. (2006). The emerging science of epigenomics. *Human Molecular Genetics, 15*(suppl_1), R95–R101. DOI: 10.1093/hmg/ddl095 PMID: 16651376

Cavanagh, C. R., Chao, S., Wang, S., Huang, B. E., Stephen, S., Kiani, S., & Akhunov, E. (2013). Genome-wide comparative diversity uncovers multiple targets of selection for improvement in hexaploid wheat landraces and cultivars. *Proceedings of the National Academy of Sciences of the United States of America, 110*(20), 8057–8062. DOI: 10.1073/pnas.1217133110 PMID: 23630259

Chalhoub, B., Denoeud, F., Liu, S., Parkin, I. A., Tang, H., Wang, X., & Wincker, P. (2014). Early allopolyploid evolution in the post-Neolithic Brassica napus oilseed genome. *Science, 345*(6199), 950–953. DOI: 10.1126/science.1253435 PMID: 25146293

Challa, S., & Neelapu, N. R. (2018). Genome-wide association studies (GWAS) for abiotic stress tolerance in plants. In *Biochemical, physiological and molecular avenues for combating abiotic stress tolerance in plants* (pp. 135-150). Academic press.

Chao, W. S., Horvath, D. P., Stamm, M. J., & Anderson, J. V. (2021). Genome-wide association mapping of freezing tolerance loci in canola (Brassica napus L.). *Agronomy (Basel), 11*(2), 233. DOI: 10.3390/agronomy11020233

Chat, V., Ferguson, R., Morales, L., & Kirchhoff, T. (2022). Ultra low-coverage whole-genome sequencing as an alternative to genotyping arrays in genome-wide association studies. *Frontiers in Genetics, 12*, 790445. DOI: 10.3389/fgene.2021.790445 PMID: 35251117

Chen, C. L., Rodiger, J., Chung, V., Viswanatha, R., Mohr, S. E., Hu, Y., & Perrimon, N. (2020). SNP-CRISPR: A web tool for SNP-specific genome editing. *G3 (Bethesda, Md.), 10*(2), 489–494. DOI: 10.1534/g3.119.400904 PMID: 31822517

Chen, H., Xie, W., He, H., Yu, H., Chen, W., Li, J., & Zhang, Q. (2014). A high-density SNP genotyping array for rice biology and molecular breeding. *Molecular Plant, 7*(3), 541–553. DOI: 10.1093/mp/sst135 PMID: 24121292

Clarke, W. E., Higgins, E. E., Plieske, J., Wieseke, R., Sidebottom, C., Khedikar, Y., & Parkin, I. A. (2016). A high-density SNP genotyping array for Brassica napus and its ancestral diploid species based on optimised selection of single-locus markers in the allotetraploid genome. *Theoretical and Applied Genetics, 129*(10), 1887–1899. DOI: 10.1007/s00122-016-2746-7 PMID: 27364915

Clarke, W. E., Parkin, I. A., Gajardo, H. A., Gerhardt, D. J., Higgins, E., Sidebottom, C., & Iniguez-Luy, F. L. (2013). Genomic DNA enrichment using sequence capture microarrays: A novel approach to discover sequence nucleotide polymorphisms (SNP) in Brassica napus L. *PLoS One*, *8*(12), e81992. DOI: 10.1371/journal.pone.0081992 PMID: 24312619

Clevenger, J., Chavarro, C., Pearl, S. A., Ozias-Akins, P., & Jackson, S. A. (2015). Single nucleotide polymorphism identification in polyploids: A review, example, and recommendations. *Molecular Plant*, *8*(6), 831–846. DOI: 10.1016/j.molp.2015.02.002 PMID: 25676455

Clevenger, J., Chu, Y., Chavarro, C., Agarwal, G., Bertioli, D. J., Leal-Bertioli, S. C., & Ozias-Akins, P. (2017). Genome-wide SNP genotyping resolves signatures of selection and tetrasomic recombination in peanut. *Molecular Plant*, *10*(2), 309–322. DOI: 10.1016/j.molp.2016.11.015 PMID: 27993622

Collard, B. C., & Mackill, D. J. (2008). Marker-assisted selection: An approach for precision plant breeding in the twenty-first century. *Philosophical Transactions of the Royal Society of London. Series B, Biological Sciences*, *363*(1491), 557–572. DOI: 10.1098/rstb.2007.2170 PMID: 17715053

Cowling, W. A., Castro-Urrea, F. A., Stefanova, K. T., Li, L., Banks, R. G., Saradadevi, R., & Siddique, K. H. (2023). Optimal contribution selection improves the rate of genetic gain in grain yield and yield stability in spring canola in Australia and Canada. *Plants*, *12*(2), 383. DOI: 10.3390/plants12020383 PMID: 36679096

Dalton-Morgan, J., Hayward, A., Alamery, S., Tollenaere, R., Mason, A. S., Campbell, E., & Batley, J. (2014). A high-throughput SNP array in the amphidiploid species Brassica napus shows diversity in resistance genes. *Functional & Integrative Genomics*, *14*(4), 643–655. DOI: 10.1007/s10142-014-0391-2 PMID: 25147024

Degen, B., & Müller, N. A. (2023). A simulation study comparing advanced marker-assisted selection with genomic selection in tree breeding programs. *G3 (Bethesda, Md.)*, *13*(10), jkad164. DOI: 10.1093/g3journal/jkad164 PMID: 37494068

Dong, C., Dalton-Morgan, J., Vincent, K., & Sharp, P. (2009). A modified TILLING method for wheat breeding. *The Plant Genome*, *2*(1), plantgenome2008.10.0012. DOI: 10.3835/plantgenome2008.10.0012

Elbasyoni, I. S., Lorenz, A. J., Guttieri, M., Frels, K., Baenziger, P. S., Poland, J., & Akhunov, E. (2018). A comparison between genotyping-by-sequencing and array-based scoring of SNPs for genomic prediction accuracy in winter wheat. *Plant Science*, *270*, 123–130. DOI: 10.1016/j.plantsci.2018.02.019 PMID: 29576064

Feizabadi, A., Noormohammadi, G., & Fatehi, F. (2021). Changes in growth, physiology, and fatty acid profile of rapeseed cultivars treated with vermicompost under drought stress. *Journal of Soil Science and Plant Nutrition, 21*(1), 200–208. DOI: 10.1007/s42729-020-00353-4

Felcher, K. J., Coombs, J. J., Massa, A. N., Hansey, C. N., Hamilton, J. P., Veilleux, R. E., & Douches, D. S. (2012). Integration of two diploid potato linkage maps with the potato genome sequence. *PLoS One, 7*(4), e36347. DOI: 10.1371/journal.pone.0036347 PMID: 22558443

Fikere, M., Barbulescu, D. M., Malmberg, M. M., Spangenberg, G. C., Cogan, N. O., & Daetwyler, H. D. (2020). Meta-analysis of GWAS in canola blackleg (Leptosphaeria maculans) disease traits demonstrates increased power from imputed whole-genome sequence. *Scientific Reports, 10*(1), 14300. DOI: 10.1038/s41598-020-71274-6 PMID: 32868838

Ganal, M. W., Durstewitz, G., Polley, A., Bérard, A., Buckler, E. S., Charcosset, A., & Falque, M. (2011). A large maize (Zea mays L.) SNP genotyping array: Development and germplasm genotyping, and genetic mapping to compare with the B73 reference genome. *PLoS One, 6*(12), e28334. DOI: 10.1371/journal.pone.0028334 PMID: 22174790

Garvin, M. R., Saitoh, K., & Gharrett, A. J. (2010). Application of single nucleotide polymorphisms to non-model species: A technical review. *Molecular Ecology Resources, 10*(6), 915–934. DOI: 10.1111/j.1755-0998.2010.02891.x PMID: 21565101

Geibel, J., Reimer, C., Weigend, S., Weigend, A., Pook, T., & Simianer, H. (2021). How array design creates SNP ascertainment bias. *PLoS One, 16*(3), e0245178. DOI: 10.1371/journal.pone.0245178 PMID: 33784304

Gent, J. I., Ellis, N. A., Guo, L., Harkess, A. E., Yao, Y., Zhang, X., & Dawe, R. K. (2013). CHH islands: De novo DNA methylation in near-gene chromatin regulation in maize. *Genome Research, 23*(4), 628–637. DOI: 10.1101/gr.146985.112 PMID: 23269663

. Goh, H. H. (2018). Integrative multi-omics through bioinformatics. *Omics applications for systems biology*, 69-80.

Gravel, S., Henn, B. M., Gutenkunst, R. N., Indap, A. R., Marth, G. T., Clark, A. G., & Wu, H. (2011). Demographic history and rare allele sharing among human populations. *Proceedings of the National Academy of Sciences of the United States of America, 108*(29), 11983–11988. DOI: 10.1073/pnas.1019276108 PMID: 21730125

Gulden, R. H., Warwick, S. I., & Thomas, A. G. (2008). The biology of Canadian weeds. 137. Brassica napus L. and B. rapa L. *Canadian Journal of Plant Science*, *88*(5), 951–996. DOI: 10.4141/CJPS07203

Guo, Z., Wang, H., Tao, J., Ren, Y., Xu, C., Wu, K., & Xu, Y. (2019). Development of multiple SNP marker panels affordable to breeders through genotyping by target sequencing (GBTS) in maize. *Molecular Breeding*, *39*(3), 1–12. DOI: 10.1007/s11032-019-0940-4

Guo, Z., Yang, W., Chang, Y., Ma, X., Tu, H., Xiong, F., & Xiong, L. (2018). Genome-wide association studies of image traits reveal genetic architecture of drought resistance in rice. *Molecular Plant*, *11*(6), 789–805. DOI: 10.1016/j.molp.2018.03.018 PMID: 29614319

Harper, A. L., Trick, M., Higgins, J., Fraser, F., Clissold, L., Wells, R., & Bancroft, I. (2012). Associative transcriptomics of traits in the polyploid crop species Brassica napus. *Nature Biotechnology*, *30*(8), 798–802. DOI: 10.1038/nbt.2302 PMID: 22820317

Heslot, N., Rutkoski, J., Poland, J., Jannink, J. L., & Sorrells, M. E. (2013). Impact of marker ascertainment bias on genomic selection accuracy and estimates of genetic diversity. *PLoS One*, *8*(9), e74612. DOI: 10.1371/journal.pone.0074612 PMID: 24040295

Hirsch, C. N., Foerster, J. M., Johnson, J. M., Sekhon, R. S., Muttoni, G., Vaillancourt, B., & Buell, C. R. (2014). Insights into the maize pan-genome and pan-transcriptome. *The Plant Cell*, *26*(1), 121–135. DOI: 10.1105/tpc.113.119982 PMID: 24488960

Huang, S., Deng, L., Guan, M., Li, J., Lu, K., Wang, H., & Hua, W. (2013). Identification of genome-wide single nucleotide polymorphisms in allopolyploid crop Brassica napus. *BMC Genomics*, *14*(1), 1–10. DOI: 10.1186/1471-2164-14-717 PMID: 24138473

Hulse-Kemp, A. M., Lemm, J., Plieske, J., Ashrafi, H., Buyyarapu, R., Fang, D. D., & Stelly, D. M. (2015). Development of a 63K SNP array for cotton and high-density mapping of intraspecific and interspecific populations of Gossypium spp. *G3 (Bethesda, Md.)*, *5*(6), 1187–1209. DOI: 10.1534/g3.115.018416 PMID: 25908569

Hurgobin, B., Golicz, A. A., Bayer, P. E., Chan, C. K. K., Tirnaz, S., Dolatabadian, A., & Edwards, D. (2018). Homoeologous exchange is a major cause of gene presence/absence variation in the amphidiploid Brassica napus. *Plant Biotechnology Journal*, *16*(7), 1265–1274. DOI: 10.1111/pbi.12867 PMID: 29205771

Jan, H. U., Abbadi, A., Lücke, S., Nichols, R. A., & Snowdon, R. J. (2016). Genomic prediction of testcross performance in canola (Brassica napus). *PLoS One*, *11*(1), e0147769. DOI: 10.1371/journal.pone.0147769 PMID: 26824924

Joughi, E. S. G., Hervan, E. M., Rad, A. S., & Noormohamadi, G. H. (2018). Fatty acid composition of oilseed rapeseed genotypes as affected by vermicompost application and different thermal regimes. *Agronomy Research (Tartu)*, *16*(1), 230–242.

Knoch, D., Meyer, R. C., Heuermann, M. C., Riewe, D., Peleke, F. F., Szymański, J., & Altmann, T. (2024). Integrated multi-omics analyses and genome-wide association studies reveal prime candidate genes of metabolic and vegetative growth variation in canola. *The Plant Journal*, *117*(3), 713–728. DOI: 10.1111/tpj.16524 PMID: 37964699

Knoll, J. E., Ramos, M. L., Zeng, Y., Holbrook, C. C., Chow, M., Chen, S., & Ozias-Akins, P. (2011). TILLING for allergen reduction and improvement of quality traits in peanut (Arachis hypogaea L.). *BMC Plant Biology*, *11*(1), 1–13. DOI: 10.1186/1471-2229-11-81 PMID: 21569438

Komar, A. (2009). *Single Nucleotide Polymorphisms*. Springer., DOI: 10.1007/978-1-60327-411-1

Koscielny, C. B., Gardner, S. W., Technow, F., & Duncan, R. W. (2020). Linkage mapping and whole-genome predictions in canola (Brassica napus) subjected to differing temperature treatments. *Crop & Pasture Science*, *71*(3), 229–238. DOI: 10.1071/CP19387

Kupiec, M., Zbikowska, A., Marciniak-Lukasiak, K., & Kowalska, M. (2020). Rapeseed oil in new application: Assessment of structure of oleogels based on their physicochemical properties and microscopic observations. *Agriculture*, *10*(6), 211. DOI: 10.3390/agriculture10060211

Kurowska, M., Daszkowska-Golec, A., Gruszka, D., Marzec, M., Szurman, M., Szarejko, I., & Maluszynski, M. (2011). TILLING-a shortcut in functional genomics. *Journal of Applied Genetics*, *52*(4), 371–390. DOI: 10.1007/s13353-011-0061-1 PMID: 21912935

Li, X., Wei, Y., Acharya, A., Jiang, Q., Kang, J., & Brummer, E. C. (2014). A saturated genetic linkage map of autotetraploid alfalfa (Medicago sativa L.) developed using genotyping-by-sequencing is highly syntenous with the Medicago truncatula genome. *G3 (Bethesda, Md.)*, *4*(10), 1971–1979. DOI: 10.1534/g3.114.012245 PMID: 25147192

Liu, Q., Guo, Y., Li, J., Long, J., Zhang, B., & Shyr, Y. (2012). Steps to ensure accuracy in genotype and SNP calling from Illumina sequencing data. *BMC Genomics*, *13*(S8), 1–8. DOI: 10.1186/1471-2164-13-S8-S8 PMID: 23281772

Malmberg, M. M., Spangenberg, G. C., Daetwyler, H. D., & Cogan, N. O. I. (2019). Assessment of low-coverage nanopore long read sequencing for SNP genotyping in doubled haploid canola (Brassica napus L.). *Scientific Reports*, *9*(1), 8688. DOI: 10.1038/s41598-019-45131-0 PMID: 31213642

Mamnabi, S., Nasrollahzadeh, S., Ghassemi-Golezani, K., & Raei, Y. (2020). Improving yield-related physiological characteristics of spring rapeseed by integrated fertilizer management under water deficit conditions. *Saudi Journal of Biological Sciences*, *27*(3), 797–804. DOI: 10.1016/j.sjbs.2020.01.008 PMID: 32127754

Mason, A. S., Higgins, E. E., Snowdon, R. J., Batley, J., Stein, A., Werner, C., & Parkin, I. A. (2017). A user guide to the Brassica 60K Illumina Infinium™ SNP genotyping array. *Theoretical and Applied Genetics*, *130*(4), 621–633. DOI: 10.1007/s00122-016-2849-1 PMID: 28220206

McCallum, C. M., Comai, L., Greene, E. A., & Henikoff, S. (2000). T argeting I nduced L ocal L esions IN G enomes (TILLING) for plant functional genomics. *Plant Physiology*, *123*(2), 439–442. DOI: 10.1104/pp.123.2.439 PMID: 10859174

McCouch, S. R., Wright, M. H., Tung, C. W., Maron, L. G., McNally, K. L., Fitzgerald, M., & Mezey, J. (2016). Open access resources for genome-wide association mapping in rice. *Nature Communications*, *7*(1), 10532. DOI: 10.1038/ncomms10532 PMID: 26842267

Muthamilarasan, M., Singh, N. K., & Prasad, M. (2019). Multi-omics approaches for strategic improvement of stress tolerance in underutilized crop species: A climate change perspective. *Advances in Genetics*, *103*, 1–38. DOI: 10.1016/bs.adgen.2019.01.001 PMID: 30904092

Nekrasov, V., Wang, C., Win, J., Lanz, C., Weigel, D., & Kamoun, S. (2017). Rapid generation of a transgene-free powdery mildew resistant tomato by genome deletion. *Scientific Reports*, *7*(1), 482. DOI: 10.1038/s41598-017-00578-x PMID: 28352080

Oliver, R. E., Tinker, N. A., Lazo, G. R., Chao, S., Jellen, E. N., Carson, M. L., & Jackson, E. W. (2013). SNP discovery and chromosome anchoring provide the first physically-anchored hexaploid oat map and reveal synteny with model species. *PLoS One*, *8*(3), e58068. DOI: 10.1371/journal.pone.0058068 PMID: 23533580

Pandey, M. K., Agarwal, G., Kale, S. M., Clevenger, J., Nayak, S. N., Sriswathi, M., & Varshney, R. K. (2017). Development and evaluation of a high density genotyping 'Axiom_ Arachis' array with 58 K SNPs for accelerating genetics and breeding in groundnut. *Scientific Reports*, *7*(1), 40577. DOI: 10.1038/srep40577 PMID: 28091575

Peng, Z., Fan, W., Wang, L., Paudel, D., Leventini, D., Tillman, B. L., & Wang, J. (2017). Target enrichment sequencing in cultivated peanut (Arachis hypogaea L.) using probes designed from transcript sequences. *Molecular Genetics and Genomics*, *292*(5), 955–965. DOI: 10.1007/s00438-017-1327-z PMID: 28492983

Ram, T., Majumder, N. D., Krishnaveni, D., & Ansari, M. M. (2007). Rice variety Dhanrasi, an example of improving yield potential and disease resistance by introgressing gene (s) from wild species (Oryza rufipogon). *Current Science*, •••, 987–992.

Raman, H., Raman, R., Coombes, N., Song, J., Prangnell, R., Bandaranayake, C., & Balasubramanian, S. (2016). Genome-wide association analyses reveal complex genetic architecture underlying natural variation for flowering time in canola. *Plant, Cell & Environment*, *39*(6), 1228–1239. DOI: 10.1111/pce.12644 PMID: 26428711

Raman, R., Diffey, S., Barbulescu, D. M., Coombes, N., Luckett, D., Salisbury, P., & Raman, H. (2020). Genetic and physical mapping of loci for resistance to blackleg disease in canola (Brassica napus L.). *Scientific Reports*, *10*(1), 4416. DOI: 10.1038/s41598-020-61211-y PMID: 32157120

Rincent, R., Charpentier, J. P., Faivre-Rampant, P., Paux, E., Le Gouis, J., Bastien, C., & Segura, V. (2018). Phenomic selection is a low-cost and high-throughput method based on indirect predictions: Proof of concept on wheat and poplar. *G3 (Bethesda, Md.)*, *8*(12), 3961–3972. DOI: 10.1534/g3.118.200760 PMID: 30373914

Roy, J., Shaikh, T. M., del Río Mendoza, L., Hosain, S., Chapara, V., & Rahman, M. (2021). Genome-wide association mapping and genomic prediction for adult stage sclerotinia stem rot resistance in Brassica napus (L) under field environments. *Scientific Reports*, *11*(1), 21773. DOI: 10.1038/s41598-021-01272-9 PMID: 34741104

Schaefer, R. J., Schubert, M., Bailey, E., Bannasch, D. L., Barrey, E., Bar-Gal, G. K., & McCue, M. E. (2017). Developing a 670k genotyping array to tag~ 2M SNPs across 24 horse breeds. *BMC Genomics*, *18*(1), 1–18. DOI: 10.1186/s12864-017-3943-8 PMID: 28750625

Shaw, G. (2013). Polymorphism and single nucleotide polymorphisms (SNP s). *BJU International*, *112*(5), 664–665. DOI: 10.1111/bju.12298 PMID: 23924425

Snowdon, R. J., & Iniguez Luy, F. L. (2012). Potential to improve oilseed rape and canola breeding in the genomics era. *Plant Breeding*, *131*(3), 351–360. DOI: 10.1111/j.1439-0523.2012.01976.x

Spindel, J. E., Dahlberg, J., Colgan, M., Hollingsworth, J., Sievert, J., Staggenborg, S. H., & Vogel, J. P. (2018). Association mapping by aerial drone reveals 213 genetic associations for Sorghum bicolor biomass traits under drought. *BMC Genomics*, *19*(1), 1–18. DOI: 10.1186/s12864-018-5055-5 PMID: 30223789

STATISTA. Worldwide Oilseed Production since 2008. Available online: https://www.statista.com/statistics/267271/worldwide-oilseed-production-since-2008 (accessed on 8 July 2020).

Strahl, B. D., & Allis, C. D. (2000). The language of covalent histonemodifications. *Nature*, *403*(6765), 41–45. DOI: 10.1038/47412 PMID: 10638745

Sun, C., Dong, Z., Zhao, L., Ren, Y., Zhang, N., & Chen, F. (2020). The Wheat 660K SNP array demonstrates great potential for marker-assisted selection in polyploid wheat. *Plant Biotechnology Journal*, *18*(6), 1354–1360. DOI: 10.1111/pbi.13361 PMID: 32065714

. Tang MinQiang, T. M., Zhang YuanYuan, Z. Y., Liu YueYing, L. Y., Tong Chao-Bo, T. C., Cheng XiaoHui, C. X., Zhu Wei, Z. W., ... & Liu ShengYi, L. S. (2019). Mapping loci controlling fatty acid profiles, oil and protein content by genome-wide association study in Brassica napus.

Tettelin, H., Masignani, V., Cieslewicz, M. J., Donati, C., Medini, D., Ward, N. L., & Fraser, C. M. (2005). Genome analysis of multiple pathogenic isolates of Streptococcus agalactiae: Implications for the microbial "pan-genome". *Proceedings of the National Academy of Sciences of the United States of America*, *102*(39), 13950–13955. DOI: 10.1073/pnas.0506758102 PMID: 16172379

. Tinker, N. A., Chao, S., Lazo, G. R., Oliver, R. E., Huang, Y. F., Poland, J. A., & Jackson, E. W. (2014). A SNP genotyping array for hexaploid oat. *The Plant Genome, 7*(3), plantgenome2014-03.

Torkamaneh, D., Laroche, J., Bastien, M., Abed, A., & Belzile, F. (2017). Fast-GBS: A new pipeline for the efficient and highly accurate calling of SNPs from genotyping-by-sequencing data. *BMC Bioinformatics*, *18*(1), 1–7. DOI: 10.1186/s12859-016-1431-9 PMID: 28049422

Varshney, R. K., Roorkiwal, M., & Sorrells, M. E. (2017). *Genomic selection for crop improvement: new molecular breeding strategies for crop improvement*. Springer International Publishing. DOI: 10.1007/978-3-319-63170-7

Vos, P. G., Uitdewilligen, J. G., Voorrips, R. E., Visser, R. G., & van Eck, H. J. (2015). Development and analysis of a 20K SNP array for potato (Solanum tuberosum): An insight into the breeding history. *Theoretical and Applied Genetics*, *128*(12), 2387–2401. DOI: 10.1007/s00122-015-2593-y PMID: 26263902

Wan, H., Cui, Y., Ding, Y., Mei, J., Dong, H., Zhang, W., & Qian, W. (2017). Time-series analyses of transcriptomes and proteomes reveal molecular networks underlying oil accumulation in canola. *Frontiers in Plant Science*, *7*, 2007. DOI: 10.3389/fpls.2016.02007 PMID: 28119706

Wang, J., Chu, S., Zhang, H., Zhu, Y., Cheng, H., & Yu, D. (2016). Development and application of a novel genome-wide SNP array reveals domestication history in soybean. *Scientific Reports*, *6*(1), 20728. DOI: 10.1038/srep20728 PMID: 26856884

Wang, J., Huang, Y., Li, X. L., & Li, H. Z. (2011). Research progress in clubroot of crucifers. *Plant prot, 37*, 153-158.

Wang, S., Wong, D., Forrest, K., Allen, A., Chao, S., Huang, B. E., & Akhunov, E. (2014). Characterization of polyploid wheat genomic diversity using a high-density 90 000 single nucleotide polymorphism array. *Plant Biotechnology Journal*, *12*(6), 787–796. DOI: 10.1111/pbi.12183 PMID: 24646323

Wang, Y., Cheng, X., Shan, Q., Zhang, Y., Liu, J., Gao, C., & Qiu, J. L. (2014). Simultaneous editing of three homoeoalleles in hexaploid bread wheat confers heritable resistance to powdery mildew. *Nature Biotechnology*, *32*(9), 947–951. DOI: 10.1038/nbt.2969 PMID: 25038773

Wang, Y., Hou, Y. P., Chen, C. J., & Zhou, M. G. (2014). Detection of resistance in Sclerotinia sclerotiorum to carbendazim and dimethachlon in Jiangsu Province of China. *Australasian Plant Pathology*, *43*(3), 307–312. DOI: 10.1007/s13313-014-0271-1

Weber, S. E., Chawla, H. S., Ehrig, L., Hickey, L. T., Frisch, M., & Snowdon, R. J. (2023). Accurate prediction of quantitative traits with failed SNP calls in canola and maize. *Frontiers in Plant Science*, *14*, 1221750. DOI: 10.3389/fpls.2023.1221750 PMID: 37936929

Williams, J. G., Kubelik, A. R., Livak, K. J., Rafalski, J. A., & Tingey, S. V. (1990). DNA polymorphisms amplified by arbitrary primers are useful as genetic markers. *Nucleic Acids Research*, *18*(22), 6531–6535. DOI: 10.1093/nar/18.22.6531 PMID: 1979162

Winfield, M. O., Allen, A. M., Burridge, A. J., Barker, G. L., Benbow, H. R., Wilkinson, P. A., & Edwards, K. J. (2016). High-density SNP genotyping array for hexaploid wheat and its secondary and tertiary gene pool. *Plant Biotechnology Journal*, *14*(5), 1195–1206. DOI: 10.1111/pbi.12485 PMID: 26466852

Wu, J., Zhao, Q., Liu, S., Shahid, M., Lan, L., Cai, G., & Zhou, Y. (2016). Genome-wide association study identifies new loci for resistance to Sclerotinia stem rot in Brassica napus. *Frontiers in Plant Science*, *7*, 1418. DOI: 10.3389/fpls.2016.01418 PMID: 27703464

Xu, C., Ren, Y., Jian, Y., Guo, Z., Zhang, Y., Xie, C., & Zou, C. (2017). Development of a maize 55 K SNP array with improved genome coverage for molecular breeding. *Molecular Breeding*, *37*(3), 1–12. DOI: 10.1007/s11032-017-0622-z PMID: 28255264

Yasin, J. K., Khan, M., Wani, S. H., Pillai, M. A., Verma, N., Pandey, P., & Singh, K. (2022). Genomics in Crop Improvement: Potential Applications, Challenges and Future Prospects. *Agricultural Biotechnology: Latest Research and Trends*, 329-349.

You, Q., Yang, X., Peng, Z., Xu, L., & Wang, J. (2018). Development and applications of a high throughput genotyping tool for polyploid crops: Single nucleotide polymorphism (SNP) array. *Frontiers in Plant Science*, *9*, 104. DOI: 10.3389/fpls.2018.00104 PMID: 29467780

Yu, H., Xie, W., Li, J., Zhou, F., & Zhang, Q. (2014). A whole-genome SNP array (RICE 6 K) for genomic breeding in rice. *Plant Biotechnology Journal*, *12*(1), 28–37. DOI: 10.1111/pbi.12113 PMID: 24034357

Zhang, H., Yin, L., Wang, M., Yuan, X., & Liu, X. (2019). Factors affecting the accuracy of genomic selection for agricultural economic traits in maize, cattle, and pig populations. *Frontiers in Genetics*, *10*, 189. DOI: 10.3389/fgene.2019.00189 PMID: 30923535

Zhou, Q., Galindo-González, L., Hwang, S. F., & Strelkov, S. E. (2021). Application of genomics and transcriptomics to accelerate development of clubroot resistant canola. *Canadian Journal of Plant Pathology*, *43*(2), 189–208. DOI: 10.1080/07060661.2020.1794541

Zhou, Y. J., Chang, Y. N., You, J. Q., Li, S. Z., Zhuang, W., & Cao, C. J. (2019). Cold-pressed canola oil reduces hepatic steatosis by modulating oxidative stress and lipid metabolism in KM mice compared with refined bleached deodorized canola oil. *Journal of Food Science*, *84*(7), 1900–1908. DOI: 10.1111/1750-3841.14504 PMID: 31183867

Zhu, Q., King, G. J., Liu, X., Shan, N., Borpatragohain, P., Baten, A., & Zhou, Q. (2019). Identification of SNP loci and candidate genes related to four important fatty acid composition in Brassica napus using genome wide association study. *PLoS One*, *14*(8), e0221578. DOI: 10.1371/journal.pone.0221578 PMID: 31442274

Zong, W., Zhong, X., You, J., & Xiong, L. (2013). Genome-wide profiling of histone H3K4-tri-methylation and gene expression in rice under drought stress. *Plant Molecular Biology*, *81*(1-2), 175–188. DOI: 10.1007/s11103-012-9990-2 PMID: 23192746

Zou, J., Hu, D., Mason, A. S., Shen, X., Wang, X., Wang, N., & Meng, J. (2018). Genetic changes in a novel breeding population of Brassica napus synthesized from hundreds of crosses between B. rapa and B. carinata. *Plant Biotechnology Journal*, *16*(2), 507–519. DOI: 10.1111/pbi.12791 PMID: 28703467

Chapter 7
Achievements and Prospects of QTL Mapping and Beneficial Gene and Allele Mining for Important Quality and Agronomic Traits in Canola Plant (Brassica napus L.)

Muhammad Muzamil
Quaid-i-Azam University, Islamabad, Pakistan

Asma Ajmal
Quaid-i-Azam University, Islamabad, Pakistan

Javed Iqbal
https://orcid.org/0000-0002-9032-2622
Department of Botany, Bacha Khan University, Charsadda, Pakistan

Tabassum Yaseen
Bacha Khan University, Charsadda, Pakistan

Ijaz Ahmed Qureshi
Ministry of National Food security and Research, Pakistan

Saira Qadir
Rawalpindi Women University, Pakistan

DOI: 10.4018/979-8-3693-8603-3.ch007

ABSTRACT

The chapter provides an overview of Canola (Brassica napus) is a vital oilseed crop valued for its high oil content, nutritional benefits and potential in sustainable energy. Key agronomic traits, including seed yield, flowering period, plant architecture, and disease resistance, are critical for maximizing production and environmental adaptability. Advances in molecular genetics, particularly QTL mapping and GWAS, have facilitated the identification of genomic regions tied to traits like disease resistance, drought and heat tolerance, and oil content, using markers like SSRs and SNPs. QTL mapping has pinpointed genetic loci for traits such as plant height, root biomass, and resistance to sclerotinia stem rot. Modern gene-editing tools like CRISPR-Cas9 allow precise modifications, improving traits like glucosinolate content and oil quality. Although QTL mapping can be slow and affected by environmental factors, the discovery of significant QTLs and beneficial alleles holds promise for enhanced agronomic performance and resilience in canola breeding.

INTRODUCTION

Brief Overview Of Canola (*Brassica napus*)

Origin and Importance of Canola as an Oilseed Crop

The allotetraploid genome (AACC) of canola (*Brassica napus*), a member of the Brassicaceae family, was produced by hybridizing Brassica rapa (AA) with Brassica oleracea (CC). It is one of the most important oilseed crops in the world because of its high oil content and nutritional value. After breeding breakthroughs removed harmful erucic acid and glucosinolates from its oil, making it safe for human consumption, brassica napus became an important crop in the middle of the 20th century (Afzal et al., 2018).

Oilseed agriculture underwent a sea change when canola was developed because it offered a new, low-saturated-fat edible oil source with desirable fatty acid profiles. Because of its mild flavor, high smoke point, and heart-healthy composition—particularly because of its high concentration of omega-3 fatty acids and low concentration of saturated fats—its refined oil is highly valued for use in cooking (Li et al., 2018). Because of its resilience to a variety of conditions, especially temperate zones, this crop has been gradually farmed in areas like Canada, Europe, and portions of Asia (Raboanatahiry et al., 2018).

Global Economic Significance in Food and Industrial Sectors

Production of canola oil is now essential to the food industry as well as non-food industries. After soybean and palm oil, canola oil is the third most important vegetable oil in the world. Grown primarily as an oilseed crop in Canada, canola oil contributes significantly to the national economy of the country. Growing middle-class populations and population growth, especially in developing nations, are driving the global demand for edible oils (Menendez et al., 2021).

Canola has attracted attention in the biofuel industry due to its high oil output and potential as a sustainable energy source. Canola oil-based biofuels are being used more frequently to make biodiesel, which lowers carbon emissions and provides a renewable substitute for fossil fuels. Canola's status as a crop of worldwide importance is cemented by its potential to service many industries, including food production, renewable energy, and even bioproducts (Afzal et al., 2018).

Furthermore, because of its high protein content, canola meal a byproduct of oil extraction is widely utilized in livestock feed, benefiting the agriculture and animal husbandry sectors financially. Future oil and protein yields could be greatly increased by the sustainable production of Brassica napus, which is supported by cutting-edge breeding techniques like genomic selection and Quantitative Trait Loci (QTL) mapping (Fattahi et al., 2018).

IMPORTANCE OF AGRONOMIC AND QUALITY TRAITS IN CANOLA

Definition and Examples of Key Agronomic Traits

Canola (Brassica napus) production and environmental adaptation are largely determined by agronomic features. These features include things like seed output, flowering period, architecture of the plant, and resistance to disease, all of which have a big impact on how well canola crops perform overall and how profitable they are.

Yield: The complex variable known as seed yield (SY) is regulated by several loci and is affected by various aspects such as silique number, biomass yield, and plant height. Given that higher seed yields directly translate into better oil output, it is regarded as the most significant agronomic characteristic (Raboanatahiry et al., 2018).

Flowering Time: Another important characteristic is flowering timing, which affects the crop's capacity to mature before unfavorable weather patterns appear. Early flowering canola variety breeding improves resilience to environmental challenges and permits increased adaptability across various climatic areas (Rahman & Kebede,

2021). The characteristics of plant architecture, like as height and branching, that are influenced by genes that regulate blooming time also have an impact on seed yield.

Disease Resistance: Many diseases, such as Plasmodiophora brassicae (clubroot) and Leptosphaeria maculans (blackleg), can affect canola. Resilience to these illnesses is essential to preserving excellent quality and output. Specific genetic areas relevant for resistance have been revealed by recent developments in Quantitative Trait Loci (QTL) mapping, providing new potential for the development of disease-resistant cultivars (Neik et al., 2020).

Importance of Quality Traits

Canola's qualitative features are just as important to its commercial success as its agronomic traits. These characteristics, which include oil content, fatty acid composition, and protein levels, are mostly connected to the makeup of seeds.

Oil Content: An important factor in determining the economic worth of canola is its oil content, hence one of the main goals of breeding efforts is to increase it. Breeders can now focus on these regions to create high-oil cultivars because QTL mapping has identified multiple loci that enhance seed oil content (Rahman & Kebede, 2021). Notably, QTLs identified on chromosome C5 have demonstrated stability in various environmental situations, making it a viable target for improving oil content.

Fatty Acid Composition: The fatty acid composition of canola oil, specifically its oleic and linoleic acid content, is essential for its industrial and nutritional benefits. The goal of breeding programs is to increase the content of oleic acid, which improves the stability and health advantages of the oil. QTLs linked to oleic acid levels have been mapped in recent research, revealing potential genes that could further enhance fatty acid composition (Zhao et al., 2019).

Protein Content: As a valuable feed for animals, canola meal, a byproduct of oil extraction, is high in protein. The goal of breeding efforts is also to improve the protein content of meals, which raises their economic value. Low glucosinolate levels have been associated with specific QTLs that increase protein content, which improves the meal's nutritional value and palatability (Rahman & Kebede, 2021).

NEED FOR GENETIC IMPROVEMENT

Challenges in Breeding (Complex Traits, Environmental Influences)

The key oilseed crop canola (Brassica napus) has several obstacles when it comes to genetic improvement because of the intricate features that determine quality, productivity, and environmental adaptation. Breeding is made more difficult by the quantitative character of most significant agronomic parameters, such as oil content, disease resistance, and seed yield. Numerous genes regulate these features, and they frequently interact with environmental factors like temperature, water availability, and soil minerals. This makes it challenging to create varieties that function well in a variety of settings on a continuous basis. For instance, environmental factors like temperature and photoperiod have a significant impact on features like blooming time and seed output, which can result in variable phenotypic results (Li et al., 2018). Additionally, the polyploid state of B. napus makes breeding even more difficult. Genetic complexity is increased by polyploidy, which causes genome duplication and many homologous gene copies that can interact in intricate ways. It becomes challenging to determine the distinct contributions of each gene or allele, especially for qualities that are governed by quantitative trait loci (QTL) (Afzal et al., 2018). The ability of Brassica napus to react to abiotic stressors such as drought and waterlogging presents another significant difficulty. Because of the impact that these environmental stressors have on seed output and root growth, it is essential to comprehend the genetic foundation of stress tolerance. This is difficult, though, because different settings frequently exhibit different phenotypic expressions of characteristics, such as root architecture or water use efficiency (Xiaoyu et al., 2020Traditional breeding techniques, which depend on phenotypic selection, are frequently insufficient to meet these issues. The challenge of preserving trait consistency in the face of changing environmental factors highlights the necessity for a more advanced method of breeding and trait selection.

The Role of Molecular Genetics and Genomics in Improving *Brassica napus*

Breeding Brassica napus has been transformed by molecular genetics and genomics, which have made it possible to overcome the drawbacks of conventional techniques. Our capacity to analyze complex traits and maximize genetic benefits has been greatly improved by the use of molecular markers, high-throughput sequencing, and sophisticated mapping methods including QTL mapping and genome-wide association studies (GWAS) (Khan et al., 2021). Finding specific genomic areas linked to desired characteristics, like as seed yield, oil content, and disease resistance, has been made easier with the use of QTL mapping. For instance, QTLs governing yield-related characteristics like seed density and thousand-seed weight have been uncovered in recent research, providing the groundwork for marker-assisted selection in breeding initiatives (Deng et al., 2019). These discoveries have increased the effectiveness of breeding programs by enabling breeders to more precisely select for desired features. Furthermore, single nucleotide polymorphisms (SNPs), which are essential for fine-mapping QTLs and identifying candidate genes responsible for trait variation, may now be identified thanks to developments in high-throughput sequencing. SNP markers, for example, have been employed to map QTLs related to fatty acid content and flowering time, assisting in the investigation of the genetic basis for these significant features (Li et al., 2018).

Understanding the intricate regulatory networks that govern trait expression has also been made possible by genomics. For instance, a network of genes involved in the control of seed oil content, a crucial characteristic in B. napus, are currently being clarified by transcriptome analysis and QTL mapping (Chao et al., 2017). Because of these developments, it is now possible to identify important regulatory genes and pathways and improve the oil content.QTL mapping and gene expression research are combined in functional genomics, which has also yielded insightful results. For example, expression QTLs (eQTLs) have been utilized to establish a connection between genetic variation and variations in gene expression, offering a more thorough comprehension of the ways in which certain genes influence phenotypic variation. Finding the genes responsible for growth-related features and flowering time, which are greatly impacted by environmental influences, has proven to be especially helpful with this technique (Li et al., 2018). Furthermore, complex traits have been successfully improved through the use of molecular breeding techniques including genomic selection (GS) and marker-assisted selection (MAS). By using molecular markers instead of just phenotypic features, MAS enables breeders to choose plants with desired qualities. This is especially helpful for characteristics like disease resistance or root architecture that are hard to quantify (Afzal et al., 2018).

BASICS OF QTL MAPPING IN CANOLA

What is QTL Mapping?

Using a genomics method called QTL (Quantitative Trait Loci) mapping, one may determine which parts of the genome are responsible for variation in complex traits like plant production, disease resistance, or blooming time. QTL mapping locates these genes based on statistical connections between phenotypic variation and genetic markers across populations (Figure 1 illustrates QTL mapping process). These loci often comprise genes that have minor to moderate effects on phenotypic traits (Raboanatahiry et al., 2018).

Figure 1. Flowchart illustrating the QTL mapping process, detailing each step from genetic mapping to application in breeding

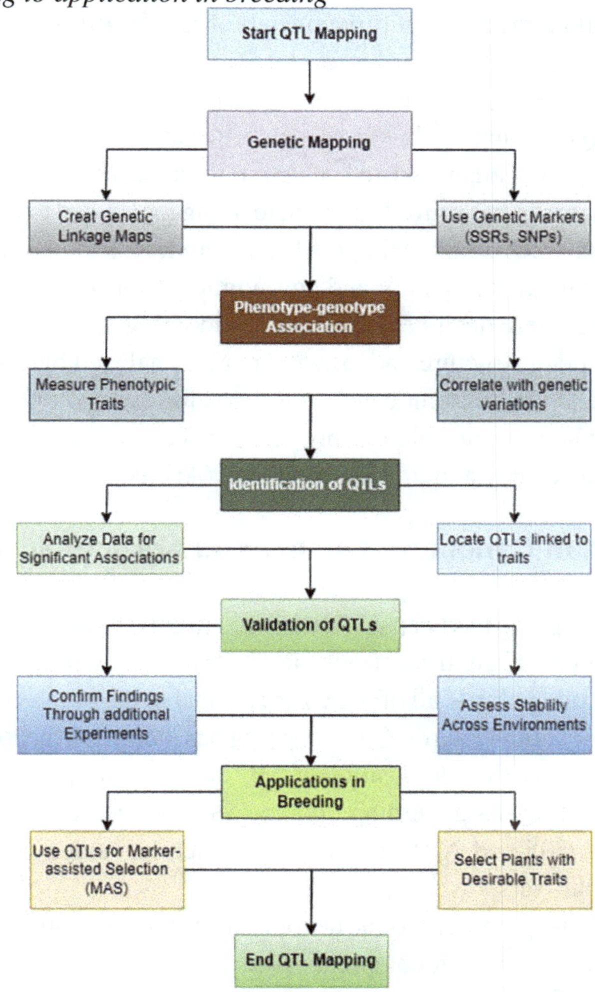

Definition of QTLs (Quantitative Trait Loci)

QTLs are characterized as genomic areas linked to particular qualities, usually quantitative ones, that vary continuously instead of categorically. These loci can affect features that are influenced by a variety of genetic and environmental factors, and they frequently span many genes. The mapping procedure locates chromosomal regions where notable correlations between particular qualities and genetic markers are noted, making it easier to pinpoint the genes in charge of characteristics like canola seed yield, disease resistance, and oil composition (Quijada et al., 2006).

Role of QTLs in Controlling Quantitative Traits

Numerous agronomic and quality parameters of canola (Brassica napus) are quantitative in nature and are regulated by several QTLs spread throughout the genome. These traits include seed yield, oil content, disease resistance, and flowering period. With the aid of QTL mapping, breeders may better understand the genetic basis of these traits and apply marker-assisted selection to increase crop performance. For instance, QTL mapping has proved useful in locating loci linked to characteristics of seed composition, such as fatty acid profiles, oil content, and protein levels, which are essential for nutritional quality and agronomic performance (Yu et al., 2016). Furthermore, characteristics like plant height, blooming time, and root vigor that are connected to plant structure and growth are also analyzed by QTL mapping. In order to generate stress-resistant canola cultivars, for example, QTL mapping has revealed specific loci responsible for root vigor, a characteristic that contributes to nutrient uptake efficiency and drought tolerance (Arifuzzaman et al., 2016).

QTL Mapping in Canola for Quality and Agronomic Traits

Canola QTL mapping has been widely utilized to determine the genetic variables governing a number of significant agronomic traits, including seed production, seed weight, and oil quality—all of which are critical to the crop's economic worth. When it comes to quality features, QTL mapping has identified the loci that increase protein content, change the composition of fatty acids, and improve oil content. For instance, significant QTLs governing oil and seed fiber content have been located on chromosomes C05 and A09, respectively, which has aided in the creation of superior canola cultivars.

Through QTL mapping, agronomic factors like disease resistance, flowering time, and seed yield have also been comprehensively examined. For example, research has shown several QTLs associated with yield and its constituents, including seed weight and number of siliques, as well as flowering time, a critical characteristic impacting crop adaptation to various settings (Delourme et al., 2018). Breeders can select for particular QTLs to increase productivity in target situations, such as low-rainfall regions or areas prone to stress conditions, thanks to the capacity to map these traits.

Significant advancements in canola breeding have been made possible by marker-assisted selection (MAS), which is powered by QTL mapping. Breeders can more efficiently improve qualities like yield and stress tolerance by concentrating on QTLs associated with advantageous alleles, which will hasten the creation of improved canola varieties. For instance, it has been determined that QTLs linked to genes related to flowering time, such as FLOWERING LOCUS T (FT), are essential for guaranteeing that canola flowers at the ideal time, balancing seed yield and environmental adaption (H. Raman et al., 2012).

Principles Of QTL Mapping

Genetic Linkage and Recombination

The tendency for genes that are next to one another on the same chromosome to be inherited together during meiosis is known as genetic linkage. Recombination is the breaking down of linkage between loci that are farther away and the mixing of alleles that happens when homologous chromosomes swap segments. The distance on a chromosome between two loci determines how quickly they recombinant. The fundamental idea behind QTL mapping is that certain genomic areas are linked to desired traits (such oil content or yield), and these regions can be found by observing how marker loci and the trait recombinant in a population that is segregating. Mapping experiments employing doubled haploid populations in canola have demonstrated this, enabling more accurate estimations of genetic impact and recombination frequencies (H. Raman et al., 2012).

Steps in QTL Mapping

Genetic Mapping: Genetic maps consist of molecular markers arranged based on their recombination frequencies. For canola, a variety of markers like Simple Sequence Repeats (SSRs), Single Nucleotide Polymorphisms (SNPs), and Restriction Fragment Length Polymorphisms (RFLPs) have been employed to create high-density linkage maps. These maps are used as a foundation to locate QTL associated with traits of interest (Ferreira et al., 1994).

Phenotype-Genotype Association: QTL mapping relies on correlating phenotypic variation (e.g., disease resistance, oil content) with the genetic variation at marker loci. In a mapping population such as recombinant inbred lines or doubled haploid lines, researchers measure traits of interest across different environmental conditions and seasons, and then analyze these data to find associations between markers and phenotypic traits (Raman et al., 2021).

Table 1. QTL mapping correlating by phenotypic variation with the genetic variation at marker loci.

Trait	QTL Identified	Chromosome	Phenotypic Variance (%)	Gene Candidates	Research Reference
Seed Color	qSC.A09	A09	10.91 - 21.83%	17 Genes (e.g., *BnaC03.TT4*)	PMC10253132
Seed Density per Silique	qSD.A9-1	A09	53.1%	56 Genes	PMC9840639
Valid Silique Length	qSL.A9	A09	47.6%	Not Specified	PMC9840639
Oil Content	None Identified	Various	Various	Known Fatty Acid Biosynthesis Genes	PMC6707581
Fatty Acid Composition	FAD2	A01, A05, C01, C03	Varies	Including *BnaFAE1*	PMC6707581
Branch Angle	qBA.C3	C3	Not Specified	54 Genes (12 differentially expressed)	Frontiers Article
Dwarf Architecture	BnDWF/DCL1	C05	Not Specified	16 Annotated Genes	BMC Plant Biology
Protein Content	Various	Various	Various	Not Specified	Various Studies
Glucosinolate Content	Various	Various	Various	Not Specified	Various Studies

Tools And Techniques For Identifying QTLs

Linkage Maps: Since linkage maps arrange markers according to recombination frequencies, they are crucial for QTL mapping. Numerous linkage maps utilizing diverse marker systems, including SSR, SNP, and Diversity Array Technology (DArT) markers, have been created for B. napus. Researchers can identify QTL that impact characteristics like flowering time, disease resistance, and seed quality with precision thanks to these high-density maps (Chen et al., 2010).

Molecular Markers: Molecular markers are DNA sequences linked to specific characteristics that are employed in the identification of genetic diversity within populations. SNP markers have proven particularly helpful in canola QTL mapping research because of their prevalence and simplicity of detection. To guarantee thorough coverage of the genome, these markers are frequently utilized in conjunction with other marker types (Liu et al., 2013).

Phenotyping and Environment: Taking into account how the environment affects phenotypic traits is a significant problem in QTL mapping. Canola plants exhibit a wide range of phenotypic plasticity due to various environmental factors.

Consequently, multi-environment trials are generally required to accurately evaluate QTL effects. Strong genotype-by-environment interactions are seen in some QTL, suggesting that the growth environment can affect how these genes are expressed (Luo et al., 2014).

Application Of QTL Mapping In Canola Improvement

The discovery of important genetic areas that control agronomic traits—a critical component of canola breeding programs—has been made easier by QTL mapping. For instance, the identification of QTL governing resistance to Leptosphaeria maculans, or blackleg disease, enables breeders to create canola cultivars that are more resistant. These QTL are frequently employed in marker-assisted selection, which facilitates the early selection of plants with desirable traits, thereby expediting the breeding process (Raman et al., 2020).

Marker Types Used In QTL Mapping

In order to connect particular chromosomal areas to phenotypic qualities, a genetic map known as QTL mapping is mostly dependent on molecular markers. The most often employed markers in canola QTL research include Simple Sequence Repeats (SSR), Single Nucleotide Polymorphisms (SNP), and Amplified Fragment Length Polymorphisms (AFLP). In addition to understanding the QTL mapping process, it is essential to recognize the various molecular markers utilized in this approach. Figure 2 illustrates the different types of molecular markers used in QTL mapping, including Simple Sequence Repeats (SSR), Single Nucleotide Polymorphisms (SNP), and Restriction Fragment Length Polymorphisms (RFLP). Each marker type has distinct characteristics and applications in breeding, which are crucial for identifying and selecting desirable traits in canola (Brassica napus).

Figure 2. Overview of molecular markers used in QTL mapping, highlighting SSRs, SNPs, and RFLPs along with their characteristics and applications

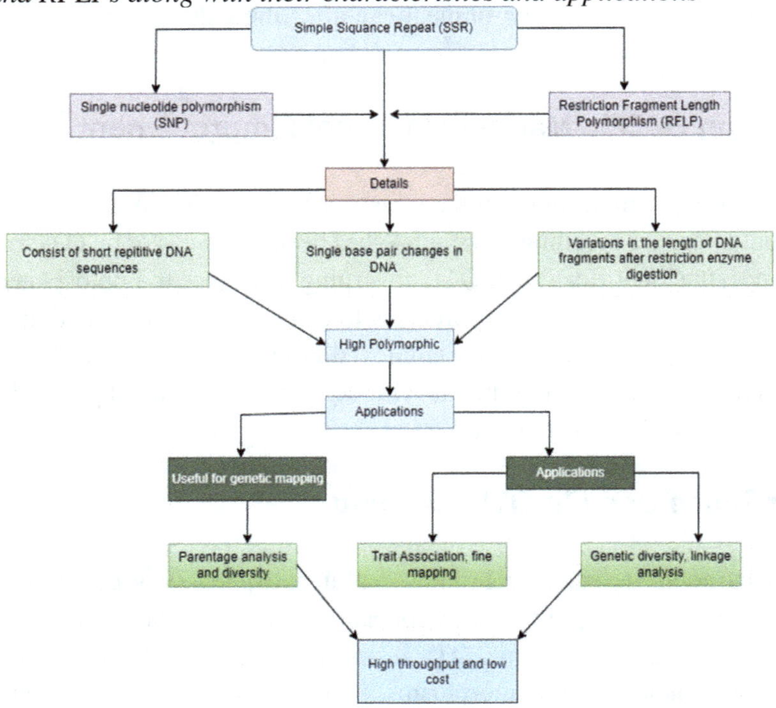

SSR (Simple Sequence Repeats)

Microsatellites, another name for SSR markers, are small DNA sequences that repeat frequently and have a high degree of polymorphism. Because they are abundant in plant genomes and co-dominant, making it possible to discriminate between heterozygous and homozygous alleles, they are frequently utilized in QTL mapping. In canola, QTLs for characteristics including plant height, flowering period, and disease resistance have been identified thanks in large part to SSR markers. For instance, a large QTL linked to semi-dwarf stature and leaf up-curling in Brassica napus was mapped using SSR markers, demonstrating how these markers aid in the knowledge of plant morphology and yield attributes (Zhao et al., 2019; Huang et al., 2019).

SNP (Single Nucleotide Polymorphisms)

SNP markers, which indicate single base-pair variations in DNA sequences, are the most common kind of genetic variation seen in genomes. SNP arrays have emerged as a crucial instrument for high-resolution QTL mapping, enabling scientists to map traits more precisely and across broader populations. For example, QTLs for canola's oleic acid content were found using a high-density SNP-based genetic map, which also revealed a novel locus that improves oil quality (Zhao et al., 2019). Furthering our knowledge of the genetic regulation of these variables, SNP markers have been essential in discovering QTLs for seed composition and agronomic traits like blooming time and silique properties (Yang et al., 2017).

AFLP (Amplified Fragment Length Polymorphisms)

AFLP markers are especially helpful for identifying polymorphisms in big, complex genomes like the canola genome because they are based on the selective amplification of DNA fragments digested by restriction enzymes. Even though SNP arrays have partially replaced AFLP markers recently, AFLP markers are still a useful tool for locating QTLs in a variety of populations. The glucosinolate concentration of canola, a crucial characteristic for enhancing meal quality in oilseed crops, has been mapped using AFLP markers in conjunction with other marker types (Jun et al., 2016).

Advantages And Limitations Of QTL Mapping

Modern plant breeding has made quantitative trait locus mapping (QTL mapping) an essential method, particularly for crops like canola (Brassica napus), where the objective is to find genetic loci affecting complex agronomic and qualitative variables. A thorough explanation of QTL mapping's benefits and drawbacks with regard to accuracy, effectiveness, and environmental dependence may be found below.

Precision in Identifying Loci Controlling Traits

The capacity of QTL mapping to identify particular genetic areas linked to complex variables like seed quality, oil content, and drought tolerance is one of its greatest advantages. Breeders can more effectively target desired alleles with this accuracy, enhancing agronomic parameters like yield and seed composition. For example, loci linked to seed oil content and other characteristics of seed quality have been successfully found in Brassica napus by QTL mapping. According to a study by Rahman and Kebede (2021), persistent QTLs for seed oil content were found on chromosome C5, indicating that genetic areas controlling essential agronomic variables may be reliably discovered by QTL mapping independent of environmental conditions.

QTL mapping has also been particularly efficient in clarifying the link between genetic loci and various phenotypes through pleiotropy. For instance, pleiotropic effects between seed color and seed composition attributes in canola were found by multi-trait QTL analysis, suggesting that a single genetic area can simultaneously affect several significant agronomic qualities (Yu et al., 2016).The precision of QTL mapping has been improved recently by molecular markers and genome-wide association studies (GWAS), making it possible to identify loci that regulate variables like yield, flowering period, and plant density more precisely (Menendez et al., 2021). This accuracy helps to comprehend the intricate genetic structure of crops and increases the effectiveness of breeding operations.

Limitations: Low Resolution, Time-Consuming, and Environment-Dependent

QTL mapping has drawbacks in spite of its benefits. It's very low resolution in comparison to other mapping methods like fine-mapping or GWAS is one of its main disadvantages. Rather of identifying the precise gene causing a problem, QTL mapping frequently finds large genomic regions; this may need to be followed up with functional genomics or fine-mapping (Fattahi et al., 2018). Furthermore, because it necessitates the development of sizable mapping populations such as doubled haploids or recombinant inbred lines QTL mapping is labor-intensive and time-consuming. It may take several generations to develop these populations, which postpones the discovery of helpful QTLs (Gacek et al., 2021). Environmental factors also have a considerable influence on QTL mapping.

Many traits can have strong environmental dependence, especially those pertaining to yield and stress tolerance. The presence of environmental variability can mask genetic signals and make it more difficult to identify stable QTLs that are adaptable to varying growth environments. By using data from many environments, multi-environment QTL analysis aims to address this, but it also lengthens the study's time and introduces another level of complexity (Zou et al., 2016; Raman et al., 2021). And last, conventional QTL mapping does not adequately take into consideration epistatic interactions between genes. The process known as epistasis, which occurs when one gene's effect is affected by another, can obscure the impact of specific loci, resulting in conclusions about the genetic basis of traits that are either partial or wrong (Zou et al., 2016).

Achievements In QTL Mapping For Canola Traits

QTLs for Agronomic Traits

A crucial technique for comprehending and optimizing Brassica napus's (canola) genetic potential is Quantitative Trait Loci (QTL) mapping. With an emphasis on recent advancements over the past seven years, this chapter describes the major breakthroughs in QTL mapping and the corresponding identification of advantageous genes and alleles for essential quality and agronomic traits in canola.

Seed Weight and Seed Number per Pod

Two key factors determining canola seed yield, seed weight (SW) and seed number per pod (SNPP), have been found to have both overlapping and separate loci by QTL mapping. Significant connections for these qualities were found in a study employing genome-wide association analysis, which also identified loci that contribute to both SW and SNPP. Significantly, it was discovered that five overlapping loci consistently influenced both parameters genetically, suggesting the possibility of improving seed weight and seed number per pod at the same time (Xin et al., 2021).

Pod Number and Yield

Another important factor influencing canola output is the number of pods. An extensive QTL analysis conducted in various conditions identified important loci influencing pod number. A total of 106 QTLs for variables that are directly correlated with yield, like plant height, branch number, and pod number, were mapped in this study. Numerous QTLs exhibited stability in various environmental circumstances, underscoring their potential for use in marker-assisted breeding aimed at enhancing yield (Zou et al., 2021).

QTL Influencing Plant Architecture in Brassica napus

Plant architecture, including traits like plant height, branch angle, and biomass, is critical for optimizing yield in canola. Key QTL findings include:

Plant Height and Branch Angle: On chromosomes A02 and A07, major QTLs governing plant height and branch angle were found; these QTLs had a substantial impact on stem architecture and lodging resistance (Shen et al., 2018).

Density Responses: Genes like TCP1 and HY5 promote growth under various plant densities, and QTLs for characteristics like biomass and seed quantity were discovered (Menendez et al., 2021).

Root and Aboveground Biomass: On chromosomes C1, C2, and C6, significant QTLs for root and aboveground biomass were found, which may help with resource acquisition and production potential (Kebede & Rahman, 2019).

Branch Angle: On A03 and C03, consistent QTLs for branch angle were discovered, emphasizing their function in raising production and light capture (Shen et al., 2017).

Integration of Flowering Intensity and Yield

Unmanned aerial vehicles (UAVs) can now be used to measure the intensity of flowering in canola fields thanks to recent advancements in phenotyping technologies. According to a study, the normalized differential yellowness index (NDYI), which measures the intensity of blooming, can be a reliable indicator of the number of pods and total seed yield. Our capacity to forecast and improve yield-related features is improved by the combination of phenotypic data and genetic information (Zhang et al., 2021).

QTLs For Stress Resistance

Drought, Salinity, and Heat Resistance QTLs

Brassica napus (canola) drought, salinity, and heat resistance genetic base has been better understood thanks to recent developments in QTL mapping. One important abiotic stressor influencing canola output is drought, especially in the reproductive stages. Numerous investigations have pinpointed QTLs associated with drought resilience, emphasizing characteristics including water use efficiency (WUE), biomass accumulation, and adaptability of seed production in water-scarce environments.

Drought Tolerance: Using a doubled haploid population, a study identified QTLs for drought tolerance and found 25 genomic areas that can explain up to 22.3% of the genotypic variance in characteristics including seed yield and the normalized deviation vegetative index (NDVI). Significant QTLs associated with genes related to blooming time and drought escape mechanisms were discovered on chromosomes A06, A10, and C04 (Raman et al., 2020).

Heat Tolerance: Heat stress has a negative impact on seed output during crucial periods, such flowering. 66 QTLs linked to characteristics including seed set and 1000-seed weight were found when canola's heat tolerance QTLs were mapped. These QTLs are essential for creating canola cultivars that are resistant to heat, especially those on chromosomes A07 and C06 (Koscielny et al., 2020).

Salinity Resistance: Through QTL mapping, the response of canola to salinity stress has been investigated, revealing important genomic areas associated with osmotic control and ion homeostasis in salt stress. Studies emphasize the significance of salinity-responsive QTLs on chromosomes A03 and A09, despite the fact that research is still in its early stages (Diouf et al., 2017).

QTLs Associated with Disease Resistance (Clubroot, Blackleg)

Blackleg Resistance: Leptosphaeria maculans is the culprit behind one of the deadliest diseases that affect canola. Studies using Quantitative Trait Loci (QTL) have shown resistance loci that are persistent in a variety of settings. A recent study identified 27 stable loci on chromosomes A02, A07, and C09 that correspond to quantitative resistance (QR) to blackleg. Breeding blackleg-resistant cultivars depends on these loci, where important candidate genes are implicated in nucleotide-binding leucine-rich repeat (LRR) receptors (Raman et al., 2018).

Clubroot Resistance: Clubroot, a serious problem in canola production, is caused by Plasmodiophora brassicae. Significant loci associated with clubroot resistance have been found on chromosomes A03 and A08 by QTL mapping. Two significant QTLs (Rcr3 and Rcr9) that confer resistance to a variety of pathotypes, including the highly pathogenic pathotype 5X, were identified in a study. The TIR-NBS-LRR proteins, which are essential for clubroot resistance, are encoded by these QTLs (Wang et al., 2022).

QTLs For Quality Traits

Oil Content and Composition (Fatty Acid Profiles, Erucic Acid Levels)

QTL mapping in Brassica napus has mostly focused on improving oil content and composition, including fatty acid profiles and erucic acid levels. Numerous investigations have discovered QTLs linked to erucic, oleic, and linoleic acids. For example, 148 significant SNP sites were shown to be associated with these features by genome-wide association studies (GWAS). These loci are primarily found in the A08 and C03 chromosomes, where candidate genes like BnaA.FAE1 and BnaC. FAE1 are important for fatty acid production (Zhu et al., 2019). According to Zhao et al. (2019), a different study identified a unique QTL on chromosome A9 that might account for up to 11.25% of the phenotypic variance for oleic acid. This suggests that the QTL has the ability to increase the amount of oleic acid in canola.

Additionally, the lowering of the unwanted fatty acid erucic acid has been discussed. QTLs for erucic acid levels on chromosomes C03 and A08 were found through mapping in various contexts, and they were substantially correlated with FAE1 alleles that regulate the synthesis of long-chain fatty acids (Zhu et al., 2019).

QTLs for Glucosinolate Reduction (An Anti-Nutritional Factor)

Reducing the amounts of glucosinolate (GSL), an anti-nutritional component in canola, is essential to raising the quality of seed meal for animal feed. Numerous QTLs linked to the decrease in GSL have been found. Notably, up to 72.87% of the phenotypic diversity in glucosinolate concentration was found to be explained by a large QTL on chromosome C02, qGSL-C2. According to Liu et al. (2019), this QTL is tightly linked to the BnaC2.MYB28 gene, a crucial regulator of GSL biosynthesis that has seen allelic variation as a result of homoeologous exchange.

Modern gene editing techniques, like CRISPR-Cas9-based glucosinolate transporter gene alterations, have effectively reduced GSL content in seeds while keeping levels high in leaves, shielding the plant from pests and enhancing the quality of seed meal (Mann et al., 2023)

Protein Content and Seed Meal Quality QTLs

Canola serves as both an oil and a source of protein, therefore increasing its protein content and enhancing the quality of its seed meal are crucial. Through QTL mapping, important areas influencing canola's protein content have been found. For example, alleles from B. oleracea dramatically boosted seed protein levels on chromosome C3, a major QTL regulating both seed protein and oil content (Rahman & Kebede, 2021).

Furthermore, QTLs that regulate the amounts of protein and glucosinolate have been identified; certain alleles that increase protein content while lowering glucosinolate levels are responsible for this. These results imply that canola seed meal's protein quality and quantity could be improved by selecting for particular QTLs (Hasan et al., 2021). Significant progress has been made in enhancing oil content, fatty acid profiles, glucosinolate reduction, and protein content as a result of recent successes in QTL mapping for canola features. In Brassica napus, the discovery of significant QTLs and candidate genes opens the door to more focused breeding and biotechnological treatments.

Table 2. QTL for stress resistance and for Quality traits.

Trait	QTL Identified	Chromosome	Key Genes Involved	Achievement in breading	Prospective for Future Research	References
Fatty Acid Profiles	qOLE1, qLIN1	A09, C03	BnaA.FAE1, BnaC.FAE1, FAD2	Improvement of oleic acid and linoleic acid profiles for better oil stability; reduced erucic acid content.	Exploration of genes controlling long-chain fatty acids and their interaction with environmental factors to improve oil stability.	Yu et al., 2016; Rahman et al., 2021
Protein Content	qPRO1, qPRO2	C03, A09	MYB28, HECTD1	Increased protein content in canola meal; improved meal quality for livestock feed.	Combined QTL mapping and transcriptomics to identify genes for enhancing protein content and reducing anti-nutritional factors.	Kebede & Rahman, 2019; Liu et al., 2019
Drought Tolerance	qDRT1, qWUE1	A06, A10, C04	FLOWERING LOCUS T, TCP1	Identification of QTLs for water-use efficiency (WUE) and seed yield stability in drought conditions.	Development of drought-tolerant varieties through genomic selection (GS) and multi-environment QTL mapping trials.	Raman et al., 2020; Diouf et al., 2017
Heat Tolerance	qHT1, qSW1	A07, C06	HY5, TCP1	QTLs identified for heat resistance during flowering and seed setting, resulting in better yield under high temperature stress.	Apply genomic selection and gene editing to enhance thermotolerance while minimizing yield losses in extreme heat conditions.	Koscielny et al., 2020

continued on following page

Table 2. Continued

Trait	QTL Identified	Chromosome	Key Genes Involved	Achievement in breading	Prospective for Future Research	References
Salinity Tolerance	qSAL1, qION1	A03, A09	FAE1, HKT1	Early identification of QTLs linked to ion homeostasis and osmotic regulation under saline conditions.	Exploration of beneficial alleles through GWAS and CRISPR to boost salinity tolerance, especially in marginal lands affected by salinity.	Diouf et al., 2017; Afzal et al., 2018
Flowering Time	qFT1, qFLT1	A02, A07	FLOWERING LOCUS T, FLC	QTLs regulating flowering time identified, contributing to the adaptability of canola in different photoperiod environments.	Incorporate environmental factors (photoperiod and temperature) in QTL mapping for precise prediction of flowering time and yield.	Shen et al., 2018; Raman et al., 2019

Case Studies: Successful QTL Mapping Projects

In canola (Brassica napus), quantitative trait loci (QTL) mapping has shown to be a reliable technique for identifying genetic mechanisms underlying important agronomic features. The effective QTL mapping initiatives that have produced notable improvements in canola trait improvement are highlighted in this section.

QTL Mapping for Water-Use Efficiency and Seed Yield

A QTL analysis was used in a study to find areas associated with canola seed yield and water-use efficiency (WUE) in an effort to increase drought resistance. It was discovered that a significant QTL on chromosome A09 controls WUE, flowering period, plant height, and seed output. Multiple traits were altered by a homoeologous exchange (HE) event that was linked to this QTL. More than 120 genes were found to be differentially expressed in this area by fine-mapping and transcriptome analysis, providing the foundation for the development of canola cultivars with increased drought resistance and yield stability (Raman et al., 2022).

Heat-Stress Tolerance in Canola

In an attempt to improve drought resilience, a QTL analysis was utilized in a study to identify regions linked to canola seed yield and water-use efficiency (WUE). A major QTL on chromosome A09 was found to regulate WUE, flowering time, plant height, and seed production. A homoeologous exchange (HE) event associated with this QTL changed multiple characteristics. By using transcriptome analysis and fine-mapping, it was discovered that over 120 genes were differentially expressed in this region. This information laid the groundwork for the creation of canola cultivars that are more resilient to drought and have stable yields (Raman et al., 2022).

Resistance to Sclerotinia Stem Rot

Canola crops are seriously threatened by sclerotinia stem rot. On chromosomes A02 and C01, substantial QTLs for resistance to this disease were found in an extensive QTL analysis encompassing several canola populations. These loci showed stability in many populations, suggesting that marker-assisted breeding programs aimed at disease resistance could benefit from using them (Shahoveisi et al., 2021).

Seed Germinability under Low Temperature

Expanding canola's cultivation range requires improving the crop's capacity to germinate in cold climates. Several loci were found in a QTL mapping investigation on low-temperature germinability; two key QTLs considerably increased seed germination rates under cold stress. According to Jiang et al. (2017), these results provide genetic resources for the creation of canola cultivars suitable for cold climes.

BENEFICIAL GENE AND ALLELE MINING IN CANOLA

Gene Discovery Techniques

In canola, gene discovery methods including transcriptomics, functional genomics, and genome-wide association studies (GWAS) have transformed our knowledge of the plant's genetic architecture and helped us identify advantageous alleles for important traits. These methods have grown to be essential for identifying allelic differences influencing agronomic and qualitative attributes, making them vital resources for crop advancement.

Overview of Gene Mining Techniques: Genome-Wide Association Studies (GWAS), Transcriptomics, and Functional Genomics

Agronomic features in canola have been linked to specific genomic areas thanks in large part to the work of genome-wide association studies (GWAS). For instance, several single nucleotide polymorphisms (SNPs) linked to flowering time, yield, and disease resistance were found in a recent GWAS investigation of canola. These correlations are important for breeding initiatives because they identify particular genomic areas that harbor advantageous alleles (Raman et al., 2019).

Transcriptomics, or the study of RNA expression, complements GWAS by offering information into the functional roles of genes. Transcriptomics is a technique that examines patterns of gene expression under various biological and environmental situations. This analysis aids in the identification of critical regulatory networks that govern critical features like metabolic efficiency and stress response. With the help of this method, primary candidate genes for canola metabolic growth variation have been effectively discovered, opening up new avenues for enhancing biomass accumulation (Knoch et al., 2023).

Functional Genomics focuses on understanding the role of genes and their interactions in controlling characteristics. This is accomplished using computational techniques in addition to experimental strategies like gene knockdown or overexpression experiments. For example, functional module extraction aids in the identification of gene clusters that cooperate during particular biological processes, improving our knowledge of features like growth optimization or disease resistance (Jha & Roy, 2020).

The process of using genomics to find advantageous alleles in canola requires combining information from several different sources, such as transcriptomics, functional genomics, and GWAS. Scientists can select genes linked to increased biomass production, stress tolerance, and disease resistance, for instance, by merging data from various platforms. In a thorough investigation, GWAS and multi-omics studies were used to identify critical alleles influencing the early build-up of biomass in canola, resulting in the discovery of primary genes like BnaDAR (Knoch et al., 2023). Furthermore, allele mining techniques like Eco-TILLING assist in finding advantageous alleles that naturally occur in gene banks and can be utilized in marker-assisted selection for crop development (Kumari et al., 2018).

Examples Of Beneficial Genes

In recent years, several key genes identified through QTL (Quantitative Trait Loci) mapping have significantly contributed to the improvement of canola traits, particularly in yield, disease resistance, and oil quality.

Yield Improvement

Numerous genes that contribute to yield-related traits have been found using QTL mapping. For example, loci for seed density, thousand-seed weight, and silique length—all of which have a direct bearing on seed yield—were mapped by a study employing SNP markers. Notably, two pleiotropic QTLs affecting seed density, silique length, and thousand-seed weight were identified on chromosome A9. With the possibility for marker-assisted selection, these findings shed light on the genetic pathways governing yield (Deng et al., 2019).

Disease Resistance

For canola resistance to important diseases including blackleg and sclerotinia stem rot (SSR), a number of QTLs have been mapped. The discovery of the Rlm3 and Rlm4 genes, which provide Leptosphaeria maculans with race-specific resistance, has greatly increased blackleg resistance. Additionally, genes encoding protein kinases involved in disease defense were linked to quantitative resistance loci (QTL) that were mapped to chromosomes A03, A07, and C04 (Raman et al., 2020). Similar to this, QTL analysis has discovered SSR resistance-related loci on chromosomes A9 and C2, providing important information for cultivating disease-resistant cultivars (Behla et al., 2017).

The combat of clubroot disease has advanced with the introduction of the CRd gene from Brassica rapa into canola cultivars, namely Zhongshuang11. Without compromising oil quality or agronomic performance, this gene offers resistance to different Plasmodiophora brassicae pathotypes (Zhan et al., 2022).

Oil Quality

One of the main breeding goals for canola is to increase its oil content. Chromosomes A10 and C5 have loci that contribute to oil content that have been mapped by QTL investigations. It has been demonstrated that the consensus QTL on A10, cqOCA10b, greatly affects oil content in various settings. Additionally, these regions contain potential genes related to lipid biosynthesis, which are essential for raising canola oil yield (Chen et al., 2017). Furthermore, QTLs for protein content and seed oil have been identified, which has led to the discovery of putative candidate genes that control oil production and enhance seed quality (Xiao et al., 2022).

Finding advantageous genes that have a major impact on canola yield enhancement, disease resistance, and oil quality has been made possible thanks in large part to QTL mapping. These observations offer a solid basis for breeding initiatives aimed at improving canola productivity and flexibility.

Mining Wild Relatives Of *Brassica Napus*

Role of Wild Relatives in Providing Beneficial Alleles

Wild cousins of Brassica napus, like Brassica oleracea and rapa, are essential in providing advantageous alleles to enhance important agronomic properties. Intraspecies cross-pollination can improve stress tolerance, disease resistance, and seed quality attributes including glucosinolate and oil content. According to recent studies, B. napus gains fresh genetic variety through the introgression of genes from wild relatives, which is essential for enhancing its production and adaptability (Li et al., 2018). Furthermore, as demonstrated by genomic comparisons, wild species contribute genes for characteristics like flowering duration, blackleg disease resistance, and tolerance to abiotic stresses like cold and drought (Lohani et al., 2019).

Cross-Species Gene Mining Efforts Using Brassica rapa and Brassica oleracea

Alleles linked to increased seed quality, yield, and disease resistance have been found thanks to cross-species gene mining between Brassica rapa and Brassica oleracea. For instance, advantageous features pertaining to seed glucosinolate content and oil composition have been transferred to B. napus through introgression from B. rapa (Liu et al., 2019). Comparable work with B. oleracea has centered on using gene-for-gene interactions including novel avirulence genes to increase resistance to illnesses like blackleg (Degrave et al., 2021). Through cross-species gene mining, the genetic basis of B. napus has been greatly increased for breeding and trait enhancement.

Applications In Breeding

Incorporation of Beneficial Genes into Breeding Programs

Oil content, seed quality, and stress resistance have all improved as a result of B. napus breeding efforts incorporating advantageous genes from wild cousins. It has been common practice to introduce these genes into elite cultivars using marker-assisted selection (MAS). For instance, studies have identified potential genes involved in oil and fatty acid biosynthesis that are introgressed from B. rapa and B. oleracea, contributing to better seed quality in B. napus (Raboanatahiry et al., 2017).

Marker-Assisted Selection (MAS) and Genomic Selection (GS) Approaches for Trait Improvement

Agronomic trait improvement in B. napus has been accelerated by the use of MAS and GS. MAS has proven very successful in identifying QTLs linked to resistance to diseases like blackleg and turnip yellows virus, as well as characteristics like flowering time and oil content (Greer et al., 2021). By using genome-wide marker data to forecast an individual's breeding value, genomic selection is further improving breeding efficiency by enabling the selection of superior genotypes even for features controlled by numerous small-effect loci (Delourme et al., 2018).

Challenges In Gene Mining

Difficulty in Identifying Small-Effect Genes

Finding small-effect genes that influence complex qualities like yield and stress tolerance is a major difficulty in the field of gene mining. Identifying these genes is frequently made more difficult by interactions with other genes or masking by larger-effect QTLs. These small-effect genes have been identified thanks to recent developments in precise mapping and transcriptome analysis, but functional validation of these genes is still challenging (Yao et al., 2020). Furthermore, by utilizing genome-wide marker data, genomic selection has demonstrated promise in capturing the impacts of these small-effect genes.

Complex Gene Interactions and Pleiotropy

Pleiotropic effects, in which a single gene influences many features, and complicated gene interactions present another hurdle. It is frequently challenging to forecast with precision how gene introgression will affect a variety of attributes because of this intricacy. For instance, there may be trade-offs in breeding programs due to genes governing the composition of seed oil also affecting growth qualities or stress tolerance. Further complicating gene mining efforts are epistatic interactions across QTLs, which have been demonstrated to play a significant effect in characteristics like biomass and flowering time (Kebede & Rahman, 201oi9).

Technological Advances Driving QTL Mapping And Gene Mining

Next-Generation Sequencing (NGS) Technologies

Advancements in next-generation sequencing (NGS) technology have had a profound impact on the field of quantitative trait loci (QTL) mapping and advantageous gene mining in Brassica napus (canola). The rapid identification of genes linked to agronomic traits like seed oil content, flowering time, and plant height is made possible by next-generation sequencing (NGS) technologies like double digest restriction-site associated DNA sequencing (ddRAD-seq), RNA sequencing (RNA-seq), and whole-genome resequencing (WGR). High-resolution QTL mapping and gene discovery are made possible by these techniques' unparalleled precision in identifying single nucleotide polymorphisms (SNPs), insertion-deletion variants (InDels), and structural rearrangements.

The polyploid nature of the Brassica napus genome, which is made up of intricate homoeologous exchanges and duplications resulting from hybridization events between Brassica rapa and Brassica oleracea, is better understood thanks to the high-density genetic linkage maps that can be produced using next-generation sequencing (NGS) (Fletcher et al., 2016). By combining WGR and QTL mapping, it is possible to identify potential genes that underlie important features like flowering time and seed oil content. These genes can then be confirmed using functional genomics techniques (Xiao et al., 2022). In a recent work, expression QTL (eQTL) data from a unique synthetic population of Brassica napus F2 were analyzed using RNA-seq. The discovery of multiple putative genes connected to growth characteristics, blooming period, and fatty acid content by the researchers emphasizes the value of NGS in explaining genetic variation.

High-Throughput Phenotyping Platforms (Advances in Precision Phenotyping for Large Populations)

Plant breeding programs' use of high-throughput phenotyping (HTP) has transformed the process of phenotyping huge populations and provided a means of bridging the genomics and phenomics divide. Technological developments in HTP platforms, especially image-based systems, enable researchers to record a wide range of plant characteristics at different phases of growth, offering a previously unheard-of degree of accuracy in tracking dynamic phenotypic changes. The use of HTP in Brassica napus (canola) has provided important new understandings of intricate characteristics such seed oil content, drought response, and salt tolerance. Quantitative trait loci (QTL) mapping and automated phenotyping can be combined in recent studies to reveal the genetic makeup of significant traits. Zhang et al. (2021) examined Brassica napus's ability to withstand salt stress using a HTP platform, finding 928.

In another pilot work, Ebersbach et al. (2022) evaluated the possibility of HTP paired with machine learning to evaluate a structured B. napus breeding population under control and drought circumstances. The research emphasized how HTP can be used to capture phenotypic variety in intricate features like growth, flower output, and canopy design. By merging RGB and NIR imaging data, this strategy provided accurate phenotypic predictions and the diagnosis of mild drought stress, highlighting the efficacy of employing machine learning models to increase precision phenotyping (Ebersbach et al., 2022). Furthermore, time-resolved phenotyping provides more profound understanding of the temporal dynamics of QTL effects by monitoring trait changes across time. Automated non-invasive HTP was used by Knoch et al. (2019) to track early development parameters like biovolume and plant.

CRISPR/Cas9 And Gene Editing

With the ability to precisely manipulate plant genomes, the CRISPR/Cas9 system has emerged as a groundbreaking tool in the field of gene editing. CRISPR/Cas9 is being investigated more and more in Brassica napus to introduce advantageous alleles on purpose, opening up new opportunities for crop improvement. Breeders may now control gene expression more effectively than ever before because to the capacity to add site-specific mutations or alleles at desired loci. This makes it possible to enhance attributes like yield, stress tolerance, and disease resistance.

Prospects of Gene Editing for Introducing Beneficial Alleles in _Brassica napus_

New pathways for the quick and accurate change of genes encoding features that are significant to agriculture have been made possible by CRISPR/Cas9 genome editing. The ability of this technique to directly insert advantageous genes into the genome without the requirement for drawn-out conventional breeding cycles is one of its most exciting features. For instance, CRISPR/Cas9 can be used to improve nutrient usage efficiency, boost stress tolerance, or raise seed oil content in canola by eliminating negative regulators or improving the activity of positive regulators of important traits. Furthermore, in polyploid organisms like B. napus, the multiplexing feature of CRISPR/Cas9—which permits the simultaneous editing of many genes—is especially helpful. This characteristic is useful for resolving issues related to the genetic redundancy found in polyploid genomes since it may be used to fine-tune complex features regulated by several loci (Afzal et al., 2018).

Case Studies of CRISPR Use in Canola Improvement

While canola is still in its early stages of CRISPR/Cas9 application, a number of case studies demonstrate the technology's potential for crop enhancement. For instance, CRISPR/Cas9 has been effectively utilized to target genes involved in the regulation of flowering time, a crucial characteristic influencing B. napus production. Researchers have optimized the crop for different growing seasons and improved production potential in a variety of environmental situations by manipulating the blooming period of the FLOWERING LOCUS C (FLC) homologs (Li et al., 2018).

Increasing the amount of seed oil in _B. napus_ is a noteworthy use of CRISPR/Cas9. A thorough QTL analysis was conducted by Xiao et al. (2022) to identify potential genes linked to the accumulation of seed oil. After editing these target genes using the CRISPR technology, oil output significantly increased. These results show that CRISPR/Cas9 has the potential to speed up breeding efforts targeted at improving commercially significant features like canola oil output (Xiao et al., 2022b). In the future, CRISPR/Cas9 technology will probably be used more in Brassica napus breeding as it develops, giving breeders effective new tools to address issues including pest pressure, rising global food demand, and climate change.

Prospects For Future Research In QTL Mapping And Gene Mining

Multi-Environment and Multi-Trait QTL Mapping

The investigation of QTL mapping in various environments, especially with fluctuating temperature and photoperiod, has great potential to improve the accuracy of QTLs that have been found. The accuracy of QTL mapping for variables like blooming time, which are influenced by both genetics and environmental factors, can be greatly increased by integrating environmental data. In Brassica napus, for example, a study found QTLs regulating blooming time under different photoperiods; this suggests that QTL interactions with environmental elements, including day duration, are critical for phenotypic expression (Rahman et al., 2018).

Another potential approach is mapping pleiotropic QTLs, which influence numerous agronomic parameters at the same time. QTLs that impact yield, seed quality, and flowering time—two characteristics that are highly relevant in crop breeding—have been identified in recent studies (Fattahi et al., 2018). Breeders will be able to increase breeding efficiency by improving numerous traits at once by identifying pleiotropic QTLs.

Precision Breeding and Digital Agriculture

A promising area for Brassica napus breeding projects is the incorporation of artificial intelligence (AI) and machine learning (ML) into precision breeding. Large datasets from phenotyping, QTL mapping, and environmental data can be analyzed by digital techniques to forecast trait performance and improve breeding tactics. In predictive breeding, where the genetic potential of lines may be predicted prior to field trials, AI-driven models can be particularly helpful in expediting the selection process (Afzal et al., 2018). By enabling breeders to track trait expression in real-time under variable field conditions, precision agriculture technologies—such as drone imaging and remote sensing—help close the gap between genotype and phenotype. For instance, under particular climatic conditions, digital techniques that monitor plant development, flowering time, and yield can improve the discovery of QTLs associated with particular features (Menendez et al., 2021).

Integration of Multi-Omics Data

Using multi-omics data genomics, transcriptomics, proteomics, and metabolomics, for example—into QTL mapping is a novel way to gain a deeper understanding of the links between genes and traits. Multi-omics analyses offer a comprehensive perspective of the molecular mechanisms behind phenotypic variation, making it possible to unravel complicated features. For example, a study conducted on B. napus found that the resolution of candidate genes linked to agronomic features such as seed composition and leaf morphology was greatly improved by integrating transcriptome data and QTL mapping (Jian et al., 2017). High-throughput sequencing and sophisticated bioinformatics pipelines are examples of genomic tools that will continue to make it easier to find new genes and alleles, which will aid in more effective crop development. A comprehensive understanding of gene function and interaction will be made possible by the integration of omics technologies, which will advance our knowledge of how various genes affect critical agronomic features (Afzal et al., 2018).

Genome-Wide Association Studies (GWAS)

Because GWAS can analyze wild populations and capture a wider range of genetic variability than traditional QTL mapping, it is becoming more and more popular. GWAS has proved helpful in finding new loci in canola (Brassica napus) that affect agronomic parameters including yield and seed oil content. For example, a study applying GWAS and QTL mapping on a population of spring-type canola accessions found unique associations for variables like biomass, silique number, and seed yield, presenting new potential for breeding improvement (Menendez et al., 2021). Exploiting advantageous alleles in various germplasm collections is another promising application of GWAS. Finding novel loci that support adaptability and yield stability in the face of shifting environmental conditions may be made easier with the capacity to connect naturally occurring genetic variation with phenotypic performance in a variety of contexts (W. Zhang et al., 2017).

Challenges and Solutions

When searching for complex features in B. napus using gene mining and QTL mapping, gene-environment interactions provide a substantial hurdle. Since environmental factors greatly affect numerous variables, most notably yield and seed quality, it is challenging to find stable QTLs in a variety of environments. On the other hand, new research has shown that it is possible to find QTLs that are stable in a variety of settings. For example, stable QTLs for seed glucosinolate content were found in B. napus and remained constant throughout China and Germany under a variety of climatic conditions (Fu et al., 2015). Another significant problem is managing massive amounts of data from high-throughput genotyping and phenotyping. This can be addressed by the development of data analysis tools like machine learning algorithms, which can process large datasets quickly and find important genetic markers linked to agronomic features. Hasan et al. (2021) state that these advances will expedite the breeding process and enable the quick development of improved canola varieties.

APPLICATIONS OF QTL MAPPING AND GENE MINING IN CANOLA BREEDING PROGRAMS

Marker-Assisted Selection (MAS)

Bypassing the conventional phenotype-based selection process and using specific genetic markers linked to desired qualities instead, Marker-Assisted Selection (MAS) is a new technique to plant breeding. MAS has been extensively used in canola (Brassica napus), particularly with the discovery of QTLs (quantitative trait loci) that regulate agronomic traits. Using QTLs associated with flowering time, which can greatly affect production stability under various environmental conditions, is one noteworthy accomplishment. QTLs for B. napus flowering time have been mapped in a number of studies, and they have been connected to genes like FLOWERING LOCUS T (FT) and FLOWERING LOCUS C (FLC). Breeding attempts to create canola cultivars with ideal blooming timings and environmental adaptability are accelerated by these findings (Raman et al., 2012).

Additionally, MAS has made it easier to modify the characteristics related to seed yield. Research utilizing QTL mapping to investigate yield-related characteristics, like plant height, silique density, and seed weight, has demonstrated similar outcomes in various settings. Certain QTLs have been found to account for as much as 20% of the variation in phenotype (Chen et al., 2007). Breeders can more effectively choose high-yielding canola varieties under particular environmental circumstances by incorporating these QTLs into breeding programs. Furthermore, breeders can target traits like oil content and disease resistance that have complex genetic bases thanks to MAS. To enhance disease resistance through MAS, QTL mapping, for instance, has shown important loci for resistance to Sclerotinia sclerotiorum, a severe pathogen in canola (Behla et al., 2017).

Benefits Of MAS

Acceleration of Breeding Cycles

By enabling selection based on genetic markers, which is quicker than waiting for phenotypic expression, MAS shortens the breeding cycle. This has proven especially helpful in areas where growing seasons are short (Quijada et al., 2006).

Efficiency in Multi-trait Selection

Breeding strategies can become more efficient when various qualities, like oil content, disease resistance, and yield, are selected for at the same time (Li et al., 2018).

Genomic Selection (GS)

Beyond the limitations of MAS, genomic selection (GS) is a more sophisticated breeding technique that forecasts breeding line success using genome-wide marker data prior to phenotypic assessment. The use of QTL data into GS models has greatly increased the accuracy of canola breeding predictions for intricate features like yield and oil quality. Utilizing high-density single nucleotide polymorphism (SNP) markers throughout the genome, genomic selection makes predictions about an individual's genetic makeup based on a training population exhibiting recognized characteristics. This eliminates the need for lengthy field trials and makes it possible to anticipate features in following generations (Zhao et al., 2016). Key agronomic trait prediction accuracy has been further improved by adding QTL information to GS models. Breeders may now more accurately anticipate yield in stressful situations like drought or nutrient shortages thanks to the integration of QTLs for traits like root vigor and flowering time into GS models (Arifuzzaman et al., 2016). Applying these GS models to canola grown in environmentally difficult areas has proven especially successful.

Integration of QTL information into GS

Studies have indicated that adding QTL-linked markers to GS models enhances the accuracy of trait selection, particularly for traits like oil content, protein concentration, and disease resistance. This is so that the inaccuracy in predicting trait values can be decreased, as QTL-linked markers offer extra genetic information (Yu et al., 2016). For instance, fatty acid profiles, which are markers for seed composition features, were incorporated into GS models to increase the accuracy of oil quality selection.

Advantages Of GS In Breeding Programs

Higher Accuracy of Selection

Compared to MAS, genomic selection allows for the more accurate selection of superior lines, particularly for traits influenced by numerous small-effect QTLs (Li et al., 2018).

Cost and Time-efficient

Breeders can choose individuals early in the breeding process thanks to GS, which eliminates the need for lengthy and costly field trials (Delourme et al., 2018).

DEVELOPMENT OF IMPROVED CANOLA VARIETIES

QTL mapping and gene mining have been combined to create superior canola varieties with improved agronomic features like higher yield, higher oil content, and better resistance to disease. QTL mapping has been effectively employed by a number of breeding initiatives to produce canola lines with increased oil output and improved environmental adaptability. Significant QTLs, for instance, were found in studies on seed yield and yield-related variables; these QTLs accounted for more than 50% of the phenotypic variance in traits including silique length and seed number (Rahaman et al., 2017). These results have had a direct impact on breeding initiatives to create canola varieties with high yields that are adaptable to different climate zones. Oil quality attributes have improved, which is another important example. The mapping of QTLs related to the composition of fatty acids has made it easier to generate canola cultivars with increased oleic acid content a desired characteristic for the production of edible oil (Li et al., 2018).

Future Prospects

It is anticipated that the ongoing use of QTL mapping and gene mining would result in the creation of canola cultivars with improved nutritional quality and stress tolerance. Breeders are in a position to discover new genes linked to significant features like disease resistance and drought tolerance thanks to the incorporation of cutting-edge genomic technologies like RNA sequencing and genome-wide association studies (GWAS) (Fletcher et al., 2016).

CONCLUSION

In recent decades, substantial advancements have occurred in QTL mapping and gene mining within Brassica napus (canola). The identification of QTLs for various significant agronomic and qualitative attributes, including seed yield, flowering time, disease resistance, and oil content, has greatly enhanced breeding efforts. Significant loci for oil content on chromosomes A10 and C5 have enhanced oil output in canola cultivars. Likewise, the identification of QTLs for resistance to diseases such as blackleg and clubroot has facilitated the creation of more robust canola cultivars. The effective incorporation of QTLs into Marker-Assisted Selection (MAS) and Genomic Selection (GS) has enabled breeders to select for numerous traits, enhancing efficiency and expediting the generation of improved varieties. QTLs for seed weight and silique number, which account for more than 50% of phenotypic variation, have proved essential in enhancing yield potential.

The forthcoming advancement in canola breeding will entail more integrative methodologies that amalgamate genomic data with environmental and phenotypic information. Multi-environment QTL mapping, along with progress in high-throughput phenotyping, will elucidate the expression of various genes across diverse climates and environmental pressures. Furthermore, the amalgamation of multi-omics data, including genomes, transcriptomics, and metabolomics, will elucidate the fundamental molecular underpinnings of intricate features such as drought tolerance and oil quality. Future breeding initiatives should also focus on leveraging new genetic techniques like CRISPR/Cas9 to incorporate advantageous alleles directly into the canola genome. This gene-editing approach could revolutionize canola breeding by accelerating the generation of cultivars with enhanced yield, nutritional quality, and stress tolerance, especially in the face of climate change. In conclusion, the achievements of QTL mapping and gene mining in canola have already provided a firm foundation for improving critical agronomic features. By combining conventional breeding approaches with new genomic tools, the future of canola production seems optimistic, with possibilities for generating high-yielding, robust, and nutritionally better varieties that fulfill the expanding worldwide demand for oilseed crops.

REFERENCES

Afzal, M., Alghamdi, S., & Rahman, M. H. (2018). Current status and future possibilities of molecular genetics techniques in Brassica napus. *Biotechnology Letters*, *40*(3), 479–492. DOI: 10.1007/s10529-018-2510-y PMID: 29344848

Afzal, M., Alghamdi, S. S., Rahman, M. H. U., Ahmad, A., Farooq, T., Alam, M., Khan, I. A., Ullah, H., Nasim, W., & Fahad, S. (2018). Current status and future possibilities of molecular genetics techniques in *Brassica napus. Biotechnology Letters*, *40*(3), 479–492. DOI: 10.1007/s10529-018-2510-y PMID: 29344848

Arifuzzaman, M., Mamidi, S., McClean, P., & Rahman, M. (2016). QTL mapping for root vigor and days to flowering in Brassica napus L. *Canadian Journal of Plant Science*, CJPS-2016-0048. Advance online publication. DOI: 10.1139/CJPS-2016-0048

Behla, R., Hirani, A., Zelmer, C., Yu, F., Fernando, W., McVetty, P., & Li, G. (2017). Identification of common QTL for resistance to Sclerotinia sclerotiorum in three doubled haploid populations of Brassica napus (L.). *Euphytica*, *213*(11), 1–16. DOI: 10.1007/s10681-017-2047-5

Behla, R., Hirani, A. H., Zelmer, C. D., Yu, F., Fernando, W. G. D., McVetty, P., & Li, G. (2017). Identification of common QTL for resistance to Sclerotinia sclerotiorum in three doubled haploid populations of Brassica napus (L.). *Euphytica*, *213*(11), 260. Advance online publication. DOI: 10.1007/s10681-017-2047-5

Chao, H., Wang, H., Wang, X., Guo, L., Gu, J., Zhao, W., Li, B., Chen, D., Raboanatahiry, N., & Li, M. (2017). Genetic dissection of seed oil and protein content and identification of networks associated with oil content in *Brassica napus. Scientific Reports*, *7*(1), 46295. Advance online publication. DOI: 10.1038/srep46295 PMID: 28393910

Chen, J., Wang, B., Zhang, Y., Yue, X., Li, Z., & Liu, K. (2017). High-density ddRAD linkage and yield-related QTL mapping delimits a chromosomal region responsible for oil content in rapeseed (Brassica napus L.). *Breeding Science*, *67*(3), 296–306. DOI: 10.1270/jsbbs.16116 PMID: 28744183

Chen, W., Zhang, Y., Liu, X., Chen, B., Tu, J., & Tingdong, F. (2007). Detection of QTL for six yield-related traits in oilseed rape (Brassica napus) using DH and immortalized F2 populations. *Theoretical and Applied Genetics*, *115*(6), 849–858. DOI: 10.1007/s00122-007-0613-2 PMID: 17665168

Degrave, A., Wagner, M., George, P., Coudard, L., Pinochet, X., Ermel, M., Gay, E., Fudal, I., Moreno-Rico, O., Rouxel, T., & Balesdent, M. (2021). A new avirulence gene of Leptosphaeria maculans, AvrLm14, identifies a resistance source in American broccoli (Brassica oleracea) genotypes. *Molecular Plant Pathology, 22*(12), 1599–1612. DOI: 10.1111/mpp.13131 PMID: 34467616

Delourme, R., Laperche, A., Bouchet, A., Jubault, M., Paillard, S., Manzanares-Dauleux, M., & Nesi, N. (2018). Genes and Quantitative Trait Loci Mapping for Major Agronomic Traits in Brassica napus L.., 41-85. https://doi.org/.DOI: 10.1007/978-3-319-43694-4_3

Delourme, R., Laperche, A., Bouchet, A., Jubault, M., Paillard, S., Manzanares-Dauleux, M., & Nesi, N. (2018). Genes and Quantitative Trait Loci Mapping for Major Agronomic Traits in Brassica napus L. In *Compendium of plant genomes* (pp. 41–85). DOI: 10.1007/978-3-319-43694-4_3

Deng, C., Liu, H., Yao, Y., Guo, S., Xiao, L., Fu, Z., & Du, D. (2019). QTL analysis of four yield-related traits for Brassica napus L. in multiple environments. *Molecular Breeding, 39*(12), 166. Advance online publication. DOI: 10.1007/s11032-019-1067-3

Deng, C., Liu, H., Yao, Y., Guo, S., Xiao, L., Fu, Z., & Du, D. (2019). QTL analysis of four yield-related traits for Brassica napus L. in multiple environments. *Molecular Breeding, 39*(12), 166. Advance online publication. DOI: 10.1007/s11032-019-1067-3

Diouf, L., Pan, Z., He, S., Gong, W., Jia, Y., Magwanga, R., Romy, K., Rashid, H., Kirungu, J., & Du, X. (2017). High-Density Linkage Map Construction and Mapping of Salt-Tolerant QTLs at Seedling Stage in Upland Cotton Using Genotyping by Sequencing (GBS). *International Journal of Molecular Sciences, 18*(12), 2622. Advance online publication. DOI: 10.3390/ijms18122622 PMID: 29206184

Dong, Z., Alam, M. K., Xie, M., Yang, L., Liu, J., Helal, M. M. U., Huang, J., Cheng, X., Liu, Y., Tong, C., Zhao, C., & Liu, S. (2021). Mapping of a major QTL controlling plant height using a high-density genetic map and QTL-seq methods based on whole-genome resequencing in *Brassica napus. G3 (Bethesda, Md.), 11*(7), jkab118. Advance online publication. DOI: 10.1093/g3journal/jkab118 PMID: 33836054

Ebersbach, J., Khan, N. A., McQuillan, I., Higgins, E. E., Horner, K., Bandi, V., Gutwin, C., Vail, S. L., Robinson, S. J., & Parkin, I. P. (2022). Exploiting High-Throughput Indoor Phenotyping to Characterize the Founders of a Structured B. napus Breeding Population. *Frontiers in Plant Science, 12*, 780250. Advance online publication. DOI: 10.3389/fpls.2021.780250 PMID: 35069637

Fattahi, F., Fakheri, B., Solouki, M., Möllers, C., & Rezaizad, A. (2018). Mapping QTL controlling agronomic traits in a doubled haploid population of winter oilseed rape (Brassica napus L.). *Journal of Genetics, 97*(5), 1389–1406. DOI: 10.1007/s12041-018-1044-3 PMID: 30555087

Ferreira, M., Williams, P. H., & Osborn, T. (1994). RFLP mapping of Brassica napus using doubled haploid lines. *Theoretical and Applied Genetics, 89*(5), 615–621. DOI: 10.1007/BF00222456 PMID: 24177938

Fletcher, R. S., Herrmann, D., Mullen, J. L., Li, Q., Schrider, D. R., Price, N., Lin, J., Grogan, K., Kern, A., & McKay, J. K. (2016). Identification of Polymorphisms Associated with Drought Adaptation QTL in *Brassica napus* by Resequencing. *G3 (Bethesda, Md.), 6*(4), 793–803. DOI: 10.1534/g3.115.021279 PMID: 26801646

Fu, Y., Lu, K., Qian, L., Mei, J., Wei, D., Peng, X., Xu, X., Li, J., Frauen, M., Dreyer, F., Snowdon, R. J., & Qian, W. (2015). Development of genic cleavage markers in association with seed glucosinolate content in canola. *Theoretical and Applied Genetics, 128*(6), 1029–1037. DOI: 10.1007/s00122-015-2487-z PMID: 25748114

Gacek, K., Bayer, P. E., Anderson, R., Severn-Ellis, A. A., Wolko, J., Łopatyńska, A., Matuszczak, M., Bocianowski, J., Edwards, D., & Batley, J. (2021). QTL Genetic Mapping Study for Traits Affecting Meal Quality in Winter Oilseed Rape (Brassica Napus L.). *Genes, 12*(8), 1235. DOI: 10.3390/genes12081235 PMID: 34440409

Greer, S., Hackenberg, D., Gegas, V., Mitrousia, G., Edwards, D., Batley, J., Teakle, G., Barker, G., & Walsh, J. (2021). Quantitative Trait Locus Mapping of Resistance to Turnip Yellows Virus in Brassica rapa and Brassica oleracea and Introgression of These Resistances by Resynthesis Into Allotetraploid Plants for Deployment in Brassica napus. *Frontiers in Plant Science, 12*, 781385. Advance online publication. DOI: 10.3389/fpls.2021.781385 PMID: 34956278

Hasan, J., Shaikh, R., Megha, S., Herrmann, D., Kebede, B., & Rahman, H. (2021). Mapping of flowering time, seed quality and clubroot resistance in rutabaga × spring canola populations and their association. Euphytica, 217. https://.oi.org/10.1007/s10681-021-02889-7

Hasan, J., Shaikh, R., Megha, S., Herrmann, D. T., Kebede, B., & Rahman, H. (2021). Mapping of flowering time, seed quality and clubroot resistance in rutabaga × spring canola populations and their association. *Euphytica, 217*(8), 160. Advance online publication. DOI: 10.1007/s10681-021-02889-7

Jha, M., & Roy, S. (2020). Functional module extraction from gene expression data using data mining techniques. *ACM SIGBioinformatics Record, 10*(1), 1–2. DOI: 10.1145/3383672.3383674

Jian, H., Yang, B., Zhang, A., Zhang, L., Xu, X., Li, J., & Liu, L. (2017). Screening of Candidate Leaf Morphology Genes by Integration of QTL Mapping and RNA Sequencing Technologies in Oilseed Rape (*Brassica napus* L.). *PLoS One*, *12*(1), e0169641. DOI: 10.1371/journal.pone.0169641 PMID: 28068426

Jiang, N., Shi, S., Shi, H., Khanzada, H., Wassan, G., Zhu, C., Peng, X., Yu, Q., Chen, X., He, X., Fu, J., Hu, L., Xu, J., Ouyang, L., Sun, X., Zhou, D., He, H., & Bian, J. (2017). Mapping QTL for Seed Germinability under Low Temperature Using a New High-Density Genetic Map of Rice. *Frontiers in Plant Science*, *8*, 1223. Advance online publication. DOI: 10.3389/fpls.2017.01223 PMID: 28747923

Jun, L., Hirani, A. H., & Zhe, L., ChunRen, W., McVetty, P., Daayf, F., & Genyi, L. (2016). QTL controlling glucosinolate content in seeds of "Brassica napus" L. *Australian Journal of Crop Science*, *10*, 152–160.

Kebede, B., & Rahman, H. (2019). QTL mapping of root and aboveground biomass in the Brassica C genome using a B. napus population carrying genome content introgressed from B. oleracea. *Molecular Breeding*, *39*(12), 169. Advance online publication. DOI: 10.1007/s11032-019-1064-6

Kebede, B., & Rahman, H. (2019). QTL mapping of root and aboveground biomass in the Brassica C genome using a B. napus population carrying genome content introgressed from B. oleracea. *Molecular Breeding*, *39*(12), 169. Advance online publication. DOI: 10.1007/s11032-019-1064-6

Knoch, D., Abbadi, A., Grandke, F., Meyer, R. C., Samans, B., Werner, C. R., Snowdon, R. J., & Altmann, T. (2019). Strong temporal dynamics of QTL action on plant growth progression revealed through high-throughput phenotyping in canola. *Plant Biotechnology Journal*, *18*(1), 68–82. DOI: 10.1111/pbi.13171 PMID: 31125482

Knoch, D., Meyer, R., Heuermann, M., Riewe, D., Peleke, F., Szymański, J., Abbadi, A., Snowdon, R., & Altmann, T. (2023). Integrated multi-omics analyses and genome-wide association studies reveal prime candidate genes of metabolic and vegetative growth variation in canola.. *The Plant journal: for cell and molecular biology*. https://doi.org/.DOI: 10.1111/tpj.16524

Knoch, D., Meyer, R., Heuermann, M., Riewe, D., Peleke, F., Szymański, J., Abbadi, A., Snowdon, R., & Altmann, T. (2023). Integrated multi-omics analyses and genome-wide association studies reveal prime candidate genes of metabolic and vegetative growth variation in canola.. *The Plant journal: for cell and molecular biology*. https://doi.org/.DOI: 10.1111/tpj.16524

Koscielny, C., Gardner, S., Technow, F., & Duncan, R. (2020). Linkage mapping and whole-genome predictions in canola (Brassica napus) subjected to differing temperature treatments. *Crop & Pasture Science*, *71*(3), 229–238. DOI: 10.1071/CP19387

Koscielny, C., Gardner, S., Technow, F., & Duncan, R. (2020). Linkage mapping and whole-genome predictions in canola (Brassica napus) subjected to differing temperature treatments. *Crop & Pasture Science*, *71*(3), 229–238. DOI: 10.1071/CP19387

Kumari, R., Kumar, P., Sharma, V., & Kumar, H. (2018). Allele Mining for Crop Improvement. *International Journal of Pure & Applied Bioscience*, *6*(1), 1456–1465. DOI: 10.18782/2320-7051.6073

Li, R., Jeong, K., Davis, J., Kim, S., Lee, S., Michelmore, R., Kim, S., & Maloof, J. (2018). Integrated QTL and eQTL Mapping Provides Insights and Candidate Genes for Fatty Acid Composition, Flowering Time, and Growth Traits in a F2 Population of a Novel Synthetic Allopolyploid Brassica napus. *Frontiers in Plant Science*, *9*, 1632. Advance online publication. DOI: 10.3389/fpls.2018.01632 PMID: 30483289

Li, R., Jeong, K., Davis, J. T., Kim, S., Lee, S., Michelmore, R. W., Kim, S., & Maloof, J. N. (2018). Integrated QTL and eQTL Mapping Provides Insights and Candidate Genes for Fatty Acid Composition, Flowering Time, and Growth Traits in a F2 Population of a Novel Synthetic Allopolyploid *Brassica napus. Frontiers in Plant Science*, *9*, 1632. Advance online publication. DOI: 10.3389/fpls.2018.01632 PMID: 30483289

Li, R., Jeong, K., Davis, J. T., Kim, S.-R., Lee, S., Michelmore, R., Kim, S., & Maloof, J. (2018). Integrated QTL and eQTL Mapping Provides Insights and Candidate Genes for Fatty Acid Composition, Flowering Time, and Growth Traits in a F2 Population of a Novel Synthetic Allopolyploid Brassica napus. *Frontiers in Plant Science*, *9*, 1632. Advance online publication. DOI: 10.3389/fpls.2018.01632 PMID: 30483289

Liu, Y., Zhou, X., Yan, M., Wang, P., Wang, H., Xin, Q., Yang, L., Hong, D., & Yang, G. (2019). Fine mapping and candidate gene analysis of a seed glucosinolate content QTL, qGSL-C2, in rapeseed (Brassica napus L.). *Theoretical and Applied Genetics*, *133*(2), 479–490. DOI: 10.1007/s00122-019-03479-x PMID: 31832742

Liu, Y., Zhou, X., Yan, M., Wang, P., Wang, H., Xin, Q., Yang, L., Hong, D., & Yang, G. (2019). Fine mapping and candidate gene analysis of a seed glucosinolate content QTL, qGSL-C2, in rapeseed (Brassica napus L.). *Theoretical and Applied Genetics*, *133*(2), 479–490. DOI: 10.1007/s00122-019-03479-x PMID: 31832742

Lohani, N., Golicz, A., Singh, M., & Bhalla, P. (2019). Genome-wide analysis of the Hsf gene family in Brassica oleracea and a comparative analysis of the Hsf gene family in B. oleracea, B. rapa and B. napus. *Functional & Integrative Genomics*, *19*(3), 515–531. DOI: 10.1007/s10142-018-0649-1 PMID: 30618014

Luo, Y. X., Luo, C., Du, D., Fu, Z., Yao, Y. M., Xu, C. C., & Zhang, H. (2014). Quantitative trait analysis of flowering time in spring rapeseed (B. napus L.). *Euphytica*, *200*(3), 321–335. DOI: 10.1007/s10681-014-1140-2

Mann, A., Kumari, J., Kumar, R., Kumar, P., Pradhan, A., Pental, D., & Bisht, N. (2023). Targeted editing of multiple homologues of GTR1 and GTR2 genes provides the ideal low-seed, high-leaf glucosinolate oilseed mustard with uncompromised defence and yield. *Plant Biotechnology Journal*, *21*(11), 2182–2195. DOI: 10.1111/pbi.14121 PMID: 37539488

Menendez, Y., Sánchez, D., Snowdon, R., Rondanini, D., & Botto, J. (2021). Unraveling the impact of plant-density genetic architecture on agronomic traits in canola. *Journal of Experimental Botany*. Advance online publication. DOI: 10.1093/jxb/erab191 PMID: 33940608

Menendez, Y. C., Sánchez, D. H., Snowdon, R., Rondanini, D. P., & Botto, J. (2021). Unraveling the impact of plant-density genetic architecture on agronomic traits in canola. *Journal of Experimental Botany*. Advance online publication. DOI: 10.1093/jxb/erab191 PMID: 33940608

Menendez, Y. C., Sanchez, D. H., Snowdon, R. J., Rondanini, D. P., & Botto, J. F. (2021). Unraveling the impact on agronomic traits of the genetic architecture underlying plant-density responses in canola. *Journal of Experimental Botany*, *72*(15), 5426–5441. DOI: 10.1093/jxb/erab191 PMID: 33940608

Neik, T. X., Amas, J., Barbetti, M., Edwards, D., & Batley, J. (2020). Understanding Host–Pathogen Interactions in Brassica napus in the Omics Era. *Plants*, *9*(10), 10. Advance online publication. DOI: 10.3390/plants9101336 PMID: 33050509

Quijada, P. A., Udall, J. A., Lambert, B., & Osborn, T. C. (2006). Quantitative trait analysis of seed yield and other complex traits in hybrid spring rapeseed (Brassica napus L.): 1. Identification of genomic regions from winter germplasm. *Theoretical and Applied Genetics*, *113*(3), 549–561. DOI: 10.1007/s00122-006-0323-1 PMID: 16767447

Raboanatahiry, N., Chao, H., Dalin, H., Pu, S., Yan, W., Yu, L., Wang, B., & Li, M. (2018). QTL Alignment for Seed Yield and Yield Related Traits in Brassica napus. *Frontiers in Plant Science*, *9*, 1127. Advance online publication. DOI: 10.3389/fpls.2018.01127 PMID: 30116254

Raboanatahiry, N., Chao, H., Guo, L., Gan, J., Xiang, J., Yan, M., Zhang, L., Yu, L., & Li, M. (2017). Synteny analysis of genes and distribution of loci controlling oil content and fatty acid profile based on QTL alignment map in Brassica napus. *BMC Genomics*, *18*(1), 776. Advance online publication. DOI: 10.1186/s12864-017-4176-6 PMID: 29025408

Rahaman, M., Mamidi, S., & Rahman, M. (2017). Association mapping of agronomic traits of canola (Brassica napus L.) subject to heat stress under field conditions. *Australian Journal of Crop Science*, *11*(09), 1094–1105. DOI: 10.21475/ajcs.17.11.09.pne512

Rahman, H., Bennett, R. A., & Kebede, B. (2018). Molecular mapping of QTL alleles of Brassica oleracea affecting days to flowering and photosensitivity in spring *Brassica napus*. *PLoS One*, *13*(1), e0189723. DOI: 10.1371/journal.pone.0189723 PMID: 29320498

Rahman, H., & Kebede, B. (2021). Mapping of seed quality traits in the C genome of Brassica napus by using a population carrying genome content of B. oleracea and their effect on other traits. *The Plant Genome*, *14*(2), e20078. Advance online publication. DOI: 10.1002/tpg2.20078 PMID: 33818008

Rahman, H., & Kebede, B. (2021). Mapping of seed quality traits in the C genome of Brassica napus by using a population carrying genome content of B. oleracea and their effect on other traits. *The Plant Genome*, *14*(2), e20078. Advance online publication. DOI: 10.1002/tpg2.20078 PMID: 33818008

Raman, H., Raman, R., Diffey, S., Qiu, Y., McVittie, B., Barbulescu, D., Salisbury, P., Marcroft, S., & Delourme, R. (2018). Stable Quantitative Resistance Loci to Blackleg Disease in Canola (Brassica napus L.) Over Continents. *Frontiers in Plant Science*, *9*, 1622. Advance online publication. DOI: 10.3389/fpls.2018.01622 PMID: 30532758

Raman, H., Raman, R., Eckermann, P., Coombes, N., Manoli, S., Zou, X., Edwards, D., Meng, J., Prangnell, R., Stiller, J., Batley, J., Luckett, D., Wratten, N., & Dennis, E. (2012). Genetic and physical mapping of flowering time loci in canola (Brassica napus L.). *Theoretical and Applied Genetics*, *126*(1), 119–132. DOI: 10.1007/s00122-012-1966-8 PMID: 22955939

Raman, H., Raman, R., Mathews, K., Diffey, S., & Salisbury, P. (2020). QTL mapping reveals genomic regions for yield based on an incremental tolerance index to drought stress and related agronomic traits in canola. *Crop & Pasture Science*, *71*(6), 562–577. DOI: 10.1071/CP20046

Raman, H., Raman, R., Pirathiban, R., McVittie, B., Sharma, N., Liu, S., Qiu, Y., Zhu, A., Killian, A., Cullis, B., Farquhar, G., Stuart-Williams, H., White, R., Tabah, D., Easton, A., & Zhang, Y. (2022). Multienvironment QTL analysis delineates a major locus associated with homoeologous exchanges for water-use efficiency and seed yield in canola. *Plant, Cell & Environment*, *45*(7), 2019–2036. DOI: 10.1111/pce.14337 PMID: 35445756

Raman, H., Raman, R., Qiu, Y., Yadav, A., Sureshkumar, S., Borg, L., Rohan, M., Wheeler, D., Owen, O., Menz, I., & Balasubramanian, S. (2019). GWAS hints at pleiotropic roles for FLOWERING LOCUS T in flowering time and yield-related traits in canola. *BMC Genomics*, *20*(1), 636. Advance online publication. DOI: 10.1186/s12864-019-5964-y PMID: 31387521

Raman, R., Diffey, S., Barbulescu, D., Coombes, N., Luckett, D., Salisbury, P., Cowley, R., Marcroft, S., & Raman, H. (2020). Genetic and physical mapping of loci for resistance to blackleg disease in canola (Brassica napus L.). *Scientific Reports*, *10*(1), 4416. Advance online publication. DOI: 10.1038/s41598-020-61211-y PMID: 32157120

Raman, R., Diffey, S., Barbulescu, D. M., Coombes, N., Luckett, D., Salisbury, P., Cowley, R., Marcroft, S., & Raman, H. (2020). Genetic and physical mapping of loci for resistance to blackleg disease in canola (Brassica napus L.). *Scientific Reports*, *10*(1), 4416. Advance online publication. DOI: 10.1038/s41598-020-61211-y PMID: 32157120

Shahoveisi, F., Oladzad, A., Mendoza, L., Hosseinirad, S., Ruud, S., & Rissato, B. (2021). *Assessing the Effect of Phenotyping Scoring Systems and SNP Calling and Filtering Parameters on Detection of QTL Associated with Reaction of Brassica napus to Sclerotinia sclerotiorum*. PhytoFrontiers., DOI: 10.1094/PHYTOFR-10-20-0029-R

Shen, Y., Xiang, Y., Xu, E., Ge, X., & Li, Z. (2018). Major Co-localized QTL for Plant Height, Branch Initiation Height, Stem Diameter, and Flowering Time in an Alien Introgression Derived Brassica napus DH Population. *Frontiers in Plant Science*, *9*, 390. Advance online publication. DOI: 10.3389/fpls.2018.00390 PMID: 29643859

Shen, Y., Yang, Y., Xu, E., Ge, X., Xiang, Y., & Li, Z. (2017). Novel and major QTL for branch angle detected by using DH population from an exotic introgression in rapeseed (Brassica napus L.). *Theoretical and Applied Genetics*, *131*(1), 67–78. DOI: 10.1007/s00122-017-2986-1 PMID: 28942459

Wang, Z., Megha, S., Kebede, B., Kav, N., & Rahman, H. (2022). Genetic and molecular analysis reveals that two major loci and their interaction confer clubroot resistance in canola introgressed from rutabaga. *The Plant Genome*, *15*(3), e20241. Advance online publication. DOI: 10.1002/tpg2.20241 PMID: 35818693

Xiao, Z., Zhang, C., Qu, C., Wei, L., Zhang, L., Yang, B., Lu, K., & Li, J. (2022). Identification of candidate genes regulating seed oil content by QTL mapping and transcriptome sequencing in Brassica napus. *Frontiers in Plant Science*, *13*, 1067121. Advance online publication. DOI: 10.3389/fpls.2022.1067121 PMID: 36570918

Xiao, Z., Zhang, C., Qu, C., Wei, L., Zhang, L., Yang, B., Lu, K., & Li, J. (2022). Identification of candidate genes regulating seed oil content by QTL mapping and transcriptome sequencing in *Brassica napus*. *Frontiers in Plant Science*, *13*, 1067121. Advance online publication. DOI: 10.3389/fpls.2022.1067121 PMID: 36570918

Xiao, Z., Zhang, C., Qu, C., Wei, L., Zhang, L., Yang, B., Lu, K., & Li, J. (2022b). Identification of candidate genes regulating seed oil content by QTL mapping and transcriptome sequencing in *Brassica napus*. *Frontiers in Plant Science*, *13*, 1067121. Advance online publication. DOI: 10.3389/fpls.2022.1067121 PMID: 36570918

Xiaoyu, D., Xu, J., He, H., Xing, Q., Shen, M., Cheng, Y., & Zhang, X. (2020). Unraveling waterlogging tolerance-related traits with QTL analysis in reciprocal intervarietal introgression lines using genotyping by sequencing in rapeseed (Brassica napus L.). *Journal of Integrative Agriculture*, *19*(8), 1974–1983. DOI: 10.1016/S2095-3119(19)62783-8

Xin, S., Dong, H., Yang, L., Huang, D., Zheng, F., Cui, Y., Wu, S., Liao, J., He, Y., Wan, H., Liu, Z., Li, X., & Qian, W. (2021). Both overlapping and independent loci underlie seed number per pod and seed weight in Brassica napus by comparative quantitative trait loci analysis. *Molecular Breeding*, *41*(6), 41. Advance online publication. DOI: 10.1007/s11032-021-01232-1 PMID: 37309442

Yang, Y., Shen, Y., Li, S., Ge, X., & Li, Z. (2017). High Density Linkage Map Construction and QTL Detection for Three Silique-Related Traits in Orychophragmus violaceus Derived Brassica napus Population. *Frontiers in Plant Science*, *8*, 1512. Advance online publication. DOI: 10.3389/fpls.2017.01512 PMID: 28932230

Yao, S., Liang, F., Gill, R., Huang, J., Cheng, X., Liu, Y., Tong, C., & Liu, S. (2020). A global survey of the transcriptome of the allopolyploid Brassica napus based on single molecule long-read isoform sequencing and Illumina-based RNA-seq data.. *The Plant journal: for cell and molecular biology*. https://doi.org/.DOI: 10.1111/tpj.14754

Yu, B., Boyle, K., Zhang, W., Robinson, S. J., Higgins, E., Ehman, L., Relf-Eckstein, J., Rakow, G., Parkin, I. P., Sharpe, A. G., & Fobert, P. R. (2016). Multi-trait and multi-environment QTL analysis reveals the impact of seed colour on seed composition traits in Brassica napus. *Molecular Breeding*, *36*(8), 111. Advance online publication. DOI: 10.1007/s11032-016-0521-8

Yu, B., Boyle, K., Zhang, W., Robinson, S. J., Higgins, E., Ehman, L., Relf-Eckstein, J.-A., Rakow, G., Parkin, I. A. P., Sharpe, A. G., & Fobert, P. R. (2016). Multi-trait and multi-environment QTL analysis reveals the impact of seed colour on seed composition traits in Brassica napus. *Molecular Breeding*, *36*(8), 111. DOI: 10.1007/s11032-016-0521-8

Zhan, Z., Shah, N., Jia, R., Li, X., Zhang, C., & Piao, Z. (2022). Transferring of clubroot-resistant locus CRd from Chinese cabbage (Brassica rapa) to canola (Brassica napus) through interspecific hybridization. *Breeding Science*, *72*(3), 189–197. DOI: 10.1270/jsbbs.21052 PMID: 36408323

Zhang, G., Zhou, J., Peng, Y., Tan, Z., Zhang, Y., Zhao, H., Liu, D., Liu, X., Li, L., Yu, L., Jin, C., Fang, S., Shi, J., Geng, Z., Yang, S., Chen, G., Liu, K., Yang, Q., Feng, H., & Yang, W. (2021). High-throughput phenotyping-based QTL mapping reveals the genetic architecture of the salt stress tolerance of *Brassica napus*. *Authorea (Authorea)*. DOI: 10.22541/au.163256646.65671725/v1

Zhang, T., Vail, S., Duddu, H., Parkin, I., Guo, X., Johnson, E., & Shirtliffe, S. (2021). Phenotyping Flowering in Canola (Brassica napus L.) and Estimating Seed Yield Using an Unmanned Aerial Vehicle-Based Imagery. *Frontiers in Plant Science*, *12*, 686332. Advance online publication. DOI: 10.3389/fpls.2021.686332 PMID: 34220907

Zhang, W., Hu, D., Raman, R., Guo, S., Wei, Z., Shen, X., Meng, J., Raman, H., & Zou, J. (2017). Investigation of the Genetic Diversity and Quantitative Trait Loci Accounting for Important Agronomic and Seed Quality Traits in Brassica carinata. *Frontiers in Plant Science*, *8*, 615. Advance online publication. DOI: 10.3389/fpls.2017.00615 PMID: 28484482

Zhao, Q., Wu, J., Cai, G., Yang, Q., Shahid, M., Fan, C., Zhang, C., & Zhou, Y. (2019). A novel quantitative trait locus on chromosome A9 controlling oleic acid content in Brassica napus. *Plant Biotechnology Journal*, *17*(12), 2313–2324. DOI: 10.1111/pbi.13142 PMID: 31037811

Zhao, Q.-H., Wu, J., Cai, G., Yang, Q., Shahid, M., Fan, C., Zhang, C., & Zhou, Y. (2019). A novel quantitative trait locus on chromosome A9 controlling oleic acid content in Brassica napus. *Plant Biotechnology Journal*, *17*(12), 2313–2324. DOI: 10.1111/pbi.13142 PMID: 31037811

Zhao, W., Wang, X., Wang, H., Tian, J., Li, B., Chen, L., Chao, H., Long, Y., Xiang, J., Gan, J., Liang, W., & Li, M. (2016). Genome-Wide Identification of QTL for Seed Yield and Yield-Related Traits and Construction of a High-Density Consensus Map for QTL Comparison in Brassica napus. *Frontiers in Plant Science*, *7*. Advance online publication. DOI: 10.3389/fpls.2016.00017 PMID: 26858737

Zhu, J., Zhang, J., Jiang, M., Wang, W., Jiang, J., Li, Y., Yang, L., & Zhou, X. (2021). Development of genome-wide SSR markers in rapeseed by next generation sequencing. *Gene*, *798*, 145798. DOI: 10.1016/j.gene.2021.145798 PMID: 34175391

Zhu, Q., King, G., Liu, X., Shan, N., Borpatragohain, P., Baten, A., Wang, P., Luo, S., & Zhou, Q. (2019). Identification of SNP loci and candidate genes related to four important fatty acid composition in Brassica napus using genome wide association study. *PLoS One*, *14*(8), e0221578. Advance online publication. DOI: 10.1371/journal.pone.0221578 PMID: 31442274

Zou, J., Zhao, Y., Liu, P., Shi, L., Wang, X., Wang, M., Meng, J., & Reif, J. C. (2016). Seed Quality Traits Can Be Predicted with High Accuracy in Brassica napus Using Genomic Data. *PLoS One*, *11*(11), e0166624. DOI: 10.1371/journal.pone.0166624 PMID: 27880793

Zou, M., Shi, T., Wang, W., Ding, G., Xu, F., & Shi, L. (2021). Identification of QTLs for Seed Yield and Yield-Related Traits in Brassica Napus Grown with Contrasting Nitrogen Supplies. https://doi.org/DOI: 10.21203/rs.3.rs-674683/v1

Chapter 8

Chloroplast and Mitochondrial Genome, Small RNA, and DNA Methylation of Canola Plant

Sowmiya C. A.
https://orcid.org/0009-0000-5876-6090
The Indian Agriculture College, Radhapuram, India

Rachel Lissy Vargheese
ICAR-Sugarcane Breeding Institute, Coimbatore, India

V. Sumithra
ICAR-Sugarcane Breeding Institute, Coimbatore, India

Shobana Sampath
https://orcid.org/0000-0002-4291-3516
Vel Tech Rangarajan Dr. Sagunthala R&D Institute of Science and Technology, India

Shoaib Khan
Abbottabad University of Science and Technology, Pakistan

ABSTRACT

The canola plant (Brassica napus L.) is a major oilseed crop globally providing high-quality oil and protein feed for livestock. However, there is limited understanding of the genetic and molecular processes that govern the plant's growth, development

DOI: 10.4018/979-8-3693-8603-3.ch008

and response to stress. This project seeks to investigate the genetic makeup of canola with a specific focus on the chloroplast and mitochondrial genomes, as well as the regulation of gene expression through small RNA and DNA methylation. It is with this in mind that we strive to elucidate the complex interaction between these components to explain the underlying mechanisms that impinge on canola productivity and its adaptability. This research shall stand as one of the major steps toward generation of high-yielding stress-tolerant and climate-resilient canola varieties, thereby further supporting the enhancement of global food security as well as sustainable agriculture.

1. INTRODUCTION

Canola belonging to the family of Brassicaceae, thrives in temperate climates, serving food and industrial purposes (McVetty & Scarth, 2012). Its oil is used in cooking, food processing, and non-food applications such as biodiesel production, while its by-products, such as canola meal, are a valuable protein source for livestock (O'Brien, 2016). The crop is also used in crop rotation systems, contributing to soil health by breaking pest cycles and enhancing nutrient availability for subsequent crops.

B. napus (rapeseed), initially resulted from natural hybridization between the two diploid species - *Brassica rapa* (2n = 2x = 20) and *Brassica oleracea* (2n = 2x = 18) (Nagaharu & Nagaharu, 1935), later led to the development of canola by selective breeding. Globally, the largest producer of canola is Canada accounting for 23% of the global production followed by China and India with 18% and 14% respectively ("USDA Foreign Agricultural Service," 2024). In India, canola is usually referred to as rapeseed and its yield can significantly vary based on various factors such as region, soil type and quality, climate and agricultural practices. Traditional rapeseed, while useful in industrial applications, was unsuitable for human consumption due to high levels of erucic acid and glucosinolates, compounds linked to potential health risks. Prolonged consumption of higher levels of erucic acid leads to the accumulation of lipids in the heart muscle fibres eventually weakening the muscle function (Galanty et al., 2023). Understanding its classical genetics provides essential insights into the traits that contribute to its agricultural value, including yield, disease resistance, and adaptability.

Traditional breeding methods primarily involve selection and crossbreeding which generally span over several generations. Plants with desirable traits such as improved oil content, disease resistance and yield are identified and are exposed to controlled pollination to develop hybrids. This meticulous process often requires multiple generations (F_1, F_2, F_3, etc.) to stabilize desired traits before commercialization. Additionally, mutation breeding has been utilized to introduce genetic variability and accelerate the development of new varieties. On the whole, the primary focus for canola breeders is mainly on enhancing oil quality by reducing undesirable compounds such as erucic acid and glucosinolates while also improving agronomic characteristics such as pest resistance and stress tolerance. The evolution of canola breeding through classical genetics and traditional methods illustrates a symbiotic relationship between historical practices and cutting-edge technologies which will be discussed in this chapter.

2. HISTORICAL DEVELOPMENT OF CANOLA

The origin of B.*napus* (rapeseed) is uncertain however it is believed to have originated in the Mediterranean region (McNaughton, 1976). Canola is a genetic variation of rapeseed developed by Canadian plant breeders specifically for its nutritional qualities, particularly its low level of saturated fat. In 1956 the nutritional aspects of rapeseed oil were questioned, especially concerning the high eicosenoic, erucic acid and glucosinolates. This led to the development of canola which started in the 1960s heavily relying on its crop genetics to minimize undesirable traits while simultaneously improving the crop's agricultural performance. The name "canola" is a combination of "Canada" and "ola," meaning oil, a proclamation of its origins and purpose. Over the years, the crop has spread to other major agricultural regions, including the United States, Europe, Australia, China and India driven by its versatility, high oil content, and wide range of uses.

Since the domestication of *B.napus*, the primary objective earlier was to minimize the erucic acid content. Extensive breeding programs were initiated to develop low erucic acid rapeseed cultivars. A natural low erucic acid variant named Liho was discovered in 1964 (Stefansson et al., 1961) and Oro was the first rapeseed cultivar developed by using Liho as one of its parents (Downey, 1990). The development of "Oro" marked one of the early milestones in creating canola with more desirable agronomic traits and better oil quality and cultivars with reduced erucic acid with less than 2% were termed LEAR cultivars (Low Erucic Acid Rapeseed). Glucosinolates are natural compounds found in rapeseed (*Brassica napus*), playing a key role in the plant's defense mechanisms against pests and diseases. They are sulfur-containing secondary metabolites which produce bioactive compounds like isothiocyanates on hydrolysis possessing both harmful and beneficial effects. In rapeseed, glucosinolates impact seed quality, particularly affecting its use in animal feed due to their potential to reduce palatability and nutrient absorption. Breeding efforts often aim to reduce glucosinolate levels in rapeseed varieties to improve their suitability for both oil production and feed use. The first "low erucic" "low glucosinolate" cultivar Tower was developed in 1974 and it was commonly called as double zero

(00) cultivars with erucic acid content less than 2% and glucosinolate content less than 30mg per g of oil (Downey, 1990). Since then, several cultivars have been bred using double zero (00) as the base material and almost all the rapeseed cultivars for human consumption were replaced by canola varieties.

Though erucic acid is undesirable for consumption, it has higher potential for industrial applications in the manufacture lubricants, surfactants and other chemical products. Similar to LEAR, High Erucic Acid Rapeseed (HEAR) were bred specifically for industrial purposes (Downey, 1990). Reston, was the first registered HEAR cultivar and since then parallel breeding efforts with an aim to develop cultivars suitable for non-food industry were carried out.

3. GENETICS OF CANOLA

3.1 Genetic Structure

Brassica napus, an allotetraploid contains two sets of homologous chromosomes from two different diploid species. A genome arises from *B.oleracea* and C genome arises from *B.campestris*, undergoes chromosome doubling to give rise to the present day *B.napus* (Figure 1).

Key features of a polyploid genome:

- Greater variation
- Enhanced adaptability
- Gene redundancy leading to gene plasticity

This polyploidy nature generally contributes to a higher degree of homeologs or duplicate genes causing functionally similar genes or sometimes they diverge and carry out specialized roles through a process called sub functionalization where a copy of a gene entirely acquires a new function. Overall, the genetic structure of canola is shaped by its polyploid nature, hybrid origin, and the extensive breeding efforts that have optimized it for modern agriculture. Canola exhibits substantial genetic diversity, which is crucial for breeding programs. Understanding the genetic variation among canola cultivars helps in selecting parents for hybridization to achieve desirable traits (Moghaieb et al., 2014).

In addition to traditional genetic factors, epigenetics plays a significant role in the regulation of gene expression in canola. Epigenetic modifications, such as DNA methylation and histone modification affect the expression of genes without disturbing the primary DNA sequence. Important traits like the development of seed, flowering time and stress tolerances are greatly influenced by these modifications.

Methylation patterns throughout the genome of *Brassica napus* during drought stress were examined and observed changes in genes linked to hormone signalling pathways and stress tolerance (Dowen et al., 2012) (Jian-ping et al., 2020).

Histone modifications such as acetylation, methylation, and phosphorylation affect chromatin structure, thus regulating gene expression. These modifications are crucial for plant development, especially for flowering time regulation, which is important for optimizing yield in canola. Finnegan et al. (2005) highlighted the importance of histone modifications in regulating the expression of the Flowering Locus C (FLC) gene, which caused delayed flowering in canola. Histone methylation of the FLC gene locus was found to be linked to environmental factors such as temperature and day length. Apart from DNA methylation and histone modification, small RNAs (sRNAs), such as microRNAs (miRNAs) and siRNAs, played pivotal roles in regulating plant stress responses, growth and development and also in yield-related traits (Li et al., 2022; J. Wang et al., 2016).

Epigenetic changes are often reversible and can be influenced by environmental factors, making them a key mechanism through which canola adapts to different conditions. The study of epigenetics in canola is still an emerging field, but it has the potential to unlock new avenues for crop improvement. Advances in epigenome editing and CRISPR-based technologies hold promise for precisely targeting and modifying these epigenetic marks to improve canola's adaptability and productivity. Epigenetics along with traditional breeding and genetic engineering, provide a powerful tool for developing superior canola varieties as globally canola continues to be a crucial crop for crop production. Hence, understanding its genetic makeup will be essential for developing improved varieties that can meet the challenges of climate change and evolving agricultural practices. Though advances in genomic tools and molecular breeding have allowed for significant improvements in yield, disease resistance, and oil quality, further research is required to ensure that canola continues to meet the demands of a changing world which will be discussed in the coming chapters.

3.2 Inheritance Pattern In Canola

The inheritance pattern in canola generally follows polygenic inheritance Mendelian genetics but some important agronomically desirable genes are influenced by complex interactions too. Canola being a dual-purpose crop for both oil and cattle feed, understanding its inheritance is essential for any breeding program. The details of traits with their respective inheritance pattern are furnished in table no 1.

Table 1. List of traits with their inheritance pattern.

No.	Trait	Inheritance pattern	Reference
1	Oil content	Polygenic	Jiefu et al. (2010)
2	Erucic acid	Two major genes	Ze-su et al. (2012)
3	Oleic acid	Two major genes	Ze-su et al. (2012)
4	Glucosinolate	Three major genes	Ze-su et al. (2012)
5	White flower	Monogenic	Huang et al. (2014)
6	Blackleg resistance	Single gene (dominant) and polygenic	Yu et al. (2005)
7	Sclerotinia	Quantitative and complex	Saharan et al. (2021)
8	White rust	One or two dominant gene	Saharan et al. (2021)
9	Clubroot	One or two dominant gene	Lammerink (1967) and Saharan et al. (2021)
10	Herbicide resistance	Single gene (dominant)	Huiming et al. (2003)

4. TRADITIONAL BREEDING IN CANOLA IMPROVEMENT

4.1 Breeding Strategies Used in Canola

Traditional breeding in canola (*Brassica napus*) involves the use of classical methods to improve the crop for desirable traits such as yield, oil content, disease resistance, and quality. Most of the methods such as selection, backcross breeding, and pedigree breeding rely on the natural genetic variation within canola and closely related species, and they have been crucial in the development of canola as a major oilseed crop. Breeders focus on developing high-yielding varieties by selecting traits that directly or indirectly affect yield, such as seed number, seed size, oil content, disease resistance, and adaptability to stress conditions (e.g., drought and heat).

4.2 Breeding for Agronomic Traits

Breeding for agronomic traits in canola (*Brassica napus*) is essential to improve characteristics such as yield, flowering time, and seed quality. Breeding programs focus on enhancing these traits through both traditional and modern techniques, ensuring that the crop meets the demands of different growing environments and markets. With the rising global demand for edible oils and biofuels, improving seed yield in *Brassica* crops is crucial for meeting these needs. Canola, a member of the *Brassica* genus, is one of the world's most important oilseed crops. Its seeds are processed to produce vegetable oil, which is used in food, industrial products, and biofuels. High seed yield directly increases the oil yield, making it a significant economic factor for farmers and industry stakeholders (Abu & Mengistu, 2023).

Adjusting flowering time and maturity is equally crucial for adapting canola to different environments and ensuring optimal yield. In some cases, early flowering varieties may escape drought, while later varieties may benefit from extended growing seasons. A more serious problem is pod shattering during winter which can cause significant yield losses. Breeding for shatter resistance helps to reduce seed loss and improves overall productivity. Selection for shatter resistance is done by identifying varieties with stronger pod walls or delayed pod dehiscence (Hossain et al., 2012). Mutagenesis has also been explored to modify genes controlling pod shattering in canola. Lodging is a common problem in canola, especially under high-yielding conditions. Breeding for lodging resistance helps maintain plant standability and facilitates mechanized harvesting where plants with stronger stems with reduced height were selected by convention method and mutagenesis to develop lodging-resistant lines (Berry et al., 2004).

Breeding for agronomic traits in canola was based on traditional methods. Economic and agronomic success in canola can be ultimately achieved by combining it with modern technologies as canola is a rising globally significant crop.

4.3 Breeding for Quality Traits

Breeding for oil content and oil quality in canola (*Brassica napus*) is a significant focus in plant breeding, driven by the growing demand for healthier and more sustainable vegetable oils. Canola oil is valued for its low levels of saturated fats and high levels of polyunsaturated fatty acids, particularly oleic acid, and its use in both the food and biofuel industries. Breeders aim to maximize oil yield without compromising the agronomic traits like seed yield and disease resistance. The focus is on increasing oleic acid content, which is healthier for consumption while reducing erucic acid and saturated fats. The relationship between erucic acid and oleic acid in *Brassica napus* is inversely correlated due to their biosynthetic pathway. By manipulating the genes responsible for fatty acid elongation (such as FAE1), breeders have been able to develop canola varieties with high oleic acid and low erucic acid content, ensuring that canola oil is safe for consumption and possesses desirable nutritional qualities (P. Wang et al., 2022). Simultaneously reducing linolenic acid improves oil stability, which is important for shelf life and cooking (Syed, 2016). There is also growing interest in altering the fatty acid profile to produce lauric and capric acid, useful in industrial applications like biofuel production. Canola breeding initially relied on traditional crossbreeding methods, selecting favourable traits over several generations. Early successes in reducing erucic acid levels were achieved by targeted mutagenesis leading to the development of first low erucic acid cultivar "ORO" (Fukai et al., 2019).

Traditionally, breeders identify and select canola plants with naturally high oil content from existing germplasm. These selected plants serve as parent lines for breeding. Breeders cross these high-oil-content plants with other elite canola lines. Offspring are evaluated for oil content and other desirable traits like disease resistance or yield. The development of early canola varieties such as 'Tower' and 'Candle' involved crossbreeding with traditional European rapeseed cultivars, which had high oil content but also high erucic acid levels. Through successive breeding and selection, low-erucic acid and high-oil-content canola varieties were developed, marking the birth of canola as a globally important oilseed crop (Iniguez-Luy & Federico, 2011). Early Polish canola breeding programs applied mass selection to improve oil content. In mass selection, large populations of canola are grown, and plants with desirable characteristics, such as high oil content, are visually selected. Seeds from these plants are harvested, and subsequent generations are grown and reselected to increase the frequency of high-oil-content plants in the population. They focused on selecting plants that produced large, oil-rich seeds. Over several generations, the average oil content of the population increased (Abbadi & Leckband, 2011).

A more promising method which led to the development of zero-erucic acid canola involved backcrossing high-oil-content donor lines with elite canola varieties. Backcrossing is used to transfer specific traits, such as high oil content, from one variety to another. Breeders cross a high-oil-content donor parent with an elite canola variety that possesses other desirable traits (e.g., disease resistance, and high yield). The progeny are repeatedly crossed back to the elite parent to retain most of the elite parent's traits while incorporating the high-oil trait. For example, Canadian breeders successfully backcrossed the 'Span' and 'Zephyr' lines to develop zero-erucic acid cultivars with high oil content (Rakow & Woods, 1987).

Mutagenesis involves exposing canola seeds or tissue to chemicals (like ethyl methanesulfonate, EMS) or radiation to induce mutations. Canadian breeders used chemical mutagenesis to develop canola varieties with significantly lower levels of erucic acid. These mutations resulted in beneficial changes, including higher oil content. Mutant plants with favourable traits were then selected and used in the breeding program. Although the primary goal was to reduce erucic acid, the process also improved oil content in some lines. Varieties like 'Tobin' and 'Midas' emerged from these efforts, showing both lower erucic acid and improved oil yield.

Hybrid development in canola is designed to develop high-yielding varieties combined with desirable agronomic traits. This greatly relies on the genetic diversity available and uses specialized techniques like male sterility to ensure effective hybridization. Male sterility is extensively used in the production of hybrid canola seeds. Male sterile lines are crossed with fertile restorer lines to produce hybrid seeds, which often exhibit heterosis (hybrid vigour), resulting in higher yields and better stress tolerance (Toriyama, 2021). In *Brassica napus*, cytoplasmic male sterility is more evident and most widely used in hybrid seed production. CMS lines are a type of male sterile lines which are governed by cytoplasmic genes especially, mitochondrial genes which leads to the development of non-viable pollen grains. Ogura, Pol, Nap, NSA and Inap are some of the widely used CMS lines in canola (Kang et al., 2017)

Traditional breeding methods like selection, crossbreeding, mass selection, backcrossing, and mutagenesis have significantly contributed to improving oil content in canola. These techniques remain important even in the era of molecular breeding, as they provide the foundation for the development of high-oil-content varieties.

4.4 Breeding for Biotic Stress Tolerance

Breeding for disease and pest resistance in canola (*Brassica napus*) is critical to improving crop yield and quality while reducing reliance on chemical pesticides. Canola faces several significant diseases and pests, predominant ones are blackleg (Y. Wang et al., 2020), clubroot (Rahman et al., 2014), aphids (Akbar et al., 2017), and cabbage seedpod weevil (Merkosky et al., 2024). Traditional breeding methods for disease and pest resistance in canola involve selecting and crossbreeding plants with natural resistance to biotic stresses such as diseases and insect pests. Large collections of canola germplasm were usually screened for resistance to specific diseases or pests (Al-Lami et al., 2023). Selected plants were exposed to the pathogen or pest, and those that show natural resistance are selected for further breeding. Disease or pest resistance was tested in both field trials and controlled environments (e.g., greenhouses or growth chambers). This ensured that the selected plants performed well under natural and controlled conditions. Once resistant plants were identified, they are crossed with high-yielding, agronomically desirable varieties. The primary goal was to combine the resistance traits of one parent with the yield and quality traits of the other. Breeders select the best offspring from these crosses over multiple generations (F_1, F_2, etc.) to maintain disease or pest resistance while enhancing other key traits such as yield, oil content, and environmental adaptability. In order to improve an existing variety, a resistant plant (donor) is crossed with an elite canola variety (recurrent parent) that lacks resistance but has other desirable traits. The resulting offspring are then repeatedly crossed back to the recurrent parent, preserving most of its traits while introducing resistance from the donor. This method helps to retain most of the high-yielding and quality traits of the recurrent parent while adding the desired resistance traits from the donor parent. A more straightforward, low-cost approach is to select plants that show resistance and other desirable traits among large populations of plants grown in fields with natural pest or disease pressure. Plants that show resistance and other desirable traits are selected for further breeding, and the process is repeated over several generations to improve overall population resistance.

Improved resistance to diseases and pests along with higher yields and better agronomic performance was generally achieved by crossing two genetically distinct inbred lines which exhibit hybrid vigour or "heterosis." Breeders can use restorer lines with known disease resistance to develop hybrids with improved pest and disease tolerance. Induced mutagenesis using radiation or chemicals brings forth new genetic variation in canola plants. Some of these mutations may confer resistance to diseases or pests. Mutants with desirable traits, such as disease resistance, are selected and used in further breeding programs.

Wild relatives of canola, such as *Brassica rapa* or *Brassica oleracea*, often have natural resistance to pests or diseases (Mei et al., 2015). These traits could be introduced into canola through crossbreeding. This process can be slow and difficult, as crossing canola with wild species may lead to undesirable traits (e.g., reduced yield or oil quality), which need to be bred out over successive generations. Breeding for disease and pest resistance in canola is essential for sustainable crop production. By employing various breeding techniques, breeders can enhance resistance to critical diseases and pests. This not only helps improve yield and quality but also reduces the need for chemical controls, supporting more environmentally friendly agricultural practices.

4.5 Breeding for Abiotic Stress

Breeding for abiotic stress tolerance in canola (*Brassica napus*) is crucial to ensuring crop productivity under varying environmental conditions. Abiotic stresses, such as drought, salinity, heat, and frost, can significantly impact canola yield and quality. Selection from germplasms, pedigree breeding, and backcross breeding are some of the methods employed to develop tolerant genotypes. Drought tolerance in canola is achieved through selecting traits like deep root systems, improved water use efficiency, and enhanced physiological traits that enable plants to maintain growth under water-limited conditions (Wu et al., 2018). Salinity stress negatively affects canola growth and yield. Breeding programs prioritized identifying and selecting genotypes that exhibit greater salinity tolerance through physiological mechanisms such as ion exclusion, osmotic adjustment, and improved antioxidant capacity. Screening methods, including hydroponic and field trials, help identify resistant varieties (Bandehagh et al., 2021). Canola being a temperate crop, heat stress during flowering can adversely affect seed set and yield. Breeding for heat tolerance involves selecting for traits that allow plants to maintain reproductive success under high temperatures (Secchi et al., 2023). Frost damage can occur during flowering and pod filling, significantly impacting yield. Breeding for frost tolerance includes selection for physiological and biochemical traits that enhance cold tolerance. Developing early-maturing varieties can help mitigate frost damage by avoiding late-season freezing (Raman et al., 2019). Focusing on traits associated with drought, salinity, heat, frost, and nutrient stress tolerance, breeders can develop resilient canola varieties that can thrive under changing environmental conditions, ultimately enhancing food security and sustainable agricultural practices.

5. FUTURE OF CANOLA BREEDING

A blend of traditional breeding methods integrated with modern technologies (Figure 2) is required to meet the increasing global demand for canola. Breeders need to focus more on the development of cultivars meeting both market demands and environmental challenges, giving importance to sustainable agriculture, climate resilience and consumer preferences. As research continues to evolve, the collaboration between scientists, farmers, and industry stakeholders will be essential in shaping the next generation of canola crops.

6. CHALLENGES IN CANOLA BREEDING

Canola breeding, like the breeding of any other crop, faces several challenges, both technical and environmental. Modern canola cultivars have a limited gene pool due to repeated selection for desirable traits, such as low levels of erucic acid and glucosinolates. This narrow genetic diversity limits the potential for further improvement and makes the crop more vulnerable to diseases and environmental stresses. This limits the plant's ability to respond to new challenges like climate change or evolving pathogens by leading to the breakdown of resistance to serious pests and diseases (Cowling, 2007). Achieving significant increases in canola yield potential is challenging due to the complexity of yield-related traits, which are influenced by multiple factors including plant architecture, photosynthetic efficiency, and seed setting. While yield increases have been achieved through conventional breeding, the gains have been relatively modest. New breeding techniques like genomic selection may offer more promise for substantial yield improvements (Ahmad et al., 2022, Ahmad et al., 2024, Ullah et al., 2023). Regarding quality traits, high-oleic canola lines have been developed, but balancing improved oil profiles with competitive yield is a breeding challenge due to negative correlations between quality and biometrical traits (McVetty & Scarth, 2002). Hybrid vigour, or heterosis, is a critical factor in boosting canola yields. However, developing stable and cost-effective hybrid breeding systems is complex and labour-intensive. Hybrid seed production systems depend on effective male sterility and fertility restoration mechanisms, which can be difficult to manage. Hybrid canola varieties have been widely adopted, but maintaining hybrid purity and seed production efficiency remains challenging (Ton et al., 2020).

Changing climate conditions, such as erratic rainfall patterns and increasing temperatures, introduce new challenges for canola breeding. Breeders need to develop varieties that are resilient to fluctuating weather patterns, but breeding for climate resilience is a time-consuming and resource-intensive process. While modern technologies such as genome editing (CRISPR-Cas9), marker-assisted selection (MAS), and genomic selection offer promise, integrating these tools into breeding programs requires substantial investment in infrastructure, knowledge, and regulatory approvals. The use of CRISPR in canola breeding is still in the experimental stages, but early trials have shown potential for targeted improvements in disease resistance and oil quality. Canola breeding faces multifaceted challenges ranging from narrow genetic diversity to the integration of advanced breeding technologies. Addressing these challenges requires continuous research, technological innovation, and collaboration between breeders, geneticists, and agronomists (Lohani et al., 2020, Ijaz et al., 2023, Ijaz et al., 2024).

7. CONCLUSION

The significant achievements visualized in canola today are the result of traditional breeding methods. Efforts were made earlier on simpler traits like erucic acid and glucosinolate content, which had significant impacts on canola's suitability as an edible oil crop. Through these traditional breeding approaches, canola evolved from a minor oilseed into a major global crop with vast economic importance. Cytoplasmic male sterility (CMS) and the exploitation of heterosis through hybrid breeding further accelerated the development of superior canola varieties, paving the way for higher yields and better stress tolerance. The development of early-maturing varieties, resistance to lodging, and adaptability to diverse climates are other achievements rooted in classical breeding practices. As canola continues to face new challenges related to climate change, disease pressures, and evolving market demands, the foundation of classical genetics will remain invaluable. Future breeding programs will increasingly rely on the rich genetic base developed through traditional methods while leveraging cutting-edge tools to meet the needs of the agricultural industry and global food security. Classical genetics and traditional breeding in canola have not only shaped the crop's past but will also continue to influence its future. Moreover, the synergy between conventional and modern breeding methods promises to ensure the ongoing success of canola as a sustainable, high-yielding, and economically important oilseed crop.

REFERENCES

Abbadi, A., & Leckband, G. (2011). Rapeseed breeding for oil content, quality, and sustainability. *European Journal of Lipid Science and Technology*, *113*(10), 1198–1206. DOI: 10.1002/ejlt.201100063

Abu, M., & Mengistu, B. (2023). *Correlation Coefficient and Path Coefficient Analysis for Seed Yield and Yield Related Traits of Rapeseed (Brassica Napus).* Genotypes.

Ahmad, Y., Haider, S., Iqbal, J., Abbasi, B. A., Yaseen, T., & Mahmood, T. (2022). The mechanisms of genome editing technologies in crop plants. In *Principles and Practices of OMICS and Genome Editing for Crop Improvement* (pp. 295–313). Springer International Publishing. DOI: 10.1007/978-3-030-96925-7_13

Ahmad, Y., Haider, S., Iqbal, J., Naseer, S., Attia, K. A., Mohammed, A. A., & Mahmood, T. (2024). In-silico analysis and transformation of OsMYB48 transcription factor driven by CaMV35S promoter in model plant–Nicotiana tabacum L. conferring abiotic stress tolerance. *GM Crops and Food: Biotechnology in Agriculture and the Food Chain*, *15*(1), 130–149. DOI: 10.1080/21645698.2024.2334476 PMID: 38551174

Akbar, W., Asif, M. U., Ismail, M., Bux, M., & Memon, R. M. (2017). Management of aphids on canola (*Brassica napus* L) through cultural practices. *Pakistan Entomologist*, *39*(2).

Al-Lami, H. F. D., You, M. P., Banga, S. S., & Barbetti, M. J. (2023). Novel resistances provide new avenues to manage alternaria leaf spot (*Alternaria brassicae*) in canola (*Brassica napus*), mustard (B. *juncea*), and other brassicaceae crops. *Plant Disease*, *107*(2), 372–381. DOI: 10.1094/PDIS-05-22-1153-RE PMID: 35802017

Bandehagh, A., Dehghanian, Z., Henry, R., & Hossain, M. A. (2021). *Salinity tolerance in canola: Insights from proteomic studies Brassica Breeding and Biotechnology*. IntechOpen., DOI: 10.5772/intechopen.96649

Berry, P. M., Sterling, M., Spink, J. H., Baker, C. J., Sylvester-Bradley, R., Mooney, S. J., Tams, A. R., & Ennos, A. R. (2004). Understanding and reducing lodging in cereals. *Advances in Agronomy*, *84*(04), 215–269. DOI: 10.1016/S1006-8104(13)60030-6

Cowling, W. A. (2007). Genetic diversity in Australian canola and implications for crop breeding for changing future environments. *Field Crops Research*, *104*(1-3), 103–111. DOI: 10.1016/j.fcr.2006.12.014

Dowen, R. H., Pelizzola, M., Schmitz, R. J., Lister, R., Dowen, J. M., Nery, J. R., Dixon, J. E., & Ecker, J. R. (2012). Widespread dynamic DNA methylation in response to biotic stress. *Proceedings of the National Academy of Sciences of the United States of America, 109*(32), E2183–E2191. DOI: 10.1073/pnas.1209329109 PMID: 22733782

Downey, R. (1990). Brassica oilseed breeding-achievements and opportunities.

Finnegan, E. J., Kovac, K. A., Jaligot, E., Sheldon, C. C., James, P. W., & Dennis, E. S. (2005). The downregulation of Flowering Locus C (FLC) expression in plants with low levels of DNA methylation and by vernalization occurs by distinct mechanisms. *The Plant Journal, 44*(3), 420–432. DOI: 10.1111/j.1365-313X.2005.02541.x PMID: 16236152

Fukai, E., Karim, M. M., Shea, D. J., Tonu, N. N., Falk, K. C., Funaki, T., & Okazaki, K. (2019). An LTR retrotransposon insertion was the cause of world's first low erucic acid *Brassica rapa* oilseed cultivar. *Molecular Breeding, 39*(1), 1–12. DOI: 10.1007/s11032-018-0916-9

Galanty, A., Grudzińska, M., Paździora, W., & Paśko, P. (2023). Erucic acid—both sides of the story: A concise review on its beneficial and toxic properties. *Molecules (Basel, Switzerland), 28*(4), 1924. DOI: 10.3390/molecules28041924 PMID: 36838911

Hossain, S., Kadkol, G., Raman, R., Salisbury, P., & Raman, H. (2012). Breeding *Brassica napus* for shatter resistance. *Plant Breeding, 14*(1), 313–332.

Huang, Z., Ban, Y. Y., Bao, R., Zhang, X. X., Xu, A. X., & Ding, J. (2014). Inheritance and gene mapping of the white flower in *Brassica napus* L. *New Zealand Journal of Crop and Horticultural Science, 42*(2), 111–117. DOI: 10.1080/01140671.2013.863211

Huiming, P., Jiefu, Z., Cunkou, Q., Jianqin, G., & Xinjun, C. (2003). Inheritance of herbicide resistance in *Brassica napus* and its utilization. *Jiangsu Journal of Agricultural Sciences, 19*(2), 81–86. DOI: 10.3390/ijms21041345

Ijaz, S., Iqbal, J., Abbasi, B. A., Tufail, A., Ullah, Z., Sharifi-Rad, J., & Uddin, S. (2023). Biofortification: Lessons from the Past and Strategies for Future Food Security. In *Legumes Biofortification* (pp. 521–545). Springer International Publishing. DOI: 10.1007/978-3-031-33957-8_23

Ijaz, S., Iqbal, J., Abbasi, B. A., Yaseen, T., Rehman, S., Kazi, M., & Mahmood, T. (2024). Role of OMICS-Based Technologies in Plant Sciences. OMICs-based Techniques for Global Food Security, 45-66.

Iniguez-Luy, F. L., & Federico, M. L. (2011). The genetics of *Brassica napus*. *Genetics and Genomics of the Brassicaceae*, 291-322. DOI: 10.1007/978-1-4419-7118-0_10

Jian-ping, W., Shu-fen, Z., Jia-cheng, Z., Jun-ping, H., Jin-hua, C., Yan-cheng, W., Kun, H., Lei, Z., Dong-guo, W., & Dong-fang, C. (2020). Genome-wide analysis of DNA methylation under drought in rapeseed (*Brassica napus* L.). *Zhongguo Youliao Zuowu Xuebao*, *42*(4).

Jiefu, Z., Cunkou, Q., Huiming, P., Song, C., Feng, C., Jianqin, G., Xinjun, C., Hui, G., & Shouzhong, F. (2010). Inheritance and QTL identification of oil content in rapeseed *Brassica napus* L.

Kang, L., Li, P., Wang, A., Ge, X., & Li, Z. (2017). A novel cytoplasmic male sterility in *Brassica napus* (inap CMS) with carpelloid stamens via protoplast fusion with Chinese woad. *Frontiers in Plant Science*, *8*, 529. DOI: 10.3389/fpls.2017.00529 PMID: 28428799

Lammerink, J. (1967). The inheritance of clubroot resistance in *Brassica napus* L. *New Zealand Journal of Agricultural Research*, *10*(1), 109–115. DOI: 10.1080/00288233.1967.10423081

Li, J., Li, Y., Wang, R., Fu, J., Zhou, X., Fang, Y., Wang, Y., & Liu, Y. (2022). Multiple Functions of MiRNAs in *Brassica napus* L. *Life (Chicago, Ill.)*, *12*(11), 1811. DOI: 10.3390/life12111811 PMID: 36362967

Lohani, N., Jain, D., Singh, M. B., & Bhalla, P. L. (2020). Engineering multiple abiotic stress tolerance in canola, *Brassica napus*. *Frontiers in Plant Science*, *11*, 3. DOI: 10.3389/fpls.2020.00003 PMID: 32161602

McNaughton, H. I. (1976). *Turpic and relatives* (Simmonds, N. W., Ed.). Longman.

McVetty, B. E., & Scarth, R. (2012). Breeding oil quality for edible and industrial applications in oilseed rape and linseed. *Technological Innovations in Major World Oil Crops, Volume 2. Perspectives*, •••, 105–122.

McVetty, P. B. E., & Scarth, R. (2002). Breeding for improved oil quality in Brassica oilseed species. *Journal of Crop Production*, *5*(1-2), 345–369. DOI: 10.1300/J144v05n01_14

Mei, J., Liu, Y., Wei, D., Wittkop, B., Ding, Y., Li, Q., Li, J., Wan, H., Li, Z., Ge, X., Frauen, M., Snowdon, R. J., Qian, W., & Friedt, W. (2015). Transfer of sclerotinia resistance from wild relative of *Brassica oleracea* into *Brassica napus* using a hexaploidy step. *Theoretical and Applied Genetics*, *128*(4), 639–644. DOI: 10.1007/s00122-015-2459-3 PMID: 25628163

Merkosky, K., Leibovitz, A., & Wong, S. (2024). Cabbage Seedpod Weevil (*Ceutorhynchus obstrictus*) Management in Canola (*Brassica napus*) across western Canada. *Plant Health Cases*(2024). DOI: 10.1079/planthealthcases.2024.0014

Moghaieb, R. E. A., Mohammed, E. H. K., & Youssief, S. S. (2014). Genetic diversity among some canola cultivars as revealed by RAPD, SSR and AFLP analyses. *3 Biotech, 4*, 403-410.

Nagaharu, U., & Nagaharu, N. (1935). Genome analysis in Brassica with special reference to the experimental formation of B. napus and peculiar mode of fertilization. *Japanese Journal of Botany*, *7*(7), 389–452.

O'Brien, D. (2016). Canola: Good protein source for dairy cattle. *Agricultural Research*, *64*(2), 1.

Rahman, H., Peng, G., Yu, F., Falk, K. C., Kulkarni, M., & Selvaraj, G. (2014). Genetics and breeding for clubroot resistance in Canadian spring canola (*Brassica napus* L.). *Canadian Journal of Plant Pathology, 36*(sup1), 122-134. DOI: 10.1080/07060661.2013.862571

Rakow, G., & Woods, D. L. (1987). Outcrossing in rape and mustard under Saskatchewan prairie conditions. *Canadian Journal of Plant Science*, *67*(1), 147–151. DOI: 10.4141/cjps87-017

Raman, H., Uppal, R. K., & Raman, R. (2019). Genetic solutions to improve resilience of canola to climate change. *Genomic designing of climate-smart oilseed crops*, 75-131. DOI: 10.1007/978-3-319-93536-2_2

Saharan, G. S., Mehta, N. K., Meena, P. D., Saharan, G. S., Mehta, N. K., & Meena, P. D. (2021). Inheritance of Disease Resistance. *Genomics of Crucifer's Host-Resistance*, 195-264.

Secchi, M. A., Fernandez, J. A., Stamm, M. J., Durrett, T., Prasad, P. V. V., Messina, C. D., & Ciampitti, I. A. (2023). Effects of heat and drought on canola (*Brassica napus* L.) yield, oil, and protein: A meta-analysis. *Field Crops Research*, *293*, 108848. DOI: 10.1016/j.fcr.2023.108848

Stefansson, B. R., Hougen, F. W., & Downey, R. K. (1961). Note on the isolation of rape plants with seed oil free from erucic acid. *Canadian Journal of Plant Science*, *41*(1), 218–219. DOI: 10.4141/cjps61-028

Syed, A. (2016). *Oxidative stability and shelf life of vegetable oils Oxidative stability and shelf life of foods containing oils and fats*. Elsevier., DOI: 10.1016/B978-1-63067-056-6.00004-5

Ton, L. B., Neik, T. X., & Batley, J. (2020). The use of genetic and gene technologies in shaping modern rapeseed cultivars (*Brassica napus* L.). *Genes, 11*(10), 1161. DOI: 10.3390/genes11101161 PMID: 33008008

Toriyama, K. (2021). Molecular basis of cytoplasmic male sterility and fertility restoration in rice. *Plant Biotechnology (Sheffield, England), 38*(3), 285–295. DOI: 10.5511/plantbiotechnology.21.0607a PMID: 34782814

Ullah, Z., Iqbal, J., Abbasi, B. A., Ijaz, S., Sharifi-Rad, J., Yaseen, T., & Mahmood, T. (2023). How Genome Editing Can Be Helpful in the Biofortification of Legumes. In *Legumes Biofortification* (pp. 207–232). Springer International Publishing. DOI: 10.1007/978-3-031-33957-8_8

USDA Foreign Agricultural Service. (2024). *Rapeseed Explorer - World Production.* Retrieved from https://ipad.fas.usda.gov/cropexplorer/cropview/commodityView.aspx?cropid=2226000

Wang, J., Jian, H., Wang, T., Wei, L., Li, J., Li, C., & Liu, L. (2016). Identification of microRNAs actively involved in fatty acid biosynthesis in developing *Brassica napus* seeds using high-throughput sequencing. *Frontiers in Plant Science, 7*, 1570. DOI: 10.3389/fpls.2016.01570 PMID: 27822220

Wang, P., Xiong, X., Zhang, X., Wu, G., & Liu, F. (2022). A review of erucic acid production in Brassicaceae oilseeds: Progress and prospects for the genetic engineering of high and low-erucic acid rapeseeds (*Brassica napus*). *Frontiers in Plant Science, 13*, 899076. DOI: 10.3389/fpls.2022.899076 PMID: 35645989

Wang, Y., Strelkov, S. E., & Hwang, S. F. (2020). Yield losses in canola in response to blackleg disease. *Canadian Journal of Plant Science, 100*(5), 488–494. DOI: 10.1139/cjps-2019-0259

Wu, W., Ma, B. L., & Whalen, J. K. (2018). Enhancing rapeseed tolerance to heat and drought stresses in a changing climate: Perspectives for stress adaptation from root system architecture. *Advances in Agronomy, 151*, 87–157. DOI: 10.1016/bs.agron.2018.05.002

Yu, F., Lydiate, D. J., & Rimmer, S. R. (2005). Identification of two novel genes for blackleg resistance in *Brassica napus. Theoretical and Applied Genetics, 110*(5), 969–979. DOI: 10.1007/s00122-004-1919-y PMID: 15798929

Ze-su, H., Paisan, L., Thitiporn, M., Ze-hui, C., Wen-dong, D., Rong, T., & Dezhen, L. (2012). Inheritance of erucic acid, glucosinolate, and oleic acid contents in rapeseed (*Brassica napus* L.). [English Edition]. *The Journal of Northeast Agricultural University, 19*(2), 1–8. DOI: 10.1016/S1006-8104(13)60030-6

APPENDIX: FIGURES

Figure 1. Evolution of Brassica napus

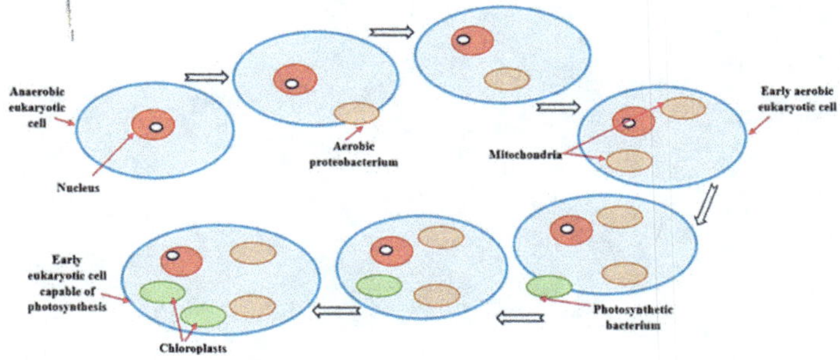

Figure 2. Modern tools for next breakthrough in Brassica napus

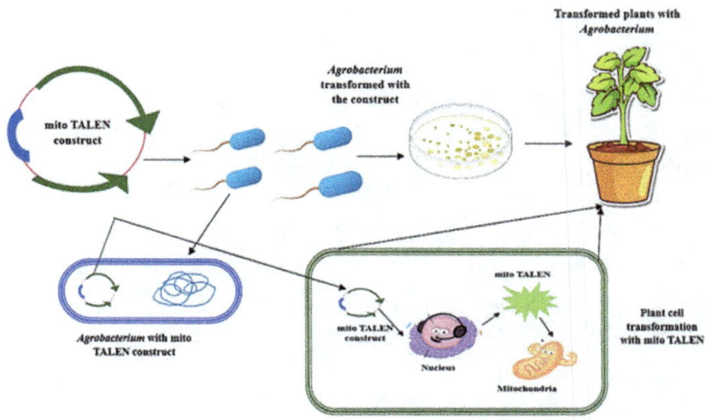

Chapter 9
Canola Plant Genetic Transformation and Gene Function Research Techniques

P. Selvakumar
https://orcid.org/0000-0002-3650-4548
Department of Science and Humanities, Nehru Institute of Technology, India

Manjunath T. C.
https://orcid.org/0000-0003-2545-9160
Rajarajeswari College of Engineering, India

ABSTRACT

Genetic transformation in canola represents a transformative approach in modern agriculture, leveraging principles of molecular biology to introduce and express new traits that enhance crop performance and functionality. Another widely used method is Particle Bombardment or Biolistics, which involves physically delivering the gene of interest into plant cells using microscopic metal particles coated with DNA. These particles are accelerated using a gene gun and penetrate the plant cell walls, introducing the genetic material into the cell. Although this method is less targeted than Agrobacterium-mediated transformation, it is valuable for species that are difficult to transform using other techniques.Principles of Selection and ScreeningAre crucial to identifying successfully transformed plants. After introducing the gene of interest, only a small fraction of the plant cells will integrate the new DNA. To isolate these cells, researchers use selectable markers—genes that confer resistance to certain herbicides or antibiotics.

DOI: 10.4018/979-8-3693-8603-3.ch009

INTRODUCTION TO GENETIC TRANSFORMATION IN CANOLA

Genetic transformation in canola represents a transformative approach in modern agriculture, leveraging principles of molecular biology to introduce and express new traits that enhance crop performance and functionality. Another widely used method is Particle Bombardment or Biolistics, which involves physically delivering the gene of interest into plant cells using microscopic metal particles coated with DNA. These particles are accelerated using a gene gun and penetrate the plant cell walls, introducing the genetic material into the cell. Although this method is less targeted than Agrobacterium-mediated transformation, it is valuable for species that are difficult to transform using other techniques.Principles of Selection and ScreeningAre crucial to identifying successfully transformed plants. After introducing the gene of interest, only a small fraction of the plant cells will integrate the new DNA. To isolate these cells, researchers use selectable markers—genes that confer resistance to certain herbicides or antibiotics. Plants that survive the selection process are likely to have incorporated the transgene. Following selection, these plants are regenerated and evaluated for the expression of the new trait and stability of the genetic modification.Disease Resistance is another critical application. Genetic transformation can introduce genes that confer resistance to common canola diseases, such as blackleg or sclerotinia stem rot. By enhancing disease resistance, transformed canola varieties reduce the need for chemical pesticides, decrease crop losses, and improve overall crop health and yield.

NUTRITIONAL ENHANCEMENT

Through genetic transformation aims to improve the nutritional profile of canola products. For instance, researchers are exploring the introduction of genes that increase the levels of beneficial fatty acids, such as omega-3 and omega-6, or enhance the oil's antioxidant content. These modifications can produce canola oils with superior health benefits, catering to consumer demands for healthier food options. Biopesticide Productionis an innovative application of genetic transformation in canola(Bonnardeaux J., 2007). By incorporating genes that produce natural pest repellents or insecticidal proteins, researchers can develop canola varieties that offer built-in pest protection. This reduces the need for chemical pesticides and supports environmentally friendly pest management practices.Regulatory and Safety ConsiderationsAre integral to the deployment of genetically transformed canola. Regulatory agencies evaluate the safety and environmental impact of transgenic crops through rigorous assessments. This includes evaluating potential risks to human health, non-target organisms, and ecosystems. Ensuring that genetic modifications are safe and beneficial requires comprehensive testing and adherence to regulatory guidelines.In summary, genetic transformation in canola represents a powerful tool for advancing agricultural practices and addressing global challenges. By introducing and optimizing specific traits, this technology enhances canola's performance, resilience, and nutritional value.

AGROBACTERIUM-MEDIATED TRANSFORMATION

Agrobacterium-mediated transformation has emerged as a pivotal tool in the genetic engineering of canola, leveraging the natural abilities of the bacterium Agrobacterium tumefaciens to introduce new genetic material into plant genomes. This method harnesses the bacterium's inherent capacity to transfer DNA into plant cells, facilitating targeted modifications that enhance canola's agronomic traits, stress tolerance, and nutritional quality(Chalhoub B., et al., 2014).This vector is engineered to include not only the gene of interest but also selectable marker genes, which are crucial for identifying successfully transformed cells. The plasmid vector is then introduced into Agrobacterium tumefaciens through a process known as transformation, resulting in a recombinant bacterium that carries the desired genetic material. In the subsequent stage is used to infect plant tissues. This is typically done using explants such as leaf disks, cotyledons, or hypocotyl segments. The bacterium attaches to the plant tissue and transfers the genetic material into the plant cells via its natural DNA transfer mechanism. Agrobacterium utilizes a specialized plasmid, known as the Ti (tumor-inducing) plasmid, which integrates the foreign DNA into the plant genome. The integration occurs through a process that closely resembles natural plant infection, where the DNA segment, or T-DNA (transfer DNA), is incorporated into the plant's chromosomal DNA.The selection process involves growing the infected tissues on selective media containing the appropriate antibiotics or herbicides. Only those cells that survive the selection pressure are likely to contain the desired transgene. Regenerating whole plantsThe next step involves from the transformed tissues. This is achieved through tissue culture(Eskin N.A.M., et al., 2003).techniques that induce the formation of shoots and roots from the selected cells. The regenerated plants are then grown to maturity and evaluated for the expression of the new trait. This involves testing for the presence of the transgene and assessing the trait's functionality and stability across generations.Agrobacterium-mediated transformation offers several advantages over other genetic modification methods. One key benefit is its ability to produce stable transformations with precise integration of the transgene into the plant genome. This stability ensures that the desired traits are reliably expressed in subsequent generations. Additionally, the method is well-suited for a wide range of plant species, including canola, which has historically been challenging to transform using other techniques.

HERBICIDE-RESISTANT CANOLA

In canola, Agrobacterium-mediated transformation has been used to achieve significant improvements in several areas. For example, the development of varieties has been a major application. By introducing genes that confer resistance to herbicides like glyphosate or glufosinate, researchers have created canola varieties that allow for more effective weed management and reduced use of chemical herbicides. This trait has transformed weed control practices, leading to higher yields and reduced production costs.Disease resistance is another crucial area where Agrobacterium-mediated transformation has made an impact. Genes conferring resistance to pathogens such as blackleg Leptosphaeria maculans or sclerotinia Sclerotinia sclerotiorum have been successfully introduced into canola(Farid M., et al., 2020). These transformations reduce the need for chemical fungicides, enhance crop health, and minimize crop losses due to disease. Looking ahead, the future of Agrobacterium-mediated transformation in canola research promises further advancements. Innovations in vector design, including the development of more efficient and precise vectors, and improvements in transformation protocols are likely to enhance the efficiency and scope of genetic modifications(Fikere M., et al., 2020). Additionally, emerging technologies such as CRISPR-Cas9 gene editing, when combined with Agrobacterium-mediated transformation, hold the potential to revolutionize the field by enabling even more precise and targeted modifications. In summary, Agrobacterium-mediated transformation stands as a powerful and transformative tool in canola research and development. Its ability to introduce and stabilize new traits has led to significant improvements in crop performance, disease resistance, stress tolerance, and nutritional quality. As technology advances, the potential applications of this method are expected to expand, continuing to drive innovations in canola cultivation and contributing to the broader goals of sustainable and productive agriculture.

BIOLISTICS AND OTHER NON-AGROBACTERIUM METHODS FOR CANOLA TRANSFORMATION

Biolistics, also known as particle bombardment, and other non-Agrobacterium methods represent alternative approaches to genetic transformation in canola, offering distinct advantages and applications beyond those provided by Agrobacterium-mediated transformation. These techniques are essential for overcoming limitations associated with Agrobacterium, such as species-specific constraints and difficulties with certain plant tissues(Gómez-Campo C., et al., 1999). Each method has unique mechanisms and applications that contribute to the advancement of genetic modification in canola.Involves the physical delivery of DNA into plant cells using high-velocity particles. In this technique, DNA-coated metal particles, usually tungsten or gold, are accelerated using a gene gun. The particles penetrate the plant cell walls and membranes, introducing the genetic material directly into the cells. This method is particularly useful for plant species that are recalcitrant to Agrobacterium transformation, including canola, and allows for the transformation of various plant tissues, including leaf discs, embryos, and callus cultures.

Biolistic

One of the primary advantages of biolistics is its versatility and applicability to a wide range of plant species. Unlike Agrobacterium-mediated methods, biolistics does not rely on bacterial infection and thus bypasses some of the limitations associated with bacterial-mediated transformation. This makes it a valuable tool for transforming canola and other crops where Agrobacterium may not be effective or practical. Additionally, biolistics can be used to introduce larger DNA constructs or multiple genes simultaneously, which is beneficial for developing complex traits. However, biolistics also has its drawbacks(Hayward A. et al., 2011). The method can result in random integration of the transgene into the plant genome, which may lead to variable expression levels and potential disruptions of endogenous genes. The process can also cause physical damage to the plant tissues, which may affect regeneration and plant health. Despite these challenges, biolistics remains a powerful tool for generating transgenic plants, particularly when other methods are not feasible. Is a method where DNA is directly injected into plant cells using fine needles. This technique is highly precise and allows for the targeted delivery of genetic material into individual cells. Microinjection is particularly useful for introducing genes into cells that are challenging to transform by other means. Despite its precision, microinjection is labor-intensive and typically suited for smaller-scale applications or research studies rather than large-scale transformations.Viral Vectors are another innovative approach for plant transformation(Katche E., et al., 2019). In this method, plant viruses are engineered to carry and deliver the gene of interest into plant cells. The virus infects the plant and introduces the genetic material into the host genome. Viral vectors can provide transient expression of the transgene, which is useful for studying gene function or generating temporary traits. However, the use of viral vectors is limited by the potential for virus spread and regulatory concerns regarding the release of transgenic plants containing viral sequences.

Sonication

The use of sound waves to induce gene transfer, is a less common but notable method. In this approach, high-frequency sound waves create microbubbles in the plant tissue, which facilitate the entry of DNA into the cells. Sonication can be used in combination with other techniques to enhance transformation efficiency(Kyriakidou M., et al., 2018). While promising, this method is still in the experimental stages and requires further optimization for broader application.Chemical Methods also offer alternative approaches to plant transformation. These methods involve using chemicals to induce DNA uptake by plant cells. For example, calcium chloride can be used to facilitate the uptake of DNA by protoplasts through a process known as chemical transformation. Although less common than other methods, chemical transformation is useful for certain applications and can be combined with other techniques to improve efficiency.Each of these non-Agrobacterium methods has its own set of advantages and limitations. Biolistics is highly versatile and applicable to many plant species, while electroporation and microinjection offer precision in gene delivery. Viral vectors provide transient expression, and sonication and chemical methods offer additional options for plant transformation. The choice of method depends on factors such as the plant species being transformed, the complexity of the genetic modifications, and the specific goals of the research or application.In summary, biolistics and other non-Agrobacterium methods for canola transformation offer valuable alternatives to traditional techniques, providing solutions for species-specific challenges and enabling a broader range of genetic modifications.

RNA Interference (RNAi) Technology for Gene Silencing in Canola

Nutritional Enhancement

Improving oil quality through RNAi involves targeting genes that affect the fatty acid composition and stability of canola oil. By silencing genes that lead to the production of undesirable fatty acids or those that affect oil oxidation, researchers can enhance the oil's quality, shelf life, and health benefits. This targeted approach allows for the fine-tuning of oil characteristics to meet specific industry and consumer needs.Challenges and considerations Associated with RNAi technology include potential off-target effects, where small RNAs may inadvertently silence genes with similar sequences. Careful design of the RNAi constructs and rigorous validation are essential to minimize these unintended consequences(Long W., et al., 2018). Additionally, the stability and efficacy of RNAi-based gene silencing need to be assessed across different canola varieties and environmental conditions to ensure consistent performance.Regulatory aspects also play a crucial role in the application of RNAi technology in agriculture. Regulatory agencies evaluate the safety and environmental impact of transgenic plants containing RNAi constructs, ensuring that the technology does not pose risks to human health, non-target organisms, or ecosystems. Compliance with regulatory requirements and transparent communication with stakeholders are essential for the successful development and commercialization of RNAi-modified canola varieties. For RNAi technology in canola include further optimization of RNAi constructs to improve silencing efficiency and specificity. Advances in RNAi delivery systems, such as nanoparticle-based methods, could enhance the precision and effectiveness of gene silencing. Additionally, combining RNAi with other genomic and breeding techniques holds promise for creating canola varieties with multiple improved traits, addressing complex challenges in crop production and quality.In summary, RNA interference (RNAi) technology offers a powerful tool for gene silencing in canola, enabling precise modulation of gene expression to achieve a range of benefits, from enhanced disease resistance and stress tolerance to improved nutritional quality (Mahadev Madgule, et al.,2023). The technology's ability to target specific genes and its applications in functional genomics and crop improvement make it a valuable asset in advancing canola research and development.

CRISPR-Cas9 Gene Editing: A Revolution in Canola Genome Engineering

Yield Improvement

Is a major goal of CRISPR-Cas9 in canola breeding. Enhancing traits related to seed production, growth efficiency, and overall plant health can lead to increased yields and better performance in various growing conditions. Editing genes involved in key physiological processes, such as flowering time, seed development, and nutrient uptake, can contribute to higher and more consistent crop yields.Gene knock-ins Where specific genes are inserted into targeted locations in the genome, represent another powerful application of CRISPR-Cas9. This technique allows researchers to introduce new traits or enhance existing ones by placing beneficial genes in precise locations. For example, inserting genes that confer herbicide resistance or improve oil quality can provide valuable traits for canola cultivation and processing(Maher L.B.W., et al., 2007). To mitigate these risks, researchers use advanced techniques such as high-fidelity Cas9 variants and comprehensive screening methods to ensure specificity and minimize unintended changes. Additionally, the efficiency of CRISPR-Cas9 delivery into plant cells and tissues, along with the stability of the edited traits across generations, are important factors to address for successful application.

VECTOR DESIGN AND CONSTRUCTION FOR CANOLA GENETIC TRANSFORMATION

Vector design and construction are pivotal processes in the genetic transformation of canola, enabling the efficient introduction and stable expression of new genes within the plant genome. The creation of effective vectors involves meticulous planning and engineering to ensure that the desired genetic modifications are achieved with precision and reliability. Vector Design begins with selecting an appropriate vector backbone, which serves as the fundamental framework for cloning and expressing the transgenes. Commonly used vector backbones include binary vectors for Agrobacterium-mediated transformation and plasmid vectors for other transformation methods like biolistics(Malmberg M.M., et al., 2018).The choice of backbone depends on the transformation technique, the desired expression system, and the specific requirements of the target gene. Include the origin of replication, selection markers, promoter sequences, and terminators. The origin of replication ensures that the vector replicates within the host cells, while selection markers are crucial for identifying transformed cells. Typically, these markers confer resistance to antibiotics or herbicides, allowing researchers to distinguish successfully transformed cells from non-transformed ones.Promoter Sequences are essential for driving the expression of the transgene. They can be selected based on their strength and specificity, with common choices including constitutive promoters like the Cauliflower Mosaic Virus (CaMV) 35S promoter, which drives high-level expression in various plant tissues, or tissue-specific promoters, which target expression to specific tissues or developmental stages(Mason A.S., et al., 2017). Terminator sequences are also included to ensure proper termination of transcription and stability of the mRNA.

Inserting the Transgene

Into the vector involves cloning the gene of interest into the vector backbone. This is typically achieved using restriction enzyme digestion and ligation or, more recently, using advanced techniques like Golden Gate Assembly or Gibson Assembly, which facilitate the precise and seamless insertion of the transgene. The choice of cloning method depends on the vector system and the complexity of the gene construct. Expression Cassettes are designed to ensure the appropriate expression of the transgene(Nadeem M.A., et al., 2018). They include the promoter, the transgene itself, and a terminator sequence. The configuration and orientation of these elements are critical for achieving the desired level of gene expression and ensuring that the gene product is functional. For instance, a strong promoter coupled with a suitable terminator ensures robust and stable expression of the transgene in canola. Also involves the integration of additional elements for specific applications. For instance, reporter genes such as green fluorescent protein (GFP) or beta-glucuronidase (GUS) can be included to monitor the expression and localization of the transgene in living plants. Gateway cloning systems can be used to facilitate the transfer of the transgene between different vectors and streamline the construction process. Additionally, homology arms may be used in vectors for targeted gene insertion through homologous recombination, providing precise control over where the transgene integrates into the plant genome.Transformation Efficiency and gene stability. Are crucial considerations during vector design. High transformation efficiency ensures that a significant proportion of plant cells are successfully transformed, which is vital for achieving a high number of transgenic plants(Nath U.K., et al., 2009). Additionally, the stability of the inserted transgene across generations is important for maintaining the desired traits in subsequent plant generations. Researchers often conduct thorough screening and molecular analysis to confirm stable integration and expression of the transgene.

Figure 1. Canola Plant

Regeneration and Selection of Transgenic Canola Plants

This medium stimulates the development of shoots from the callus tissue. Shoot induction is a crucial step as it marks the transition from undifferentiated callus to differentiated plant structures.RootingOnce shoots have developed, they are transferred to a rooting medium to promote root formation. This step ensures that the regenerated shoots can develop a functional root system, which is essential for the plant's survival and nutrient uptake once transferred to soil.Acclimatization Regenerated plants are then gradually acclimatized to soil conditions(Nath U.K., et al., 2016). This process involves transferring the plants from the controlled tissue culture environment to a greenhouse or growth chamber where they are gradually exposed to ambient conditions. Acclimatization is crucial for the successful transition from in vitro to in vivo growth, ensuring that the plants adapt to the external environment and continue to grow and develop properly.Selection of transgenic plants is essential to identify those that have successfully integrated the transgene and are expressing it stably. Selection typically involves:Observing the phenotype of the regenerated plants provides additional confirmation of successful transformation. This involves evaluating the plants for the expression of the desired traits, such as resistance to diseases, tolerance to environmental stresses, or improved nutritional profiles. Phenotypic analysis helps to correlate the molecular data with observable traits and ensures that the genetic modifications are having the intended effects. Rigorous screening and validation processes are necessary to address these challenges and ensure the development of high-quality transgenic plants.In conclusion, the regeneration and selection of transgenic canola plants are critical steps in plant biotechnology that involve the conversion of transformed cells into whole plants and the identification of those expressing the desired traits(Oluwadare O., et al., 2019). This process requires a combination of tissue culture techniques, selection strategies, and molecular and phenotypic analyses to ensure the successful development of genetically modified canola.

Genomic and Epigenomic Analysis of Transgenic Canola

Genomic and epigenomic analysis of transgenic canola provides critical insights into the molecular changes induced by genetic modifications and their implications for plant function and stability. These analyses are essential for understanding the integration, expression, and regulation of transgenes, as well as for ensuring the safety and efficacy of genetically modified crops. By employing advanced genomic and epigenomic tools, researchers can thoroughly evaluate the effects of genetic transformations and optimize canola varieties for improved performance and trait expression.Genomic Analysis involves the examination of the plant's entire genome to assess the effects of transgene integration and to identify any unintended genetic changes(Paramasivam Selvakumar, et al., 2023). The key components of genomic analysis include:Accurate characterization of the transgene insertion site is crucial for understanding its impact on the host genome. Techniques such as Southern blotting and PCR are used to confirm the presence and copy number of the transgene. Next-generation sequencing (NGS) provides a more comprehensive view, allowing researchers to map the exact location of the transgene within the genome and to identify any potential rearrangements or integrations into non-target regions.

Integration of Genomic and Epigenomic Data

Allows for a comprehensive understanding of how genetic modifications affect transgenic canola. Combining data from genomic sequencing with epigenomic profiling provides a holistic view of the interactions between the transgene, the host genome, and the regulatory networks. This integrated approach helps in identifying potential off-target effects, assessing the functional impact of genetic changes, and ensuring that the transgene is expressed as intended without disrupting critical regulatory pathways.Challenges in Genomic and Epigenomic Analysis include the complexity of analyzing large-scale data and the potential for subtle changes in gene expression and regulation. Advanced bioinformatics tools and statistical methods are required to interpret the data accurately and to distinguish between biologically relevant changes and technical noise (Palmer C.E.D., et al., 2011). Additionally, the variability between individual plants and environmental conditions can complicate the analysis, necessitating careful experimental design and replication. In genomic and epigenomic analysis involve the development of more refined and high-throughput techniques. Single-cell genomics and epigenomics are emerging fields that offer the potential for more detailed and nuanced understanding of gene expression and regulation at the cellular level. Additionally, advancements in synthetic biology and genome editing technologies will continue to drive improvements in the design and analysis of transgenic crops, enabling more precise and controlled modifications.In summary, genomic and epigenomic analysis of transgenic canola provides essential insights into the effects of genetic modifications on the plant genome and its regulation. By employing advanced techniques to study transgene integration, expression, and epigenetic regulation, researchers can ensure the stability, safety, and efficacy of genetically modified canola(Schiessl S.V., et al., 2019). These analyses are crucial for optimizing crop traits and advancing the development of canola varieties with improved performance and resilience.

GENE EXPRESSION ANALYSIS: QUANTITATIVE PCR AND TRANSCRIPTOMIC

Gene expression analysis, employing techniques such as quantitative PCR (qPCR) and transcriptomics, is crucial for understanding how genes are regulated and expressed in various biological contexts. These methods provide insights into gene activity, expression levels, and the functional implications of genetic modifications, particularly in the study of transgenic plants like canola. This normalization process accounts for variations in sample input and experimental conditions, providing accurate and reliable data on gene expression.Comparative Analysis between qPCR and transcriptomics offers complementary insights. While qPCR provides precise quantitative data for specific genes, transcriptomics offers a global view of gene expression across the entire genome. Combining these approaches allows researchers to validate findings from transcriptomics with targeted qPCR assays, enhancing the robustness and reliability of the data.Applications of gene expression analysis are vast and varied. In transgenic canola research, qPCR and transcriptomics are used to evaluate the expression of introduced genes, assess their impact on endogenous gene networks, and study the effects of genetic modifications on plant development, stress responses, and productivity(Qu C., wt al., 2017.. For example, qPCR can quantify the expression of a transgene to determine its activity level, while RNA-Seq can reveal how the transgene affects the overall gene expression landscape, identifying any unintended consequences or regulatory changes. In gene expression analysis include ensuring the accuracy and reproducibility of results. For qPCR, careful design of primers, optimization of reaction conditions, and proper selection of reference genes are essential to obtain reliable data. In transcriptomics, challenges include handling large volumes of data, dealing with sequencing errors, and interpreting complex expression patterns. Advanced bioinformatics tools and statistical methods are required to process and analyze transcriptomics data, and the integration of data from multiple sources can enhance the interpretation and validation of findings.In conclusion, gene expression analysis through quantitative PCR and transcriptomics is a powerful combination of techniques that provides valuable insights into gene regulation and expression(Scarth R.,et al., 1988). qPCR offers precise quantification of specific genes, while transcriptomics provides a comprehensive overview of gene expression across the transcriptome. Together, these methods enhance our understanding of genetic modifications, improve the development of transgenic canola varieties, and contribute to advances in plant biotechnology and functional genomics.

Protein Function Analysis: Biochemical and Cellular Approaches

Protein function analysis is essential for understanding the roles of proteins in biological processes and their implications in genetic modifications, such as those introduced in transgenic canola. This analysis can be carried out using various biochemical and cellular approaches, each providing unique insights into protein activity, interactions, and effects on cellular functions. By combining these methods, researchers can obtain a comprehensive understanding of how proteins contribute to the physiological and molecular characteristics of plants.Biochemical Approaches focus on characterizing the chemical properties, interactions, and activities of proteins in vitro(Paramasivam Selvakumar, et al., 2023). These methods typically involve:Protein Purification To study a protein's function, it must first be isolated in a pure form. Techniques such as affinity chromatography, gel filtration, and ion exchange chromatography are used to separate proteins based on their size, charge, or specific interactions with ligands. Purified proteins are essential for detailed functional assays and structural studies. For proteins that have enzymatic activity, enzyme assays measure their catalytic functions. These assays involve incubating the enzyme with a substrate and measuring the conversion of substrate to product. This approach provides information on enzyme kinetics, substrate specificity, and inhibition mechanisms. For example, if a transgenic canola plant expresses a protein with a novel enzymatic activity, enzyme assays can determine how efficiently the protein performs its function compared to the wild-type.Protein-Protein Interactions Techniques such as co-immunoprecipitation (Co-IP),yeast two-hybrid screening, and bioluminescence resonance energy transfer (BRET) are used to identify and analyze interactions between proteins(Scarth R.,et al., 1988). Understanding these interactions helps elucidate protein complexes and signaling pathways. For instance, a protein of interest in transgenic canola might interact with other regulatory proteins, influencing its function and the overall cellular response.Cellular Approaches investigate how proteins function within the context of living cells and tissues.

Protein Localization Studies

Techniques such as fluorescence microscopy and immunohistochemistry allow researchers to visualize the localization of proteins within cells. Fluorescence microscopy involves tagging proteins with fluorescent markers or tags and observing their distribution in live or fixed cells. Immunohistochemistry uses antibodies to detect specific proteins in tissue sections, revealing their cellular and subcellular localization.Functional Assays Cellular assays assess the functional impact of proteins on cellular processes. For example, reporter assays use reporter genes (such as GFP or luciferase) linked to a promoter of interest to measure the activity of transcription factors or other regulatory proteins(Scheben A., et al., 2017). Cell viability assays and proliferation assays evaluate the effects of protein expression on cell health and growth.Genetic Modifications Knockout and knockdown approaches, such as CRISPR-Cas9 and RNA interference (RNAi), are used to assess the impact of removing or reducing the expression of specific proteins. These genetic modifications help determine the physiological roles of proteins by observing changes in cellular or organismal. This comprehensive approach involves analyzing the entire protein content of a cell or tissue. Techniques like mass spectrometry are used to identify and quantify proteins, providing insights into their abundance, post-translational modifications, and interactions. Proteomics can reveal how the expression of a transgene affects the overall proteome and cellular processes.Challenges in protein function analysis include ensuring the specificity and sensitivity of assays, dealing with the complexity of protein interactions, and accurately interpreting results within the cellular context. Variability in protein expression levels, cellular conditions, and experimental techniques can influence the outcomes, necessitating careful experimental design and validation.Future Directions in protein function analysis involve advancements in technology and methodology (Spasibionek S., et al., 2020). Developments in single-cell proteomics and high-resolution imaging techniques promise to provide more detailed insights into protein function at the cellular and molecular levels. Additionally, the integration of omics approaches, such as proteogenomics, which combines proteomics with genomic and transcriptomic data, offers a more holistic view of protein functions and interactions.In conclusion, protein function analysis through biochemical and cellular approaches provides essential insights into the roles and activities of proteins in biological systems.

Gene Function Validation using Mutagenesis and Complementation

Gene function validation is a critical process in molecular biology, enabling researchers to understand the roles of specific genes in various biological processes. Two fundamental techniques used in this validation are mutagenesis and complementation(Selvakumar, P.2022). These methods allow scientists to investigate gene function by observing the effects of gene alterations and determining if normal function can be restored by introducing wild-type genes.Mutagenesis involves introducing mutations into a gene to assess its function. This can be achieved through various methods, such as chemical mutagenesis, where chemicals induce random mutations, or site-directed mutagenesis, where specific changes are introduced at targeted locations in the gene. The resulting mutant organisms or cells are then analyzed for phenotypic changes. These changes can provide insights into the gene's role, as observed deficiencies or alterations in biological functions often correlate with the loss or alteration of the gene's activity. For instance, in model organisms like yeast, bacteria, or mice, specific phenotypic traits, such as growth defects or disease symptoms, can reveal the gene's function. Complementation can be carried out through various methods, including plasmid transformation in bacteria, transfection in eukaryotic cells, or gene replacement in genetically modified organisms. This approach helps verify that the observed effects are directly attributable to the mutated gene, rather than being caused by unrelated factors.

High-Throughput Phenotyping and Genotyping for Canola Breeding

High-throughput phenotyping and genotyping are transformative technologies in the field of canola breeding, offering unprecedented opportunities to enhance crop improvement programs. These approaches streamline the process of evaluating and selecting plants with desirable traits, thereby accelerating the development of new canola varieties with superior performance, disease resistance, and stress tolerance (Selvakumar, P.,2023). Technologies like drone-based imaging, hyperspectral cameras, and automated phenotyping platforms enable breeders to monitor and analyze plants in real-time across large experimental plots. By capturing detailed phenotypic data, high-throughput phenotyping allows for a more comprehensive understanding of how different genotypes express specific traits under varying environmental conditions.The integration of high-throughput phenotyping into canola breeding programs offers several advantages. Firstly, it enhances the accuracy and efficiency of trait evaluation by reducing the time and labor required for manual measurements. This efficiency is crucial for handling the large number of genotypes typically involved in breeding programs. Secondly, it enables breeders to identify and select plants with subtle or complex traits that may be difficult to measure with traditional methods. For example, high-throughput phenotyping can detect variations in drought tolerance or disease resistance by assessing plant responses to controlled stress conditions. The combination of high-throughput phenotyping and genotyping facilitates the development of genotype-by-environment (GxE) studies, which explore how genetic and environmental factors interact to influence plant traits. These studies are crucial for understanding how canola varieties perform under different environmental conditions, such as varying soil types, moisture levels, or temperatures. By integrating phenotypic and genotypic data, breeders can identify genotypes that exhibit stable performance across diverse environments, leading to the development of resilient and adaptable canola varieties. GS involves developing predictive models based on large-scale genotypic and phenotypic data, which can then be used to select the best candidates for breeding. This approach reduces the reliance on field trials, accelerates the breeding cycle, and improves the efficiency of developing new canola varieties with targeted traits.The implementation of high-throughput phenotyping and genotyping in canola breeding programs also aligns with the growing emphasis on precision agriculture. Companies and researchers must carefully review existing patents to avoid infringement and may need to obtain licenses for technologies developed by others. This process involves conducting freedom-to-operate (FTO) analyses and negotiating licensing agreements. The patent landscape in agricultural biotechnology is dynamic, with ongoing developments and new applications emerging regularly, which necessitates vigilant IP management.Trademarks and plant breeders' rights (PBR) also play roles in the commercialization of GM canola. Trademarks protect the branding of GM canola products, ensuring that consumers can identify and differentiate them in the marketplace. Plant breeders' rights, on the other hand, offer protection to new plant varieties by granting exclusive rights to propagate and market them. This form of protection supports the development of new canola cultivars by providing breeders with incentives to develop and commercialize innovative varieties.Regulatory issues are equally critical in the field of canola genetic transformation, as GM crops are subject to rigorous safety assessments and approval processes before they can be cultivated and marketed. Regulatory frameworks vary by country but generally include a series of evaluations to assess the safety and efficacy of GM canola.

Bioinformatics Tools for Canola Genomic and Gene Function Analysis

Bioinformatics tools have become indispensable for advancing our understanding of canola genomics and gene function, providing critical insights into genetic variation, gene expression, and functional annotation. These tools facilitate the analysis of large-scale genomic data, helping researchers to identify key genes, unravel complex genetic networks, and accelerate the breeding of improved canola varieties.Functional Genomics tools enable researchers to investigate gene function through various methods(Sugumaran K. M, D., et al.,2024). BLAST (Basic Local Alignment Search Tool) is used for sequence alignment and functional annotation by comparing canola gene sequences against known databases to predict gene functions. KEGG (Kyoto Encyclopedia of Genes and Genomes) and GO (Gene Ontology) databases provide information on gene pathways and functional categories, helping to elucidate the roles of genes in metabolic pathways and biological processes. Functional prediction tools like InterPro and Pfam offer insights into protein domains and functional motifs that are critical for understanding gene functions. Tools such as Cytoscape enable the visualization of complex gene and protein interaction networks, while STRING and BioGRID databases provide information on known interactions and functional relationships between proteins. Systems biology helps in understanding how genes interact within biological pathways and how these interactions influence plant traits and responses..In summary, bioinformatics tools are essential for advancing our understanding of canola genomics and gene function. As bioinformatics technologies continue to evolve, their integration with emerging tools and methodologies will further enhance our ability to explore and manipulate the canola genome, paving the way for innovations in crop breeding and genetic research.

Intellectual Property and Regulatory Issues in Canola Genetic Transformation

Intellectual property (IP) and regulatory issues play a significant role in the field of canola genetic transformation, impacting the development, commercialization, and management of genetically modified (GM) canola varieties. These concerns encompass a range of aspects, from the protection of biotechnological innovations to the compliance with legal and safety standards that govern the use of GM crops. Intellectual Property (IP) rights are crucial for protecting innovations in canola genetic transformation. This protection typically involves patents, trademarks, and plant breeders' rights (PBR). Patents grant exclusive rights to inventors for novel and non-obvious genetic modifications or transformation technologies(Vijayakumar G, et al.,2024). For instance, patents may cover specific genetic constructs, transformation methods, or traits conferred to canola plants, such as herbicide resistance or improved oil composition. IP protection encourages investment in research and development by providing legal recourse against unauthorized use of proprietary technologies and ensuring that innovators can recoup their investments. In canola genetic transformation can be complex due to the need to navigate overlapping claims and ensure freedom to operate(Wei D., Cui Y., et al.,2017).

ENVIRONMENTAL RISK ASSESSMENTS

Are a core component of regulatory evaluations. These assessments examine the potential impacts of GM canola on biodiversity, ecosystem functions, and non-target organisms. For example, regulators assess whether GM canola may crossbreed with wild relatives or other crops, potentially leading to unintended ecological consequences(Yuan Y.,et al.,2017). Measures are often required to mitigate such risks, including confinement strategies and monitoring programs.Food safety assessments ensure that GM canola poses no undue risks to human health. This involves evaluating the nutritional composition of GM canola, as well as assessing potential allergenicity or toxicity of novel proteins introduced through genetic transformation. The safety data submitted by developers is scrutinized through rigorous testing and review processes to ensure that GM canola is safe for consumption. Are regulatory requirements that ensure transparency in the food supply chain. In many jurisdictions, GM canola products must be labeled to inform consumers about their genetic modification. Additionally, traceability systems are implemented to track GM canola throughout its lifecycle, from production to market, ensuring that GM crops can be easily identified and monitored.International trade also intersects with IP and regulatory issues in canola genetic transformation(Hayward A. et al., 2011). Different countries have varying regulations and standards for GM crops, which can affect trade and market access. Companies involved in the international trade of GM canola must navigate diverse regulatory landscapes and comply with both domestic and international standards. This may include adherence to agreements such as the Cartagena Protocol on Biosafety, which governs the transboundary movement of GMOs.Ethical and societal considerations.Also play a role in the regulation of GM canola. Public perception of genetic modification and concerns about potential impacts on agriculture and ecosystems can influence regulatory policies and market acceptance. Engaging with stakeholders, including farmers, consumers, and environmental groups, is important for addressing these concerns and ensuring that GM canola is developed and used responsibly(Nath U.K., et al., 2009).In summary, intellectual property and regulatory issues are integral to the field of canola genetic transformation, influencing the protection of innovations, safety assessments, and market access.

FUTURE DIRECTIONS: EMERGING TECHNIQUES AND APPLICATIONS IN CANOLA RESEARCH

The future of canola research is poised to be significantly transformed by emerging techniques and applications, driven by advancements in genomics, biotechnology, and data analytics. These innovations promise to enhance our understanding of canola biology, improve crop traits, and address global agricultural challenges with unprecedented precision and efficiency(Paramasivam Selvakumar, et al., 2023).High-Throughput Phenotyping continues to evolve with the integration of remote sensing technologies and artificial intelligence (AI). Drones equipped with multispectral and hyperspectral sensors, combined with AI-driven image analysis, enable rapid and detailed assessment of plant traits over large areas. Metabolomics and Proteomics are emerging fields that complement genomics by providing comprehensive insights into the biochemical processes within canola plants. Metabolomics involves the large-scale study of metabolites, such as sugars, lipids, and secondary metabolites, which can reveal the physiological and biochemical status of the plant. Proteomics focuses on the analysis of the plant proteome, offering insights into protein expression, modifications, and interactions. Together, these approaches help to elucidate the metabolic pathways and regulatory networks that underpin key traits in canola. Is a key focus for future canola research(Palmer C.E.D., et al., 2011). With climate change posing significant challenges to agriculture, developing canola varieties that can withstand extreme weather conditions, such as drought, flooding, and temperature fluctuations, is critical. Research in this area involves identifying and incorporating stress-tolerant traits into canola varieties through genetic modification and traditional breeding techniques.Genomic Selection (GS) and Phenomic Selection (PS) are innovative approaches that combine genomic and phenotypic data to predict and select for desirable traits more effectively. GS uses genetic markers to predict the breeding value of canola lines, while PS integrates high-throughput phenotypic data to refine these predictions. These methods enhance the efficiency of breeding programs by enabling the selection of superior lines based on comprehensive data.

Bioinformatics and Computational Tools

Continue to advance, providing more sophisticated methods for analyzing and interpreting complex biological data. Tools such as machine learning algorithms and network analysis software enhance the ability to predict gene function, identify key regulatory elements, and model biological processes. These tools facilitate the integration of diverse datasets and the extraction of actionable insights for canola improvement.Ethical and Societal Considerations are increasingly important as new technologies are developed and implemented. Engaging with stakeholders, including farmers, consumers, and policymakers, ensures that the benefits of technological advancements are realized while addressing potential concerns related to GM crops, environmental impact, and food security(Schiessl S.V., et al., 2019). Public dialogue and transparent communication are essential for the responsible development and adoption of new canola technologies.In summary, the future directions in canola research are characterized by the integration of advanced genomic techniques, high-throughput phenotyping, and emerging technologies such as synthetic biology and precision agriculture. These innovations promise to enhance our understanding of canola biology, improve crop traits, and address the challenges of modern agriculture. As research progresses, the combination of these techniques with ethical considerations and stakeholder engagement will be crucial for advancing the field and achieving sustainable and productive canola cultivation.

CONCLUSION

The integration of sophisticated gene function research techniques, such as transcriptomic and proteomic analyses, has furthered our comprehension of the complex gene networks governing canola's growth and response to environmental stresses. Despite these advancements, several issues persist that need addressing for further progress. The complexity of canola's genome, coupled with its polyploid nature, poses significant challenges in gene functional analysis and trait development. The advent of high-throughput sequencing technologies and advanced bioinformatics tools offers promising avenues to unravel these complexities. Furthermore, the regulatory and ethical considerations surrounding genetically modified organisms (GMOs) require continuous scrutiny to ensure safe and responsible application of these technologies.Future research should focus on integrating multi-omics approaches to gain a holistic view of gene function and interactions within the canola genome. Collaborative efforts between geneticists, agronomists, and bioinformaticians will be essential to tackle the intricacies of gene expression and regulation. Additionally, advancements in genome-editing technologies should be harnessed to create precise genetic modifications with minimal off-target effects. The exploration of gene editing in combination with traditional breeding methods could yield canola varieties that not only meet agricultural demands but also contribute to sustainable farming practices. In summary, while the current techniques in canola plant genetic transformation and gene function research have paved the way for significant progress, ongoing innovation and interdisciplinary collaboration are crucial for overcoming existing challenges. Companies and researchers must carefully review existing patents to avoid infringement and may need to obtain licenses for technologies developed by others. This process involves conducting freedom-to-operate (FTO) analyses and negotiating licensing agreements. The patent landscape in agricultural biotechnology is dynamic, with ongoing developments and new applications emerging regularly, which necessitates vigilant IP management.Trademarks and plant breeders' rights (PBR) also play roles in the commercialization of GM canola. Trademarks protect the branding of GM canola products, ensuring that consumers can identify and differentiate them in the marketplace. Plant breeders' rights, on the other hand, offer protection to new plant varieties by granting exclusive rights to propagate and market them. This form of protection supports the development of new canola cultivars by providing breeders with incentives to develop and commercialize innovative varieties.Regulatory issues are equally critical in the field of canola genetic transformation, as GM crops are subject to rigorous safety assessments and approval processes before they can be cultivated and marketed. Regulatory frameworks vary by country but generally include a series of evaluations to assess the safety and efficacy of GM canola.

REFERENCES

Bonnardeaux, J. (2007). *Uses for Canola Meal*. Department of Agriculture and Food.

Chalhoub, B., Denoeud, F., Liu, S., Parkin, I. A. P., Tang, H., Wang, X., Chiquet, J., Belcram, H., Tong, C., Samans, B., Corréa, M., Da Silva, C., Just, J., Falentin, C., Koh, C. S., Le Clainche, I., Bernard, M., Bento, P., Noel, B., & Wincker, P. (2014). Early allopolyploid evolution in the post-Neolithic *Brassica napus* oilseed genome. *Science*, *345*(6199), 950–953. DOI: 10.1126/science.1253435 PMID: 25146293

Eskin, N. A. M., & Przybylski, R. (2003). RAPE SEED OIL/CANOLA. In Caballero, B. (Ed.), *Encyclopedia of Food Sciences and Nutrition* (2nd ed., pp. 4911–4916). Academic Press. DOI: 10.1016/B0-12-227055-X/01349-3

Farid, M., Yang, R. C., Kebede, B., & Rahman, H. (2020). Evaluation of Brassica oleracea accessions for resistance to Plasmodiophora brassicae and identification of genomic regions associated with resistance. *Genome*, *63*(2), 91–101. DOI: 10.1139/gen-2019-0098 PMID: 31600449

Fikere, M., Barbulescu, D. M., Malmberg, M. M., Spangenberg, G. C., Cogan, N. O. I., & Daetwyler, H. D. (2020). Meta-analysis of GWAS in canola blackleg (Leptosphaeria maculans) disease traits demonstrates increased power from imputed whole-genome sequence. *Scientific Reports*, *10*(1), 14300. DOI: 10.1038/s41598-020-71274-6 PMID: 32868838

Gómez-Campo, C., & Prakash, S. (1999). 2 Origin and domestication. In Gómez-Campo, C. (Ed.), *Developments in Plant Genetics and Breeding* (Vol. 4, pp. 33–58). Elsevier.

Hayward, A. (2011). Introduction-Oilseed Brassicas. In Edwards, D., Batley, J., Parkin, I., & Kole, C. (Eds.), *Genetics, Genomics and Breeding of Oilseed Brassicas*. Taylor & Francis Group.

Katche, E., Quezada-Martinez, D., Katche, E. I., Vasquez-Teuber, P., & Mason, A. S. (2019). Interspecific Hybridization for Brassica Crop Improvement. *Crop Breeding, Genetics and Genomics*, *1*, e190007. DOI: 10.20900/cbgg20190007

Kyriakidou, M., Tai, H. H., Anglin, N. L., Ellis, D., & Strömvik, M. V. (2018). Current Strategies of Polyploid Plant Genome Sequence Assembly. *Frontiers in Plant Science*, *9*, 1660. DOI: 10.3389/fpls.2018.01660 PMID: 30519250

Li, F.-W., & Harkess, A. (2018). A guide to sequence your favorite plant genomes. *Applications in Plant Sciences*, *6*(3), e1030. DOI: 10.1002/aps3.1030 PMID: 29732260

Long, W., Hu, M., Gao, J., Chen, S., Zhang, J., Cheng, L., & Pu, H. (2018). Identification and Functional Analysis of Two New Mutant BnFAD2 Alleles That Confer Elevated Oleic Acid Content in Rapeseed. *Frontiers in Genetics*, *9*, 399. Advance online publication. DOI: 10.3389/fgene.2018.00399 PMID: 30294343

Mahadev Madgule, N. (2023). Mechanical properties and microstructure of activated TIG welded similar joints of Inconel alloys by desirability approaches. *Materials Today: Proceedings*, *77*, 528–533. DOI: 10.1016/j.matpr.2022.12.250

Maher, L. B. W., Salisbury, P., Debonte, L., & Deng, X. (2007).(Organising Committee of: Wuhan). High Oleic, low linolenic (HOLL) specialty canola development in Australia; *Proceedings of the 12th International Rapeseed Congress*; Wuhan, China. 23–30 March; pp. 22–24.

Malmberg, M. M., Barbulescu, D. M., Drayton, M. C., Shinozuka, M., Thakur, P., Ogaji, Y. O., Spangenberg, G. C., Daetwyler, H. D., & Cogan, N. O. I. (2018). Cogan N.O.I. Evaluation and Recommendations for Routine Genotyping Using Skim Whole Genome Re-sequencing in Canola. *Frontiers in Plant Science*, *9*, 1809. Advance online publication. DOI: 10.3389/fpls.2018.01809

Mason, A. S., Higgins, E. E., Snowdon, R. J., Batley, J., Stein, A., Werner, C., & Parkin, I. A. (2017). A user guide to the Brassica 60K Illumina Infinium™ SNP genotyping array. *Theoretical and Applied Genetics*, *130*(4), 621–633. DOI: 10.1007/s00122-016-2849-1 PMID: 28220206

McVetty, P. B. E., & Duncan, R. W. (2016).Reference Module in Food Science. Elsevier; Amsterdam, The Netherlands. Canola/Rapeseed: Genetics and Breeding.

Nadeem, M. A., Nawaz, M. A., Shahid, M. Q., Doğan, Y., Comertpay, G., Yıldız, M., Hatipoğlu, R., Ahmad, F., Alsaleh, A., Labhane, N., Özkan, H., Chung, G., & Baloch, F. S. (2018). DNA molecular markers in plant breeding: Current status and recent advancements in genomic selection and genome editing. *Biotechnology, Biotechnological Equipment*, *32*(2), 261–285. DOI: 10.1080/13102818.2017.1400401

Nath, U. K., Kim, H.-T., Khatun, K., Park, J.-I., Kang, K.-K., & Ill-Sup, N. (2016). Modification of Fatty Acid Profiles of Rapeseed (Brassica napus L.) Oil for Using as Food, Industrial Feed-Stock and Biodiesel. *Plant Breeding and Biotechnology*, *4*(2), 123–134. DOI: 10.9787/PBB.2016.4.2.123

Nath, U. K., Wilmer, J. A., Wallington, E. J., Becker, H. C., & Möllers, C. (2009). Increasing erucic acid content through combination of endogenous low polyunsaturated fatty acids alleles with Ld-LPAAT + Bn-fae1 transgenes in rapeseed (Brassica napus L.). *Theoretical and Applied Genetics*, *118*(4), 765–773. Advance online publication. DOI: 10.1007/s00122-008-0936-7 PMID: 19050848

Oluwadare, O., Highsmith, M., & Cheng, J. (2019). An Overview of Methods for Reconstructing 3-D Chromosome and Genome Structures from Hi-C Data. *Biological Procedures Online*, *21*(1), 7. DOI: 10.1186/s12575-019-0094-0 PMID: 31049033

Palmer, C. E. D., & Keller, W. A. (2011). Future Prospects for Brassica Oilseed Improvement through Genomics. In Edwards, D., Batley, J., Parkin, I., & Kole, C. (Eds.), *Genetics, Genomics and Breeding of Oilseed Brassicas* (pp. 1–13). Taylor & Francis Group.

Qu, C., Jia, L., Fu, F., Zhao, H., Lu, K., Wei, L., Xu, X., Liang, Y., Li, S., Wang, R., & Li, J. (2017). Genome-wide association mapping and Identification of candidate genes for fatty acid composition in Brassica napus L. using SNP markers. *BMC Genomics*, *18*(1), 232. DOI: 10.1186/s12864-017-3607-8 PMID: 28292259

Scarth, R., McVetty, P. B. E., Rimmer, S. R., & Stefansson, B. R. (1988). STELLAR LOW LINOLENIC-HIGH LINOLEIC ACID SUMMER RAPE. *Canadian Journal of Plant Science*, *68*(2), 509–511. DOI: 10.4141/cjps88-061

Scheben, A., Batley, J., & Edwards, D. (2017). Genotyping-by-sequencing approaches to characterize crop genomes: Choosing the right tool for the right application. *Plant Biotechnology Journal*, *15*(2), 149–161. DOI: 10.1111/pbi.12645 PMID: 27696619

Schiessl, S. V., Quezada-Martinez, D., Tebartz, E., Snowdon, R. J., & Qian, L. (2019). The vernalisation regulator FLOWERING LOCUS C is differentially expressed in biennial and annual Brassica napus. *Scientific Reports*, *9*(1), 14911. DOI: 10.1038/s41598-019-51212-x PMID: 31624282

Selvakumar, P. (2022). *Phytochemical and Pharmacological Profile Review of Bryophyllum Pinnatum. Biomedical and Biotechnology Research Journal (BBRJ) 6(3)*. Jul–Sep., DOI: 10.4103/bbrj.bbrj_126_22

Selvakumar, P. (2023). Fiber Reinforced Polymer Composites: Classification, Chemical Treatment, Mechanical and Tribological Properties. *Research Journal of Chemistry and Environment*, *27*(1), 111–119. DOI: 10.25303/2701rjce1110119

Selvakumar, P. (2023). Novel Drug Target with Diverse Therapeutic Potential in Cancer Therapy. *Pharmacy Times*, *55*(1), 11–14.

Selvakumar, P. (2023). Biomedical applications of polymers in biosensors, cancer vaccines and drug delivery systems. *Biopolimery i Kletka*, *39*(4), 245–256. DOI: 10.7124/bc.000AA0

Selvakumar, P., Mariappan, P., Shunmugam, S., Paul, R., & Jabez, S. (2024). Removal of chemicals from effluent using photobioreactor technology to improve environmental and health impacts. *Zeitschrift für Physikalische Chemie*, *238*(7), 1295–1302. Advance online publication. DOI: 10.1515/zpch-2023-0450

Selvakumar, P., Palanisamy, S. K., Cinthaikinian, S., Palanisamy, V., Mariappan, R., & Selvakumar, P. K. (2024). Biosensors and its diverse applications in healthcare systems. *Zeitschrift für Physikalische Chemie*. Advance online publication. DOI: 10.1515/zpch-2023-0406

Spasibionek, S., Mikołajczyk, K., Ćwiek-Kupczyńska, H., Piętka, T., Krótka, K., Matuszczak, M., Nowakowska, J., Michalski, K., & Bartkowiak-Broda, I. (2020). Bartkowiak-Broda Marker assisted selection of new high oleic and low linolenic winter oilseed rape (Brassica napus L.) inbred lines revealing good agricultural value. *PLoS One*, *15*(6), e0233959. DOI: 10.1371/journal.pone.0233959 PMID: 32497146

Sugumaran, K. M. D., Pandiyarajan V., Y. B S, Selvakumar P. and M. S, (2024). "Analysis of Electromyography Signals for Control of Mechanical Prosthesis using Machine Learning Techniques," 2024 2nd International Conference on Intelligent Data Communication Technologies and Internet of Things (IDCIoT), Bengaluru, India, 2024, pp. 1059-1067, DOI: 10.1109/IDCIoT59759.2024.10467707

Vijayakumar, G., Rajkumar, M., Rajiv Chandar, N., Selvakumar, P., & Duraisamy, R. (2024). Environmentally friendly TDS removal from waste water by electrochemical ion exchange batch-type recirculation (EIR) technique. *Environmental Science. Water Research & Technology*, *10*(4), 826–835. DOI: 10.1039/D3EW00793F

Wei, D., Cui, Y., He, Y., Xiong, Q., Qian, L., Tong, C., Lu, G., Ding, Y., Li, J., Jung, C., & Qian, W. (2017). A genome-wide survey with different rapeseed ecotypes uncovers footprints of domestication and breeding. *Journal of Experimental Botany*, *68*(17), 4791–4801. DOI: 10.1093/jxb/erx311 PMID: 28992309

Chapter 10
Abiotic Resistance of Canola in the Functional Genomics Era

Fadime Karabulut
https://orcid.org/0000-0001-5186-2303
Hizan Vocational School, Bitlis Eren University, Bitlis, Turkey

Mohammad Faizan
Maulana Azad National Urdu University, India

Amin Fathi
https://orcid.org/0000-0002-7539-0053
Islamic Azad University, Amol, Iran

ABSTRACT

Abiotic stresses have a negative impact on the growth, development and productivity of canola plants which results in large losses in crop yield. The complex mechanisms underlying plants' ability to withstand abiotic stress are becoming easier to understand recent tremendous advancements in genomics. Increased levels of ROS, which are highly reactive and toxic and impact photosynthetic capacity, antioxidant enzyme activities, protein, carbohydrate and lipid biosynthesis are also observed in canola plants under abiotic stress. Candidate genes, microRNAs and transcription factors (TFs) involved in particular plant processes can be identified using transgenic techniques. The molecular and physiological processes influencing plant tolerance and productivity are thus better understood. Transgenic breeding could therefore prove to be a useful technique for reducing Canola's exposure to

DOI: 10.4018/979-8-3693-8603-3.ch010

abiotic stress. This chapter discusses the viability and effective uses of transgenic breeding to increase plant productivity and resistance to abiotic stress, as well as future developments in this field.

1. INTRODUCTION

Abiotic stress conditions are something that plants regularly face throughout their lives, and to combat the detrimental effects of these unfavourable conditions, they have developed a number of defence strategies. However, the way in which plants react to abiotic stressors is a highly complex mechanism that results in changes in transcription, translation, and post-translation, leading to a comprehensive change at both metabolic and biochemical levels, resulting in morphological and physiological adaptation, as evidenced by numerous years of scientific study. Furthermore, transcriptome, proteome, and metabolome have come to characterise all cellular activity in the years that have passed, and they are now referred to as "-omics," which is a term for all data acquired using high-throughput technologies. As new technologies emerge, the field of omics has expanded to include lipidomics, ionomics, and phenomics in addition to transcriptomics, proteomics, and metabolomics. Systems biology is now utilising all of the data collected with omics technologies thanks to bioinformatics techniques. Abiotic stress response in plants is better understood from new angles thanks to the development of omics technologies, given the complexity of changes in cellular metabolism (Faizan et al., 2024; Karabulut et al., 2024, Ahmad et al., 2022). A comprehensive perspective of biological systems is offered by high-throughput technologies known as omics. Numerous crop plants now have their genome sequences available thanks to technological advancements. The identification of useful alleles and novel resistance genes from the wild relatives of current crops is aided by this, as is the understanding of intricate networks and gene functions. For use in molecular breeding, sophisticated genotyping systems yield elite agronomically suited cultivars with desired characteristics. Huge collections of germ plasm from different species all over the world can be used for high-density genotyping and whole genome resequencing, and they represent an amazing natural diversity with untapped genetic potential (Singh et al., 2020). Genome-wide association studies (GWAS) are then conducted using high-throughput phenotyping and population-wide genotyping data to obtain insight into marker-trait associations (MTAs), quantitative trait loci (QTLs), quantitative trait nucleotides (QTNs), and genomic estimated breeding value (GEBVs) for the selection of desired traits (Huynh et al., 2018). Additionally, it offers information about potential source donors for superior haplotypes and desired traits that can be used in genomics-assisted breeding (GAB) and pre-breeding (Varshney et al., 2019). Furthermore, different GAB techniques, for example, marker-assisted recurrent selection (MARS), genomic selection (GS), and marker-assisted selection (MAS), can aid in the integration or accumulation of genomic loci that contribute to desirable traits (Singh et al., 2017). Jenko et al. (2015) describe promotion of alleles through Peptide-Assisted Genome Editing (PAGE) as a method of Peptide-Assisted Genome Editing that can also be used to modify genes with defined QTNs. In order to reduce the generation time and hasten the release of new varieties, fast breeding is being combined with Peptide-Assisted Genome Editing and molecular breeding. When it comes to production systems, canola can be grown in a range of settings from low-input to highly intensive, thanks to its high plasticity and long flowering window. Soil moisture retention is the primary determinant of canola production. Canola is sown by farmers in a dry manner in preparation for later downpours. The yields of canola vary annually. In limited conditions, they respond to abiotic stresses with less than 1 tonne/ha, and in irrigated conditions, 6–8 tonnes/ha. This is a broad range of responses. Drought and heat are acknowledged as the two biggest global constraints on crop productivity (Smith and De Smet 2012, Ahmad et al., 2024).

A number of subspecies of *Brassica rapa* (*B. rapa*) (2n=20, AA) are known to produce edible oils and leafy, root, and stem vegetables for human consumption. Additionally, according to Cheng et al. (2014), it symbolises the beginning of the Brassica 'A' genome and supports the Brassica allopolyploid oil crops *Brassica juncea* (*B. juncea*) (AABB) and *Brassica napus* (*B. napus*) (AACC). Because of this, *B. rapa* shows a great deal of promise as a model for genomic and evolutionary studies of species in the genus Brassica (Zhang et al., 2015). *B. rapa* can be distinguished based on how well it resists cold temperatures by looking at its overwintering rate and cold resistance performance. While most cold-resistant cultivars can overwinter in the areas east of Wushaoling Mountain, only exceptionally and strongly cold-resistant cultivars can do so in the Hexi Corridor (Sun et al., 2007). China's cold and arid northern regions are now an important winter canola seed production zone owing to the development and breeding of *B. rapa* cultivars, the problem of oilseed crop overwintering has been resolved. As a result, it boosts economic returns and crop density. ROS are crucial to plant life because they control how cells react to external stimuli such as hydrogen peroxide (H_2O_2), superoxide anions (O_2^-), and hydroxyl radicals (OH-), which can oxidise proteins and lipids and damage DNA and other cellular components. Most cold-tolerant plants have evolved antioxidant defence mechanisms to shield themselves from the excessive accumulation of ROS. These defence mechanisms include the use of antioxidants like catalase (CAT), superoxide dismutase (SOD), and ascorbate peroxidase (APX), ascorbic acid (Gilroy et al., 2016). Plants have a peroxidase superfamily member called APX, which is involved in processes that scavenge ROS (Apel and Hirt, 2004). In order to scavenge H_2O_2 from plant chloroplasts and cytoplasm, plants have evolved APX, an essential enzyme in the ascorbate-glutathione cycle. It scavenges H_2O_2 produced in plants by using ascorbic acid as an electron donor, strengthening and boosting resistance to oxidative stress (Ding et al., 2019, Ali et al., 2022, Haider et al., 2021). According to a study on Arabidopsis, information about past cold stress can be transmitted over time through the regulation of plastid APX, even in the absence of cold adaptation. Following the rice APXa gene's cloning, scientists found that the promoter, located 81 bp upstream of the TATA box, contains the minimal heat shock factor binding motif 5'-nGAAnnTTCn-3. This discovery provides rice seedlings with protection from the cold and validates heat shock-mediated APX gene expression (Sato et al., 2001). Studies on B. *rapa*'s APX regulation of cold adaptation are, nevertheless, scarce. Nowadays, improving agronomic practices through improvement programs and management program improvements is the foundation for Brassica oilseed and mustard crop yield and yield stability. With the inclusion of molecular resources such as breeding pedigrees, trait-associated loci identification in breeding programs, and germplasm characterisation, all these programs are gaining tremendous traction. Studies show that crossing hybrid *Brassica rapa* and/or *Brassica oleracea* can be used to quickly establish high-yielding, short-lived *B. napus* lines that can yield up to 4.6 times higher yields (Karim et al., 2014). But it's not always easy to focus on a particular phenotypic trait, like yield or oil seed content. Crop phenotype and genotype are not always associated. Although substantial genomic resources, such as reference genomes or pangenomes, are available, the underlying genes for quantitative resistance have not yet been identified due to its multigene nature (Amas et al., 2021; Raman et al., 2012). Furthermore, phenotypic characteristics frequently dictate the viability of an area. This is so because abiotic stressors like heavy metals, salinity, pH, temperature, and dehydration have a major impact on yield security.

This chapter focusses on transgenic breeding's potential, successful applications, and future prospects for enhancing canola productivity and resistance to abiotic stress..

2. GENES RESPONSIVE TO ABIOTIC STRESS IN THE CANOLA

The term “genomics” refers to the study of the genome, which contains enough data to describe the chemical, physiological, and biological processes of an organism as well as the composition, sequence, and function of individual genes (Gilliham et al., 2017). First-generation sequencing, which began in the 1970s, and third-generation sequencing, which saw swift advancements in genome sequencing technology, represent the evolutionary history of genomics. (El-Metwally et al., 2014). The identification of genes and their functions associated with stress stimuli is aided by functional genomics (Wang et al., 2018). Climate-resilient crop production is aided by our understanding of gene expression and regulation with genome-wide complex stress-responsive traits (Gilliham et al., 2017). Further research on plants can be conducted using transcriptomics, proteomics, metabolomics, and online genomic data in conjunction with genome engineering (CRISPR/Cas) system (Shen et al., 2019). Because abiotic stresses are persistent and have detrimental effects on agriculture globally, there is interrelatedness among various traits (Raza et al., 2020). Thus, the development of plant varieties resistant to abiotic stress has been the primary goal of molecular breeding. Gene functions related to plant stresses have previously been widely utilised through crop identification and manipulation to create genotypes that are stress-tolerant, as well as the application of genomic and genetic techniques like transcriptome analysis, QTL, GWAS, microarray technology, and MAS (Bancroft et al., 2011, Ijaz et al., 2023). The development of plants resistant to abiotic stress is being made possible by recent developments in high-throughput technologies. Modern techniques for genetic improvement, such as Peptide-Assisted Genome Editing (regularly clustered short palindromic repeats, or CRISPR/Cas system) and omics approaches, are being combined with the accessibility of genomic tools for gene documentation and Canola trait research to expand molecular breeding capabilities. This gene's expression pattern suggests that leaves and seeds react to drought in different ways. Under drought stress, this gene belongs to a novel class of gene family that widely accumulates in vegetative parts (Downing et al., 1992).

Whole plant genomes can now be sequenced thanks to advancements in next-generation technologies. Additionally, the availability of whole genome sequences makes it possible to identify all putative genes and regulatory elements on the genome using bioinformatics techniques. Furthermore, using genetic and physical mapping, functional genomics, comparative genomics, and structural genomics can be used to find genes on the genome, identify the gene or genes' function, and confirm the genome structure in different organisms or organs. For the purpose of demonstrating the roles and functions of every gene found on the genome in cellular metabolism, functional genomics is essential (Rhee and Mutwil, 2014). While it is possible to sequence the genome and identify potential genes and regulatory elements, further research is necessary to assign a functional role in cellular metabolism. This research should focus on protein function, subcellular localisation, transcription, and transcription variation in various organs, developmental stages, and environmental conditions. Additionally, potential alternative splicing products should be taken into consideration. Functional genomics is the most significant method for comprehending the functional makeup of the plant genome in this regard (Figure 1).

Figure 1. Mechanisms of abiotic stress signalling

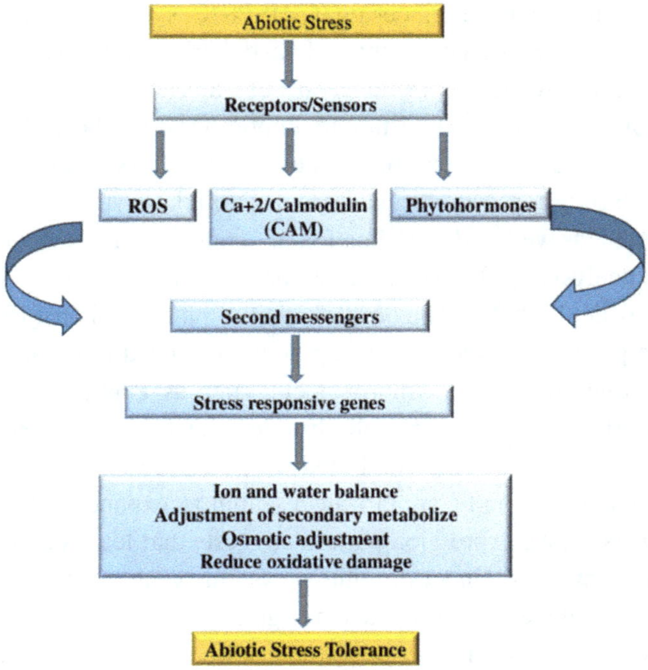

Single nucleotide polymorphisms (SNPs) are found in agriculture using a comparative genomics approach, which is made possible by genomics. Finally, determining SNPs or QTLs associated with a desired phenotype is the aim. SNPs are very common in the genome, but they usually do not produce an observable phenotype in plants because of the high levels of environmental stress they face. This is particularly important when studying abiotic stress because it's necessary to observe the desired phenotype—here, resistance to abiotic stresses—in different environmental conditions in order to rule out variations in the desired trait caused by the interaction of environment and genotype (G x E). Therefore, GWAS and genotyping by sequencing (GBS) techniques have been widely used to identify SNPs linked to field crops' tolerance to abiotic stresses. In large-scale breeding studies, MAS can be used to develop a marker for selection when tolerance and SNP(s) have a strong association (Kang et al., 2016). An estimated hundred SNPs or more, along with QTLs linked to abiotic stress tolerance, are typically reported by the majority of these studies. Only a small number of studies, though, validate SNPs and QTLs using separate genotypes. Genes or regulatory regions with altered nucleotide sequences that could potentially boost plant vigour in harsh environments are identified by GWAS-identified SNPs linked to abiotic stress tolerance. Strong environmental interactions can cause phenotypic changes, which can make it difficult to demonstrate the functional role of the same SNPs or QTLs in the selection of tolerant genotypes in individual plants. The primary disadvantage of employing these SNPs and QTLs as reliable markers in breeding strategies to enhance abiotic stress tolerance is this. It is critical to recognise that multiple abiotic stressors will manifest simultaneously as a predicted effect of global climate change. Plants will eventually have to adapt to high salinity, for example, as water availability decreases. Similarly, rising temperatures cause evaporation to accelerate, which further reduces water availability. Agriculture is therefore concerned about yield loss even when a crop is heat-sensitive and drought-tolerant. The effects of a variety of abiotic stresses on cropland are therefore the main focus of research to identify SNPs or QTLs (Mangin et al. 2017, Ijaz et al., 2024, Ullah et al., 2023).

3. Progress In Genomics By Comparison

Plants bred in challenging environments yield the highest possible yield by stopping the plant's vital life cycle. Crop resistance to abiotic stress rises as a result. Participatory breeding in conjunction with evolution makes it more effective in overcoming environmental challenges (Raza et al., 2019). For instance, the three genomes A, B, and C are crossed to create the Brassica species AB, AC, and BC. On the other hand, there is no cross-breeding on the B genome chromosome. Nevertheless, self-pollination results in the removal of the genome B chromosome (Xiao et al. 2010). *B. napus* is growing in importance as a crop across borders as a result of its exceptional breeding success. Primitive breeding methods have shown to be beneficial for boosting plant production since the beginning of agricultural breeding. On the other hand, genetic methods combine our understanding to produce a particular organism through breeding. One major challenge is boosting oil production and quality in oilseed crops like canola under unfavourable conditions. Elite genotypes are created by direct gene diversification through the use of combination and traditional breeding techniques. Under unfavourable environmental circumstances, these genotypes can adapt (Nelson et al. 2004). Researchers are working to create canola varieties that typically yield more plants and have better oil quality. Using herbicide resistance, for instance, Canadian breeders of canola seed have successfully produced a number of hybrids with enhanced resistance to abiotic stresses and improved seed superiority (McVetty and Duncan, 2015). Likewise, from 2005 to 2014, China recorded about 215 canola varieties that were better in terms of yield, plant count, oil content, and resistance to abiotic stresses (Hu et al. 2014). As temperature (high or low) and drought stresses become global issues, another study claims that MAS is using molecular techniques and tools, such as genetics or other genetic engineering methods, to support traditional breeding efforts aimed at increasing yield and confirming global food security (Raza et al. 2020). Furthermore, MAS is still useful in developing novel varieties of canola seed that have increased resistance to a range of abiotic stresses because it has been successfully applied to seed breeding strategies (He et al., 2014). Thanks to NGS, whole plant genome sequencing and resequencing are progressing at a rapid pace, leading to the discovery of new molecular markers. Abiotic stressor responses in canola may be influenced by these markers, which may include aspirant genes, QTLs, TFs, and microRNAs. Molecular markers can be used to regulate allelic cooperation behaviours. Determining how many loci are more strongly correlated with seed mass, colour, and crop options is also helpful. These characteristics might differ between varieties (Raman et al., 2016). Modern biotechnological techniques also reveal plant development and offer deeper insights.

These molecular techniques have applications in the field of food crops in the twenty-first century as an integrative and renewing discipline (Varshney et al. 2007). But because of its close kinship with the experimental plant Arabidopsis, studies on *B. napus* are known to be more valuable. Additionally, better outcomes can be achieved by utilising genomic tools for Brassica crops. There appears to be a relationship gap between genotypic and phenotypic data that makes genome-assisted breeding limited. An analysis of physiological parameters and metabolite profiles comprise phenotyping. With regard to intricate characteristics like seed development, oil content, and control crop resilience against the negative consequences of climate change, these high-throughput profiles can offer comprehensive information. Consequently, *B. napus*'s higher yield characters are manipulated. Moreover, transcriptomes and genomic variation associated with economically significant traits can be easily identified thanks to high-throughput phenotyping techniques (Table 1).

Table 1. Reference genomes useful for omics research for Brassica

Genome	Approach	Principal Results	References
B. nigra accession CGN7651	ONT, Hi-C	The centromeric regions exhibited a hotspot of ALE-type retroelements, indicating their significant contribution to the centromere divergence of B. nigra.	Perumal, et al. 2020
B. oleracea cultivar “C-8”	PacBio, Illumina HiSeq, transcriptomics	Within the Brassica genus, cauliflower is the most recent variety to evolve. In comparison to other B. oleracea species, it has a higher number of repetitive sequences.	Sun, et al. 2019
991 B. napus worldwide accessions	WGRS, SNP genotyping	Regions with selective sweeps that are enriched in stress response-related genes	Wu, et al. 2019
588 B. napus worldwide accessions	WGRS, SNP genotyping	A number of potential genes have been found to contribute to the biotic stress response through sub-genomic-specific selection.	Lu, et al. 2019
Population mapping of inbred lines “C1234” (resistant) and “C1184” (susceptible) for cabbage B. oleracea var. capitata	WGRS, QTL mapping	Black rot resistance in B. oleracea is linked to 21 putative NBS-LRR genes.	Lee, et al. 2015
B. rapa cultivar “Chiifu-401-42”	Optical mapping, Hi-C, PacBio	Improved repeat reads, centromere locations, and additional gene annotations in these challenging regions were all included in V3.0. greater amount of TEs annotated	Zhang, et al. 2018
B. napus winter cultivar “Express 617”	Illumina, ONT, Optical mapping, PacBio HiSeq,	Broken-point sequence at homologous exchange regions resolved	Lee, et al. 2020
B. juncea–B. fruticulosa introgression lines	GBS, GWAS	The majority of the 20 candidate genes are found on B. juncea's A subgenome.	Atri, et al. 2019

Conventional comparative genomics investigations employ a sequence-based methodology, analysing entire genomes to discern patterns and distinctions in the composition and capabilities of genomic sequences. The basis of this strategy is the shared functions of sequence identities that are conserved across species. Although coding regions have been studied in great detail, comparative genomics has not been able to identify less conserved regulatory regions as targets. Comparing the evolutionary divergence of various plant groups and identifying their distinctive functional signatures has been a successful application of this method (Butler et al., 2017). Comparative genomics is enabling researchers to expand their ideas about biochemistry, metabolism, physiology, and genetics by providing access to more genomic resources. Fundamentally, the key to comparative genomics' success in clarifying plant structure, regulatory pathways, and genome annotation is the availability of genome sequence data for plants from all major evolutionary lineages. Studies reveal that, despite advancements, sequencing might not be enough to compare the organisation and structure of genomes across different species (Lewin et al., 2009). Accordingly, the tool is becoming more widely used in comparative regulomics, which uses expression analysis studies to find pairs of orthologues within a gene family that have highly conserved expression (called expressionologues) (Patel et al., 2012). Scientists expression-based data is always used in addition to sequence-based information to identify expressionologues (Ingvarsson et al., 2016). Furthermore, data describing chromatin structure, TF binding, co-expression data, and epigenetic markers further enrich comparative genomics. To transform data into knowledge that is biologically meaningful, a number of computational tools for plant comparative genomics are being developed (Aoki et al., 2016; Lee et al., 2015). These tools will have a significant impact on studies related to ecology, conservation, and evolution as well as the identification of previously unidentified genes (new transcripts), entirely new RNA classes, and unidentified regulatory mechanisms. With the advancement of technology, this method's applications are being expanded to the field of agriculture in order to improve plant crops by identifying allelic differences between species that are linked to agronomic traits (Figure 2).

Figure 2. Enhancement of plants by incorporating omics techniques

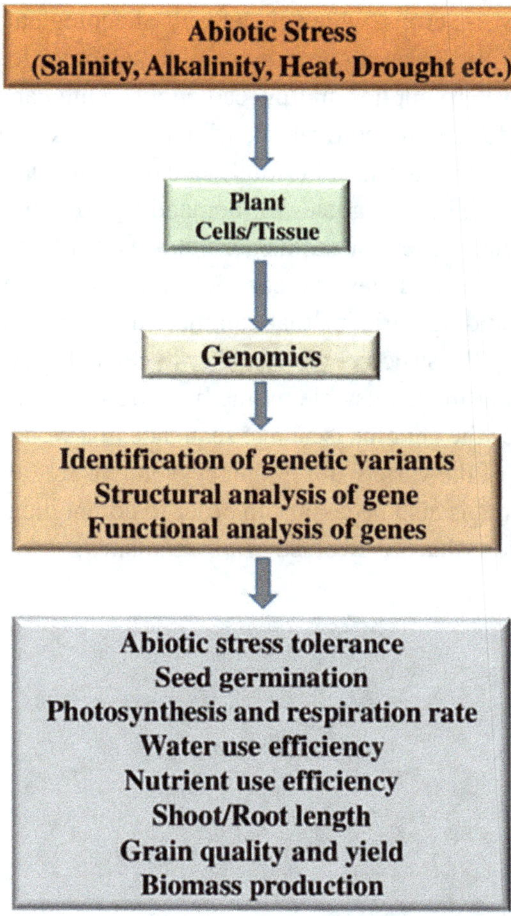

4. CONCLUSION AND FUTURE DIRECTIONS

Integrating data from omics studies is a necessary next step towards comprehending complex traits related to abiotic stress tolerance. Finding the important variables influencing tolerance and/or susceptibility to abiotic stress can also be accomplished by comparing the genotypes of tolerant and susceptible individuals. What's problematic, though, is figuring out which genotype is more sensitive or tolerant of another genotype. Investigating omics analyses to prime tolerance to abiotic stressors is also necessary, as this could reveal information about metabolic adaptation to environmental cues. Crop improvement and the understanding of the functional basis of allelic diversity for various traits and genetic bottlenecks resulting from human and ecological niches can be facilitated by the abundance of germplasm found in naturally diverse crops, which are a result of ecological, evolutionary, and domestication processes. Breeding programs fail to select for low rainfall environments with limited water supplies and instead choose elite canola varieties under the ideal conditions for canola cultivation, which are areas with medium to high rainfall. Because of this, high-value elite alleles are probably biassed in the breeding germplasm used today. The best alleles for demanding conditions should be taken from related species or a variety of canola germplasms. With the help of tools from plant breeding, genomics, and classical genetics, the diverse canola germplasm set has produced high-yielding varieties that can meet the world's needs for food and energy. Understanding the structure of several traits related to adaptation and canola productivity is being made possible by the breakdown of genetic loci and their functional roles. Since the genome is a huge sequence that won't be useful until it is decoded, knowledge of the variety of regulatory elements, regulation mechanisms, and expression patterns is also necessary in order to comprehend the various macromolecular interactions that are involved in an organism's growth and development.

REFERENCES

Ahmad, Y., Haider, S., Iqbal, J., Abbasi, B. A., Yaseen, T., & Mahmood, T. (2022). The mechanisms of genome editing technologies in crop plants. In *Principles and Practices of OMICS and Genome Editing for Crop Improvement* (pp. 295–313). Springer International Publishing. DOI: 10.1007/978-3-030-96925-7_13

Ahmad, Y., Haider, S., Iqbal, J., Naseer, S., Attia, K. A., Mohammed, A. A., & Mahmood, T. (2024). In-silico analysis and transformation of OsMYB48 transcription factor driven by CaMV35S promoter in model plant–Nicotiana tabacum L. conferring abiotic stress tolerance. *GM Crops and Food: Biotechnology in Agriculture and the Food Chain, 15*(1), 130–149. DOI: 10.1080/21645698.2024.2334476 PMID: 38551174

Ali, I., Salah, K. B. H., Sher, H., Ali, H., Ullah, Z., Ali, A., & Mahmood, T. (2022). Drought stress enhances the efficiency of floral dip method of Agrobacterium-mediated transformation in Arabidopsis thaliana. *Brazilian Journal of Biology, 84*, e259326. DOI: 10.1590/1519-6984.259326 PMID: 35703626

Amas, J., Anderson, R., Edwards, D., Cowling, W., & Batley, J. (2021). Status and advances in mining for blackleg (Leptosphaeria maculans) quantitative resistance (QR) in oilseed rape (*Brassica napus*). *Theoretical and Applied Genetics, 134*(10), 3123–3145. DOI: 10.1007/s00122-021-03877-0 PMID: 34104999

Aoki, Y., Okamura, Y., Tadaka, S., Kinoshita, K., & Obayashi, T. (2016). ATTED-II in 2016: A plant coexpression database towards lineagespecific coexpression. *Plant & Cell Physiology, 57*(1), e5. DOI: 10.1093/pcp/pcv165 PMID: 26546318

Apel, K., & Hirt, H. (2004). REACTIVE OXYGEN SPECIES: Metabolism, Oxidative Stress, and Signal Transduction. *Annual Review of Plant Biology, 55*(1), 373–399. DOI: 10.1146/annurev.arplant.55.031903.141701 PMID: 15377225

Atri, C., Akhatar, J., Gupta, M., Gupta, N., Goyal, A., Rana, K., Kaur, R., Mittal, M., Sharma, A., Singh, M. P., Sandhu, P. S., Barbetti, M. J., & Banga, S. S. (2019). Molecular and genetic analysis of defensive responses of Brassica juncea—B. Fruticulosa introgression lines to Sclerotinia infection. *Scientific Reports, 9*(1), 17089. DOI: 10.1038/s41598-019-53444-3 PMID: 31745129

Bancroft, I., Morgan, C., Fraser, F., Higgins, J., Wells, R., Clissold, L., Baker, D., Long, Y., Meng, J., Wang, X., Liu, S., & Trick, M. (2011). Dissecting the genome of the polyploid crop oilseed rape by transcriptome sequencing. *Nature Biotechnology, 29*(8), 762–766. DOI: 10.1038/nbt.1926 PMID: 21804563

Butler, J. B., Vaillancourt, R. E., Potts, B. M., Lee, D. J., King, G. J., Baten, A., Shepherd, M., & Freeman, J. S. (2017). Comparative genomics of Eucalyptus and Corymbia reveals low rates of genome structural rearrangement. *BMC Genomics*, *18*(1), 397. DOI: 10.1186/s12864-017-3782-7 PMID: 28532390

Cheng, F., Wu, J., & Wang, X. (2014). Genome Triplication Drove the Diversification of Brassica Plants. *Horticulture Research*, *1*(1), 14024. DOI: 10.1038/hortres.2014.24 PMID: 26504539

Ding, Y., Shi, Y., & Yang, S. (2019). Advances and Challenges in Uncovering Cold Tolerance Regulatory Mechanisms in Plants. *The New Phytologist*, *222*(4), 1690–1704. DOI: 10.1111/nph.15696 PMID: 30664232

Downing, W. L., Mauxion, F., Fauvarque, M. O., Reviron, M.-P., De Vienne, D., Vartanian, N., & Giraudat, J. (1992). A Brassica napus transcript encoding a protein related to the Künitz protease inhibitor family accumulates upon water stress in leaves, not in seeds. *The Plant Journal*, *2*(5), 685–693. DOI: 10.1111/j.1365-313X.1992.tb00137.x PMID: 1302628

El-Metwally, S., Ouda, O. M., & Helmy, M. (2014). *Next generation sequencing technologies and challenges in sequence assembly* (Vol. 7). Springer Science & Business. DOI: 10.1007/978-1-4939-0715-1

Faizan, M., Sultan, H., Alam, P., Karabulut, F., Cheng, S. H., Rajput, V. D., Minkina, T., Hayat, S., Khan, M. N., & Nie, L. (2024). Melatonin and its cross-talk with other signaling molecules under abiotic stress. *Plant Stress*, *11*, 100410. DOI: 10.1016/j.stress.2024.100410

Gilliham, M., Able, J. A., & Roy, S. J. (2017). Translating knowledge about abiotic stress tolerance to breeding programmes. *The Plant Journal*, *90*(5), 898–917. DOI: 10.1111/tpj.13456 PMID: 27987327

Gilroy, S., Białasek, M., Suzuki, N., Górecka, M., Devireddy, A. R., Karpiński, S., & Mittler, R. (2016). ROS, Calcium, and Electric Signals: Key Mediators of Rapid Systemic Signaling in Plants. *Plant Physiology*, *171*(3), 1606–1615. DOI: 10.1104/pp.16.00434 PMID: 27208294

Haider, S., Iqbal, J., Naseer, S., Yaseen, T., Shaukat, M., Bibi, H., & Mahmood, T. (2021). Molecular mechanisms of plant tolerance to heat stress: Current landscape and future perspectives. *Plant Cell Reports*, *40*(12), 1–25. DOI: 10.1007/s00299-021-02696-3 PMID: 33890138

He, J., Zhao, X., Laroche, A., Lu, Z. X., Liu, H., & Li, Z. (2014). Genotyping-bysequencing (GBS), an ultimate marker-assisted selection (MAS) tool to accelerate plant breeding. *Frontiers in Plant Science*, *5*, 484. DOI: 10.3389/fpls.2014.00484 PMID: 25324846

Hu, Q., Hua, W., Yin, Y., Zhang, X., Liu, L., Shi, J., Zhao, Y., Qin, L., Chen, C., & Wang, H. (2017). Rapeseed research and production in China. *The Crop Journal*, *5*(2), 127–135. DOI: 10.1016/j.cj.2016.06.005

Huynh, B. L., Ehlers, J. D., Huang, B. E., Muñoz-Amatriaín, M., Lonardi, S., Santos, J., Ndeve, A., Batieno, B. J., Boukar, O., Cisse, N., Drabo, I., Fatokun, C., Kusi, F., Agyare, R. Y., Guo, Y.-N., Herniter, I., Lo, S., Wanamaker, S. I., Xu, S., & Roberts, P. A. (2018). A multi-parent advanced generation inter-cross (MAGIC) population for genetic analysis and improvement of cowpea (*Vigna unguiculata* L. Walp.). *The Plant Journal*, *93*(6), 1129–1142. DOI: 10.1111/tpj.13827 PMID: 29356213

Ijaz, S., Iqbal, J., Abbasi, B. A., Tufail, A., Ullah, Z., Sharifi-Rad, J., & Uddin, S. (2023). Biofortification: Lessons from the Past and Strategies for Future Food Security. In *Legumes Biofortification* (pp. 521–545). Springer International Publishing. DOI: 10.1007/978-3-031-33957-8_23

Ijaz, S., Iqbal, J., Abbasi, B. A., Yaseen, T., Rehman, S., Kazi, M., & Mahmood, T. (2024). Role of OMICS-Based Technologies in Plant Sciences. OMICs-based Techniques for Global Food Security, 45-66.

Ingvarsson, P. K., Hvidsten, T. R., & Street, N. R. (2016). Towards integration of population and comparative genomics in forest trees. *The New Phytologist*, *212*(2), 338–344. DOI: 10.1111/nph.14153 PMID: 27575589

Jenko, J., Gorjanc, G., Cleveland, M. A., Varshney, R. K., Whitelaw, C. B., Woolliams, J. A., & Hickey, J. M. (2015). Potential of promotion of alleles by genome editing to improve quantitative traits in livestock breeding programs. *Genetics, Selection, Evolution : GSE*, *47*(1), 55. DOI: 10.1186/s12711-015-0135-3 PMID: 26133579

Kang, Y. J., Lee, T., Lee, J., Shim, S., Jeong, H., Satyawan, D., Kim, M. Y., & Lee, S. H. (2016). Translational genomics for plant breeding with the genome sequence explosion. *Plant Biotechnology Journal*, *14*(4), 1057–1069. DOI: 10.1111/pbi.12449 PMID: 26269219

Karabulut, F., & Khan, A. Abd_Allah, E. F., Parray, J. A. and Hashem, A. (2024). Omics and other biotechnological tools for biopesticide and microbial bioremediation. In Microbiome-Assisted Bioremediation (pp. 103-128). Academic Press.

Karim, M. M., Siddika, A., Tonu, N. N., Hossain, D. M., Meah, M. B., Kawanabe, T., Fujimoto, R., & Okazaki, K. (2014). Production of high yield short duration Brassica napus by interspecific hybridization between B. oleracea and B. rapa. *Breeding Science*, *63*(5), 495–502. DOI: 10.1270/jsbbs.63.495 PMID: 24757390

Lee, H., Chawla, H. S., Obermeier, C., Dreyer, F., Abbadi, A., & Snowdon, R. (2020). Chromosome-scale assembly of winter oilseed rape Brassica napus. *Frontiers in Plant Science*, *11*, 496. DOI: 10.3389/fpls.2020.00496 PMID: 32411167

Lee, J., Izzah, N. K., Jayakodi, M., Perumal, S., Joh, H. J., Lee, H. J., Lee, S.-C., Park, J. Y., Yang, K.-W., Nou, I.-S., Seo, J., Yoo, J., Suh, Y., Ahn, K., Lee, J. H., Choi, G. J., Yu, Y., Kim, H., & Yang, T.-J. (2015). Genome-wide SNP identification and QTL mapping for black rot resistance in cabbage. *BMC Plant Biology*, *15*(1), 32. DOI: 10.1186/s12870-015-0424-6 PMID: 25644124

Lee, T., Yang, S., Kim, E., Ko, Y., Hwang, S., Shin, J., Shim, J. E., Shim, H., Kim, H., Kim, C., & Lee, I. (2015). AraNet v2: An improved database of co-functional gene networks for the study of Arabidopsis thaliana and 27 other nonmodel plant species. *Nucleic Acids Research*, *43*(Database issue), D996–D1002. DOI: 10.1093/nar/gku1053 PMID: 25355510

Lewin, H. A., Larkin, D. M., Pontius, J., & O'Brien, S. J. (2009). Every genome sequence needs a good map. *Genome Research*, *19*(11), 1925–1928. DOI: 10.1101/gr.094557.109 PMID: 19596977

Lu, K., Wei, L., Li, X., Wang, Y., Wu, J., Liu, M., Zhang, C., Chen, Z., Xiao, Z., Jian, H., Cheng, F., Zhang, K., Du, H., Cheng, X., Qu, C., Qian, W., Liu, L., Wang, R., Zou, Q., & Li, J. (2019). Whole-genome resequencing reveals Brassica napus origin and genetic loci involved in its improvement. *Nature Communications*, *10*(1), 1154. DOI: 10.1038/s41467-019-09134-9 PMID: 30858362

Mangin, B., Casadebaig, P., Cadic, E., Blanchet, N., Boniface, M. C., Carrère, S., Gouzy, J., Legrand, L., Mayjonade, B., Pouilly, N., André, T., Coque, M., Piquemal, J., Laporte, M., Vincourt, P., Munos, S., & Langlade, N. B. (2017). Genetic control of plasticity of oil yield for combined abiotic stresses using a joint approach of crop modelling and genome-wide association. *Plant, Cell & Environment*, *40*(10), 2276–2291. DOI: 10.1111/pce.12961 PMID: 28418069

McVetty, P. B. E., & Duncan, R. W. (2015). *Canola, rapeseed, and mustard: for biofuels and bioproducts. Industrial crops. Handbook of plant breeding*. Springer.

Nelson, R. J., Naylor, R. L., & Jahn, M. M. (2004). The role of genomics research in improvement of" orphan" crops. *Crop Science*, *44*(6), 1901–1904. DOI: 10.2135/cropsci2004.1901

Patel, R. V., Nahal, H. K., Breit, R., & Provart, N. J. (2012). BAR expressolog identification: Expression profile similarity ranking of homologous genes in plant species. *The Plant Journal*, *71*(6), 1038–1050. DOI: 10.1111/j.1365-313X.2012.05055.x PMID: 22607031

Perumal, S., Koh, C. S., Jin, L., Buchwaldt, M., Higgins, E. E., Zheng, C., Sanko, D., Robinson, S. J., Kagale, S., & Navabi, Z.-K.. (2020). A high-contiguity Brassica nigra genome localizes active centromeres and defines the ancestral Brassica genome. *Nature Plants*, *6*(8), 929–941. DOI: 10.1038/s41477-020-0735-y PMID: 32782408

Raman, H., Raman, R., Coombes, N., Song, J., Prangnell, R., Bandaranayake, C., Tahira, R., Sundaramoorthi, V., Killian, A., Meng, J., Dennis, E. S., & Balasubramanian, S. (2016). Genome-wide association analyses reveal complex genetic architecture underlying natural variation for flowering time in canola. *Plant, Cell & Environment*, *39*(6), 1228–1239. DOI: 10.1111/pce.12644 PMID: 26428711

Raman, R., Taylor, B., Marcroft, S., Stiller, J., Eckermann, P., Coombes, N., Rehman, A., Lindbeck, K., Luckett, D., Wratten, N., Batley, J., Edwards, D., Wang, X., & Raman, H. (2012). Molecular mapping of qualitative and quantitative loci for resistance to Leptosphaeria maculans causing blackleg disease in canola (*Brassica napus* L.). *Theoretical and Applied Genetics*, *125*(2), 405–418. DOI: 10.1007/s00122-012-1842-6 PMID: 22454144

Raza, A., Ashraf, F., Zou, X., Zhang, X., & Tosif, H. (2020). Plant adaptation and tolerance to environmental stresses: mechanisms and perspectives. In *Plant ecophysiology and adaptation under climate change: mechanisms and perspectives I* (pp. 117–145). Springer. DOI: 10.1007/978-981-15-2156-0_5

Raza, A., Charagh, S., Razzaq, A., Javed, R., Khan, R. S. A., & Hasanuzzaman, M. (2020). Brassicaceae plants response and tolerance to drought stress: physiological and molecular interventions. In *The plant family Brassicaceae* (pp. 229–261). Springer. DOI: 10.1007/978-981-15-6345-4_7

Raza, A., Razzaq, A., Mehmood, S. S., Zou, X., Zhang, X., Lv, Y., & Xu, J. (2019). Impact of climate change on crops adaptation and strategies to tackle its outcome: A review. *Plants*, *8*(2), 34. DOI: 10.3390/plants8020034 PMID: 30704089

Rhee, S. Y., & Mutwil, M. (2014). Towards revealing the functions of all genes in plants. *Trends in Plant Science*, *19*(4), 213–221. DOI: 10.1016/j.tplants.2013.10.006 PMID: 24231067

Sato, Y., Murakami, T., Funatsuki, H., Matsuba, S., Saruyama, H., & Tanida, M. (2001). Heat Shock-mediated APX Gene Expression and protection against Chilling Injury in rice Seedlings. *Journal of Experimental Botany*, *52*(354), 145–151. DOI: 10.1093/jexbot/52.354.145 PMID: 11181723

Shen, W., Li, H., Teng, R., Wang, Y., Wang, W., & Zhuang, J. (2019). Genomic and transcriptomic analyses of HD-Zip family transcription factors and their responses to abiotic stress in tea plant (*Camellia sinensis*). *Genomics*, *111*(5), 1142–1151. DOI: 10.1016/j.ygeno.2018.07.009 PMID: 30031053

Singh, R. K., Prasad, A., Muthamilarasan, M., Parida, S. K., & Prasad, M. (2020). Breeding and biotechnological interventions for trait improvement: Status and prospects. *Planta*, *252*(4), 54. DOI: 10.1007/s00425-020-03465-4 PMID: 32948920

Singh, R. K., Sahu, P. P., Muthamilarasan, M., Dhaka, A., & Prasad, M. (2017). Genomics-assisted breeding for improving stress tolerance of graminaceous crops to biotic and abiotic stresses: Progress and prospects. In Senthil-Kumar, M. (Ed.), *Plant tolerance to individual and concurrent stresses* (pp. 59–81). Springer., DOI: 10.1007/978-81-322-3706-8_5

Smith, S., & De Smet, I. (2012). Root system architecture: Insights from Arabidopsis and cereal crops. *Philosophical Transactions of the Royal Society of London. Series B, Biological Sciences*, *367*(1595), 1441–1452. DOI: 10.1098/rstb.2011.0234 PMID: 22527386

Sun, D., Wang, C., Zhang, X., Zhang, W., Jiang, H., Yao, X., Liu, L., Wen, Z., Niu, G., & Shan, X. (2019). Draft genome sequence of cauliflower (*Brassica oleracea* L. var. botrytis) provides new insights into the C genome in Brassica species. *Horticulture Research*, *6*(1), 82. DOI: 10.1038/s41438-019-0164-0 PMID: 31645943

Sun, W., Ma, W., Lei, J., Liu, Q., Yang, R., & Wu, J.. (2007). Study on Adaptation and Introduction Possibility of Winter Rapeseed to Dry and Cold Areas in Northwest China. *Zhongguo Nong Ye Ke Xue*, *40*, 2716–2726. DOI: 10.3321/j.issn:0578-1752.2007.12.008

Ullah, Z., Iqbal, J., Abbasi, B. A., Ijaz, S., Sharifi-Rad, J., Yaseen, T., & Mahmood, T. (2023). How Genome Editing Can Be Helpful in the Biofortification of Legumes. In *Legumes Biofortification* (pp. 207–232). Springer International Publishing. DOI: 10.1007/978-3-031-33957-8_8

Varshney, R., (2007). Extending the repertoire of microsatellite markers for genetic linkage mapping and germplasm screening in chickpea. *Journal of SAT Agricultural Research*, *5*, 1–3.

Varshney, R. K., Pandey, M. K., Bohra, A., Singh, V. K., Thudi, M., & Saxena, R. K. (2019). Toward the sequence-based breeding in legumes in the post-genome sequencing era. *Theoretical and Applied Genetics*, *132*(3), 797–816. DOI: 10.1007/s00122-018-3252-x PMID: 30560464

Wang, P., Su, L., Gao, H., Jiang, X., Wu, X., Li, Y., Zhang, Q., Wang, Y., & Ren, F. (2018). Genome-wide characterization of bHLH genes in grape and analysis of their potential relevance to abiotic stress tolerance and secondary metabolite biosynthesis. *Frontiers in Plant Science*, *9*, 64. DOI: 10.3389/fpls.2018.00064 PMID: 29449854

Wu, D., Liang, Z., Yan, T., Xu, Y., Xuan, L., Tang, J., Zhou, G., Lohwasser, U., Hua, S., Wang, H., Chen, X., Wang, Q., Zhu, L., Maodzeka, A., Hussain, N., Li, Z., Li, X., Shamsi, I. H., Jilani, G., & Jiang, L. (2019). Whole-genome resequencing of a worldwide collection of rapeseed accessions reveals the genetic basis of ecotype divergence. *Molecular Plant*, *12*(1), 30–43. DOI: 10.1016/j.molp.2018.11.007 PMID: 30472326

Xiao, Y., Chen, L., Zou, J., Tian, E., Xia, W., & Meng, J. (2010). Development of a population for substantial new type Brassica napus diversified at both A/C genomes. *Theoretical and Applied Genetics*, *121*(6), 1141–1150. DOI: 10.1007/s00122-010-1378-6 PMID: 20556596

Zhang, J., Li, G., Li, H., Pu, X., Jiang, J., Chai, L., Zheng, B., Cui, C., Yang, Z., Zhu, Y., & Jiang, L. (2015). Transcriptome Analysis of Interspecific Hybrid between Brassica napusand B. Rapa Reveals Heterosis for Oil Rape Improvement. *International Journal of Genomics*, *2015*, 1–11. DOI: 10.1155/2015/230985 PMID: 26448924

Zhang, L., Cai, X., Wu, J., Liu, M., Grob, S., Cheng, F., Liang, J., Cai, C., Liu, Z., Liu, B., Wang, F., Li, S., Liu, F., Li, X., Cheng, L., Yang, W., Li, M., Grossniklaus, U., Zheng, H., & Wang, X. (2018). Improved Brassica rapa reference genome by single-molecule sequencing and chromosome conformation capture technologies. *Horticulture Research*, *5*(1), 50. DOI: 10.1038/s41438-018-0071-9 PMID: 30131865

Chapter 11
Transcriptomics of Canola Plants

Ilham Khan
Quaid-i-Azam University, Islamabad, Pakistan

Munazza Jabeen
Quaid-i-Azam University, Islamabad, Pakistan

Zarqa Riaz
Quaid-i-Azam University, Islamabad, Pakistan

Junaid Shehzad
Quaid-i-Azam University, Islamabad, Pakistan

Najeeba Parre Paker
https://orcid.org/0009-0001-4501-3345
Quaid-i-Azam University, Islamabad, Pakistan

Banzeer Ahsan Abbasi
https://orcid.org/0000-0003-3036-4481
Rawalpindi Women University, Pakistan

Ghazala Mustafa
Quaid-i-Azam University, Islamabad, Pakistan

ABSTRACT

This chapter provides detailed insight into the complex transcriptomic landscape of an important oilseed crop, canola. It provided an overview of the economic importance of canola, some progress in genomic resources followed by reviews of canola genome and transcriptome analyses. It discusses insights from transcriptomics on canola development and explains molecular mechanisms that characterize growth and differentiation. This chapter examines the response of canola to abiotic and biotic stresses and examines how transcriptomic analysis is used to inform breeding strategies to address environmental challenges. In addition, the role of transcriptomics in breeding and genetic improvement to create improved varieties with improved characteristics is emphasized. Finally, this chapter describes recent

DOI: 10.4018/979-8-3693-8603-3.ch011

advances in transcriptomics technologies, such as high-throughput sequencing and bioinformatics tools, which play important role in leading us further towards understanding the biology of canola and its adaptation to a changing environment.

1. INTRODUCTION

Canola, belonging to the Brassicaceae family, is now one of the most important oilseed crops worldwide because of its very high-quality oil content and nutritional value. This crop is widely grown for its seeds, which have about 40% oil and 15% protein content. Therefore, it plays a very significant role in edible oil supply and animal nutrition (Razavizadeh, 2015; Assis et al., 2020). The production of this crop has increased over the last few decades. It has come to be recognized as the second most produced oilseed crop globally after soybean mainly (Koh et al., 2016). This is due to an increase in acreage unofficially, which has grown suddenly because of its favorable fatty acid composition with low levels of saturated fats, hence attracting health-conscious consumers.

Besides also being one of the nutritionally valued crops, beyond nutritional value, canola has a lot to do with agricultural systems, especially in rotations, since it contributes to soil and reduces certain pests (Kruger et al., 2016). On the other hand, regarding economy, this crop has a huge role in many countries' agriculture, among the first being Canada, since it is one of the key commodities exported by it (Eldin et al., 2018; Assis et al., 2020). Differences in canola cultivar variability also provide possibilities for breeding programs with the aim of enhancing specific traits correlated with oil quality, disease resistance, and abiotic stress tolerance (Rahman, 2013; Ton et al., 2020). Besides oil production, by-products from canola, such as canola meal, are used in animal feed and other industrial applications (Nikvash et al., 2010).

Transcriptomics has become a key aspect under focus where the transcripts of RNA developed by the genome under conditions are studied. It has thus imposed importance on the research for the understanding of gene expression in crops like canola. Transcriptomic information provides insight into dynamic changes in the expression of genes which might be a result of environmental stimuli in the forms of biotic or abiotic stresses (Jing, 2018; Lohani et al., 2020). Application of high-throughput sequencing has made the identification of similar differentially expressed genes responsible for any phenotypic trait achievable, thereby explaining the molecular basis of such a wide range of plant responses to stressors (Zhang et al., 2014: Jia et al., 2015). Such knowledge is important in the formulation of methods aimed at improving crop resilience and productivity both in present times and considering future climate change and agricultural requirements (Lohani et al., 2020).

Because of the potential to improve vital agronomic traits, the importance of transcriptomic studies in canola is underlined. Indeed, improved transcriptomic approaches have shown that the genetic loci responsible for disease resistance could be identified, most importantly against diseases by a pathogen such as *Leptosphaeria maculans*, which causes blackleg disease in canola (Raman et al., 2012; Larkan et al., 2012; Larkan et al., 2014). The development of cultivars with enhanced disease resistance will minimize dependency on chemical fungicides and offer more feasible alternatives for the pursuit of sustainable agriculture through the application of genetics known about the resistance (Howlett et al., 2015). Transcriptomic analyses have also been used in the improvement of oil quality where genes associated with the biosynthesis or metabolism of lipids are identified and serve as entities for genetic modification or selective breeding (Kittipol et al., 2019; Ton et al., 2020).

In addition, transcriptomic studies showed highlights on the mechanisms of stress tolerance in canola, under specific conditions of salinity and drought (Shahbaz et al., 2013; Jing, 2018). The identification of such stress-responsive genes with their regulatory networks is of prime importance in developing the canola varieties that are productive under adversities for food security amidst climate variability (Lohani et al., 2020). This ensures an increased understanding of the complex interactions of genes, proteins, and metabolites involved in the canola pathway by integrating the transcriptomic data with proteomic and metabolomic approaches, hence facilitating comprehensive breeding strategies (Jia et al., 2015).

The present chapter outlines the multi-dimensional importance of canola as an essential oilseed crop and describes the revolutionary power in transcriptomic studies upgrading agricultural aspects of the plant. Integration of transcriptomic information to classically applied breeding methodologies opens encouraging perspectives in the pursue of disease- and abiotic stresses-tolerant high-yielding varieties of canola. In any case, with increased demand for canola the world over, the integration of improved genomic technologies will characterize future production methods in pursuit of sustainability in the agricultural system.

2. CANOLA GENOME AND TRANSCRIPTOME

It is known that canola is an allotetraploid and, consequently, its genome derived from hybridization between two diploid parents, *Brassica rapa* and *Brassica oleracea*. As an allopolyploidy species, the genome of canola has a very intriguing organization, as it harbors two main subgenomes, usually referred to as A and C, with many structurally variable genes (Havlíčková et al., 2017). Allotetraploidy in canola has a high effect on its genetic composition because homoeologous exchanges during meiosis allow genetic diversity and adaptiveness to environmental stressors (Chalhoub et al., 2014; Gabur et al., 2018). Extensive structural variation also prevails in the genome, being one part of the differential expression of genes that are responsible for important agronomic traits such as glucosinolate content and biosynthesis of oil in canola (Mason et al., 2018). They include variants which may be a bit inconvenient in the interpretation of transcriptomic data, since the different expression level can be widely changed between two sub-genomes-leading to functional divergence. Moreover, its entire evolutionary history from domestication to further selection included dynamic changes within the genome, to this day, according to agricultural and environmental changes.

Furthermore, genomic complexity in canola is accentuated by the presence of numerous QTLs associated with economically important traits such as disease resistance and yield. Characterization and identification of these QTLs are among the most vital aspects of breeding programs toward the improvement of canola cultivars (Saad et al., 2021). Indeed, integration of the transcription analysis with genomic data advances such understanding of the genetic basis of those traits, helping in the enhancement of resilience and productivity within canola development.

Different techniques are used for transcriptome profiling in canola, but RNA-Seq and microarrays are the prominent ones. RNA-Seq revolutionized the field of transcriptomics by allowing high-throughput and cost-effective quantification of gene expression across the whole genome. This approach gives the researcher an opportunity to capture the canola transcriptome complexity by detecting novel transcripts, alternative splicing events, and fusion genes that couldn't be determined by using traditional microarray approaches (Malmberg et al., 2017).

Although presently less in use after the advent of RNA-Seq, microarray still finds a place in transcriptome analysis, especially when one is interested in targeted studies related to particular gene sets. These platforms allow for the analysis of parallel thousands of genes for their expression levels and provide an overview of the regulatory networks that control plant responses against biotic and abiotic stresses (Saad et al., 2021). In techniques of transcriptome profiling, a particular method is usually chosen concerning the question being addressed, available resources, and resolution of gene expression data. Except for RNA-Seq and microarrays, other methodologies have been employed in transcriptomic studies, such as GBS, to further understand patterns of gene expression with respect to genetic variation.

GBS enables the analysis of both genetic markers and transcript levels to be performed on a sample for a more complete view of the genotype-to-phenotype relationship in canola. It will be of particular value in breeding programs because the improvements of cultivars under these conditions will depend largely on the clear understanding of the genetic basis of traits (Mason et al., 2018). The transcriptomic canola studies contribute to breeding and genetic studies. Basically, the elucidation of the genetic architecture of such important traits facilitates and enables the development of more effective breeding strategies leveraging genetic diversity within canola germplasm. For instance, expression profiling using genes related to the resistance of diseases will help breeders in their choice of parental lines to develop superior cultivars with a view to arriving at significant resistance against the pathogen *Leptosphaeria maculans*, among others (Zou et al., 2021).

Integration of transcriptomic data with genomic information enables acceleration of the process of identifying candidate genes for traits of interest, hence targeted approaches to genetic improvement. Besides, the capability to determine the level of gene expression to be associated with a specific phenotype allows breeders to make more informed selections on which line should be advanced in their programs (Saad et al., 2021). This accelerates the pace of the entire breeding cycle and enhances the efficiency of the processes involved in the production of canola. This allotetraploid nature of canola faces both challenging and enabling perspectives toward the understanding of gene expression and its consequence for breeding. While transcriptomic technologies are in continuous evolution, they will surely play an important role in improving our understanding of the genetics of canola and the sustainability of the production systems soon.

3. TRANSCRIPTOMIC INSIGHTS INTO CANOLA DEVELOPMENT

Since transcriptomics studies in canola invariably deal with critical knowledge regarding the molecular mechanisms which govern various developmental phases, such as seed development, flowering, and vegetative growth, this chapter synopsizes the contemporary understanding of the pattern of gene expression and regulatory networks that underline these processes, bringing into the limelight key genes and their role in canola development. In canola, seeds develop through a very active and differentiated process, which has been described to involve the following: fertilization, seed maturity, and oil biosynthesis (Figure 1). Several key genes participating in these described processes have been identified by transcriptomic analyses. Including those involved in the encoding of fatty acid desaturases and acyl-CoA synthetases, generally believed to be important in oil biosynthesis during the determination of the fatty acid composition of canola seed (Wang et al., 2017). All these genes are highly regulated during maturation, where their expression is either activated or repressed in relation to developmental and environmental advice through certain transcription factors (Raman et al., 2023). Recent works underlined the role of auxins in seed development, showing that the inoculation with auxin-producing bacteria may improve root architecture and, at last, seed yield (Raimi & Adeleke, 2023). The interaction between the canola root and beneficial microbes can enhance nutrient uptake and stress tolerance, which can further impact seed development and oil content (Suchan et al., 2020). Transcriptomic information combined with functional studies identified that auxin signaling pathways take part in coordinating processes of seed growth and maturation, playing a critical role in the final yield (Su et al., 2017). This transition in canola from vegetative to reproductive stages is controlled by the elaborate interaction of environmental signals with genetic factors. Transcriptomic investigations have identified a set of key genes responsible for flowering time, such as the FLOWERING LOCUS T and EARLY FLOWERING genes. Such genes are known to integrate photoperiod and temperature signals, hence influencing the timing of flowering in response to environmental conditions (Raman et al., 2020).

It also reveals that with a long photoperiodic signal, FT expression promotes flowering; meanwhile, the ELF genes are known to regulate the timing of expression according to the circadian rhythm (Raman et al., 2022). QTL mapping was also done to explain the fertility inherited characteristics of rapeseed, and loci responsible for reproductive traits subsequently underlined seed set and pod development (Asaduzzaman, 2024). These results point out the importance of the understanding of underlying genetics behind the architecture of flowering and reproductive machinery that might inform breeding strategies toward improving yield stability under variable environmental conditions (Umesh, 2023). Vegetative growth in canola involves developing leaves and roots, which are quite important in nutrient acquisition and overall plant health. Transcriptomic comparisons within leaf tissues along the growth stages of plants have shown differential expression in genes associated with photosynthesis, nutrient transport, and stress responses (Williams and Vail, 2021). For instance, the expression of genes related to nitrogen assimilation is highly important during the early vegetative stage, since canola has a high demand for nitrogen to achieve good vegetative growth (Kiran et al., 2019).

Taking into consideration the other important factor in canola development, it is its root architecture, which particularly impacts on a plant attaining good access to water and nutrient availability. RSA has been noted as highly variable and affected by the environment, especially among the nutrient levels of the soil and the moisture levels of the soil (Stanic et al., 2021). Major regulatory genes identified in RSA through transcriptomic profiling involve those related to hormonal pathways, such as auxin signaling and strigolactone pathways, which again play a major role in the adjustment of root growth and root branching (Hu et al., 2018). There is immense potential for interaction between root exudates and soil microorganisms to prove highly important in shaping the root architecture for better nutrient uptake (Pan et al., 2017). Integration of transcriptomic data from various stages of development and tissues provides a wide perspective into the biology of canola (Arifuzzaman et al., 2019). Identification of shared regulatory networks underlying phenomena such as seed development, flowering, and vegetative growth underlines the interconnectivity of these very aspects.

Advanced genomic techniques, like GWAS, have helped find their genetic variants that causate traits of interest and further inform targeted breeding programs (Lamaoui et al., 2018). We would have resulting insights from such transcriptomic studies that go beyond basic research into canola breeding and agronomy (Raza et al., 2020). This is possible because breeders verifying the reason behind traits like drought tolerance, nutrient use efficiency, and disease resistance on a molecular basis could develop canola varieties that suit changing conditions in the environment and agricultural practices. Further, the transcriptomic exploration of plant-microbe interactions has given way to new prospects for enhancing crop resilience and productivity by adopting sustainable methods of agriculture. Therefore, transcriptomic surveys have so far advanced knowledge about the development in canola by illustrating complex gene expression patterns and regulatory networks that underline seed maturation, flowering, and vegetative growth. The insight obtained from these studies furthers not only basic knowledge regarding the biology of plants but also has important implications for the improvement of breeding strategies and agricultural practices in canola. The demand for canola as a source for edible oil is gaining, along with its biofuel attributes; hence, the integration of transcriptomic data on other omics approaches would be the most promising towards resilient, high-yielding canola varieties.

Figure 1. Transcriptomic insights into Canola development. This figure shows the active gene expression profiles in Canola throughout the developing stages, from seedling to flowering and seed maturity. Key TFs are underlined and specify organ-specific responses in roots, leaves, stems, and reproductive organs

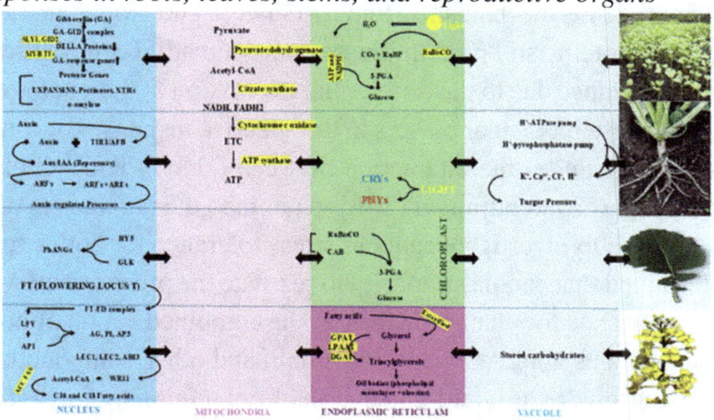

Abbreviations: *ARFs (Auxin response factors), AREs (Antioxidant response elements), LEC (Leafy cotyledon), ABI3 (Abscisic acid insensitive 3), WRI1 (Wrinkled1), AP-1 (Activator protein 1), LFY (Leafy), PhANGs (Photosynthesis-associated nuclear genes), GLK (Golden2-Like), GPAT (Glycerol-sn-3-phosphate acyltransferase), LPAAT (Lysophosphatidic Acid Acyltransferase), DGAT (Diacylglycerol acyltransferase).*

4. RESPONSE OF CANOLA TO ABIOTIC STRESS

Canola is one of the major oilseed crops, and its response to abiotic stressors such as drought, salinity, temperature extremes, and nutrient deficiencies are all very important in respect to a changing climate and increasing requirements for oilseed crops. For this reason, there is a desperate need to understand the transcriptomic responses of canola to such stresses that provide insight into underlying mechanisms regulating plant resilience and adaptation (Figure 2). The chapter compiles findings from a wide range of studies explaining the transcriptomic response of canola to drought and salinity stress, temperature stresses, and nutrient deficiencies (Table 1). Drought stress significantly affects the growth and yield of canola, and several works have reported the morphological and physiological changes that canola varieties undergo during water stress. For instance, Khodabin et al. (2020) showed that under drought stress conditions, drought-tolerant canola varieties grow in height, root length, and overall biomass more than sensitive varieties, which proofs that selecting the hardy genotype would be an advantage for its cultivation under arid conditions. On the other hand, Rezayian et al. (2023) reported that canola is sensitive to drought at all stages of its growth; hence, it requires a specific breeding strategy in line with improving drought tolerance.

Salinity stress has also been regarded as one of the major factors that influence canola production at especially early vegetative stages. It is reported that salinity reduces photosynthetic rates and ion exchange, which affects growth and development in plants (Stafen et al., 2018). Canola has been rated as a moderately salt-tolerant crop, but its yield drops drastically with increased salinity levels (Jing, 2018). Transcription profiling of canola under salinity stress has revealed the complex regulatory network involving genes responsible for ion transport, osmotic adjustment, and antioxidant defense mechanisms (Shokri-Gharelo and Noparvar, 2018). Some calcium-binding proteins and protein kinases have already been implicated in improved salinity tolerance due to the differential expression of their certain members, thus implicating these molecular players in the adaptive responses of canola (Zhang et al., 2014). This includes the interaction between drought and salinity stress. In this respect, Yanagawa and Fujimaki (2013) examined the parameters of the model of root water uptake to investigate canola's stress tolerance for both stresses, and it has been determined that canola's capacity to regulate the processes of water uptake is of crucial importance for plant growth after the combined action of both stresses.

Besides, exogenous applications of melatonin and other compounds have been able to improve growth and yield under drought conditions; thus, indicating the possible agronomic practices that could be employed in mitigating stress impacts (Rafique et al., 2023). Another very important aspect of the adaptability in canola is its response to temperature extremes, whether it is heat or cold. The very high temperatures normally affect canola during its growth stage and reduce its seed yield and oil content. Ahmad et al. 2024 reported that under high-temperature stress, plants showed reduced height and relative water content, which resulted in significant reduction in seed yield. The transcriptomic profiling done in canola under heat stress identified important genes linked to the synthesis of heat shock proteins and antioxidant defense, thus playing an essential role in protecting cellular integrity during thermal stresses (Koscielny et al., 2020). On the other hand, cold stress is also one of the challenges in canola production, especially in those countries where temperature fluctuation is very prevalent. Frost stress cross-stress tolerance in canola has been documented by Fordyce et al. (2018), where plants with cold treatment can be set off to become more resilient to the succeeding abiotic stresses.

Cold tolerance in canola is imparted by the regulation in the expression and activity of those transcription factors that, in turn, activate the transcription of genes involved in the adaptive response induced by low temperature (Zhu et al. 2016). Nutrient deficiency, especially in some of the key components or elements such as nitrogen, phosphorus, and potassium, would dramatically alter negative development and production in canola (Raman et al., 2020). Nutrient stress has been indicated through transcriptomic analyses genes upregulated in canola associated with nutrient uptake and metabolism. In relation, most potassium transporters were expressed regarding osmotic stresses, hence proved that potassium is so vital in cellular homeostasis under nutrient-deficient conditions (Osakabe et al., 2013). Integration of transcriptomic data and physiology tests deciphered the complex interaction among nutrient availability and abiotic stress responses (Gao et al., 2018). For instance, photosynthetic efficiency in canola is more responsive to water deficit and applied nitrogen compared with other oilseed crops (Raman et al., 2022). This underlines the importance of managing nutrients in environments where stressors are prevalent. More so, QTLs that are involved in water-use efficiency and nutrient uptake can be identified, which will further facilitate the development of canola cultivars adapted to the soil nutrient-poor condition. The transcriptomic responses of canola to abiotic stresses brought on by drought, salinity, extreme temperatures, and nutrient deficiencies have unraveled a complex interaction of genetic, physiological, and biochemical mechanisms involved in plant resilience. These are useful to understand in view of proper breeding strategy and agronomical practices for the improvement of canola regarding changed environmental variables. Therefore, future studies should be geared towards the integration of transcriptomic data with phenotype assessments to identify key regulatory networks and pathways that comprise potential targets for further improvement in stress tolerance in canola.

Table 1. Transcriptomics response of Canola to abiotic stress.

S. No	Stress	Transcription Factors	Function
1	Heat	ERF	Ethylene-responsive pathways, Heat Tolerance
2	Drought	NAC	Drought Tolerance, ABA-responsive Genes
3	Heat	bHLH, NAC, ERF	Increase Tolerance
4	Heat	ERF, HSFs, bHLH, MYB, C_2H_2 and NAC	Stress-Specific Gene Regulation
5	Heat, Drought	MYB	Secondary Metabolism
6	Heat	NAC103	ROS Accumulation and Cell Death
7	Drought	NAC	Enhances Drought Tolerance, ABA-Responsive Genes
8	Heat	bHLH, MYB and C2H2	Heat Tolerance
9	Drought, Salinity	WRKY	ROS Scavenging Pathways, Abiotic Stress Tolerance
10	Salt, Drought and Heat	ERF	Heat Shock
11	Cold	Zinc-finger	Cellular stability, Cold Tolerance
12	Salinity, Cold	bZIP	Salinity Cold and Tolerance, ABA Signaling Pathway
13	Drought, Salt	ARF	Auxin Signaling Pathways
14	High, Light	bHLH	Protection from Photodamage
15	Heat	HSF	HSP and Antioxidants
16	Drought	ERF/ AP2	Stress Tolerance and Osmotic Pressure

Source: Ozhuner et al., 2013; Niu et al., 2014; Sarkar et al., 2014; Wang et al., 2017; Qian et al., 2019; Chen et al., 2022

5. RESPONSE OF CANOLA TO BIOTIC STRESS

The specific genetic, biochemical, and physiological mechanisms of both the canola-pathogen and canola-pest interactions involve complex interactions that constitute the response in plants against biotic stress. Canola is also susceptible to several pathogens including fungi, bacteria, and viruses--that cause serious losses in yield as well as in quality. The induction of several genes along with signaling cascades becomes very crucial for its defenses against such biotic stresses (Figure 2). The WRKY family of transcription factors, for example, has been implicated in the response of the plant to various biotic and abiotic stresses (Jiang et al., 2014; Becker et al., 2017). Transcripts studies have identified the upregulation of several of these WRKY genes during pathogen infection, hence enhancing resistance machinery in plants. This type of gene expression is induced by seeds, or it may be chemically induced through many signal molecules such as salicylic acid and JA, which act as critical modules in most plant defense response pathways. Among the fungal pathogens, blackleg caused by *Leptosphaeria maculans* is one of the most economically important diseases affecting the canola crops (Becker et al., 2017).

Several receptor genes and signaling genes have been identified to be upregulated in resistant lines of canola and are considered important for resistance against the disease. Such genes will provide information on the molecular basis for resistance and identify targets for breeding programs aimed at improving disease resistance in canola (Zou and Fernando, 2023). Moreover, the combined action of biotic and abiotic stress further complicates the response mechanisms. In this regard, it has been reported that the availability of nitrogen affects the expression of several biotic stress response genes, indicating that nutrient management plays a significant role in disease susceptibility (Farjad et al., 2018). Such interplay indicates that nutrient optimization may lead to enhanced canola resilience against pathogens.

Figure 2. Response of Canola to Biotic/Abiotic Stress. This figure shows the molecular and cellular responses of canola to biotic and abiotic stress. Upon detecting stress, canola stimulates receptors in plasma membrane, initiating downstream signaling pathways such as ROS production Ca^{2+} signaling, and UPR. Which regulates transcriptional responses including production of antioxidants, activation of stress-responsive genes, and BiP to maintain protein homeostasis, protect cells from damage, and increase stress resistance

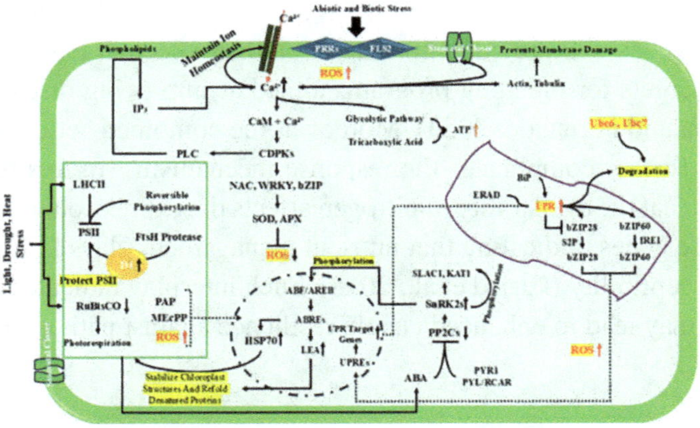

Abbreviations: *PRRs (Pattern recognition receptors), FLS2 (Flagellin-Sensing 2), CaM (Calmodulin), CDPKs (Calcium-dependent protein kinases), PLC (Phospholipase C), IP_3 (Inositol trisphosphate), LHCII (Light-harvesting complex II), PSII (Photosystem II), PAP (3'-phosphoadenosine-5'-phosphate), MEcPP (Methylerythritol cyclodiphosphate), HSPs (Heat shock proteins), PYL (Pyrabactin Resistance1-Like), RCAR (Regulatory components of ABA receptors), PP2Cs (Protein phosphatase 2Cs), SnRK2 (SNF1-related protein kinase 2), AREB (ABA-responsive element-binding factors), LEA (Late embryogenesis abundant), UPR (Unfolded protein response), S2P (Site-2 protease), IRE1 (Inositol-requiring enzyme 1), BiP (Binding immunoglobulin protein), ERAD (ER-associated degradation).*

Besides pathogens, canola is also attacked by insect pests causing serious damage to it. A few defense mechanisms that canola uses against insect herbivores have been unraveled by transcriptomic studies. For instance, genes involved in the biosynthesis of secondary metabolites like glucosinolates are commonly induced following pest attacks. These metabolites may act as deterrents to herbivores and may even be involved in attracting natural enemies of pests (Cornelsen, 2023). The role of JA in triggering plant responses to insect herbivory has been vastly studied. This is a major signaling molecule that plays an important role in triggering defense pathways resulting in the production of protective compounds and proteins (Müller and Munné-Bosch, 2015). Which is complicated by the fact that various pests cause different transcriptomic responses in canola, referring to specific defense strategies, depending on the type of herbivore.

Recent advances in multi-omics techniques have facilitated an integrated understanding of the battery of defense responses that canola mounts against pests. The integration of transcriptomic, proteomic, and metabolomic data in such investigations by Sarkar et al. 2023, thus allowed the identification of regulatory networks and key pathways involved in pest resistance. This holistic approach not only enhances our understanding of plant defense but also informs breeding strategies toward the development of pest-resistant canola varieties.

The interaction between biotic and abiotic stresses is an important area of research, especially now that these are becoming acute under fluctuating climatic conditions. Several works have indicated that abiotic stresses such as drought and salinity can impact the reaction of the host plant toward biotic stresses. For example, drought stress has been shown to increase the vulnerability of canola to some pathogens, hence management alone would keep crops healthy (Suzuki et al., 2014). The crosstalk between signaling pathways arising from abiotic and biotic stresses further complicates the response mechanisms in plants. For example, the activation of stress-responsive genes may interfere with drought-responsive gene expression, or negatively affect the expression of defense genes against pathogens, hence compromising plant immune response (Nejat and Mantri, 2017). Essentially, the reaction of canola to biotic stress comes through the complex network of genes and signaling pathways triggered by its pathogens and pests. Identification of key regulatory genes, mostly from the WRKY family of transcription factors, draws attention to potential targets that can be used for breeding to demonstrate increased disease resistance.

More recent works, which have included transcriptomic studies, have enlightened the mechanisms developing in the process of pest resistance, underlining in a very strong way the role of secondary metabolites and hormonal signaling in plant defense. This complicated interaction between biotic and abiotic stresses further makes the response mechanisms operate at a level that requires a holistic approach toward crop management, taking into consideration the multidimensionality of those plant stress responses. Continued research into this area will be of utmost importance for the development of resilient canola varieties against pathogens and environmental stresses.

6. TRANSCRIPTOMICS FOR CANOLA BREEDING AND GENETIC IMPROVEMENT

Canola is one of the most economically important oilseeds crops due to its high content and quality of oil, hence a vital component of world agricultural production. The transcriptomics era has completely revolutionized canola breeding and genetic improvement by facilitating the identification of candidate genes for desirable traits, enhancing stress tolerance, and applying molecular markers within the breeding programs. Improvement in the content and quality of oil is considered one of the most critical goals in canola breeding programs. Indeed, transcriptomic analyses have contributed to the identification of candidate genes involved in the biosynthesis of oil. For example, studies have shown that certain gene expressions related to triacylglycerol biosynthesis are critical to the development in oil accumulation in developing seeds of canola (Wan et al., 2017). Besides, the integration of transcriptomic data with metabolic profiling showed molecular networks governing oil accumulation, providing insight into genetic regulation that governed oil yield (Wan et al., 2017).

On the other hand, GWAS revealed QTL featuring oil content. Indeed, such investigations have unraveled the complex genetic architecture underlying oil production traits, and breeders are able to effectively select for high oil content (Fletcher et al., 2016). High-throughput sequencing technologies have accelerated the discovery of genetic markers associated with oil quality traits such as fatty acid composition, which is of paramount importance to meet the market demands for healthier oil products. Resistance to abiotic stresses like drought and salinity contributes towards yield stability in canola under variable climatic conditions. The transcriptomic approach has been employed in deciphering the molecular mechanism of tolerance of canola to abiotic stresses (Ulfat et al., 2020).

In this regard, scientific studies have been able to prove that under salt stress, a certain gene expression profile could be activated by improved ion transporters and antioxidants that promote salt tolerance (Shokri-Gharelo and Noparvar, 2018). Moreover, the integration of transcriptomic data with physiological assessments has provided functional insight in how canola plants respond under different types of environmental stresses. Under drought stress, evidence has been found in studies for differential expressions that are associated with photosynthesis and the signaling pathways of stresses, which are considered important in developing stress-resistant varieties (Gul et al., 2022). Transcriptomic analyses are applied for the identification of stress-responsive genes, and hence, targeted breeding strategies have been deployed to enhance the resilience in canola under abiotic stresses. The application of transcriptomics in the development of molecular markers revolutionized canola breeding practices. High-density SNP mapping and the integration of transcriptomic data pave the way for locating genetic markers associated with important agronomical traits, facilitating marker-assisted selection MAS, as developed by Li et al. (2020).

Using this approach, breeders will be in a better position to select the character at seedling stages, hence saving themselves from consuming time, which is usually long during conventional breeding. Genomic selection is one of the potent tools in canola breeding, using transcriptomic information for the prediction of genetic potential by breeding lines. Studies showed that including information from transcriptomics into GS models increases the accuracy of yield and oil content predictions among others (Knoch et al., 2021).

The increased power of the genetic architecture understanding of the studied traits is afforded by genomic and transcriptomic data combinations. These will enable breeders to make more informed decisions in their selection processes (Ton et al., 2020). Further, multi-omics approaches, which integrate the transcriptomic, genomic, and metabolomic data, improved the prediction of the hybrid performance in canola. This holistic approach will provide a better understanding of the interplay among genes, metabolites, and environmental factors leading to the development of superior canola varieties. Transcriptomics has been well incorporated into the process of breeding and genetic improvement in canola to identify candidate genes for oil content enhancement, development of varieties resistant to stresses, and application of molecular markers for efficient breeding practices. Since research keeps taking twists, contributing through transcriptomics toward sustainable production of canola in the face of global changing challenges has become promising. Future studies further need to be done on functional validation of the identified genes and exploration of technologies on gene editing to hasten the development of improved varieties of canola.

7. TECHNOLOGICAL ADVANCES IN CANOLA TRANSCRIPTOMICS

This will considerably speed up the pace of progress in this field with the development of next-generation sequencing technologies. The introduction of NGS revolutionarily changed the mode in which gene expression and transcript variation have been studied in rapeseed. High-throughput sequencing enables a global view that captures completeness of transcripts. This will enable the identification of differentially expressed genes, alternative splicing events, and novel transcripts that are important to be cognizant of regarding understanding the genetic basis that governs traits like resistance and stress tolerance in canola (Lowe et al., 2017; Wang et al., 2019). This now enables researchers to investigate the subtleties of gene regulation and expression using large data banks from various developmental stages and environmental conditions thanks to unprecedented depths in RNA sequencing (Oikonomopoulos et al., 2020). Further, the combination of RNA sequencing with bioinformatics tools has facilitated the intensive modeling and analysis of transcriptomic data using gene networks and pathways involved in the response of canola to stresses, both biotic and abiotic (Lowe et al., 2017). Some of the examples include the transcriptomic studies on the underlying molecular mechanisms of resistance in canola against various pathogens like *Leptosphaeria maculans*.

This provides valuable information which will be useful during breeding programs for enhanced diseases' resistance, as stated by Cantila et al. (2020) and Fikere et al.

(2020). In addition, NGS application in canola enabled further identification of genetic markers connected with the most important agronomically valuable traits and, correspondingly, allowed the acceleration of breeding by means of marker-assisted selection (Knoch et al., 2021). Another promising technology that holds much potential to unveil our knowledge of canola biology is single-cell transcriptomics. This approach enables the examination of gene expression at the level of single cells; therefore, it provides information on cell-specific functions and heterogeneity within plant tissues (Aldridge and Teichmann, 2020). This technology can be advantageously employed in canola for studying the dynamics of gene expression during seed development, stress responses, and interactions with beneficial microbes. The dissection of the transcriptomic profiles of each cell type enables identification of the key regulatory genes and pathways responsible for the important physiological processes, such as oil accumulation and stress tolerance (Gul et al., 2022).

Single-cell technologies have the potential, if used in canola transcriptomics, to disclose novel insights into the cellular mechanisms underpinning complex traits. For example, identifying the spatial and temporal expression patterns of genes associated with oil biosynthesis could eventually lead to the development of new varieties with improved oil content and quality (Wan et al., 2017). On the other hand, single-cell transcriptomics can define rare cells that may have highly important functions related to plant immunity responses and stress tolerance, hence informing breeding strategies aiming at improving the existing resilient environmental effects. Coupled with NGS and single-cell transcriptomics, other technological advancements that will help shape the future of research in canola include CRISPR-based editing of genes and multi-omics approaches. CRISPR technology allows for precise editing in the genome of canola, thereby enabling researchers to explore the functional role of some candidate genes in biological processes (Zalatan et al., 2015). This capability is particularly useful for dissecting the genetic basis of traits important to canola productivity in response to climatic variation. Multi-omics approaches, in which transcriptomics is integrated with genomics, proteomics, and metabolomics, provide overall understanding of canola biological systems. However, integration of data across more omics layers can give comprehensive insights into interaction amongst genes, proteins, and metabolites in the process of plant development and stress responses (Ton et al., 2020). For instance, transcriptomic data integrated into proteomic analyses have identified critical nodes in metabolic pathways controlling oil biosynthesis in canola-a process whose efficient regulation depends on coordination at the levels of gene transcription, translation, and protein function (Wan et al., 2017).

Advances in canola transcriptomics enhance our knowledge on plant biology and promote innovation in agriculture. The knowledge obtained from transcriptomic studies is consequently feeding into the breeding programs targeted at enhancing canola cultivars for adaptation to changing environmental conditions and improving nutrition (Cantila et al., 2020). While demand for sustainable agricultural practices is increasingly encouraged, transcriptomics will be vital in canola research to help address food security and environmental sustainability challenges over the next few years. In summary, advances in canola transcriptomics with NGS and single-cell transcriptomic tools open a new horizon in plant research studies. These technologies provide new insight into the genetic and molecular mechanisms regulating growth, development, and responses against environmental stresses in canola. Where there are researchers continuing to embrace this, then the basis of improving canola varieties through informed breeding strategies not only will remain strong but will continue to improve into the future, further helping with the sustainability-productivity nexus of this important crop.

8. CHALLENGES AND FUTURE DIRECTIONS

This chapter summarizes the challenges of data integration and analysis, functional validation of transcriptomic data, and prospects to enhance productivity and resilience in the canola crop. Transcriptomic large-scale integration and data analysis also have challenges in terms of the enormity and complexity of the data generated into transcriptomics. Since transcriptomics depends on high-throughput sequencing technologies, the resulting datasets are huge and complex; hence, efficient analysis will require strong computational tools (Rohart et al., 2017). With the aid of bioinformatics tools, such integration has been performed on a variety of omics datasets, thus making it able to find only robust biomarkers across independent studies. In general, ensuring such utilities can handle the heterogeneity of different data types and formats remains a challenge that often acts as a barrier to comprehensive analysis (Zhang et al., 2022). Moreover, the interpretation of transcriptomic data itself is complex, as genetic profiles need to be related to a phenotypic outcome. This requires proficiency not only in the biological context in which the genes work but also the environmental factors affecting gene expression.

Climate variability alone, for instance, has been shown to explain a large part of global crop yield variability, suggesting that in any analysis of transcriptomic data given by Ray et al. (2015), environmental variables must be taken into consideration. Future work, therefore, needs to be directed more at developing integrative frameworks which can put together efficiently the transcriptomic data with environmental and phenotypic information to improve understanding of plant response against stressors. Indeed, transcriptomic studies have greatly contributed to the elucidation of gene expression patterns. However, for translating these data into practice, the functionality of the findings must be highly validated. New gene editing technologies, such as CRISPR-Cas9, enable the performance of functional genomics directly by manipulating gene functions and assessing the resulting phenotypic changes (Li et al., 2021). This could help validate the roles of certain genes identified through transcriptomic analysis and hence give more insight into their contribution to the traits like drought tolerance or disease resistance. For example, transcriptomic data have been shown to identify candidate genes implicated in abiotic stress responses such as drought and salinity. However, in the absence of functional validation, the importance of the genes remains speculative (Das et al., 2020).

Future research should therefore be focused on integrating transcriptomic data with functional genomics approaches for determining the functions of genes in canola and other crops (Aksoy et al., 2015). This can be carried out in concert with verification of gene function and information about the underlying molecular mechanisms using sophisticated RNA interference and transgenic approaches. Judging from the prospect of using this technology in viewpoints to help improve crop productivity and resilience, transcriptomics in canola research holds a bright future. The other main emphasis is on multi-omics combined approaches, employing transcriptomics together with proteomics and metabolomics techniques to create deeper insights into plant biology (Zhang et al., 2022). The integrating approach can assess the complex crosstalk occurring within divergent biological pathways and can identify novel targets for crop improvement. To this effect, the appearance of abiotic stresses such as drought and heat in canola crops should also be studied for more tolerant varieties, as climate change still presents a challenge to this effect on agricultural productivity. Recent research has focused on the understanding of molecular mechanisms underpinning plant responses to such stress; such an understanding is explainable through transcriptomic analysis (Priest et al., 2014).

The identification of stress-responsive genes and pathways will, therefore, enable the development of more tolerant varieties of abiotic stresses, hence contributing to food security under variable climatic conditions (Jagadish et al., 2020). Gene editing technologies also have great potential to accelerate the development of canola varieties with more desirable traits. With knowledge garnered through transcriptomics studies, researcher-targeted gene editing can be utilized to further advance traits like oil content, disease resistance, and stress tolerance (Li et al., 2021). It not only rationalizes breeding but also ensures precision in the modification of traits, which reduces the possibility of an adverse effect associated with traditional breeding techniques.

Data integration and its analysis due to complexity, followed by the validation of gene functions through advanced genomics techniques, and investigation into multi-omics approaches, will help researchers lead the way to major steps toward improvement in canola crop productivity and resilience.

9. CONCLUSION

Most of the key aspects described involve transcriptomics dealing with the effects of environmental stresses, growth traits, and diseases in general. High-throughput sequencing and transcriptome analysis have been employed to gain substantial insight into the adaptation of canola against environmental challenges such as salinity, drought, and pathogen attacks. The integration of transcriptomic data with physiological and biochemical analyses has provided substantial insight into the complex regulatory networks driving plant responses under abiotic and biotic stresses. Another important finding presented in this chapter involves the determination of genes up-regulated or down-regulated after the imposition of salt stress. Several studies demonstrated that salinity induces changes in the ion transport of canola plants, osmotic adjustment, and biosynthesis of compatible and protective metabolites. Transcriptional studies have supported the view that some of the key genes regulating these processes are upregulated under salt stress conditions, thus proving their critical role in the enhancement of Salt stress. For instance, photosynthesis and antioxidant defense-related genes have been shown to be highly expressed in salt-tolerant cultivars of canola, hence showing a coordinated response which maintains cellular homeostasis while reducing oxidative stress.

The chapter further discusses the heterosis phenomenon of canola hybrids, whereby transcriptomic profiling has unraveled the genetic basis driving higher biomass and yield over their parents. It identified that the early establishment of photosynthesis and biosynthesis of auxin are some of the critical factors contributing to hybrid vigor. By comparing transcriptomes across a range of canola hybrids, researchers have been able to identify specific genes differentially expressed during the early growth stages of seedlings that play an important role in optimizing growth and productivity. Besides abiotic stress responses, the chapter also brought forth the importance of plant-pathogen interaction with the use of transcriptomic studies. Identification of resistance genes against pathogens such as *Leptosphaeria maculans* has been one of the key focuses of research. Many defense-related genes and signal pathways that are important in the induction of effective immunity have come into view from the transcriptomic analyses. It also contributes valuable basic knowledge about resistance mechanisms of diseases in canola that will contribute to more efficient and effective breeding of resistant cultivars once applied.

This chapter also focuses on the potential of integrating transcriptomics with other omics technologies, such as metabolomics and proteomics, to provide an overview of how plants respond to environmental stress. An integrative approach has, so far, enabled the identification of key metabolic pathways and regulatory networks that function in concert during adaptation to stresses. With such advanced techniques, researchers will know about more complex interactions of genes and metabolites with environmental factors toward better crop resilience and productivity. In other words, transcriptomics of canola plants is a dynamically developing area of science which can give huge opportunities not only for improving knowledge about plant biology but for agriculture in general. Insights obtained from transcriptomic studies not only enrich our basic understanding of the responses of plants to stress but also offer a rationale for developing new strategies for crop improvement. Therefore, with more research being constantly developed, the application of transcriptomics in canola and other crops would no doubt be important to address these issues on climate change, food security, and sustainable agriculture.

REFERENCES

Ahmad, M., Waraich, E. A., Zulfiqar, U., Yong, J. W. H., Ishfaq, M., din, K., Ullah, A., Abbas, A., Awan, M. I., Moussa, I. M., & Elshikh, M. S. (2024). Thiourea improves yield and quality traits of brassica napus l. by upregulating the antioxidant defense system under high temperature stress. *Scientific Reports*, *14*(1), 12195. Advance online publication. DOI: 10.1038/s41598-024-62257-y PMID: 38806561

Aksoy, E., Demirel, U., Öztürk, Z., Çalışkan, S., & Çalışkan, M. (2015). Recent advances in potato genomics, transcriptomics, and transgenicsunder drought and heat stresses: A review. *Turkish Journal of Botany*, *39*, 920–940. DOI: 10.3906/bot-1506-25

Aldridge, S., & Teichmann, S. (2020). Single cell transcriptomics comes of age. *Nature Communications*, *11*(1), 4307. Advance online publication. DOI: 10.1038/s41467-020-18158-5 PMID: 32855414

Arifuzzaman, M., Oladzad, A., McClean, P., & Rahman, M. (2019). Shovelomics for phenotyping root architectural traits of rapeseed/canola (brassica napus l.) and genome-wide association mapping. *Molecular Genetics and Genomics*, *294*(4), 985–1000. DOI: 10.1007/s00438-019-01563-x PMID: 30968249

Asaduzzaman, A. (2024). Genotype-by-environment interaction and stability of canola (<em>brassica napus </em>l) for weed suppression through improved interference. https://doi.org/DOI: 10.20944/preprints202408.0227.v1

Assis, R., Silva, F., Carneiro, L., Silva, G., Cruz, S., & Machado, C. (2020). Canola agronomic performance in three sowing dates in jataí-go. *Bioscience Journal*, *36*, 36–47. DOI: 10.14393/BJ-v36n0a2020-48258

Becker, M., Zhang, X., Walker, P., Wan, J., Millar, J., Khan, D., Granger, M. J., Cavers, J. D., Chan, A. C., Fernando, D. W. G., & Belmonte, M. (2017). Transcriptome analysis of the brassica napus–leptosphaeria maculans pathosystem identifies receptor, signaling and structural genes underlying plant resistance. *The Plant Journal*, *90*(3), 573–586. DOI: 10.1111/tpj.13514 PMID: 28222234

Cantila, A., Saad, N., Amas, J., Edwards, D., & Batley, J. (2020). Recent findings unravel genes and genetic factors underlying leptosphaeria maculans resistance in brassica napus and its relatives. *International Journal of Molecular Sciences*, *22*(1), 313. DOI: 10.3390/ijms22010313 PMID: 33396785

Chalhoub, B., Denœud, F. S. L., Parkin, I., Tang, H., Wang, X., & Wincker, P. (2014). Early allopolyploid evolution in the post-neolithic brassica napus oilseed genome. *Science*, *345*(6199), 950–953. DOI: 10.1126/science.1253435 PMID: 25146293

Chen, H., Wang, Y., Liu, J., Zhao, T., Yang, C., Ding, Q., & Wang, D. (2022). Identification of WRKY transcription factors responding to abiotic stresses in Brassica napus L. *Planta*, *255*(1), 1–17. DOI: 10.1007/s00425-021-03733-x PMID: 34837557

Cornelsen, J., Ort, N. W. W., Gabert, R. K., Epp, I., & Rempel, C. B. (2023). Current and potential pest threats for canola in the canadian prairies. *Pest Management Science*, *80*(5), 2220–2234. DOI: 10.1002/ps.7858 PMID: 37899491

Das, R., Pradhan, S., & Parida, A. (2020). De-novo transcriptome analysis unveils differentially expressed genes regulating drought and salt stress response in panicum sumatrense. *Scientific Reports*, *10*(1), 21251. Advance online publication. DOI: 10.1038/s41598-020-78118-3 PMID: 33277539

Eldin, S., Ziena, H., Khair, S., & Rozan, M. (2018). Canola seed meal as a potential source of natural antioxidant. *Alexandria Science Exchange Journal*, *39*(OCTOBER-DECEMBER), 615–619. DOI: 10.21608/asejaiqjsae.2018.19582

Farjad, M., Rigault, M., Pateyron, S., Martin-Magniette, M., Krapp, A., Meyer, C., & Fagard, M. (2018). Nitrogen limitation alters the response of specific genes to biotic stress. *International Journal of Molecular Sciences*, *19*(11), 3364. DOI: 10.3390/ijms19113364 PMID: 30373239

Fikere, M., Barbulescu, D., Malmberg, M., Spangenberg, G., Cogan, N., & Daetwyler, H. (2020). Meta-analysis of gwas in canola blackleg (leptosphaeria maculans) disease traits demonstrates increased power from imputed whole-genome sequence. *Scientific Reports*, *10*(1), 14300. Advance online publication. DOI: 10.1038/s41598-020-71274-6 PMID: 32868838

Fletcher, R., Herrmann, D., Mullen, J., Li, Q., Schrider, D., Price, N. & McKay, J. (2016). Identification of polymorphisms associated with drought adaptation qtl inbrassica napusby resequencing. G3 Genes|genome|genetics, 6(4), 793-803. https://doi.org/DOI: 10.1534/g3.115.021279

Fordyce, S., Torrion, J., Lachowiec, J., Miller, P., Biddulph, B., & Carr, P. (2018). Field evidence for frost-induced cross-stress tolerance in canola. *Agricultural & Environmental Letters*, *3*(1), 180020. Advance online publication. DOI: 10.2134/ael2018.05.0020

Gabur, I., Chawla, H., Liu, X., Kumar, V., Faure, S., Tiedemann, A., & Obermeier, C. (2018). Finding invisible quantitative trait loci with missing data. *Plant Biotechnology Journal*, *16*(12), 2102–2112. DOI: 10.1111/pbi.12942 PMID: 29729219

Gao, L., Caldwell, C., & Jiang, Y. (2018). Photosynthesis and growth of camelina and canola in response to water deficit and applied nitrogen. *Crop Science*, *58*(1), 393–401. DOI: 10.2135/cropsci2017.07.0406

Gul, H., Ulfat, M., Zafar, Z., Haider, W., Ali, Z., Manzoor, H., & Athar, H. (2022). Photosynthesis and salt exclusion are key physiological processes contributing to salt tolerance of canola (brassica napus l.): Evidence from physiology and transcriptome analysis. *Genes*, *14*(1), 3. DOI: 10.3390/genes14010003 PMID: 36672744

Havlíčková, L., He, Z., Wang, L., Langer, S., Harper, A., Kaur, H., Broadley, M. R., Gegas, V., & Bancroft, I. (2017). Validation of an updated associative transcriptomics platform for the polyploid crop species brassica napus by dissection of the genetic architecture of erucic acid and tocopherol isoform variation in seeds. *The Plant Journal*, *93*(1), 181–192. DOI: 10.1111/tpj.13767 PMID: 29124814

Howlett, B., Marcroft, S., & Wouw, A. (2015). Evolution of virulence in fungal plant pathogens: Exploiting fungal genomics to control plant disease. *Mycologia*, *107*(3), 441–451. DOI: 10.3852/14-317 PMID: 25725000

Hu, L., Robert, C., Cadot, S., Zhang, X., Ye, M., Li, B., Manzo, D., Chervet, N., Steinger, T., van der Heijden, M. G. A., Schlaeppi, K., & Erb, M. (2018). Root exudate metabolites drive plant-soil feedback on growth and defense by shaping the rhizosphere microbiota. *Nature Communications*, *9*(1), 2738. Advance online publication. DOI: 10.1038/s41467-018-05122-7 PMID: 30013066

Jagadish, S., Pal, M., Sukumaran, S., Parani, M., & Siddique, K. (2020). Heat stress resilient crops for future hotter environments. *Plant Physiology Reports*, *25*(4), 529–532. DOI: 10.1007/s40502-020-00559-9

Jia, H., Shao, M., He, Y., Guan, R., Chu, P., & Jiang, H. (2015). Proteome dynamics and physiological responses to short-term salt stress in brassica napus leaves. *PLoS One*, *10*(12), e0144808. DOI: 10.1371/journal.pone.0144808 PMID: 26691228

Jiang, Y., Duan, Y., Yin, J., Ye, S., Zhu, J., Zhang, F., & Luo, K. (2014). Genome-wide identification and characterization of the populus wrky transcription factor family and analysis of their expression in response to biotic and abiotic stresses. *Journal of Experimental Botany*, *65*(22), 6629–6644. DOI: 10.1093/jxb/eru381 PMID: 25249073

Jing, X. (2018). Peer review #1 of "molecular response of canola to salt stress: insights on tolerance mechanisms (v0.1)". https://doi.org/DOI: 10.7287/peerj.4822v0.1/reviews/1

Khodabin, G., Tahmasebi-Sarvestani, Z., Rad, A., & Modarres-Sanavy, S. (2020). Effect of drought stress on certain morphological and physiological characteristics of a resistant and a sensitive canola cultivar. *Chemistry & Biodiversity*, *17*(2), e1900399. Advance online publication. DOI: 10.1002/cbdv.201900399 PMID: 31909554

Kiran, A., Wakeel, A., Snowdon, R., & Friedt, W. (2019). Genetic dissection of root architectural traits by qtl and genome-wide association mapping in rapeseed (brassica napus). *Plant Breeding*, *138*(2), 184–192. DOI: 10.1111/pbr.12665

Kittipol, V., He, Z., Wang, L., Doheny-Adams, T., Langer, S., & Bancroft, I. (2019). Genetic architecture of glucosinolate variation in brassica napus. *Journal of Plant Physiology*, *240*, 152988. DOI: 10.1016/j.jplph.2019.06.001 PMID: 31255878

Knoch, D., Werner, C., Meyer, R., Riewe, D., Abbadi, A., Lücke, S., Snowdon, R. J., & Altmann, T. (2021). Multi-omics-based prediction of hybrid performance in canola. *Theoretical and Applied Genetics*, *134*(4), 1147–1165. DOI: 10.1007/s00122-020-03759-x PMID: 33523261

Koh, J., Barbulescu, D., Salisbury, P., & Slater, A. (2016). Pterostilbene is a potential candidate for control of blackleg in canola. *PLoS One*, *11*(5), e0156186. DOI: 10.1371/journal.pone.0156186 PMID: 27213274

Koscielny, C., Gardner, S., Technow, F., & Duncan, R. (2020). Linkage mapping and whole-genome predictions in canola (brassica napus) subjected to differing temperature treatments. *Crop & Pasture Science*, *71*(3), 229–238. DOI: 10.1071/CP19387

Kruger, D., Fourie, J., & Malan, A. (2016). The effect of cover crops and their management on plant-parasitic nematodes in vineyards. *South African Journal of Enology and Viticulture*, *36*(2). Advance online publication. DOI: 10.21548/36-2-953

Lamaoui, M., Jemo, M., Datla, R., & Bekkaoui, F. (2018). Heat and drought stresses in crops and approaches for their mitigation. *Frontiers in Chemistry*, *6*, 26. Advance online publication. DOI: 10.3389/fchem.2018.00026 PMID: 29520357

Larkan, N., Lydiate, D., Parkin, I., Nelson, M., Epp, D., Cowling, W., Rimmer, S. R., & Borhan, M. (2012). The brassica napus blackleg resistance gene lepr3 encodes a receptor-like protein triggered by the leptosphaeria maculans effector avrlm1. *The New Phytologist*, *197*(2), 595–605. DOI: 10.1111/nph.12043 PMID: 23206118

Larkan, N., Lydiate, D., Yu, F., Rimmer, S., & Borhan, M. (2014). Co-localisation of the blackleg resistance genes rlm2 and lepr3 on brassica napus chromosome a10. *BMC Plant Biology*, *14*(1), 387. Advance online publication. DOI: 10.1186/s12870-014-0387-z PMID: 25551287

Li, H., Feng, H., Guo, C., Yang, S., Huang, W., Xiong, X., & Yang, W. (2020). High-throughput phenotyping accelerates the dissection of the dynamic genetic architecture of plant growth and yield improvement in rapeseed. *Plant Biotechnology Journal, 18*(11), 2345–2353. DOI: 10.1111/pbi.13396 PMID: 32367649

Li, S., Zhang, C., Li, J., Yan, L., Wang, N., & Xia, L. (2021). Present and future prospects for wheat improvement through genome editing and advanced technologies. *Plant Communications, 2*(4), 100211. DOI: 10.1016/j.xplc.2021.100211 PMID: 34327324

Lohani, N., Jain, D., Singh, M., & Bhalla, P. (2020). Engineering multiple abiotic stress tolerance in canola, brassica napus. *Frontiers in Plant Science, 11*, 3. Advance online publication. DOI: 10.3389/fpls.2020.00003 PMID: 32161602

Lowe, R., Shirley, N., Bleackley, M., Dolan, S., & Shafee, T. (2017). Transcriptomics technologies. *PLoS Computational Biology, 13*(5), e1005457. DOI: 10.1371/journal.pcbi.1005457 PMID: 28545146

Malmberg, M., Pembleton, L., Baillie, R., Drayton, M., Sudheesh, S., Kaur, S., & Cogan, N. (2017). Genotyping-by-sequencing through transcriptomics: Implementation in a range of crop species with varying reproductive habits and ploidy levels. *Plant Biotechnology Journal, 16*(4), 877–889. DOI: 10.1111/pbi.12835 PMID: 28913899

Mason, A., Chauhan, P., Banga, S., Banga, S., Salisbury, P., Barbetti, M., & Batley, J. (2018). Agricultural selection and presence–absence variation in spring-type canola germplasm. *Crop & Pasture Science, 69*(1), 55–64. DOI: 10.1071/CP17161

Müller, M., & Munné-Bosch, S. (2015). Ethylene response factors: A key regulatory hub in hormone and stress signaling. *Plant Physiology, 169*(1), 32–41. DOI: 10.1104/pp.15.00677 PMID: 26103991

Nejat, N., & Mantri, N. (2017). Plant immune system: Crosstalk between responses to biotic and abiotic stresses the missing link in understanding plant defense. *Current Issues in Molecular Biology*, •••, 1–16. DOI: 10.21775/cimb.023.001 PMID: 28154243

Nikvash, N., Kraft, R., Kharazipour, A., & Euring, M. (2010). Comparative properties of bagasse, canola and hemp particle boards. *Holz als Roh- und Werkstoff, 68*(3), 323–327. DOI: 10.1007/s00107-010-0465-3

Niu, F., Wang, B., Wu, F., Yan, J., Li, L., Wang, C., & Jiang, Y. Q. (2014). Canola (Brassica napus L.) NAC103 transcription factor gene is a novel player inducing reactive oxygen species accumulation and cell death in plants. *Biochemical and Biophysical Research Communications*, *454*(1), 30–35. DOI: 10.1016/j.bbrc.2014.10.057 PMID: 25450358

Oikonomopoulos, S., Bayega, A., Fahiminiya, S., Djambazian, H., Bérubé, P., & Ragoussis, J. (2020). Methodologies for transcript profiling using long-read technologies. *Frontiers in Genetics*, *11*, 606. Advance online publication. DOI: 10.3389/fgene.2020.00606 PMID: 32733532

Osakabe, Y., Arinaga, N., Umezawa, T., Katsura, S., Nagamachi, K., Tanaka, H., & Yamaguchi-Shinozaki, K. (2013). Osmotic stress responses and plant growth controlled by potassium transporters inarabidopsis. *The Plant Cell*, *25*(2), 609–624. DOI: 10.1105/tpc.112.105700 PMID: 23396830

Ozhuner, E., Eldem, V., Ipek, A., Okay, S., Sakcali, S., Zhang, B., & Unver, T. (2013). Boron stress responsive microRNAs and their targets in barley. *PLoS One*, *8*(3), e59543. DOI: 10.1371/journal.pone.0059543 PMID: 23555702

Pan, W., Maaz, T., Madsen, I., Reese, M., Hammac, W., Wysocki, D., & Brown, J. (2017). 4r nitrogen management when integrating canola into semi-arid wheat. *Crops & Soils*, *50*(3), 16–66. DOI: 10.2134/cs2017.50.0309

Priest, H., Fox, S., Rowley, E., Murray, J., Michael, T., & Mockler, T. (2014). Analysis of global gene expression in brachypodium distachyon reveals extensive network plasticity in response to abiotic stress. *PLoS One*, *9*(1), e87499. DOI: 10.1371/journal.pone.0087499 PMID: 24489928

Qian, Y., Ren, Q., Zhang, J., & Chen, L. (2019). Transcriptomic analysis of the maize (Zea mays L.) inbred line B73 response to heat stress at the seedling stage. *Gene*, *692*, 68–78. DOI: 10.1016/j.gene.2018.12.062 PMID: 30641208

Rafique, N., Ilyas, N., Aqeel, M., Raja, N., Shabbir, G., Ajaib, M., & Ansari, M. (2023). Interactive effects of melatonin and salicylic acid on brassica napus under drought condition. *Plant and Soil*. Advance online publication. DOI: 10.1007/s11104-023-05942-7

Rahman, H. (2013). Review: Breeding spring canola (brassica napusl.) by the use of exotic germplasm. *Canadian Journal of Plant Science*, *93*(3), 363–373. DOI: 10.4141/cjps2012-074

Raimi, A., & Adeleke, R. (2023). 16s rrna gene-based identification and plant growth-promoting potential of cultivable endophytic bacteria. *Agronomy Journal, 115*(3), 1447–1462. DOI: 10.1002/agj2.21241

Raman, H., Raman, R., McVittie, B., Borg, L., Diffey, S., Yadav, A., & Farquhar, G. (2020). Genetic and physiological bases for variation in water use efficiency in canola. *Food and Energy Security*, *9*(4), e237. Advance online publication. DOI: 10.1002/fes3.237

Raman, H., Raman, R., Pirathiban, R., McVittie, B., Sharma, N. S. L., & Zhang, Y. (2022). Multienvironment qtl analysis delineates a major locus associated with homoeologous exchanges for water-use efficiency and seed yield in canola. *Plant, Cell & Environment*, *45*(7), 2019–2036. DOI: 10.1111/pce.14337 PMID: 35445756

Raman, H., Shamaya, N., Pirathiban, R., McVittie, B., Raman, R., Cullis, B., & Easton, A. (2023). Quantitative trait loci for genotype and genotype by environment interaction effects for seed yield plasticity to terminal water-deficit conditions in canola (brassica napus l.). *Plants*, *12*(4), 720. DOI: 10.3390/plants12040720 PMID: 36840067

Raman, R., Taylor, B., Marcroft, S., Stiller, J., Eckermann, P., Coombes, N., Rehman, A., Lindbeck, K., Luckett, D., Wratten, N., Batley, J., Edwards, D., Wang, X., & Raman, H. (2012). Molecular mapping of qualitative and quantitative loci for resistance to leptosphaeria maculans causing blackleg disease in canola (brassica napus l.). *Theoretical and Applied Genetics*, *125*(2), 405–418. DOI: 10.1007/s00122-012-1842-6 PMID: 22454144

Ray, D., Gerber, J., MacDonald, G., & West, P. (2015). Climate variation explains a third of global crop yield variability. *Nature Communications*, *6*(1), 5989. Advance online publication. DOI: 10.1038/ncomms6989 PMID: 25609225

Raza, W., Wang, J., Jousset, A., Friman, V., Mei, X., Wang, S., . . . Shen, Q. (2020). Bacterial community richness shifts the balance between volatile organic compound-mediated microbe–pathogen and microbe–plant interactions. Proceedings of the Royal Society B Biological Sciences, 287(1925), 20200403. https://doi.org/DOI: 10.1098/rspb.2020.0403

Razavizadeh, R. (2015). Protein pattern of canola (brassica napus l.) changes in response to salt and salicylic acid in vitro. *Biological Letters*, *52*(1-2), 19–36. DOI: 10.1515/biolet-2015-0012

Rezayian, M., Niknam, V., & Arabloo, M. (2023). Iron nanoparticle regulate succinate dehydrogenase activity in canola plants under drought stress. *Scientific Reports, 13*(1), 9628. Advance online publication. DOI: 10.1038/s41598-023-36105-4 PMID: 37316501

Rohart, F., Gautier, B., Singh, A., & Cao, K. (2017). Mixomics: An r package for 'omics feature selection and multiple data integration. *PLoS Computational Biology, 13*(11), e1005752. DOI: 10.1371/journal.pcbi.1005752 PMID: 29099853

Saad, N., Severn-Ellis, A., Pradhan, A., Edwards, D., & Batley, J. (2021). Genomics armed with diversity leads the way in brassica improvement in a changing global environment. *Frontiers in Genetics, 12*, 600789. Advance online publication. DOI: 10.3389/fgene.2021.600789 PMID: 33679880

Sarkar, A., Kisiała, A., Adhikary, D., Basu, U., Emery, R., Rahman, H., & Kav, N. (2023). Silicon ameliorates clubroot responses in canola (brassica napus): A "multi-omics"-based investigation into possible mechanisms. *Physiologia Plantarum, 175*(2), e13900. Advance online publication. DOI: 10.1111/ppl.13900 PMID: 36992551

Sarkar, N. K., Kim, Y. K., & Grover, A. (2014). Coexpression network analysis associated with call of rice seedlings for encountering heat stress. *Plant Molecular Biology, 84*(1-2), 125–143. DOI: 10.1007/s11103-013-0123-3 PMID: 23975147

Shahbaz, M., Noreen, N., & Perveen, S. (2013). Triacontanol modulates photosynthesis and osmoprotectants in canola (brassica napusl.) under saline stress. *Journal of Plant Interactions, 8*(4), 350–359. DOI: 10.1080/17429145.2013.764469

Shokri-Gharelo, R., & Noparvar, P. (2018). Molecular response of canola to salt stress: Insights on tolerance mechanisms. *PeerJ, 6*, e4822. DOI: 10.7717/peerj.4822 PMID: 29844974

Stafen, C., Pedrolo, A., Schroeder, L., Viana, V., Busanello, C., Rosa, M., & Pegoraro, C. (2018). Seed germination and seedling growth of canola treated with abiotic factors that can affect growth and development. *International Journal of Vegetable Science, 24*(6), 574–582. DOI: 10.1080/19315260.2018.1453909

Stanic, M., Hickerson, N., Arunraj, R., & Samuel, M. (2021). Gene-editing of the strigolactone receptor bnd14 confers promising shoot architectural changes in brassica napus (canola). *Plant Biotechnology Journal, 19*(4), 639–641. DOI: 10.1111/pbi.13513 PMID: 33219600

Su, Z., Wang, T., Shrivastava, N., Chen, Y., Liu, X., Sun, C., & Lou, B. (2017). Piriformospora indica promotes growth, seed yield and quality of brassica napus l. *Microbiological Research, 199*, 29–39. DOI: 10.1016/j.micres.2017.02.006 PMID: 28454707

Suchan, D., Bergsveinson, J., Manzon, L., Pierce, A., Kryachko, Y., Korber, D., & Yost, C. (2020). Transcriptomics reveal core activities of the plant growth-promoting bacterium delftia acidovorans ray209 during interaction with canola and soybean roots. *Microbial Genomics*, *6*(11). Advance online publication. DOI: 10.1099/mgen.0.000462 PMID: 33151138

Suzuki, N., Rivero, R., Shulaev, V., Blumwald, E., & Mittler, R. (2014). Abiotic and biotic stress combinations. *The New Phytologist*, *203*(1), 32–43. DOI: 10.1111/nph.12797 PMID: 24720847

Ton, L., Neik, T., & Batley, J. (2020). The use of genetic and gene technologies in shaping modern rapeseed cultivars (brassica napus l.). *Genes*, *11*(10), 1161. DOI: 10.3390/genes11101161 PMID: 33008008

Ulfat, M., Athar, H., Khan, Z., & Kalaji, H. (2020). Rnaseq analysis reveals altered expression of key ion transporters causing differential uptake of selective ions in canola (brassica napus l.) grown under nacl stress. *Plants*, *9*(7), 891. DOI: 10.3390/plants9070891 PMID: 32674475

Umesh, M. (2023). Growth and physiological response of canola (brassica napus l.) tointeractive effect of temperature, moisture and nitrogen stressesunder controlled environment. Journal of Oilseeds Research, 40 (Specialissue). https://doi.org/DOI: 10.56739/5fxsvn54

Wan, H., Cui, Y., Ding, Y., Mei, J., Dong, H., Zhang, W., & Qian, W. (2017). Time-series analyses of transcriptomes and proteomes reveal molecular networks underlying oil accumulation in canola. *Frontiers in Plant Science*, *07*. Advance online publication. DOI: 10.3389/fpls.2016.02007 PMID: 28119706

Wang, B., Kumar, V., Olson, A., & Ware, D. (2019). Reviving the transcriptome studies: An insight into the emergence of single-molecule transcriptome sequencing. *Frontiers in Genetics*, *10*, 384. Advance online publication. DOI: 10.3389/fgene.2019.00384 PMID: 31105749

Wang, P., Yang, C., Chen, H., Song, C., Zhang, X., & Wang, D. (2017). Transcriptomic basis for drought-resistance in Brassica napus L. *Scientific Reports*, *7*(1), 40532. DOI: 10.1038/srep40532 PMID: 28091614

Wang, X., Chen, Y., Thomas, C., Ding, G., Xu, P., Shi, D., & Shi, L. (2017). Genetic variants associated with the root system architecture of oilseed rape (brassica napus l.) under contrasting phosphate supply. *DNA Research : an International Journal for Rapid Publication of Reports on Genes and Genomes*, *24*(4), 407–417. DOI: 10.1093/dnares/dsx013 PMID: 28430897

Williams, S., Vail, S., & Arcand, M. M. (2021). Nitrogen use efficiency in parent vs. hybrid canola under varying nitrogen availabilities. *Plants*, *10*(11), 2364. DOI: 10.3390/plants10112364 PMID: 34834725

Yanagawa, A., & Fujimaki, H. (2013). Tolerance of canola to drought and salinity stresses in terms of root water uptake model parameters. *Journal of Hydrology and Hydromechanics*, *61*(1), 73–80. DOI: 10.2478/johh-2013-0009

Zalatan, J., Lee, M., Almeida, R., Gilbert, L., Whitehead, E., Russa, M., & Lim, W. (2015). Engineering complex synthetic transcriptional programs with crispr rna scaffolds. *Cell*, *160*(1-2), 339–350. DOI: 10.1016/j.cell.2014.11.052 PMID: 25533786

Zhang, H., Liu, W., Zhang, Y., Deng, M., Niu, F., Yang, B., & Jiang, Y. (2014). Identification, expression and interaction analyses of calcium-dependent protein kinase (cpk) genes in canola (brassica napusl.). *BMC Genomics*, *15*(1), 211. Advance online publication. DOI: 10.1186/1471-2164-15-211 PMID: 24646378

Zhang, H., Yang, B., Liu, W., Li, H., Wang, L., Wang, B., & Jiang, Y. (2014). Identification and characterization of cbl and cipk gene families in canola (brassica napus l.). *BMC Plant Biology*, *14*(1), 8. Advance online publication. DOI: 10.1186/1471-2229-14-8 PMID: 24397480

Zhang, R., Zhang, C., Yu, C., Dong, J., & Hu, J. (2022). Integration of multi-omics technologies for crop improvement: Status and prospects. *Frontiers in Bioinformatics*, *2*, 1027457. Advance online publication. DOI: 10.3389/fbinf.2022.1027457 PMID: 36438626

Zhu, M., Monroe, J., Suhail, Y., Villiers, F., Mullen, J., Pater, D., & Assmann, S. (2016). Molecular and systems approaches towards drought-tolerant canola crops. *The New Phytologist*, *210*(4), 1169–1189. DOI: 10.1111/nph.13866 PMID: 26879345

Zou, Z., & Fernando, W. (2023). Overexpression of bnnac19 in brassica napus enhances resistance to leptosphaeria maculans, the blackleg pathogen of canola. *Plant Pathology*, *73*(1), 104–114. DOI: 10.1111/ppa.13791

Zou, Z., Liu, F., & Fernando, D. (2021). Genome-wide identification and analysis of the valine-glutamine motif-containing gene family inbrassica napusand functional characterization ofbnmks1in response toleptosphaeria maculans. *Phytopathology*, *111*(2), 281–292. DOI: 10.1094/PHYTO-04-20-0134-R PMID: 32804045

Chapter 12
Metabolomics of Canola Plants

Ilham Khan
Quaid-i-Azam University, Islamabad, Pakistan

Najeeba Parre Paker
https://orcid.org/0009-0008-7335-9344
Quaid-i-Azam University, Islamabad, Pakistan

Junaid Shehzad
Quaid-i-Azam University, Islamabad, Pakistan

Banzeer Ahsan Abbasi
https://orcid.org/0000-0003-3036-4481
Rawalpindi Women University, Pakistan

Farhat Gul
Quaid-i-Azam University, Islamabad, Pakistan

Ghazala Mustafa
Quaid-i-Azam University, Islamabad, Pakistan

ABSTRACT

Canola is one of the major oilseed crops because of its nutritional and industrial applications. This chapter describes metabolomics of canola and represents truly complex metabolic process compared to plant growth, developmental stages and stressed conditions. The canola metabolome includes wide array of metabolites; fatty acids, phenolics and glucosinolates important in seed development and for the health of the plant in general. A clear vision of metabolomic variation during seed development explains the biochemical mechanisms underpinning yield and quality attributes. Conversely, metabolic changes in canola are essential to its response through resilience and adaptation mechanisms toward both abiotic and biotic stresses. The application of metabolomics in canola breeding facilitates identification of biomarkers and traits related to stress tolerance and oil quality improvement. Besides that, metabolic engineering could improve agronomical performance and nutritional quality, opening more perspectives toward sustainable crop improvement under changing environmental conditions.

DOI: 10.4018/979-8-3693-8603-3.ch012

1. INTRODUCTION

Canola is one of the important oilseed crops grown in many parts of the world due to its high-quality edible oil. Its importance can be outlined by its adaptation to a wide range of climatic conditions, especially in temperate and subtropical regions, where it has attained much importance in agricultural practice. Canola basically demonstrates economic viability due to its oil-rich composition of unsaturated fatty acids, making the crop suitable for both cooking oil and food products. Besides, canola meal, which originates after oil is extracted from canola seed, is a good source of protein in feed, hence applicable in animal husbandry (Garcia-Aloy et al., 2019). Canola cultivation also faces quite a few challenges, as it is very susceptible to several diseases like Blackleg, Clubroot, and Sclerotinia Stem Rot that may badly affect yield and quality. Climate change has only enhanced the severity of these diseases, which threatens global food security (Neik et al., 2020). Understanding the biochemical processes underlying the growth and resilience in canola is central to devising effective management strategies. That is where metabolomics incorporates identification of metabolic pathways and profiles characteristic of canola under different environmental conditions and stressors.

Metabolomics is the comprehensive study of metabolites within a biological system and hence it plays a major role in the elucidation of biochemical processes in canola. The metabolic profiling analysis helps researchers to understand how canola responds under a plethora of abiotic and biotic stresses, including nutrient limitations and pathogen attacks. For instance, sulfur deficiency has been one of the nutrient deficiencies that have widely reported to cause drastic physiological and proteomic changes in canola hence affecting its growth with respect to plant development and seed quality, such as oil quality (D'Hooghe et al., 2013). This underpins the importance of sulfur to the metabolic dynamics of canola, since this crop is quite responsive to this nutrient. Metabolomics is currently at the forefront in the study of the interactions between canola and pathogens. For example, it has been reported that the infection of canola by Pseudomonas aeruginosa results in the biosynthesis of phytoalexins, crucial compounds in the plant defense mechanism (Cook et al., 2022). This interaction underlines the involvement of secondary metabolites in plant immunity and their possible use in breeding programs toward resistance against diseases in canola. This view has been further elaborated in the integration of metabolomics with other omics, such as genomics and transcriptomics.

For example, studies have employed combined metabolomic and transcriptomic analyses to study flower color variation in canola, and the results show very clearly the complex interplay that occurs between gene expression and metabolite accumulation. In fact, such integrative approaches go toward enhancing our understanding of the biology of canola and also offer a way to targeted breeding for improved traits

(Cui, 2024). Besides the prospects of disease resistance and nutritional utilization, metabolomics has also been applied to investigate environmental stresses in canola. It has been indicated by studies that transgenic canola lines exhibiting increased expression of certain genes had changed their metabolite profiles and further altered their tolerance to low-temperature stress and pathogens (Nokhrina et al., 2014). This demonstrates the power of metabolomics in identifying key metabolic pathways that confer a stress phenotype and thus inform breeding programs for developing climate-resilient canola varieties. More importantly, metabolomics enhances nutritional quality in canola oil. Recent advances in metabolic engineering have focused on a genetic modification of the fatty acid composition of canola oil to make it healthy and friendly (Miao et al., 2021; Rauf et al., 2023).

This is achieved by first determining metabolic pathways through which biosynthesis of fatty acids occur, followed by manipulation of the metabolic pathways to produce an oil, exhibiting increased nutritional values-for instance, high levels of omega-3 fatty acids. Application of metabolomics in canola research is not limited to the understanding of stress responses and improvement in oil quality but also extends to the exploration of allelopathic properties of canola. Allelopathy is a chemical interaction between plants which can inhibit or stimulate the growth of neighboring plants. Different metabolomic studies have pointed out that canola has allelopathic potential, which can be harnessed for our benefit regarding instituting sustainable weed management practices (Barrales-Cureño et al., 2022). This aspect of the research on canola brings into play the many facets of benefit attendant on coming to understand its metabolomic landscape.

Canola, more correctly Brassica napus, is one of the most economically and nutritionally important oilseed crops. Introducing metabolomics into the study of canola has a profound consequence for the insight it gives into the biochemical processes of this plant, relating to its growth and development and potential stress responses, besides nutritional value. Application of metabolomics will, therefore, be quite central in the development of novel approaches to canola cultivation and use, as the agricultural face keeps changing due to climate change and issues of food security.

2. METABOLOME OF CANOLA PLANTS

The canola metabolome encompasses a vast range of both primary and secondary metabolites, which are highly critical in plant physiology, stress response, and interaction with the environment. Primary metabolites consist of organic compounds involved in major metabolic pathways related to basic cellular functions and energy metabolism. These are glucosinolates, flavonoids, and phenolic compounds (Table

1). These compounds participate in the plant's defense mechanisms and allelopathy-commonly understood as chemical inhibition of one plant species by another. Their study is very important both for understanding canola ecological interactions and for possible agricultural applications in enhancing crop resilience against pests and diseases (Asaduzzaman et al., 2015; Tian et al., 2016). The studies have determined that in most of the cultivars, the profile of metabolites is colossally variable, hence relating them to their respective allelopathic properties. Under laboratory conditions, some canola genotypes possess allelochemicals with the potency to inhibit the germination of weed species and hence act as a natural method for controlling weeds. Its potential for allelopathy was mainly because of the important secondary metabolites, which were quite variable among cultivars.

Understanding this variation may lead toward the selection of varieties with high yield but with enhanced allelopathic traits and thus contribute toward more sustainable agricultural production (Barrales-Cureño et al., 2022). Therefore, the choice of analytical techniques immediately becomes important for metabolomics, requiring such complex profiling of canola metabolome. Among those, two predominant techniques utilized in metabolomic analyses include gas chromatography-mass spectrometry and liquid chromatography-mass spectrometry. While GC-MS is effective in the case of volatile and semi-volatile compounds, polar and thermally labile metabolites are best separated by LC-MS (Hanifah et al., 2018; Zeki et al., 2020). A combination of methods can thus allow the detection of a wider range of metabolites, since different methods may respectively have their strength in analyzing different classes of compounds. In addition, nuclear magnetic resonance spectroscopy is non-destructive and can provide information about the structure of the metabolites with a minimum amount of sample preparation. These approaches taken together enable comprehensive investigation of metabolic pathways and interactions in the canola plant (Emwas et al., 2019; Wishart et al., 2022). More recently, metabolomics research in canola extended the use of this to the study of the plant responses against biotic stresses, such as pathogens.

Indeed, it was shown that canola infected with *Plasmodiophora brassicae* the pathogen responsible for clubroot disease-displayed an overall change in their metabolic profile, including significant accumulations of phytoalexins, which are indeed antimicrobial compounds produced during attacks by pathogens (Cook et al., 2022; Adhikary et al., 2022). This underlines the role of metabolomics in unraveling the biochemical pathways in the mechanisms of plant defense and in pointing out possible biomarkers for disease resistance. The integration of transcriptomic and proteomic data along with metabolomic analysis will provide an in-depth overview of the molecular alteration occurring in the canola plants under different stressful conditions. The authors have integrated the data of transcriptome, proteome, and metabolome to characterize the early responses of canola challenged with clubroot infection and observed that substantial changes occurred in metabolic pathways associated with disease resistance (Adhikary, 2024). Importantly, this multi-omics approach has not only enriched the knowledge of the interaction between plants and pathogens but also helped in formulating certain breeding strategies for resistance in canola against specific diseases.

Apart from the health of a plant and agricultural productivity, metabolomics is related to nutritional quality in canola oil. A metabolomics profile used in canola seeds may alter the fatty acid composition, the tocopherols, and other bioactive constituting the oil to an epileptic one for human health (Tian et al., 2016). Thus, advanced metabolomics will enable the identification and selection of canola varieties that will produce oil with superior nutritional value, hence contributing to food security and public health. In other words, the metabolome of the canola plant is highly diversified into primary and secondary metabolites that contribute to growth, development, and respond to environmental cues. Such metabolites and their function in plant physiology and stress responses could be deeply unraveled by the sophisticated technique of metabolomics involving GC-MS, LC-MS, and NMR. This continues to be an exciting area of study as the dimensionality of canola metabolome is realized, with application toward the improvement of agricultural practices, crop hardiness, and nutritional quality in canola oil.

Table 1. Metabolites present in different parts of Canola plant.

S. No	Compound Name	Plant Part				
1	**Malic acid**	**Root**	**Stem**	**Leaf**	**Inflorescence**	**–**
2	Benzoyl choline	–	–	–	–	Seed
3	Phenylalanine	–	–	Leaf	–	–
4	Isocitric acid	Root	Stem	Leaf	Inflorescence	–
5	Kaempferol hydroxyferuloyl dihexoside	–	–	Leaf	–	–
6	Tryptophan	–	–	Leaf	–	–
7	Kaempferol caffeoyl dihexoside	–	–	–	Inflorescence	–
8	Sinapoyl choline hexoside	–	–	–	–	Seed
9	Quercetin trihexoside	–	Stem	Leaf	Inflorescence	–
10	Gluconic acid	Root	Stem	Leaf	Inflorescence	–
11	Ferulic acid	–	–	–	Inflorescence	–
12	Quercetin sinapoyl trihexoside	–	Stem	Leaf	Inflorescence	–
13	Isorhamnetin dihexoside	–	–	–	Inflorescence	–
14	Kaempferol sinapoyl dihexoside	–	Stem	Leaf	–	–
15	Quercetin dihexoside	–	–	–	Inflorescence	–

continued on following page

Table 1. Continued

S. No	Compound Name	Plant Part				
1	**Malic acid**	**Root**	**Stem**	**Leaf**	**Inflorescence**	**–**
16	Sinapic acid	–	–	–	Inflorescence	–
17	Isorhamnetin dihexoside	–	Stem	Leaf	Inflorescence	–
18	Kaempferol feruloyl trihexoside	–	–	Leaf	Inflorescence	–
19	Sinapoyl choline	–	–	–	–	Seed
20	Quercetin dihexoside	–	Stem	Leaf	Inflorescence	–
21	Kaempferol feruloyl dihexoside	–	–	Leaf	–	–
22	Kaempferol-3-O-sophoroside	–	–	Leaf	–	–
23	Rutin	–	–	Leaf	Inflorescence	–
24	Isorhamnetin dihexoside isomer	–	Stem	–	Inflorescence	–
25	Kaempferol-3-O-β-D-rutinoside	–	–	Leaf	–	–
26	Feruloyl choline ferulyl ester	–	–	–	–	Seed
27	Isorhamnetin dihexoside isomer	–	Stem	–	Inflorescence	–
28	Sinapic acid hexoside	–	Stem	–	–	–
29	Kaempferol-3-O-hexoside	–	–	Leaf	Inflorescence	–
30	Miquelianin	–	–	–	Inflorescence	Seed
31	Isoquercitrin	–	–	–	Inflorescence	–
32	Kaempferol sinapoyl dihexoside isomer	–	–	Leaf	–	–
33	Isorhamnetin-3-O-hexoside	–	Stem	Leaf	Inflorescence	–
34	Ferulic acid isomer	–	–	Leaf	–	–
35	Kaempferol feruloyl dihexoside	–	–	Leaf	Inflorescence	–
36	3,4-Dimethoxy cinnamoyl choline	–	–	–	–	Seed
37	Sinapoyl choline feruloyl ester	–	–	–	–	+
38	Dihydroxy-oxo-12- octadecenoic acid	–	–	Leaf	–	–
39	Sinapoyl malic acid	–	–	–	–	Seed
40	Isorhamnetin	–	Stem	Leaf	Inflorescence	–
41	Sinapoyl choline feruloyl ester isomer	–	–	–	–	Seed
42	Hydroxy octadecadienoic acid isomer	Root	–	–	–	Seed
43	Quercetin	–	–	–	Inflorescence	–
44	Naringenin	–	–	–	Inflorescence	–
45	Hydroxy octadecadienoic acid	Root	–	–	–	Seed
46	Kaempferol	–	–	–	Inflorescence	–
47	Myristic acid	–	Stem	Leaf	Inflorescence	–
48	Linoleic acid	–	–	–	–	Seed

continued on following page

Table 1. Continued

S. No	Compound Name	Plant Part				
1	Malic acid	Root	Stem	Leaf	Inflorescence	–
49	Hydroxy octadecanedioic acid	Root	–	–	–	Seed
50	Oleic acid	Root	–	Leaf	Inflorescence	Seed
51	Hydroxy octadecanedioic acid isomer	Root	–	–	–	Seed
52	Stearic acid	Root	Stem	Leaf	Inflorescence	Seed
53	Hydroxy octadecatrienoic acid	Root	–	–	–	–
54	Hydroxy palmitic acid	–	–	Leaf	Inflorescence	–
55	Erucic acid	–	–	–	–	Seed
56	Palmitic acid	Root	–	–	–	Seed
57	Linolenic acid	–	–	–	–	Seed

Source: Farag et al., 2013

3. METABOLOMICS IN SEED DEVELOPMENT

Specificity of the rationale for metabolomics in the development of seeds, especially regarding pathways of oil biosynthesis and profiling of metabolites across different stages of seed maturation, is at the core of their complex processes. Synthesis of fatty acids and lipids is a key process affecting energy reserve accumulation within seeds, which becomes very important during germination and the establishment of the seedling. This response will explain in detail oil biosynthesis pathways, metabolite identification associated with the production of fatty acids and lipids, and metabolite profiles during seed maturation to which special emphasis will be given to their influence on oil quality. Oil biosynthesis in seeds mainly consists of triacylglycerols, which are synthesized from fatty acids. This process is initiated in the plastids where the synthesis of fatty acids is mediated by FAS complexes.

These catalyze the condensation of acetyl-CoA and malonyl-CoA, forming long-chain fatty acids taken to the ER for assembly into TAG (Kim et al., 2012). Of these, ABC transporters such as AtABCA9 play a significant role in supplying fatty acids from plastids to the ER, where they are esterified with glycerol-3-phosphate to form TAGs (Cai et al., 2021). Besides these transport mechanisms, regulation of lipid metabolism is provided by various transcription factors, including WRINKLED1 (WRI1), which orchestrates the expression of genes involved in the biosynthesis of fatty acids and the accumulation of TAGs. The WRINKLED1 transcription factor controls the expression of genes involved in the synthesis of fatty acids and the accumulation of TAG (Grimberg et al., 2015). The metabolic transition in seed development often involves a switch from photosynthetic metabolism into oil biosynthesis. This would, in turn, indicate the importance of carbon source allocation from various plant tissues, such as leaves and stems, to developing seeds, as mentioned by Vanhercke et al. (2013) and Tan et al. (2015).

The dynamic interrelationship between these and other metabolic pathways, including glycolysis and the tricarboxylic acid cycle, provides all the precursors required for lipid synthesis. Such flexibility in the metabolic network has been evidenced by the conversion of starch to fatty acids in some Arabidopsis mutants during seed maturation in the presence of variable carbon sources. (Li et al., 2015; Lu et al., 2018). This flexibility is important for optimizing oil content in seeds under different environmental conditions. Seed Maturation and Metabolite Profiling Seed maturation is a rather complex process and is joined by significant changes in the metabolite composition of seeds, which directly influences the quality of the oils. The seeds, while developing through a maturation phase, undergo several biochemical transformations, leading to the accumulation of storage compounds like oils, proteins, and carbohydrates. Metabolite profiling has shown that seed fatty acid composition is highly dynamic across the development stages and a subset of these metabolites' associates with oil quality improvement (Song et al., 2013). For instance, frame metabolite profiles have been analyzed in developing seeds of Brassica napus to identify key metabolites correlating with seed weight and oil content. Such initiating dynamic changes in metabolites-including phosphatidic acid and a variety of fatty acids-during seed development are indicative of a metabolic shift in seeds as they transition from a growth phase into a storage phase.

More recently, the funiculus has been identified regarding nutrient transport during seed development, emphasizing that it is crucial for supplying necessary metabolites to account for the quality of seeds (Li et al., 2019). It has also been compared in different rice cultivars during seed development, where certain amino acids and fatty acids are associated with better germination and seedling vigor (Tan et al., 2016). The identification of such metabolites can give information on the metabolic pathways that control the quality of seeds, thereby highlighting potential targets for metabolic engineering toward oil production improvement in crops. Factors of the Environment Affecting Quality of Oil. It is dedication and oxidative stress due to all such environmental factors that affect the viability of seeds and their oil quality negatively. The generation and accumulation of ROS in the seeds during aging have been reported to induce lipid peroxidation, which impairs seed integrity and diminishes the quality of the contained oil (Hu et al., 2014; Wiebach et al., 2019). Antioxidants have been trying to improve seed viability and maintain oil quality during their storage by minimizing this oxidative damage (Adetunji et al., 2021).

Besides, the interaction between lipid metabolism and environmental stress responses is critical in the process of maintaining seed quality (Ahmad et al., 2022a;

Ahmad et al., 2024a). For example, it was found that the expression of phospholipase Dα was related to desiccation tolerance in seeds, with its activity thus modulating compositional changes in lipids and membrane stability during seed maturation (Hu et al., 2014; Chen et al., 2017). Understanding these interactions may, therefore, be of paramount importance in developing strategies toward improved seed oil quality across different environmental conditions. Metabolomics applied to seed development provides important information on the biochemical pathway of oil biosynthesis and its complex factors affecting seed quality. The whole range of metabolites at each maturation stage brings critical information on metabolic shifts and changes happening in oil production. The demands of vegetable oils are continuous, and translation of this knowledge through metabolic engineering, coupled with better agricultural practices, will become necessary for improvement in yield and quality of oil in crops.

4. ABIOTIC STRESS RESPONSES

Abiotic stress, especially drought and salinity, poses a serious threat to agricultural yield and crops like canola. Its response due to such conditions would include complex metabolomics, comprising different levels of biochemical pathways that could help in maintaining intracellular homoeostasis and promote survival under stressful conditions (Figure 1). This might be a typical response where changes take place in the levels of primary and secondary metabolites, reflecting their important roles for osmoregulation, antioxidant defense, and overall stress tolerance in plants (Öztürk et al., 2020; Haider et al., 2021a; Haider et al., 2022a). Drought stress triggers a sequence of physiological and biochemical alterations in canola, leading to the accumulation of specific metabolites that aid this plant during stressed conditions due to lack of water (Wang et al., 2022). One of the most studied metabolites in the context of the drought stress response is the amino acid proline. It is an Osmoprotectant means that it maintains protein and cellular stability, while at the same time it takes part in the osmotic balance during drought. The increase of proline accumulation together with other free amino acids, such as asparagine and arginine, represent general responses to drought stress, aiming at facilitating recovery and tolerance upon subsequent rehydration. In addition, secondary metabolism was also highly induced by the drought treatment, such as in biosynthesis of phenolics and flavonoids, further supporting the protection mechanisms against oxidative stress (Rezayian and Ebrahimzadeh, 2019; Naheed et al., 2021).

Besides indicating drought, salinity stress induces deep metabolic changes in canola. High levels of salinity promote ion toxicity and osmotic stress that may contribute significantly to impeding plant growth and development. Compatible solutes like glycine betaine and proline accumulate in response to salinity, while playing a role in mitigating the adverse effects of high salt concentrations (Rossi et al., 2016; Cui et al., 2018). Moreover, it has been suggested that salinity stress conditions require the induction of antioxidant pathways, in which metabolites such as phenolic compounds participate in detoxifying ROS. Ahmad et al. (2021) reported that several metabolic responses involve plants under salinity and drought stress. These conditions get even more complicated because a plant faces the interaction between these two stresses (Patel et al., 2022). A comparison of the metabolomics profile with that under conditions of abiotic stress demonstrated significant changes in levels of various metabolites, such as sugars, organic acids, and amino acids. For example, drought stress has been noted in several studies to increase the levels of soluble sugars, acting not only as the energy source but also taking part in osmotic adjustment (Mibei et al., 2017; Das et al., 2017, Haider et al., 2021b). Organic acids also accumulate under drought and salinity, reflecting changes in metabolic pathways to afford stress tolerance (Kráľová et al., 2012; Obata and Fernie, 2012). These metabolic changes are important in the context of maintenance of cellular integrity and function under stressful conditions.

It also causes temperature stress, both low and high to induce specific metabolomic responses in canola. For example, heat stress may induce increased respirational rates or changes in energy metabolism which result in the accumulation of certain metabolites to aid survival under a high-temperature environment (Wu et al., 2020). On the other hand, cold stress may trigger the synthesis of cryoprotectants, which are certain sugars and amino acids responsible for allowing the stabilization of plant cellular structures and prevention or delay of ice formation in the plant tissues. The interaction between temperature with other abiotic stresses such as drought and salinity further complicate the metabolic responses and requires a profound understanding of these interactions as one of the main preconditions for the elaboration of stress management techniques (Khodabin et al., 2020). Such use of metabolomics to study the responses has provided substantial information on biochemical pathways in stress tolerance mechanisms. Advanced techniques of analysis, while identifying and quantifying the wide range of metabolites, have facilitated further elucidation on the metabolic shift occurring under stress conditions in abiotic stresses (Jorge et al., 2015, Haider et al., 2022b).

This knowledge is of prime importance in breeding programs for the development of stress-resistant canola varieties because it helps in the identification of key metabolites and pathways to be targeted for improvement. The changes in metabolomics of canola plants under drought, salinity, and temperature stress are multivariable and include complex interactions from primary and secondary metabolites. Such changes are prerequisites for the adaptability of the plant under extreme conditions of the environment. This understanding enhances our knowledge not only about plant physiology but also about agricultural practice with respect to its various modes of enhancing crop resilience against climate change.

5. BIOTIC STRESS RESPONSES

Biotic stress responses are very crucial for the survival and productivity of plants, including canola. These responses are largely mediated through secondary metabolites-very important bioactive organic compounds involved in the defense mechanisms against pathogens and pests (Figure 1). This chapter, therefore, chronicles the metabolites involved in such defense mechanisms and places emphasis on phytoalexins and other defense compounds that canola plants use in combating biotic stresses. One class of secondary metabolite synthesized de novo in response to biotic stress, including pathogen infection, is phytoalexins. These have antimicrobial activity, which has been considered important in the defense of plants (Pedras et al., 2011; Cook et al., 2022). The diversity in canola includes camalexin, brassinin, and other phytoalexins produced against a wide array of pathogens, including *Pseudomonas aeruginosa* and fungal infections. Their biosynthesis is usually controlled by a signaling pathway involving the phytohormones JA and ethylene, which is induced after an attack by the pathogen, as mentioned by Pérez-Llorca et al. (2023) and Meng & Zhang (2013).

Canola has been induced to express, for example, camalexin biosynthetic genes through interaction with *Pseudomonas aeruginosa*, this in a way elaborates on the importance of phytoalexins in plant artillery (Cook et al., 2020). Besides phytoalexins, a suite of secondary metabolites is produced by canola plants which include glucosinolates after all; these are sulfur-containing compounds which take part in the replies against biotic stresses too. Glucosinolates and their breakdown products have been found to deter herbivores, interfere with the growth of pathogens (Hanschen et al. 2014; Ahanger et al. 2020). Depending on the nature of the attack made by herbivores, enzymatic breakdown of glucosinolate allows the release of such bioactive compounds that may trigger plant defense mechanisms as attributed to signaling by JA.

The given interaction draws the intricate interrelations between various secondary metabolic classes in the plant defense. Recent studies indicate that silicon is important in the improvement of canola's secondary metabolite production. Silicon improves the accumulation of different phenolics and glucosinolates, showing peculiar antimicrobial properties and enhancement of overall stress tolerance in plants (Li et al., 2018; Sarkar et al., 2023). The application of silicon improves not only physical entities of plant tissues but also metabolic processes developing those produced compounds in defense. Such is the case when silicon-mediated stress tolerance has been related to the change in lignin deposition, cell wall composition, and more generally necessary to resist pathogen invasion (Hashemi et al., 2010; Lahlali et al., 2017; Ahmad et al., 2022b). Further, the integration of different omics approaches, including transcriptomics and metabolomics, gave way to more detailed information on metabolic responses against biotic stresses in canola. These high-throughput techniques enable the characterization of complex regulatory networks governing the biosynthesis of secondary metabolites against pathogen and pest attacks (Wang et al., 2020; Kumari, 2024). For instance, some examples of transcriptomic analyses have shown differential expressions of genes corresponding to the biosynthesis of phytoalexins and other related defense metabolites under biotic stress conditions (Cook et al., 2020; Vo et al., 2021). In fact, this comprehensive understanding of plant defense mechanisms is urgently needed for developing superior varieties with enhanced resistance against various biotic stresses.

ROS contribution to the signaling pathway of biotic stress responses is also not negligible. ROS generation happens upon infection of pathogens, which is crucial for the induction of internal defense reactions such as phytoalexin biosynthesis and strengthening of cell walls (Jaspers and Kangasjärvi, 2010; Meng and Zhang, 2013). Their accrual may, therefore, trigger a chain reaction of signal transduction events that mobilize the activation of defense genes and secondary metabolite production, enhancing the resilience of crops against biotic stresses. The canola plant protection mechanism against the attack of biotic stress is complex and involves a wide array of secondary metabolites, namely phytoalexins and glucosinolates. All these metabolites interact in a complex way with the regulatory functions of signaling molecules such as JA and ethylene, underlining the complexity of the plant defense mechanisms. In addition, the use of omics technologies and silicon application to enhance the production of metabolites open quite promising perspectives for further research to improve crop resilience under biotic stresses.

Figure 1. Comprehensive metabolomic pathway illustrating the response of canola plants to abiotic/biotic stress, highlighting mechanisms employed to mitigate stress. Abbreviations: PAL (Phenylalanine ammonia-lyase), C4H (Cinnamate 4-hydroxylase), 4CL (4-coumarate-CoA ligase), CHS (Chalcone synthase), CHI (Chalcone isomerase), CCR (Cinnamoyl-CoA reductase), CAD (Cinnamyl alcohol dehydrogenase), PRRs (Pathogen Recognition Receptors), PAMPs (Pathogen-associated molecular patterns), PTI (PAMP-triggered immunity), RLKs (Receptor-like kinases), ROS (Reactive oxygen species), MAPKs (Mitogen-Activated Protein Kinases), AOS (Allene oxide synthase), PLA2 (Phospholipase A2) 13-HPOT (13-hydroperoxyoctadecatrienoic acid), OPDA (12-oxo-phytodienoic acid), CYP79 (cytochrome P450 monooxygenases)

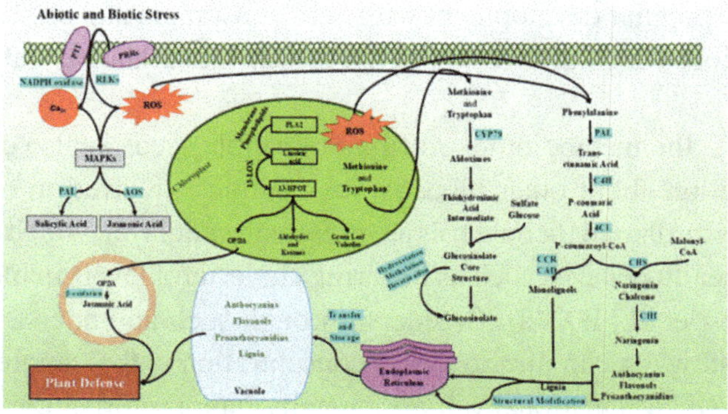

6. METABOLOMICS FOR CANOLA BREEDING

Metabolomics now stands as one of the new approaches in plant breeding. Among these crop improvements, that of canola is of utmost importance. The identification of key metabolites will enable breeders to further enhance oil composition, its nutritional value, and resilience against environmental stressors (Lin et al., 2013, Ijaz et al., 2023). The fatty acid profile of canola is excellent, rich in monounsaturated and polyunsaturated acids. Despite the large amount of health-related literature existing, for example, cardiovascular improvement associated with the consumption of canola oil, it is well-documented. Metabolomics could enable the identification of those specific metabolites responsible for these health benefits and would allow breeders to select for canola varieties based on optimized profiles of oil.

Recent works developed the possibility to use, as a source of dietary omega-3 fatty acids, canola oil after supplementation with oils rich in alpha-linolenic acid (ALA) (Nguyen et al., 2017). Metabolic pathways responsible for increased ALA content in canola seeds were identified using metabolomic approaches. According to Popovici et al. (2022), the incorporation of lichen extracts favored its antioxidant action on canola oil, probably delaying its rancidity and giving it an even better shelf life. In this respect, it is good to refer to the fact that metabolic profiling might enhance not only the nutritional value of canola oil but also its stability and usability in food applications.

Breeding high-oleic-acid-content varieties of canola has recently been on the increase, in view of the sustainability of the oil at high-temperature cooking and its biodiesel uses (Spasibionek et al., 2020). Conversely, metabolomics analyses can reveal the genetic and biochemical pathways responsible for oleic acid accumulation and therefore provide breeders with more efficient selection for those traits. It has become a more popular development with both consumers and producers alike due to its formulation, which affords improved health and oxidative stability (Jones et al., 2014).

Besides FAs, the metabolomic profiling of canola seeds is also capable of providing information about other bioactive compounds involved in oil quality. For instance, phytosterols and tocopherols are present in canola oil, which are believed to show various health benefits such as lowering cholesterol levels in blood (Adeyemi et al., 2016; Ijaz et al., 2024a). Breeders, through such an understanding of their biosynthetic pathways, can improve their concentration in the-canola oil and hence further improve their nutritional value. Generally, canola cultivation faces different abiotic stresses-drought and salinity-and biotic stress in the form of disease (Neik et al., 2020). Metabolomics opens views on creating resistant species of canola to decipher the changed metabolic states occurring in plants after environmental challenges. For example, some metabolites accounting for resistance to stress may be used at due time in breeding programs for improving tolerance. From this view, some metabolites such as proline and several phenolic compounds are involved in plant responses to various stress conditions (Richardson et al., 2017, Ahmad et al., 2024b, Ijaz et al., 2024b). This will enable breeders using metabolome techniques to screen the levels of metabolites across diverse canola genotypes under specified conditions of stress. This information can hence be used to select for varieties that show higher concentrations of stress-related metabolites, hence generally enhancing their resilience.

This, in addition, can help the breeding programs increase in precision through the integration of metabolomics with genomic data. Breeders will, therefore, be able to apply marker-assisted selection techniques that would identify specific genetic markers associated with desirable metabolic traits located within the genome (Spasibionek et al., 2020). Hence, helping to hasten the process of developing varieties of canola resistant to stress, which in turn saves time and enhances the success rate of breeding programs in general. In fact, the effect of the environment on metabolic profiles is well documented in the case of canola. For example, the variation in soil nutrients can well affect the compositional characteristics of the fatty acids in canola oil (Gao et al., 2010). If such interactions are known using metabolomic analyses, breeders can employ them to deduce the best ways of growing any species and select varieties that could thrive well under challenging environmental conditions. This is furthered by the fact that metabolomics can even attain the characteristics that promote stress resistance without necessarily using the actual traits through the conventional breeding systems. For example, secondary metabolites such as glucosinilates provide knowledge concerning their active roles in plant resistance to pests and diseases (Gomaa et al., 2018). This, in turn, can be used to select canola varieties that can contain higher levels of such compounds, hence increasing their general hardiness, as well as reducing their dependence on chemical means of pest control.

Soon, metabolomics will be combined with other omics such as genomics and proteomics in the breeding of canola. Most definitely, this will ensure that knowledge on interaction across genes, proteins, and metabolites in the canola plant will be better. This knows how to enable breeders to handle canola varieties that meet consumer needs for quality oil and easily cope with environmental stress. This is further enhanced by the potential of high-performance analytical techniques like mass spectrometry and NMR, which will further enable the power of metabolomics in research on canola (Richardson et al., 2017). These technologies have enabled the detailed profiling of metabolites involved in biochemical pathways that dictate oil composition and stress responses. As these techniques become increasingly available, their adoption into breeding programs cannot but further accelerate the process of developing better varieties. Performing metabolomics in concert with the new gene editing and synthetic biology methodologies could provide a new paradigm-shifting opportunity for the discipline of canola breeding. This will allow for very specific changes to metabolic pathways of oil biosynthesis and, hence, enable the breeder to develop new canola varieties with superior and value-added oil composition, with additional tolerance to sudden and extreme environmental stresses (MacIntosh et al., 2021). New approaches hold great promise for the response to challenges of climate change and increased global demand for sustainable agricultural production. Consequently, metabolomics presents a strong tool that will help in the enhancement of canola breeding. Key metabolites whose identification is linked with oil quality and stress resistance will enable the breeder to develop the most acceptable variety of canola, characterized by meeting nutritional requirements while at the same time being resilient under environmental challenges. In so doing, metabolomics, combined with other "omics" technologies and state-of-the-art breeding techniques, will enable the development of the next generation of canola varieties to become sustainable while being profitable.

7. METABOLIC ENGINEERING OF CANOLA

Genetic engineering of canola is a high castle in agricultural biotechnology, where improvement in the desired traits like oil yield, disease resistance, and abiotic stress are predicted. Targeting Key Pathways Metabolic engineering geared at manipulating biosynthetic pathways in canola has been one of the points of focus for researchers. The general aims have been to improve oil yield and plant resilience. Canola is an allotetraploid species with two sets of chromosomes derived from different ancestral species, *Brassica rap*a, and *Brassica oleracea* (Chu et al., 2020). This genetic complexity presents a challenge and an opportunity when carrying out metabolic engineering. Metabolic engineering also conveys overproduction of desirable compounds such as omega-3 fatty acids in transgenic crops and enhances the tolerance of plants against environmental stresses through the targeting of specific metabolic pathways.

One of the main pathways targeted in modification is the biosynthetic pathway for fatty acids. Traditionally, canola oil is valued for containing very low levels of saturated fat and high levels of unsaturated fatty acids, especially oleic acid (Han et al., 2022). However, recent metabolic engineering has enabled the introduction of genes involved in the synthesis of LC-PUFAs such as EPA and DHA. Delivery of genes from various marine microorganisms has transformed canola into a line capable of producing these essential fatty acids that play a crucial role in human health and aquaculture (Weldon, 2023). It will enrich not only the nutrient levels in canola oil but also promise sustainability for fish oil as an aqua feed. More related to metabolic engineering, perhaps, is the development of herbicide resistance traits, such as those that confer glyphosate resistance in canola.

For instance, it has been observed that the glyphosate acetyl transferase transgene GAT4621 is effectively incorporated into canola and confers excellent weed control with no effects on the crop (McClinchey et al., 2016). This not only increases the yield by reducing competition amongst weeds but also makes the management of canola crops easier, hence more popular. In canola, metabolic engineering has improved oil yield and added herbicide tolerance, but most of the work has focused on enhancing disease resistance. Adding resistance genes would significantly enhance the host plant's defense capabilities against a broad array of pathogens, reducing the application of chemical fungicides. For example, methods of synthetic biology have been used to develop canola varieties resistant to fungal diseases that threaten crop yields all over the world (Sargent et al., 2022). By using gene editing technologies, scientists can make precise changes in canola's genome, enhancing its natural immune responses and developing more robust crop varieties. Synthetic Biology Approaches Synthetic biology has emerged recently as a powerful tool in metabolic engineering, enabling the design and construction of new biological parts and systems in canola.

This approach allows for the more precise manipulation of metabolic pathways to achieve novel properties not attainable by conventional breeding (Wulimin et al., 2013). Applications of synthetic biology to canola might play a critical role in crop production by improving the trait for oil composition, stress tolerance, and disease resistance. The most promising aspect of synthetic biology in canola is that with this technology, it is possible to design novel promoters that may potentially drive gene expression in a tissue-specific manner. For example, isolation of native promoters from Brassica species has been carried out and could be used to regulate transgene expression in canola (Buzdin and Патрушев, 2021). This specificity is very important to ensure the introduced characteristics appear only in the desired tissues and minimize the possible unintended effects on the growth and development of the plant. It is possible to use such native promoters for assuring finer expression of genes responsible for oil biosynthesis, thus in turn giving a higher yield and quality of oil (Beszterda and Nogala-Kałucka, 2019). Besides, the use of synthetic biology in concert with the state-of-the-art genome editing technologies like CRISPR/Cas9 opened more vistas for metabolic engineering applications in canola. These technologies will be able to provide precise changes in the canola genome, thus allowing several desirable traits of this crop to be introduced into new varieties in just one generation.

For example, with the use of this technology, scientists would be able to enhance oil yield, take over disease resistance, and further improve abiotic stress tolerance while targeting the overall multiple genes within pathways that control said traits. Advanced biotechnology massively reduces the time used for breeding work while offering multiplex capability and hence presents an opportunity to come up with new varieties with value addition in canola (Hernández-Terán et al., 2017). Applications of synthetic biology further extend to metabolic engineering for the improved production of bioactive compounds in canola. Engineering metabolic pathways responsible for producing these active ingredients can improve their levels in canola oil, hence improving nutritionally. This is particularly placed in developing functional foods that are expected to offer health benefits above basic nutrition (Sargent et al., 2022). The production of the bioactive compounds in canola would classify it as a commodity crop that is headed for value addition to functional foods. New emerging technologies in metabolic engineering have the potential to change the face of canola production by solving modern agriculture's challenges. Most of the significant technologies involve advanced genomic tools that are applied for the identification and characterization of genes involved in oil biosynthesis.

High-throughput sequencing and transcriptomic analyses have provided insights into the diversities of complex regulation with respect to oil production in canola (Chu et al., 2020). Indeed, this kind of understanding opens the avenue toward identifying those key genes which could be targeted for modification to achieve improved yield and quality in the case of canola oil. Further, the use of systems biology approaches allows the integration of genomic, transcriptomic, and metabolomic data into the elaboration of holistic models of canola metabolism. In this respect, these models can be used to predict the effect of genetic interventions on the metabolic network, a fact which will allow for the optimization of engineering strategies (Sargent et al., 2022). By running simulations of the metabolic responses in canola against different genetic interventions, researchers can then optimize their methodology for lab conditions before taking those into the field. AI algorithms can analyze large datasets from genomic and phenotypic studies for patterns or correlations that may not be visible under traditional data analysis. By embracing AI, it will be easier for researchers to speed up the discovery of beneficial traits and develop effective breeding programs. This might ultimately allow for the release of improved performing varieties of canola. Public perception and regulatory barriers to GMOs in many parts of the world vary from country to country, and therefore are still one of the major obstacles to commercialization of the engineered varieties of canola (Bawa & Anilakumar, 2012). Consumer acceptance of the safety and environmental impact of GMOs is an essential base upon which the commercial success of the new engineered varieties depends. Besides, the occurrence of random phenotypic changes due to genetic modifications raises concerns about the potential ecological consequences of releasing these crops into the wild (Hernández-Terán et al., 2017). Thus, besides broad risk assessments, long-term studies will be necessary about the impacts of transgenic canola on biodiversity and ecosystem health. Optimization of pathways and enhancing genetic modification efficiencies are some of the current areas of research. Metabolic engineering in canola encompasses approaches to improve oil yield, confer disease resistance capacity, and enhance tolerance to stresses. By targeting key biosynthetic pathways and powered by synthetic biology, researchers could change the face of canola production to meet consumer and environmental requirements.

8. TECHNOLOGICAL ADVANCES IN METABOLOMICS

Metabolomics, or the comprehensive study of metabolites within biological systems, has notably benefited from key advances in analytical technologies enabling high-throughput profiling of metabolites. These advances are crucial in understanding complex biological processes and applying these to precision medicine, agriculture, and biotechnology (Gowda and Raftery, 2015; Beger et al., 2016). High-throughput metabolomics has been made possible using highly efficient analytical systems, namely MS and NMR spectroscopy; these enable processing many metabolites from any given biological sample with high sensitivity and speed (Figure 2). Of these,

MS has become the cornerstone for metabolomic studies owing to its sensitivity and capability for complex mixture analysis.

Advances in MS technology, such as the development of new ionization techniques and new data acquisition methods, have made it possible to observe thousands of metabolites in a single run (Kumar et al., 2017). For instance, LC-MS provided much easier and more conscious separation and identification of metabolites. Thus, the researchers viewed a fuller metabolome (Hong et al., 2016; Katam et al., 2022). Large-scale investigations-such as the metabolic response of crops to environmental stressors or the identification of a metabolic signature related to diseases-are now both critical and possible with such capability diseases (Razzaq et al., 2019; Oh et al., 2023). Besides, NMR is a nondestructive approach for metabolomics study in which comprehensive structural information of the metabolites could be provided.

Even though NMR is normally less sensitive than MS, its quantitative data presentation without the extensive requirement in sample preparation makes them irreplaceable in certain cases (Microbial Metabolomics, 2016). When these advantages complement by each other while combining the techniques in MS and NMR, this combination can enhance the robustness of metabolomic analyses (Sumner et al. 2015; Beger et al. 2016). This integration added immense pace to this field through the introduction of high-throughput metabolomics along with the bioinformatics tool. Advanced computational methods were employed to manage and analyze huge data from such analytical platforms. For example, one of the online tools that was in wide use is MetaboAnalyst, which allows facilitation of statistical analysis and visualization in metabolomic data and thus allows meaningful biological interpretation from such complex datasets (Xia et al., 2015; Choudhury et al., 2020). The biological significance interpretation while there are changes in metabolomics is important and it integrates technology with informatics. It also searches for biomarkers related to diseases or stress responses in plants (Jendoubi, 2021). Among other powerful approaches towards understanding biological phenomena at systems levels is the integration of metabolomics with other omics disciplines, such as genomics, transcriptomics, and proteomics. With this strategy of multi-omics, one enables studies of the complicated interaction between metabolites, genes, and proteins and may thus reach an understanding of biological processes at a mechanistic level (Lu and Zhan, 2018; Rattray et al., 2018).

Genomics provides insight into genetic architecture, whereas transcriptomics gives views into the expression patterns of genes involved in metabolic pathways. Together with metabolomic profiles, these would yield insights into how genetic variation affects metabolic phenotypes (Guo et al., 2020; Bongaerts et al., 2022). Although metabolomics integrated into genomic data facilitates the identification of genetic variants for metabolic traits in crops, thus developing better breeding strategies, different studies have outlined. This will be an integrative approach, particularly valuable in precision agriculture, and can explain the genetic and metabolic basis of traits for crop improvement (Hong et al., 2016; Razzaq et al., 2019). Transcriptomics follows this integrated path as well. In this way, by analyzing levels of gene expression that accompany metabolite concentrations, researchers could unearth the most significant regulatory pathways and metabolic networks that respond to changed environmental conditions or other stressors (Oh et al., 2023). For example, this can be applied in the case of plant responses to abiotic stressors at either drought or salinity, which may be associated merely with changes in gene expression and changes in metabolite profiles via a multi-omics approach, giving an overview of the adaptive mechanisms operating in the plant (Katam et al., 2022). Proteomics, focusing on the study of proteins and their functions, therefore complements metabolomics because it can shed light on which functional consequences arise due to metabolic changes (Jendoubi, 2021). Integration of proteomic data into metabolomic profiles shows how metabolic changes affect protein expression and activity, further aiding cellular processes. Such a whole-system level perspective cuts across the complexities in biological systems and allows targeted intervention, as is necessary in medicine, agriculture, among other sectors (Ritchie et al., 2015).

Fully integrated multi-omics approaches through machine learning and other advanced statistical methods are being widely applied. Such approaches will disclose the patterns and associations in the data and facilitate the identification of biomarkers and therapeutic targets in an easily manageable way (Barupal et al., 2018; Hooft et al., 2020). This integration with diverse omics data will improve, without any doubt, our understanding of biological systems and enhance the precision of interventions in health and agriculture, as metabolomics continues to evolve. In brief, rapid technological advances in the field of metabolomics, high-throughput analytical platforms, and integration with other omics disciplines have affected our perception of biological systems beyond our imagination. The potential profile of many metabolites that enables the inference gathered from genomics, transcriptomics, and proteomics facilitates the development of a cohesive model for the study of complicated biological processes. That these technologies are continuing to be refined promises to open new ways of research and applications in precision medicine, and crop improvement.

Figure 2. Recent technological advancements in plant metabolomics

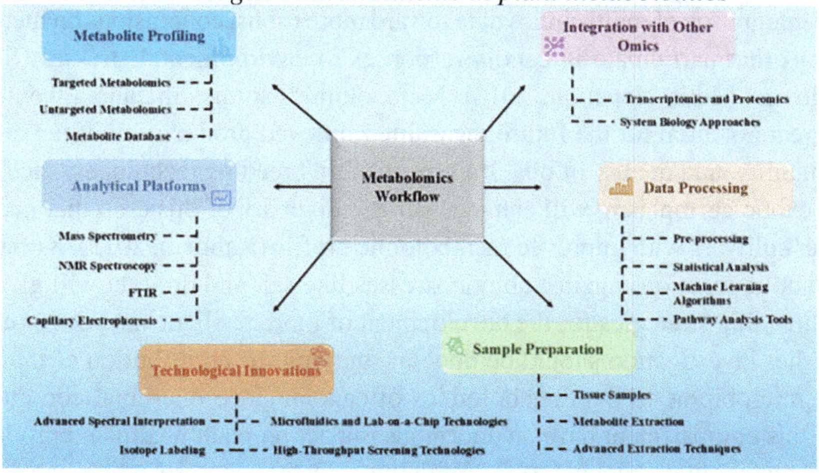

9. CHALLENGES AND FUTURE DIRECTIONS

The complexity in the canola metabolome has introduced big challenges in the identification and quantification of the huge array of metabolites within this crop. Such include a wide range of metabolites such as fatty acids, tocopherols, and phytosterols, which require advanced techniques for their profiling. These studies further showed that the key role of extraction techniques resulted in a higher content of tocopherols in cold-pressed canola oil when compared with refined oils (Azadmard-Damirchi et al., 2010; Ghazani et al., 2014). This variability complicates the establishment of a comprehensive metabolomic profile, since in each process different techniques can result in the loss of useful compounds; as in the case of canolol, which is often below detection limits in refined versions of oil. The presence of other secondary metabolites, such as glucosinolates and isothiocyanates, whose content is influenced by the plant growth stage and conditions, further complicates the metabolome (Gurusinghe et al., 2019).

Other challenges of canola metabolomics involve data interpretation and integration. High throughput metabolomic analyses produce a very large volume of information, which requires advanced computational tools to analyze. Current methodologies cannot resolve datasets of different natures such as genomic, transcriptomic, and proteomic data besides the metabolomic data itself (Shokri-Gharelo and Noparvar, 2018). For example, recent works emphasize that in canola, stress responses interact particularly with lipid metabolism, and understanding such relations is relevant for improving crop resilience and oil quality (Wang et al., 2022).

For this reason, there is a need to improve various bioinformatics tools that could enable integration of multi-omics data toward more robust conclusions on metabolic pathways that participate in canola responses to environmental stresses (Shokri-Gharelo and Motie-Noparvar, 2018). Metabolomics studies in canola would then offer great potential for the future regarding improved productivity of the crop, its sustainability, and quality of oils. Improvement in breeding techniques, such as the use of exotic germplasm, will enhance genetic diversity of this crop that may lead to better cultivars with improved metabolomic profile (Rahman, 2013). Knowledge of metabolic reactions against abiotic stresses like salt and drought will give light to the breeding strategies for the development of more resilient varieties of canola.

Further investigation should be done in studying the contribution of the rhizosphere microbiome to the metabolomics of canola, since microbial communities within this compartment have an enormous impact on plant health and production of plant metabolites (Nichol, 2023). With this information, researchers can develop specific methods that would help improve the nutritional quality of the canola oil without necessarily compromising agronomic concerns (Lay et al., 2018).

In short, the challenges associated with canola metabolome complexity, improvement of data interpretation, and future research directions all add up to emphasize the multitasking nature of canola metabolomics. Without a shadow of a doubt, the way forward toward further understanding of this important crop and enhancing its contributions toward food security and nutrition lies in addressing these issues.

10. CONCLUSION

Metabolomics, as a high-throughput analytical approach to functional genomics, has vastly improved our understanding of the biochemical phenotype of the canola plant, including profiles of complex metabolic pathways and their responses to various environmental stimuli. Most of the canola metabolomics studies have focused on secondary metabolites, given their crucial role in plant defense mechanisms, stress responses, and overall plant health. Identification and quantification of such metabolites highlighted how canola plants respond to abiotic stresses such as drought and salinity, besides temperature variations. Further, this chapter identified the potential use of metabolomics in the enhancement of breeding programs toward resilient varieties with improved nutritional profiles.

Further, the integration of metabolomic data with genomic and transcriptomic information allowed a more holistic understanding of the regulatory networks controlling metabolic processes in canola. This multi-omics approach will not only improve our knowledge regarding plant biology but may also provide opportunities for various biotechnological applications, such as the engineering of canola plants toward improved yields and higher quality production of oil. The chapter also focused on the methodological developments in metabolomic analyses, including high-throughput technologies and advanced data analysis techniques. Such innovations have allowed researchers to generate large datasets from which predictive modeling and systems biology approaches may benefit further, enriching further the knowledge of plant metabolomics.

In other words, the metabolomics of canola plants is a dynamic area of study that keeps on changing very fast and holds great potential as far as agricultural sustainability and enhancement of food security are concerned. The data discussed herein in this chapter are of critical value to basic knowledge on plant metabolism and provide a portal to applications regarding crop productivity and management. Future studies should continue interrogating the interaction of complicated metabolites with environmental factors and possibly use such information in the development of canola varieties that could address challenges resulting from changing climate and increasing global food demand.

REFERENCES

Adetunji, A., Varghese, B., & Pammenter, N. (2021). Effects of exogenous application of five antioxidants on vigour, viability, oxidative metabolism and germination enzymes in aged cabbage and lettuce seeds. *South African Journal of Botany, 137*, 85–97. DOI: 10.1016/j.sajb.2020.10.001

Adeyemi, K., Sabow, A., Ebrahimi, M., Samsudin, A., & Sazili, A. (2016). Fatty acid composition, cholesterol and antioxidant status of infraspinatus muscle, liver and kidney of goats fed blend of palm oil and canola oil. *Italian Journal of Animal Science, 15*(2), 181–190. DOI: 10.1080/1828051X.2016.1158081

Adhikary, D., Kisiała, A., Sarkar, A., Basu, U., Rahman, H., Emery, R., & Kav, N. (2022). Early-stage responses toplasmodiophora brassicaeat the transcriptome and metabolome levels in clubroot resistant and susceptible oilseedbrassica napus. *Molecular Omics, 18*(10), 991–1014. DOI: 10.1039/D2MO00251E PMID: 36382681

Adhikary, D., Mehta, D., Kisiala, A., Basu, U., Uhrig, R. G., Emery, R. J. N., Rahman, H., & Kav, N. N. V. (2024). Proteome- and metabolome-level changes during early stages of clubroot infection in brassica napus canola. *Molecular Omics, 20*(4), 265–282. DOI: 10.1039/D3MO00210A PMID: 38334713

Adhikary, D., Mehta, D., Uhrig, R., Rahman, H., & Kav, N. (2022). A proteome-level investigation into plasmodiophora brassicae resistance in brassica napus canola. *Frontiers in Plant Science, 13*, 860393. Advance online publication. DOI: 10.3389/fpls.2022.860393 PMID: 35401597

Ahanger, M., Bhat, J., Siddiqui, M., Rinklebe, J., & Ahmad, P. (2020). Integration of silicon and secondary metabolites in plants: A significant association in stress tolerance. *Journal of Experimental Botany, 71*(21), 6758–6774. DOI: 10.1093/jxb/eraa291 PMID: 32585681

. Ahmad, Y., Haakim, Z., Iqbal, J., Abbasi, B. A., Mahmood, T., & Kazi, M. (2024b). Technological Innovations for Abiotic Stress Resistance in Horticultural Crops. OMICs-based Techniques for Global Food Security, 233-244.

Ahmad, Y., Haider, S., Iqbal, J., Abbasi, B. A., Yaseen, T., & Mahmood, T. (2022a). The mechanisms of genome editing technologies in crop plants. In *Principles and Practices of OMICS and Genome Editing for Crop Improvement* (pp. 295–313). Springer International Publishing. DOI: 10.1007/978-3-030-96925-7_13

Ahmad, Y., Haider, S., Iqbal, J., Abbasi, B. A., Yaseen, T., & Mahmood, T. (2022b). The mechanisms of genome editing technologies in crop plants. In *Principles and Practices of OMICS and Genome Editing for Crop Improvement* (pp. 295–313). Springer International Publishing. DOI: 10.1007/978-3-030-96925-7_13

Ahmad, Y., Haider, S., Iqbal, J., Naseer, S., Attia, K. A., Mohammed, A. A., & Mahmood, T. (2024a). In-silico analysis and transformation of OsMYB48 transcription factor driven by CaMV35S promoter in model plant–Nicotiana tabacum L. conferring abiotic stress tolerance. *GM Crops and Food: Biotechnology in Agriculture and the Food Chain, 15*(1), 130–149. DOI: 10.1080/21645698.2024.2334476 PMID: 38551174

Ahmad, Z., Anjum, S., Skalický, M., Waraich, E., Tariq, R., Ayub, M., & Sabagh, A. (2021). Selenium alleviates the adverse effect of drought in oilseed crops camelina (camelina sativa l.) and canola (brassica napus l.). *Molecules (Basel, Switzerland), 26*(6), 1699. DOI: 10.3390/molecules26061699 PMID: 33803724

Asaduzzaman, M., Pratley, J., An, M., Luckett, D., & Lemerle, D. (2015). Metabolomics differentiation of canola genotypes: Toward an understanding of canola allelochemicals. *Frontiers in Plant Science, 5*. Advance online publication. DOI: 10.3389/fpls.2014.00765 PMID: 25620971

Azadmard-Damirchi, S., Habibi-Nodeh, F., Hesari, J., Nemati, M., & Achachlouei, B. (2010). Effect of pretreatment with microwaves on oxidative stability and nutraceuticals content of oil from rapeseed. *Food Chemistry, 121*(4), 1211–1215. DOI: 10.1016/j.foodchem.2010.02.006

Barrales-Cureño, H., Herrera-Cabrera, B., Montiel-Montoya, J., López-Valdez, L., Salgado-Garciglia, R., Ocaño-Higuera, V., Sánchez-Herrera, L. M., Lucho-Constantino, G. G., & Zaragoza-Martínez, F. (2022). Metabolomics studies of allelopathy: A review. *Revista Colombiana De Ciencias Químico Farmacéuticas, 51*(1). Advance online publication. DOI: 10.15446/rcciquifa.v51n1.102693

Barupal, D., Fan, S., & Fiehn, O. (2018). Integrating bioinformatics approaches for a comprehensive interpretation of metabolomics datasets. *Current Opinion in Biotechnology, 54*, 1–9. DOI: 10.1016/j.copbio.2018.01.010 PMID: 29413745

Bawa, A., & Anilakumar, K. (2012). Genetically modified foods: Safety, risks and public concerns—a review. *Journal of Food Science and Technology, 50*(6), 1035–1046. DOI: 10.1007/s13197-012-0899-1 PMID: 24426015

Beale, D. J., Kouremenos, K. A., & Palombo, E. A. (2016). *Microbial metabolomics*. Springer International Publishing., DOI: 10.1007/978-3-319-46326-1

Beger, R., Dunn, W., Schmidt, M., Gross, S., Kirwan, J., Cascante, M., Brennan, L., Wishart, D. S., Oresic, M., Hankemeier, T., Broadhurst, D. I., Lane, A. N., Suhre, K., Kastenmüller, G., Sumner, S. J., Thiele, I., Fiehn, O., & Kaddurah-Daouk, R. (2016). Metabolomics enables precision medicine: "a white paper, community perspective". *Metabolomics*, *12*(9), 149. Advance online publication. DOI: 10.1007/s11306-016-1094-6 PMID: 27642271

Beszterda, M., & Nogala-Kałucka, M. (2019). Current research developments on the processing and improvement of the nutritional quality of rapeseed (brassica napus l.). *European Journal of Lipid Science and Technology*, *121*(5), 1800045. Advance online publication. DOI: 10.1002/ejlt.201800045

Bongaerts, M., Bonte, R., Demirdas, S., Huidekoper, H., Langendonk, J., Wilke, M., de Valk, W., Blom, H. J., Reinders, M. J. T., & Ruijter, G. (2022). Integration of metabolomics with genomics: Metabolic gene prioritization using metabolomics data and genomic variant (cadd) scores. *Molecular Genetics and Metabolism*, *136*(3), 199–218. DOI: 10.1016/j.ymgme.2022.05.002 PMID: 35660124

Buzdin, A., & Патрушев, M. (2021). Will plant genome editing play a decisive role in "quantum-leap" improvements in crop yield to feed an increasing global human population? *Plants*, *10*(8), 1667. DOI: 10.3390/plants10081667 PMID: 34451712

Cai, G., Wang, G., Kim, S., Li, J., Zhou, Y., & Wang, X. (2021). Increased expression of fatty acid and abc transporters enhances seed oil production in camelina. *Biotechnology for Biofuels*, *14*(1), 49. Advance online publication. DOI: 10.1186/s13068-021-01899-w PMID: 33640013

Chen, H., Yu, X., Zhang, X., Yang, L., Huang, X., Zhang, J., Pritchard, H. W., & Li, W. (2017). Phospholipase dα1-mediated phosphatidic acid change is a key determinant of desiccation-induced viability loss in seeds. *Plant, Cell & Environment*, *41*(1), 50–63. DOI: 10.1111/pce.12925 PMID: 28152567

Choudhury, R., Beezley, J., Davis, B., Tomeck, J., Gratzl, S., Golzarri-Arroyo, L., & O'Connell, T. (2020). Viime: Visualization and integration of metabolomics experiments. *Journal of Open Source Software*, *5*(54), 2410. DOI: 10.21105/joss.02410 PMID: 33768193

Chu, U., Kumar, S., Sigmund, A., Johnson, K., Li, Y., Vongdeuane, P., & Jones, T. (2020). Genotype-independent transformation and genome editing of brassica napus using a novel explant material. *Frontiers in Plant Science*, *11*, 579524. Advance online publication. DOI: 10.3389/fpls.2020.579524 PMID: 33133118

Cook, J., Douglas, G., Zhang, J., Glick, B., Langille, M., Liu, K., & Cheng, Z. (2020). Transcriptomic profiling of brassica napus responses to pseudomonas aeruginosa. *Innate Immunity*, *27*(2), 143–157. DOI: 10.1177/1753425920980512 PMID: 33353474

Cook, J., Hui, J., Zhang, J., Kember, M., Berrué, F., Zhang, J., & Cheng, Z. (2022). Production of quorum sensing-related metabolites and phytoalexins during pseudomonas aeruginosa–brassica napus interaction. *Microbiology*, *168*(8). Advance online publication. DOI: 10.1099/mic.0.001212 PMID: 35980361

Cui, C., Zhang, K., Chai, L., Zheng, B., Zhang, J., Jiang, J., Tan, C., Li, H., Chen, D., & Jiang, L. (2024). Unraveling the mechanism of flower color variation in brassica napus by integrated metabolome and transcriptome analyses. *Frontiers in Plant Science*, *15*, 1419508. Advance online publication. DOI: 10.3389/fpls.2024.1419508 PMID: 38933465

Cui, F., Sui, N., Duan, G., Liu, Y., Yan, H., Liu, S., & Li, G. (2018). Identification of metabolites and transcripts involved in salt stress and recovery in peanut. *Frontiers in Plant Science*, *9*, 217. Advance online publication. DOI: 10.3389/fpls.2018.00217 PMID: 29520289

D'Hooghe, P., Escamez, S., Trouverie, J., & Avice, J. (2013). Sulphur limitation provokes physiological and leaf proteome changes in oilseed rape that lead to perturbation of sulphur, carbon and oxidative metabolisms. *BMC Plant Biology*, *13*(1), 23. Advance online publication. DOI: 10.1186/1471-2229-13-23 PMID: 23391283

Das, A., Rushton, P., & Rohila, J. (2017). Metabolomic profiling of soybeans (glycine max l.) reveals the importance of sugar and nitrogen metabolism under drought and heat stress. *Plants*, *6*(2), 21. DOI: 10.3390/plants6020021 PMID: 28587097

Emwas, A., Roy, R., McKay, R., Tenori, L., Saccenti, E., Gowda, G., Raftery, D., Alahmari, F., Jaremko, L., Jaremko, M., & Wishart, D. (2019). Nmr spectroscopy for metabolomics research. *Metabolites*, *9*(7), 123. DOI: 10.3390/metabo9070123 PMID: 31252628

Farag, M. A., Sharaf Eldin, M. G., Kassem, H., & Abou el Fetouh, M. (2013). Metabolome classification of Brassica napus L. organs via UPLC–QTOF–PDA–MS and their anti-oxidant potential. *Phytochemical Analysis*, *24*(3), 277–287. DOI: 10.1002/pca.2408 PMID: 23055344

Gao, J., Thelen, K., Min, D., Smith, S., Hao, X., & Gehl, R. (2010). Effects of manure and fertilizer applications on canola oil content and fatty acid composition. *Agronomy Journal*, *102*(2), 790–797. DOI: 10.2134/agronj2009.0368

Garcia-Aloy, M., Hulshof, P., Estruel-Amades, S., Osté, M., Lankinen, M., Geleijnse, J., de Goede, J., Ulaszewska, M., Mattivi, F., Bakker, S. J. L., Schwab, U., & Andrés-Lacueva, C. (2019). Biomarkers of food intake for nuts and vegetable oils: An extensive literature search. *Genes & Nutrition*, *14*(1), 7. Advance online publication. DOI: 10.1186/s12263-019-0628-8 PMID: 30923582

Ghazani, S., García-Llatas, G., & Marangoni, A. (2014). Micronutrient content of cold-pressed, hot-pressed, solvent extracted and rbd canola oil: Implications for nutrition and quality. *European Journal of Lipid Science and Technology*, *116*(4), 380–387. DOI: 10.1002/ejlt.201300288

Gomaa, W., Mosaad, G., & Yu, P. (2018). On a molecular basis, investigate association of molecular structure with bioactive compounds, anti-nutritional factors and chemical and nutrient profiles of canola seeds and co-products from canola processing: Comparison crusher plants within canada and within china as well as between canada and china. *Nutrients*, *10*(4), 519. DOI: 10.3390/nu10040519 PMID: 29690527

Gowda, G., & Raftery, D. (2015). Can nmr solve some significant challenges in metabolomics? *Journal of Magnetic Resonance (San Diego, Calif.)*, *260*, 144–160. DOI: 10.1016/j.jmr.2015.07.014 PMID: 26476597

Grimberg, Å., Carlsson, A., Marttila, S., Bhalerao, R., & Hofvander, P. (2015). Transcriptional transitions in nicotiana benthamiana leaves upon induction of oil synthesis by wrinkled1 homologs from diverse species and tissues. *BMC Plant Biology*, *15*(1), 192. Advance online publication. DOI: 10.1186/s12870-015-0579-1 PMID: 26253704

Guo, X., Sarup, P., Jensen, J., Orabi, J., Kristensen, N., Mulder, F., Jahoor, A., & Jensen, J. (2020). Genetic variance of metabolomic features and their relationship with malting quality traits in spring barley. *Frontiers in Plant Science*, *11*, 575467. Advance online publication. DOI: 10.3389/fpls.2020.575467 PMID: 33193515

Gurusinghe, S., Brooks, T., Barrow, R., Zhu, X., Thotagamuwa, A., Dennis, P., Gupta, V. V. S. R., Vanniasinkam, T., & Weston, L. (2019). Technologies for the selection, culture and metabolic profiling of unique rhizosphere microorganisms for natural product discovery. *Molecules (Basel, Switzerland)*, *24*(10), 1955. DOI: 10.3390/molecules24101955 PMID: 31117282

Haider, S., Iqbal, J., Naseer, S., Shaukat, M., Abbasi, B. A., Yaseen, T., & Mahmood, T. (2022b). Unfolding molecular switches in plant heat stress resistance: A comprehensive review. *Plant Cell Reports*, *41*(3), 775–798. DOI: 10.1007/s00299-021-02754-w PMID: 34401950

Haider, S., Iqbal, J., Naseer, S., Yaseen, T., Shaukat, M., Bibi, H., & Mahmood, T. (2021a). Molecular mechanisms of plant tolerance to heat stress: Current landscape and future perspectives. *Plant Cell Reports*, *40*(12), 1–25. DOI: 10.1007/s00299-021-02696-3 PMID: 33890138

Haider, S., Iqbal, J., Shaukat, M., Naseer, S., & Mahmood, T. (2021b). The epigenetic chromatin-based regulation of somatic heat stress memory in plants. *Plant Gene*, *27*, 100318. DOI: 10.1016/j.plgene.2021.100318

Haider, S., Raza, A., Iqbal, J., Shaukat, M., & Mahmood, T. (2022a). Analyzing the regulatory role of heat shock transcription factors in plant heat stress tolerance: A brief appraisal. *Molecular Biology Reports*, *49*(6), 5771–5785. DOI: 10.1007/s11033-022-07190-x PMID: 35182323

Han, L., Silvestre, S., Sayanova, O., Haslam, R., & Napier, J. (2022). Using field evaluation and systematic iteration to rationalize the accumulation of omega-3 long-chain polyunsaturated fatty acids in transgenic camelina sativa. *Plant Biotechnology Journal*, *20*(9), 1833–1852. DOI: 10.1111/pbi.13867 PMID: 35656640

Hanifah, A., Maharijaya, A., Putri, S., Laviña, W., & Sobir, . (2018). Untargeted metabolomics analysis of eggplant (solanum melongena l.) fruit and its correlation to fruit morphologies. *Metabolites*, *8*(3), 49. DOI: 10.3390/metabo8030049 PMID: 30200482

Hanschen, F., Lamy, E., Schreiner, M., & Rohn, S. (2014). Reactivity and stability of glucosinolates and their breakdown products in foods. *Angewandte Chemie*, *53*(43), 11430–11450. DOI: 10.1002/anie.201402639 PMID: 25163974

Hashemi, A., Abdolzadeh, A., & Sadeghipour, H. (2010). Beneficial effects of silicon nutrition in alleviating salinity stress in hydroponically grown canola,brassica napusl., plants. *Soil Science and Plant Nutrition*, *56*(2), 244–253. DOI: 10.1111/j.1747-0765.2009.00443.x

Hernández-Terán, A., Wegier, A., Benítez, M., Lira, R., & Escalante, A. (2017). Domesticated, genetically engineered, and wild plant relatives exhibit unintended phenotypic differences: A comparative meta-analysis profiling rice, canola, maize, sunflower, and pumpkin. *Frontiers in Plant Science*, *8*, 2030. Advance online publication. DOI: 10.3389/fpls.2017.02030 PMID: 29259610

Hong, , Yang, L., Zhang, D., & Shi, J. (2016). Plant Metabolomics: An Indispensable System Biology Tool for Plant Science. *International Journal of Molecular Sciences*, *17*(6), 767. Advance online publication. DOI: 10.3390/ijms17060767 PMID: 27258266

Hong, J., Yang, L., Zhang, D., & Shi, J. (2016). Plant metabolomics: An indispensable system biology tool for plant science. *International Journal of Molecular Sciences, 17*(6), 767. DOI: 10.3390/ijms17060767 PMID: 27258266

Hooft, J., Mohimani, H., Bauermeister, A., Dorrestein, P., Duncan, K., & Medema, M. (2020). Linking genomics and metabolomics to chart specialized metabolic diversity. *Chemical Society Reviews, 49*(11), 3297–3314. DOI: 10.1039/D0CS00162G PMID: 32393943

Hu, C., Shi, J., Quan, S., Cui, B., Kleeßen, S., Nikoloski, Z., & Zhang, D. (2014). Metabolic variation between japonica and indica rice cultivars as revealed by non-targeted metabolomics. *Scientific Reports, 4*(1), 5067. Advance online publication. DOI: 10.1038/srep05067 PMID: 24861081

Ijaz, S., Iqbal, J., Abbasi, B. A., Tufail, A., Ullah, Z., Sharifi-Rad, J., & Uddin, S. (2023). Biofortification: Lessons from the Past and Strategies for Future Food Security. In *Legumes Biofortification* (pp. 521–545). Springer International Publishing. DOI: 10.1007/978-3-031-33957-8_23

Ijaz, S., Iqbal, J., Abbasi, B. A., Yaseen, T., Rehman, S., Kazi, M., & Mahmood, T. (2024a). Role of OMICS-Based Technologies in Plant Sciences. OMICs-based Techniques for Global Food Security, 45-66.

Ijaz, S., Iqbal, J., Abbasi, B. A., Yaseen, T., Rehman, S., Kazi, M. & Mahmood, T. (2024b). Role of OMICS-Based Technologies in Plant Sciences. OMICs-based Techniques for Global Food Security, 45-66.

Jaspers, P., & Kangasjärvi, J. (2010). Reactive oxygen species in abiotic stress signaling. *Physiologia Plantarum, 138*(4), 405–413. DOI: 10.1111/j.1399-3054.2009.01321.x PMID: 20028478

Jendoubi, T. (2021). Approaches to integrating metabolomics and multi-omics data: A primer. *Metabolites, 11*(3), 184. DOI: 10.3390/metabo11030184 PMID: 33801081

Jones, P., Senanayake, V., Pu, S., Jenkins, D., Connelly, P., Lamarche, B., Couture, P., Charest, A., Baril-Gravel, L., West, S. G., Liu, X., Fleming, J. A., McCrea, C. E., & Kris-Etherton, P. (2014). Dha-enriched high–oleic acid canola oil improves lipid profile and lowers predicted cardiovascular disease risk in the canola oil multicenter randomized controlled trial. *The American Journal of Clinical Nutrition, 100*(1), 88–97. DOI: 10.3945/ajcn.113.081133 PMID: 24829493

Jorge, T., Rodrigues, J., Caldana, C., Schmidt, R., Dongen, J., Thomas-Oates, J., & António, C. (2015). Mass spectrometry-based plant metabolomics: Metabolite responses to abiotic stress. *Mass Spectrometry Reviews*, *35*(5), 620–649. DOI: 10.1002/mas.21449 PMID: 25589422

Katam, R., Lin, C., Grant, K., Katam, C., & Chen, S. (2022). Advances in plant metabolomics and its applications in stress and single-cell biology. *International Journal of Molecular Sciences*, *23*(13), 6985. DOI: 10.3390/ijms23136985 PMID: 35805979

Khodabin, G., Tahmasebi-Sarvestani, Z., Rad, A., & Modarres-Sanavy, S. (2020). Effect of drought stress on certain morphological and physiological characteristics of a resistant and a sensitive canola cultivar. *Chemistry & Biodiversity*, *17*(2), e1900399. Advance online publication. DOI: 10.1002/cbdv.201900399 PMID: 31909554

Kim, S., Yamaoka, Y., Ono, H., Kim, H., Shim, D., Maeshima, M., Martinoia, E., Cahoon, E. B., Nishida, I., & Lee, Y. (2012). Atabca9 transporter supplies fatty acids for lipid synthesis to the endoplasmic reticulum. *Proceedings of the National Academy of Sciences of the United States of America*, *110*(2), 773–778. DOI: 10.1073/pnas.1214159110 PMID: 23269834

Kráľová, K., Jampílek, J., & Ostrovský, I. (2012). Metabolomics - useful tool for study of plant responses to abiotic stresses. *Ecological Chemistry and Engineering. S*, *19*(2), 133–161. DOI: 10.2478/v10216-011-0012-0

Kumar, R., Bohra, A., Pandey, A., Pandey, M., & Kumar, A. (2017). Metabolomics for plant improvement: Status and prospects. *Frontiers in Plant Science*, *8*, 1302. Advance online publication. DOI: 10.3389/fpls.2017.01302 PMID: 28824660

Kumari, M., Yagnik, K. N., Gupta, V., Singh, I. K., Gupta, R., Verma, P. K., & Singh, A. (2024). Metabolomics-driven investigation of plant defense response against pest and pathogen attack. *Physiologia Plantarum*, *176*(2), e14270. Advance online publication. DOI: 10.1111/ppl.14270 PMID: 38566280

Lahlali, R., Song, T., Chu, M., Yu, F., Kumar, S., Karunakaran, C., & Peng, G. (2017). Evaluating changes in cell-wall components associated with clubroot resistance using fourier transform infrared spectroscopy and rt-pcr. *International Journal of Molecular Sciences*, *18*(10), 2058. DOI: 10.3390/ijms18102058 PMID: 28954397

Lay, C., Bell, T., Hamel, C., Harker, K., Mohr, R., Greer, C., Yergeau, É., & St-Arnaud, M. (2018). Canola root–associated microbiomes in the canadian prairies. *Frontiers in Microbiology*, *9*, 1188. Advance online publication. DOI: 10.3389/fmicb.2018.01188 PMID: 29937756

Li, K., Wang, D., Gong, L., Lyu, Y., Guo, H., Chen, W., & Luo, J. (2019). Comparative analysis of metabolome of rice seeds at three developmental stages using a recombinant inbred line population. *The Plant Journal*, *100*(5), 908–922. DOI: 10.1111/tpj.14482 PMID: 31355982

Li, L., Hur, M., Lee, J., Zhou, W., Song, Z., Ransom, N., Demirkale, C. Y., Nettleton, D., Westgate, M., Arendsee, Z., Iyer, V., Shanks, J., Nikolau, B., & Wurtele, E. (2015). A systems biology approach toward understanding seed composition in soybean. *BMC Genomics*, *16*(S3), S9. Advance online publication. DOI: 10.1186/1471-2164-16-S3-S9 PMID: 25708381

Li, Z., Song, Z., Zhang, Y., Hao, Q., Song, A., Liu, L., & Liang, Y. (2018). Silicon enhancement of estimated plant biomass carbon accumulation under abiotic and biotic stresses. a meta-analysis. *Agronomy for Sustainable Development*, *38*(3), 26. Advance online publication. DOI: 10.1007/s13593-018-0496-4

Lin, L., Allemekinders, H., Dansby, A., Campbell, L., Durance-Tod, S., Berger, A., & Jones, P. (2013). Evidence of health benefits of canola oil. *Nutrition Reviews*, *71*(6), 370–385. DOI: 10.1111/nure.12033 PMID: 23731447

Lu, M., & Zhan, X. (2018). The crucial role of multiomic approach in cancer research and clinically relevant outcomes. *The EPMA Journal*, *9*(1), 77–102. DOI: 10.1007/s13167-018-0128-8 PMID: 29515689

Lu, S., Sturtevant, D., Aziz, M., Jin, C., Li, Q., Chapman, K., & Guo, L. (2018). Spatial analysis of lipid metabolites and expressed genes reveals tissue-specific heterogeneity of lipid metabolism in high- and low-oil brassica napus l. seeds. *The Plant Journal*, *94*(6), 915–932. DOI: 10.1111/tpj.13959 PMID: 29752761

MacIntosh, S., Shaw, M., Connelly, M., & Yao, Z. (2021). Food and feed safety of ns-b5øø27-4 omega-3 canola (brassica napus): A new source of long-chain omega-3 fatty acids. *Frontiers in Nutrition*, *8*, 716659. Advance online publication. DOI: 10.3389/fnut.2021.716659 PMID: 34660659

McClinchey, S., Gillespie, J., Fisher, T., Taylor, K., Challender, M., & Schmidt, D. (2016). Quality characteristics of glyphosate-tolerant canola cultivars containing a glyphosate acetyltransferase transgene (event dp-ø73496-4). *Crop Science*, *56*(4), 1736–1746. DOI: 10.2135/cropsci2015.12.0744

Meng, X., & Zhang, S. (2013). Mapk cascades in plant disease resistance signaling. *Annual Review of Phytopathology*, *51*(1), 245–266. DOI: 10.1146/annurev-phyto-082712-102314 PMID: 23663002

Miao, H., Zeng, W., Wang, J., Zhang, F., Sun, B., & Wang, Q. (2021). Improvement of glucosinolates by metabolic engineering in brassica crops. *aBIOTECH*, *2*(3), 314–329. DOI: 10.1007/s42994-021-00057-y PMID: 36303883

Mibei, E., Owino, W., Ambuko, J., Giovannoni, J., & Onyango, A. (2017). Metabolomic analyses to evaluate the effect of drought stress on selected african eggplant accessions. *Journal of the Science of Food and Agriculture*, *98*(1), 205–216. DOI: 10.1002/jsfa.8458 PMID: 28573744

Microbial Metabolomics (2016) DOI: 10.1007/978-3-319-46326-1

Morsi, N., Hashem, O., El-Hady, M., Abd-Elkrem, Y., El-temsah, M., Galal, E., & Abdelkader, M. (2023). Assessing drought tolerance of newly developed tissue-cultured canola genotypes under varying irrigation regimes. *Agronomy (Basel)*, *13*(3), 836. DOI: 10.3390/agronomy13030836

Naheed, R., Aslam, H., Kanwal, H., Farhat, F., Gamar, M., Al-Mushhin, A., & Hessini, K. (2021). Growth attributes, biochemical modulations, antioxidant enzymatic metabolism and yield in brassica napus varieties for salinity tolerance. *Saudi Journal of Biological Sciences*, *28*(10), 5469–5479. DOI: 10.1016/j.sjbs.2021.08.021 PMID: 34588857

Neik, T., Amas, J., Barbetti, M., Edwards, D., & Batley, J. (2020). Understanding host–pathogen interactions in brassica napus in the omics era. *Plants*, *9*(10), 1336. DOI: 10.3390/plants9101336 PMID: 33050509

Nguyen, D., Le, V., Nguy n, Q., Malau-Aduli, B., Nichols, P., & Malau-Aduli, A. (2017). Omega–3 long-chain fatty acids in the heart, kidney, liver and plasma metabolite profiles of australian prime lambs supplemented with pelleted canola and flaxseed oils. *Nutrients*, *9*(8), 893. DOI: 10.3390/nu9080893 PMID: 28817082

Nichol, J., Ribano, A. K. B., Hickerson, N. M. N., Ali, N., Jamois, F., & Samuel, M. A. (2023). Plant growth regulator extracts from seaweeds promote plant growth and confer drought tolerance in canola (brassica napus). *Plant Signaling & Behavior*, *18*(1), 2267222. Advance online publication. DOI: 10.1080/15592324.2023.2267222 PMID: 37903454

Nokhrina, K., Ray, H., Bock, C., & Georges, F. (2014). Metabolomic shifts in brassica napus lines with enhancedbnplc2expression impact their response to low temperature stress and plant pathogens. *GM Crops and Food: Biotechnology in Agriculture and the Food Chain*, *5*(2), 120–131. DOI: 10.4161/gmcr.28942 PMID: 24787279

Obata, T., & Fernie, A. (2012). The use of metabolomics to dissect plant responses to abiotic stresses. *Cellular and Molecular Life Sciences : CMLS*, *69*(19), 3225–3243. DOI: 10.1007/s00018-012-1091-5 PMID: 22885821

Oh, S., Imran, M., Kim, E., Park, S., Lee, S., Park, H., & Ryu, T. (2023). Approach strategies and application of metabolomics to biotechnology in plants. *Frontiers in Plant Science*, *14*, 1192235. Advance online publication. DOI: 10.3389/fpls.2023.1192235 PMID: 37636096

Öztürk, M., Ünal, B., García-Caparrós, P., Khursheed, A., Gul, A., & Hasanuzzaman, M. (2020). Osmoregulation and its actions during the drought stress in plants. *Physiologia Plantarum*, *172*(2), 1321–1335. DOI: 10.1111/ppl.13297 PMID: 33280137

Patel, J., Khandwal, D., Choudhary, B., Ardeshana, D., Jha, R., Tanna, B., Yadav, S., Mishra, A., Varshney, R. K., & Siddique, K. (2022). Differential physio-biochemical and metabolic responses of peanut (arachis hypogaea l.) under multiple abiotic stress conditions. *International Journal of Molecular Sciences*, *23*(2), 660. DOI: 10.3390/ijms23020660 PMID: 35054846

Pedras, M., Yaya, E., & Glawischnig, E. (2011). The phytoalexins from cultivated and wild crucifers: Chemistry and biology. *Natural Product Reports*, *28*(8), 1381. DOI: 10.1039/c1np00020a PMID: 21681321

Pérez-Llorca, M., Pollmann, S., & Müller, M. (2023). Ethylene and jasmonates signaling network mediating secondary metabolites under abiotic stress. *International Journal of Molecular Sciences*, *24*(6), 5990. DOI: 10.3390/ijms24065990 PMID: 36983071

Popovici, V., Bucur, L., Gîrd, C., Rambu, D., Calcan, S., Cucolea, E., Costache, T., Ungureanu-Iuga, M., Oroian, M., Mironeasa, S., Schröder, V., Ozon, E.-A., Lupuliasa, D., Caraiane, A., & Badea, V. (2022). Antioxidant, cytotoxic, and rheological properties of canola oil extract of usnea barbata (l.) weber ex f.h. wigg from călimani mountains, romania. *Plants*, *11*(7), 854. DOI: 10.3390/plants11070854 PMID: 35406834

Rahman, H. (2013). Review: Breeding spring canola (brassica napusl.) by the use of exotic germplasm. *Canadian Journal of Plant Science*, *93*(3), 363–373. DOI: 10.4141/cjps2012-074

Rattray, N., Deziel, N., Wallach, J., Khan, S., Vasiliou, V., Ioannidis, J., & Johnson, C. (2018). Beyond genomics: Understanding exposotypes through metabolomics. *Human Genomics*, *12*(1), 4. Advance online publication. DOI: 10.1186/s40246-018-0134-x PMID: 29373992

Rauf, S., Fatima, S., & Ortíz, R. (2023). Modification of fatty acid profile and oil contents using gene editing in oilseed crops for a changing climate. *GM Crops and Food: Biotechnology in Agriculture and the Food Chain, 14*(1), 1–12. DOI: 10.1080/21645698.2023.2243041 PMID: 37551783

Razzaq, A., Sadia, B., Raza, A., Hameed, M., & Saleem, F. (2019). Metabolomics: A way forward for crop improvement. *Metabolites, 9*(12), 303. DOI: 10.3390/metabo9120303 PMID: 31847393

Rezayian, M., & Ebrahimzadeh, H. (2019). Different effects of calcium and penconazole on primary and secondary metabolites of brassica napus under drought. *Physiology and Molecular Biology of Plants : an International Journal of Functional Plant Biology, 25*(2), 497–509. DOI: 10.1007/s12298-018-00634-4 PMID: 30956431

Richardson, C., Hennebelle, M., Otoki, Y., Zamora, D., Yang, J., Hammock, B., & Taha, A. (2017). Lipidomic analysis of oxidized fatty acids in plant and algae oils. *Journal of Agricultural and Food Chemistry, 65*(9), 1941–1951. DOI: 10.1021/acs.jafc.6b05559 PMID: 28157307

Ritchie, M., Holzinger, E., Li, R., Pendergrass, S., & Kim, D. (2015). Methods of integrating data to uncover genotype–phenotype interactions. *Nature Reviews. Genetics, 16*(2), 85–97. DOI: 10.1038/nrg3868 PMID: 25582081

Rossi, L., Zhang, W., Lombardini, L., & Ma, X. (2016). The impact of cerium oxide nanoparticles on the salt stress responses of brassica napus l. *Environmental Pollution, 219*, 28–36. DOI: 10.1016/j.envpol.2016.09.060 PMID: 27661725

Sargent, D., Conaty, W., Tissue, D., & Sharwood, R. (2022). Synthetic biology and opportunities within agricultural crops. *Journal of Sustainable Agriculture and Environment, 1*(2), 89–107. DOI: 10.1002/sae2.12014

Sarkar, A., Kisiała, A., Adhikary, D., Basu, U., Emery, R., Rahman, H., & Kav, N. (2023). Silicon ameliorates clubroot responses in canola (brassica napus): A "multi-omics"-based investigation into possible mechanisms. *Physiologia Plantarum, 175*(2), e13900. Advance online publication. DOI: 10.1111/ppl.13900 PMID: 36992551

Shokri-Gharelo, R., & Motie-Noparvar, P. (2018). Molecular response of canola to salt stress: insights on tolerance mechanisms. https://doi.org/DOI: 10.7287/peerj.preprints.26716v1

Shokri-Gharelo, R., & Noparvar, P. (2018). Molecular response of canola to salt stress: Insights on tolerance mechanisms. *PeerJ, 6*, e4822. DOI: 10.7717/peerj.4822 PMID: 29844974

Song, Q., Li, Q., Liu, Y., Zhang, F., Ma, B., Zhang, W., Man, W.-Q., Du, W.-G., Wang, G.-D., Chen, S.-Y., & Zhang, J. (2013). Soybean gmbzip123 gene enhances lipid content in the seeds of transgenic arabidopsis plants. *Journal of Experimental Botany*, *64*(14), 4329–4341. DOI: 10.1093/jxb/ert238 PMID: 23963672

Spasibionek, S., Mikołajczyk, K., Ćwiek-Kupczyńska, H., Pietka, T., Krotka, K., Matuszczak, M., Nowakowska, J., Michalski, K., & Bartkowiak-Broda, I. (2020). Marker assisted selection of new high oleic and low linolenic winter oilseed rape (brassica napus l.) inbred lines revealing good agricultural value. *PLoS One*, *15*(6), e0233959. DOI: 10.1371/journal.pone.0233959 PMID: 32497146

Sumner, L., Lei, Z., Nikolau, B., & Saito, K. (2015). Modern plant metabolomics: Advanced natural product gene discoveries, improved technologies, and future prospects. *Natural Product Reports*, *32*(2), 212–229. DOI: 10.1039/C4NP00072B PMID: 25342293

Tan, H., Xiang, X., Tang, J., & Wang, X. (2016). Nutritional functions of the funiculus in brassica napus seed maturation revealed by transcriptome and dynamic metabolite profile analyses. *Plant Molecular Biology*, *92*(4-5), 539–553. DOI: 10.1007/s11103-016-0530-3 PMID: 27539000

Tan, H., Xie, Q., Xiang, X., Li, J., Zheng, S., Xu, X., & Ye, W. (2015). Dynamic metabolic profiles and tissue-specific source effects on the metabolome of developing seeds of brassica napus. *PLoS One*, *10*(4), e0124794. DOI: 10.1371/journal.pone.0124794 PMID: 25919591

Tian, H., Lam, S., & Shui, G. (2016). Metabolomics, a powerful tool for agricultural research. *International Journal of Molecular Sciences*, *17*(11), 1871. DOI: 10.3390/ijms17111871 PMID: 27869667

Tripathi, P., Rabara, R., Reese, R., Miller, M., Rohila, J., Subramanian, S., & Rushton, P. (2016). A toolbox of genes, proteins, metabolites and promoters for improving drought tolerance in soybean includes the metabolite coumestrol and stomatal development genes. *BMC Genomics*, *17*(1), 102. Advance online publication. DOI: 10.1186/s12864-016-2420-0 PMID: 26861168

Vanhercke, T., Tahchy, A., Liu, Q., Zhou, X., Shrestha, P., Divi, U., & Petrie, J. (2013). Metabolic engineering of biomass for high energy density: Oilseed-like triacylglycerol yields from plant leaves. *Plant Biotechnology Journal*, *12*(2), 231–239. DOI: 10.1111/pbi.12131 PMID: 24151938

Vo, K., Rahman, M., Rahman, M., Trinh, K., Kim, S., & Jeon, J. (2021). Proteomics and metabolomics studies on the biotic stress responses of rice: An update. *Rice (New York, N.Y.), 14*(1), 30. Advance online publication. DOI: 10.1186/s12284-021-00461-4 PMID: 33721115

Wang, W., Pang, J., Zhang, F., Sun, L., Yang, L., Fu, T., & Siddique, K. (2022). Salt-responsive transcriptome analysis of canola roots reveals candidate genes involved in the key metabolic pathway in response to salt stress. *Scientific Reports, 12*(1), 1666. Advance online publication. DOI: 10.1038/s41598-022-05700-2 PMID: 35102232

Wang, X., Li, Y., Wang, X., Li, X., & Dong, S. (2022). Physiology and metabonomics reveal differences in drought resistance among soybean varieties. *Botanical Studies, 63*(1), 8. Advance online publication. DOI: 10.1186/s40529-022-00339-8 PMID: 35332430

Wang, Y., Liu, Q., Du, L., Hallerman, E., & Li, Y. (2020). Transcriptomic and metabolomic responses of rice plants to cnaphalocrocis medinalis caterpillar infestation. *Insects, 11*(10), 705. DOI: 10.3390/insects11100705 PMID: 33076419

Weldon, A., Davis, D. A., Rhodes, M., Morey, A., Iassonova, D., & Roy, L. A. (2023). Use of genetically modified canola oil as a replacement for fish oil in practical diets for whiteleg shrimp litopeneaus vannamei reared in green water conditions. *Aquaculture Research, 2023*, 1–12. DOI: 10.1155/2023/2999827

Wiebach, J., Nagel, M., Börner, A., Altmann, T., & Riewe, D. (2019). Age-dependent loss of seed viability is associated with increased lipid oxidation and hydrolysis. *Plant, Cell & Environment, 43*(2), 303–314. DOI: 10.1111/pce.13651 PMID: 31472094

Wishart, D., Cheng, L., Copié, V., Edison, A., Eghbalnia, H., Hoch, J., Gouveia, G. J., Pathmasiri, W., Powers, R., Schock, T. B., Sumner, L. W., & Uchimiya, M. (2022). Nmr and metabolomics—A roadmap for the future. *Metabolites, 12*(8), 678. DOI: 10.3390/metabo12080678 PMID: 35893244

Wu, L., El-Mezawy, A., & Shah, S. (2013). Isolation and evaluation of three novel native promoters in Brassica napus. *Botany, 91*(6), 414–419. DOI: 10.1139/cjb-2012-0245

Wu, W., Duncan, R., & Ma, B. (2020). The stage sensitivity of short-term heat stress to lodging-resistant traits and yield determination in canola (brassica napus l.). *Journal Agronomy & Crop Science, 207*(1), 74–87. DOI: 10.1111/jac.12464

Xia, J., Sinelnikov, I., Han, B., & Wishart, D. (2015). Metaboanalyst 3.0—Making metabolomics more meaningful. *Nucleic Acids Research, 43*(W1), W251–W257. DOI: 10.1093/nar/gkv380 PMID: 25897128

Zeki, Ö., Eylem, C., Reçber, T., Kır, S., & Nemutlu, E. (2020). Integration of gc–ms and lc–ms for untargeted metabolomics profiling. *Journal of Pharmaceutical and Biomedical Analysis*, *190*, 113509. DOI: 10.1016/j.jpba.2020.113509 PMID: 32814263

Chapter 13
Future Perspectives in the Omics Era for Canola Breeding

Muhammad Zeeshan Arshed
University of the Punjab, Pakistan

Ammara Saleem
University of the Punjab, Pakistan

Asma Zulfiqar
University of the Punjab, Pakistan

Muhammad Aijaz Ahmad
University of the Punjab, Pakistan

ABSTRACT

This chapter explores the possibilities of multi-omics technologies in catalyzing canola breeding for sustainable agriculture. From genomics to transcriptomics, proteomics and metabolomics, omics techniques offer a unified view of genetic and molecular regulation of important agronomic traits. Genomic technologies are essential in screening yield related QTL, oil quality and stress tolerance traits. Proteomics and metabolomics are another level up from genomics and transcriptomics as they show intricate protein–protein and metabolic networks that regulate plant growth, development and stress tolerance. Further, functional genomics styles like CRISPR-Cas9, RNA interference provide new approaches targeting genome modification for canola improvement, disease resistance and efficient nutrient utilization under climate change. Further, author presents potential work direction using multi-omics data

DOI: 10.4018/979-8-3693-8603-3.ch013

with advanced platforms; high-throughput phenotyping and artificial intelligence that will culminate to development of improved canola varieties to enhance food security and industrial utility.

1. INTRODUCTION

Canola (*Brassica napus* L. var. oleifera), an oilseed crop from the Brassicaceae family, is an improved variety of rapeseed (*Brassica napus* L.) (Estevez et al., 2014). Canola" as a term is used to refer to a particular type of rapeseed that has been bred to contain not more than 2% erucic acid as well as not more than 30 micromoles of Glucosinolates per 50 grams of defatted meal. This category comprises different cultivars of the species *Brassica napus*, *Brassica rapa* and *Brassica juncea* and it is widely used in cooking oils and many food products because of its health roles (Tang et al., 2018). Canola is second only to soybean among the oilseed-producing crops and is grown in many countries. It is however only used domestically as an oilseed crop, a characteristic that seems to have been domesticated only some 400-500 years back. In the course of these years, it has become one of the most important crops in the world and canola oil is one of the most popular types of vegetable oils (Zhou et al., 2017). Canola, an international crop that belongs to the Brassica group, is one of the most important oilseed crops that has boosted agricultural economies especially in Canada, Australia and China and in some parts of Europe (McVetty et al., 2016). Canola is a rich source of oil with low content of saturated fats and high content of omega 3 fatty acids, the byproduct of canola oil, canola meal also has great value since it is used as animal feed with high protein content (Welter et al., 2016). Some breeding programs targeting canola improvement have recorded tremendous developments in yield, diseases resistance and quality of the oil. Nevertheless, such endeavours are confronted with several challenges that present various complexities that call for dynamism especially in the case of climate change, new diseases, and constantly changing needs of sustainable agriculture (Yadava et al., 2012). On the other hand, the ever advancing field of omics technologies and genetic engineering provided new prospects for improvement of canola, which may form its future. The global demand for canola (*Brassica napus*) continues to rise due to its versatility as a source of oil and protein, making it a critical crop for food security and agricultural sustainability (So & Duncan, 2021). Potential threats such as climate change, new pest and diseases, and the requirement for higher yield, better quality, and drought tolerance require new developments. Yield is one of the main objectives of plant breeding (Chapman et al., 2012). When the breeder takes time to select good plant varieties and crossbreed them, then the plant varieties that would give more yield per acre can be developed. For instance, advances in breeding have given birth to improved varieties of basic crops, these include wheat, rice, canola as well as maize. Further, breeding also make the crops better in quality and this make nutrients rich, tasty, and have longer shelf life.

Hence genomics, transcriptomics proteomics, metabolomics and phenomics are transforming canola breeding since they provide systems biology view of the genetic and biochemical regulation of quantitative traits of significance to agriculture. This chapter discusses the bottleneck in canola breeding, prospects in canola breeding, future direction of canola breeding in the omics era, that is, multi-omics strategy, utilizing genetic variability, and prospects for climate sensitive agriculture.

2. CHALLENGES IN CANOLA BREEDING

Canola breeding faces several challenges despite advancements in biotechnological tools. Breeders must overcome increasing resistance to abiotic stresses such as drought, salinity, and extreme temperatures. Additionally, the need to enhance disease resistance against emerging pathogens remains a significant hurdle. Another constraint is low genetic variation in the breeding material, which is solved with the help of innovative methods such as gene modification and marker-associated selection. In addition, the unification of high yield varieties while retaining oil quality and environmental concerns makes breeding programs a bit complicated (Chauhan et al., 2011). Accomplishing these challenges requires further optimization of biotechnological approaches and successful combination of omics approaches to adapt the agronomical demands. Here are briefed challenges:

2.1. Limited Genetic Variation

The big issue that has been observed about breeding of canola is that it has a limited number of genes in the species. In the long run, breeding for such attributes as yield and quality of the oil has led to crops with a small genetic base. This makes the crop accommodate something like diseases, pests, climate etc., that may not be of any benefit to the crop (Ton et al., 2020).

The absence of variation also constrains the possibility of introducing new characters as for instance drought or heat resistance which is very crucial due to the changing climate. Increasing the genetic potential of canola is also a goal, but it is also problematic in its way. The introduction of new genetic material from its wild relatives or from related species sometimes leads to such negative characters as low yield, poor adaptability, low quality of the seed oil, etc. This means that it becomes obligatory for breeders to carry out long breeding cycles to try to eradicate these undesirable features and the process is very expensive and lengthy (Jovovic et al., 2020).

2.2. Disease and Pest Resistance (Abiotic and Biotic Stress)

Like all the cruciferous crops, Canola is prone to various diseases and pests and the most significant of which include: Blackleg (*Leptosphaeria maculans*), Club root (*Plasmodiophora brassicae*), Sclerotinia stem rot (*Sclerotinia sclerotiorum*). Sometimes they are capable of reducing yield or even completely destroying a crop (Mansouripour, 2018). Disease management is all about coming up with disease resistant varieties and the diseases themselves are dynamic. For instance, although blackleg-resistant varieties of canola have been developed, the disease has since evolved to other, even more severe types hence less resistance to the disease by the developed varieties. Climate change is changing the conditions under which canola is produced and these conditions have to be taken into consideration by breeding programs (Kaur et al., 2009).

One of the major challenges for disease control in canola is managing blackleg disease, caused by the fungal pathogen *Leptosphaeria maculans*. While resistant canola varieties offer the most promising control method, the high evolutionary potential of *L. maculans* allows it to overcome resistance genes in just a few years. For example, the blackleg resistance gene *Rlm3* provided effective resistance in Canadian canola varieties until the early 2000s when fungal populations adapted (Zhang & Fernando, 2017).

Drought and high temperature have increased and low temperature stress is relatively frequent and canola is known to be sensitive to those stresses at flowering and podding stages. Heat stress for instance can lead to reduced pollination and seed formation; hence a big blow to yield (Shafighi et al., 2021). Creation of new canola varieties resistant to these severe conditions is rather difficult as abiotic stress tolerance is generally determined by quantitative factors and involves multiple genes. Furthermore, increasing one stress tolerance trait, for instance, drought tolerance, may result to reduction in the other important traits like oil content or yield. As a consequence, the breeders are to optimize stress tolerance on one hand and high yield and quality of the produced oil on the other.

2.3. Sustainable Agriculture and Environmental Concerns

As the world tries to adopt sustainable agricultural practices, canola breeding has been confronted with the task of minimizing the impact of this crop on environment. Farmers are being forced to minimize the application of synthetic fertilizers and pesticides that are considered as culprits to environmental pollution (Shah &

Wu, 2019). But then again, canola is a high input crop and attempts at minimizing the use of chemicals by organic agriculture results to low yields and susceptibility to pests and diseases. Breeding for traits that improve sustainability of production systems is one approach to this challenge for example, nutrient use efficiency. Varieties that use little nitrogen, or those which are able to better utilize the available nitrogen and other nutrients would help rid canola of the need for fertilizers while at the same time maintaining the environmentally unfriendly effects of growing canola. But, selection for such characters is challenging and the required genes are yet to be defined and used optimally (Raman et al., 2019).

2.4. Oil Quality Improvement

Even though canola has a high quality oil, the consumers need more healthy oil than before. Breeders are challenge with a continuous enhancement of oil characteristics, especially with regard to its fatty acid (omega-3) profile where there is a need to have high levels of PUFA, particularly omega-3 while having low levels of saturated fatty acids (Tocher et al., 2019). One challenge in improving canola oil is reducing its saturated fat content, which currently stands at 7%, while simultaneously increasing the levels of beneficial omega-3 polyunsaturated fatty acids, such as alpha-linolenic acid (9–11%). Additionally, enhancing the concentration of phytosterols, which currently make up about 0.9% by weight, presents another challenge for developing a more nutritionally valuable oil (Gaber et al., 2018).

More importantly, other uses of canola oil in industrial processes like biodiesel production need different compositions of oils which makes the breeding process challenging (Babadi et al., 2022). In particular, it remains difficult to reach these improvements without losing yield or disease resistance. However, the attempt to create transgenic traits to improve the quality of oil has aroused much resistance from the authorities and the public in many countries thus making it difficult for genetically modified canola varieties to be accepted in the market.

3. OPPORTUNITIES AND ADVANTAGES OF MULTI-OMICS TECHNIQUES IN CROP AND CANOLA IMPROVEMENT

Multi-omics approaches, which integrate data from genomics, transcriptomics, proteomics, and metabolomics, are revolutionizing canola research and breeding. These powerful techniques provide unprecedented insights into the complex biology of this important oilseed crop.

Here are some of the key opportunities and advantages:

3.1. Elaborate Study of Biology of Canola

Multi-omics is an integrated and systematic study of canola through genetics, transcriptomics, proteomics and metabolomics. This integrated view is required for assessment of polygenic features and stress reactions expressions (Yang et al., 2021). One major constraint facing growers of canola is clubroot disease that is caused by *Plasmodiophora brassicae* resulting to dismal yields in Brassica crops. It is good to cultivate and sustain resistant canola types since the pathogen is known to overcome the resistance to prevail in the longer run to combat the pathogen (Sarkar et al., 2023). A deep understanding of the canola response to clubroot disease and other biotic and abiotic stresses is given by integrated analysis of multi-omics data. Breeding for resilience to climate change is possible in canola where interlinking of multi-omics strategies is expected to offer many opportunities. Multi-omics may afford the opportunity to link each pathway or metabolite from fatty acid synthesis through the biosynthesis of secondary metabolites in canola (Ashraf et al., 2022). This knowledge leads to the production of improved canola cultivars that are capable of resistance of environmental dynamics, quality of the oil and nutritional worth of the products. This information is very useful for engineering of improved oil quality and quantity in seed through optimization of particular enzymes and regulatory genes.

3.2. Agricultural Breeding for Sustainable Crops Production using Omics Science

It is widely assumed that through the technology of omics technologies (OTs), we will have the capability to enhance the breeding of food crops as far as yield, nutritional quality, and abiotic stress resistance is concerned. For example, it is useful for identification of genes linked to desirable characteristics like yield, disease resistance and nutritive qualities, by using the genomic data produced from OTs. Integration of multi-omics technologies like genomics, proteomics and metabolomics coupled with techniques like genome editing and high through-put phenotyping these new paradigms remain open in the case of canola breeding, according to Mmbando of the year 2024 (Mmbando, 2024). It has been noted that such approaches can assist in a faster pace of efficiency of the new varieties of canola providing nutrient and water use efficient, lessen utilization of fertilizers and water for irrigation improving the practice of the sustainable agriculture (Riar et al., 2020).

Prospects include efficiency gains from the application of omics technologies (OTs) for appropriate creation of required GM varieties. Such technologies benefit in assessing the safety, accessibility, and ecological influence of GM crops, in developing appreciably more stress-tolerant canola forms. This can lead to higher yields, and making of products that are sustainable under the changing climate and other unfavourable weather conditions (Benevenuto et al., 2023). Such an approach due to multi-omics is instrumental in correctly defining the exact genetic markers and pathways associated with such traits as drought tolerance, disease tolerance and quality of the oil yielded. By incorporating biotic stress tolerance traits in crop varieties, the genetic information means that scientists can identify the particular gene or genetic indicators of that particular tolerance. This information can then be utilized in omics-driven approaches (Figure 1) like genetic engineering and marker-assisted selection to develop crops with enhanced resistance to biotic stresses, such as pests and diseases.

Figure 1. Overview of dome important challenges and opportunities of OMICS

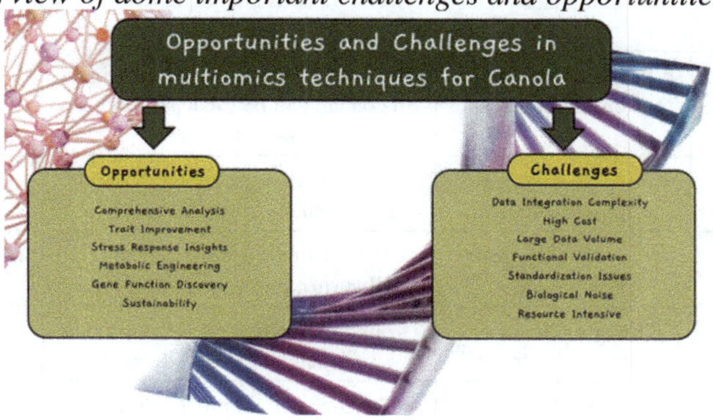

The Table 1 below highlights the role of omics approaches in breeding programs targeting both biotic (e.g., pests, diseases) and abiotic (e.g., drought, heat) stresses in various crops, showcasing how these technologies enhance precision breeding for desirable traits. Various crops, such as wheat, maize, and cotton, have undergone breeding programs to improve their resilience against factors like drought, disease, and pests, showcasing the vital need for continued breeding efforts in the face of global challenges. Conclusively, synergy between multi-omics, genome editing, and other technologies will transcend the canola breeding sector to produce high yielding and stress resistant sustainable varieties of canola. Such a comprehensive approach will be instrumental in the management of future challenges in agriculture and enhancement of the functional utilization of canola in food security and other industries.

Table 1. Omics-Driven Breeding Objectives and Improvements in Crops for Biotic and Abiotic Stress Resistance

Crop	Breeding Objective	Improvement Achieved	References
Rice (*Oryza sativa*)	Disease resistance, yield improvement	High-yielding, disease-resistant varieties like IR8	(Jamaloddin, Mahender, et al., 2021)
Wheat (*Triticum spp.*)	Drought tolerance, rust resistance	Varieties like Borlaug's dwarf wheat with increased yield and rust resistance	(Hussain et al., 2017)
Maize (*Zea mays*)	Drought resistance, pest resistance	Development of Bt maize for pest control, drought-tolerant varieties	(Oyekunle et al., 2023)
Tomato (*Solanum lycopersicum*)	Disease resistance, fruit quality improvement	Varieties resistant to tomato yellow leaf curl virus and improved shelf life	(Singh et al., 2019)
Soybean (*Glycine max*)	Pest resistance, oil content improvement	Development of soybean with higher oil content and resistance to soybean cyst nematode	(J. Wang et al., 2022)
Potato (*Solanum tuberosum*)	Late blight resistance, yield improvement	Blight-resistant varieties like 'Bionica', improved yield	(Berindean et al., 2024)
Cotton (*Gossypium spp.)*	Fiber quality, pest resistance	Bt cotton for resistance to bollworms, improved fiber quality	(Mora-Oberlaender et al., 2022)
Peanut (*Arachis hypogaea*)	Drought tolerance, disease resistance	Drought-tolerant varieties, resistance to groundnut rosette disease	(Herselman et al., 2004)
Barley (*Hordeum vulgare*)	Stress tolerance, malting quality improvement	Varieties with improved drought and heat tolerance, better malting properties	(Mahalingam, 2017)
Sorghum (*Sorghum bicolor*)	Drought tolerance, pest resistance	Drought-resistant varieties for arid regions, pest resistance improvement	(Verma et al., 2022)
Canola (*Brassica napus*)	Oil quality, disease resistance	Varieties with improved oil quality and resistance to blackleg disease	(Potter et al., 2016)
Banana (*Musa spp.)*	Disease resistance, fruit quality improvement	Varieties resistant to Panama disease and improved fruit shelf life	(Nansamba et al., 2020)
Sugarcane (*Saccharum spp.*)	Yield improvement, disease resistance	Development of high-yielding, disease-resistant hybrids	(Wei et al., 2022)
Cassava (*Manihot esculenta*)	Virus resistance, starch content improvement	Varieties resistant to cassava mosaic virus, higher starch content	(Ntui et al., 2024)
Apple (*Malus domestica*)	Disease resistance, fruit quality	Varieties resistant to scab disease and with improved flavor and storage potential	(Pereira-Lorenzo et al., 2018)

4. PRECISION BREEDING AND METABOLIC ENGINEERING FOR IMPROVED OIL QUALITY AND YIELD IN CROPS

Omics technologies (OTs) are critical in enhancing yield and quality of crops and nutritional values, nutrient content, and anti-nutrient content of foods (Figure 2). Proteomics and metabolomics for instance, can be used to accurately describe nutrient content of a given crop (Jamaloddin, Maliha, et al., 2021). The researchers are meaning to identify the concentration of the different vitamins, minerals, amino acids and other nutritional needs in crops through the metabolites and proteins. Using Multi-omics data for precision breeding and metabolic engineering of canola is changing the dynamics of canola breeding by improving both the quality and the yield of the oil (Cao et al., 2022). Further, the omics technologies (OTs) also support the generation of biofortified crops and these crops are developed to have higher nutrient density (Yang et al., 2021). After getting the information of the genes and the metabolic routes that are implicated in assembly of nutrients and storage, then it becomes reasonably uncomplicated to conduct plant breeding for enhanced nutritional status. Multi-omics in precision breeding are incorporated into MAS and GS, enhancing the accuracy of the prediction of traits such as seed yield, biomass, and oil content. Such models, which include genomic selection models, transcriptomic models, proteomic models, and metabolomic models, are better than old techniques that apply genetic marker selection (Cao et al., 2022). It indicates that with multi-omics data for selection, breeders can do better/faster on traits for genetic progress. Similarly, metabolic engineering targets the enhancement of biosynthetic paths of oil in the canola seed, as well as the quality of the oil generated. When the detailed metabolomic profiles are obtained in conjunction with the genetic data, researchers have a basis for targeting particular genes involved in the synthesis of fatty acids and triacylglycerols (TAGs) (Vanhercke et al., 2019). For instance, 13C-metabolic flux analysis (MFA) techniques used in the laboratory enable scientists to monitor carbon distribution in various central metabolic pathways; this can be used to determine which metabolite flux can be regulated in order to enhance oil generation (Antoniewicz, 2021). In this regard, the technique CRISPR/Cas9 is often employed to increase the efficiency of the enzymes, or eliminate restrictive metabolic pathways, overall increasing yield and oil quality (Jiang & Doudna, 2017). Furthermore, there is a "push-pull" regulation system that means transcription factors, which stimulate fatty acid production, are overproduced, while enzymes, that declaim TAGs, are underproduced. It also enhances the quality of canola oil for the food industry while providing industrial uses in other areas including renewable biofuel and bioplastics (Andrews et al., 2016).

Figure 2. Role of Omics technologies (OTs) in enhancing yield, oil and quality of canola

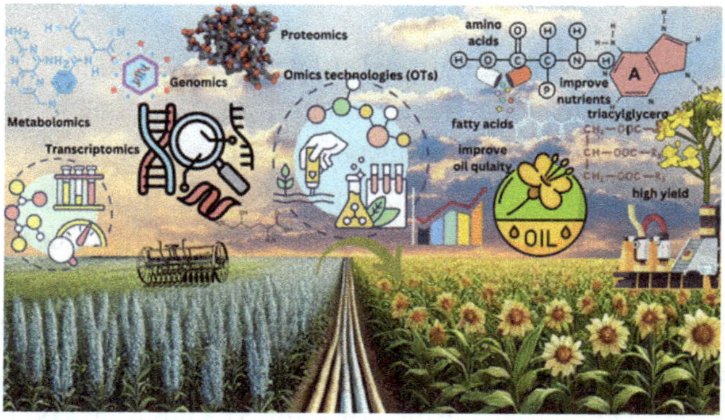

5. OMICS STRATEGIES IN CANOLA RESEARCH AND BREEDING

5.1. Genomic Insights, Functional Genomics, and Molecular Breeding in *Brassica Napus*

It is quite unusual that *Brassica napus* (canola) does not exist in natural environments or stands as wild population. Therefore, the tendency is to investigate genetic differentiation for cultivars and breeding lines originated from different countries and breeding programs (Malik, 2023). As a consequence, in order to discover the genetic polymorphism present in *B. napus* germplasm resources, various molecular techniques such as RFLP, SSR, AFLP, RAPD and isozyme markers have been used by several researchers including Qu et al., (2012). The stock variability is important in any breeding exercise since it defines the differentiation in plants depending on various characteristics. The Brassica genus is comprised of approximately 100 species, and demonstrate considerable genetic variation (Qu et al., 2012). Genetic variability is crucial for the success of any breeding program, as it reflects the divergence in plants based on various traits. The Brassica genus consists of around 100 species that exhibit significant variability (Raza et al., 2020). Previous study described the cytogenetic relationships among rapeseed species, identifying that the three allotetraploid species *B. napus*, *B. juncea*, and *B. carinata* are amphidiploids derived from the diploid species *B. nigra*, *B. rapa*, and *B. oleracea*. *Brassica napus* (AnAnCnCn, 2n = 38) is the second most grown oilseed crop in the world which is used as a source of edible vegetable oil (Gu et al., 2024; Kumar et al., 2022). This all tetraploid species came into existence about 7500 years back due to hybridization between *Brassica rapa* with genome ArAr, 2n = 20 and *Brassica oleracea* with genome CoCo, 2n = 18. *B. napus* has a diploid genome and its genome is diploid and comprises of two sub-genomes An and Cn which is derived from the genomes of its diploid ancestors (Wang, 2016). The latest genomic analysis of populations of *B. napus*, *B. rapa* and *B. oleracea* has suggested that the An subgenome of *B. napus* might have evolved from *B. rapa* var. rapa. To date, no wild germplasm of *B. napus* has been found in the wild state, hence a cultigen which has a complex evolutionary history associated with its progenitors (Gazave et al., 2016). Winter-type oilseed rapes (WORs) are believed to be the first and least developed type that occurred in the early stage of domestication and cultivation of rapeseed. Later, two other ecotypes, spring type oilseed rapes (SORs), and semi-winter type oilseed rapes (SWORs) developed. Through process of domestication to different climatic condition and through the process of chromosome introgression from ancestral species and related brassica species over the last thousand years, these ecotypes evolved and became adapted to different environments. Such an evolutionary process shows the ability of *Brassica napus* to grow in a wide range of environments (Wu et al., 2019).

The genome assembly statistics and transposable element (TE) rates for various *Brassica napus* species and cultivars are summarized as follows. The total genome size of the Darmorbzh Genome Assembly v4.1 was reported to be 850.29 Mb, with an anchored size of 645 Mb, an anchorage rate of 75.86%, and a TE rate of 12.5% (Chalhoub et al., 2014). For the *B. napus* 'ZS11' genome assembly, the total genome size was 976.05 Mb, with an anchored size of 854.78 Mb, an anchorage rate of 87.60%, and a TE rate of 14.9% (Sun et al., 2017). Darmorbzh_v8.1 had a total genome size of 850.29 Mb, with 799 Mb anchored, an anchorage rate of 93.97%, and no reported TE rate (Bayer et al., 2017). The Tapidor v6.3 genome assembly reported a total size of 636.32 Mb, with 616.7 Mb anchored, an anchorage rate of 96.92%, and a TE rate of 96.92% (Bayer et al., 2017). Ningyou 7 had a total genome size of 993.78 Mb, with 892 Mb anchored, an anchorage rate of 89.76%, and a TE rate of 56.9% (Zhao et al., 2016). The ZS11 genome assembly reported a total size of 1008.16 Mb, with 960.82 Mb anchored, an anchorage rate of 95.30%, and a TE rate of 55.6%. For the Westar assembly, the total size was 1007.28 Mb, with 942.07 Mb anchored, an anchorage rate of 93.53%, and a TE rate of 55.8% (Gu et al., 2024; Song et al., 2020). No2127 had a total genome size of 1011.42 Mb, with 932.4 Mb anchored, an anchorage rate of 92.19%, and a TE rate of 56.9% . Zheyou7 reported a total genome size of 1015.12 Mb, with 923.81 Mb anchored, an anchorage rate of 91.00%, and a TE rate of 56.4% (Gu et al., 2024; Song et al., 2020). Gangan had a total genome size of 1033.32 Mb, with 933.99 Mb anchored, an anchorage rate of 90.39%, and a TE rate of 56.6%. Shengli's genome assembly reported a total size of 1001.45 Mb, with 929.75 Mb anchored, an anchorage rate of 92.84%, and a TE rate of 56.3%. Tapidor had a total genome size of 1013.60 Mb, with 932.47 Mb anchored, an anchorage rate of 92.00%, and a TE rate of 56.6% (Gu et al., 2024; Song et al., 2020). Quinta reported a total genome size of 1003.23 Mb, with 932.68 Mb anchored, an anchorage rate of 92.97%, and a TE rate of 56.3% (Song et al., 2020). Express617_v1 had a total genome size of 925 Mb, with 764.62 Mb anchored, an anchorage rate of 82.66%, and a TE rate of 43.6% (Lee et al., 2020; Tavares et al., 2014). Darmorbzh_v10 reported a total genome size of 923.8 Mb, with 866.92 Mb anchored, an anchorage rate of 93.84%, and a TE rate of 54.0%. Finally, the ZS11_PB genome assembly had a total genome size of 921.54 Mb, with 884.92 Mb anchored, an anchorage rate of 96.03%, and no reported TE rate (Chen et al., 2010; Chen et al., 2021).

5.2. QTL Mapping in *Brassica Napus* Breeding

Accomplishments of multi-omics integrating with GWAS and QTL (Table 2), have defined genes, proteins, and metabolites governing important agronomic traits including oil content, disease resistance, and drought tolerance that has revolutionized the process of developing canola (Mahmood et al., 2022). GWAS has catalogued over several thousand of these loci, but, due to LD in large candidate regions, exact identification of underlying causal genes is difficult. This is the rationale for defining multi-omics, comprising genomic, transcriptomic, proteomic, and metabolomics, data combining data from more omics layers allows to select the genes of interest and to decrease uncertainty of trait localization in QTL regions (Li & Ritchie, 2021). It has been established that the seed yield in open pollinated and hybrid lines of *Brassica napus* involves several genomic regions which can be identified using QTL mapping mainly with mapping populations from European winter, Chinese semi-winter, and resynthesized *B. napus* (Rahman et al., 2017). It has concluded other QTL alleles on chromosomes A10, C4 and C5 which are highest genres to enhance the seed yield. These QTLs originating from multiple germplasm sources account for a significant proportion of phenotypic variation; some of the genomic regions harbor multiple closely linked QTLs suggesting their significant impact on agronomically important traits such as seed yield (Zhao et al., 2016). The discovered co-location of QTL for traits indicates that both A and C genome alleles from exotic germplasm, including resynthesized *B. napus*, can be used to increase seed yield in spring *B. napus* cultivars. However, the expr imbalance effects of exotic QTL alleles on seed yield potential are quantitatively different due to interactions with the genetic background in which these alleles originated (Kim et al., 2012). The first something to recognize is that while some alleles increase yield in one genetic background, others may be inconsequential or even detrimental because of interactions between them (Kim et al., 2012). This is because traits with such patterns of genetic control are difficult to analyze and even more so given that the interactions underlying high seed yield hardly come with simple additivity where new QTL alleles are introduced. Sustained long-term investigation and systematic breeding and propagation, such as molecular plant breeding, is crucial to the incorporation of these alleles into local elite lines, which in beneficially increasing genetic variation, will improve the probability of high yield in both op and hybrid canola cultivars (Rauf et al., 2010). On the other hand, to confirm the roles of these candidate genes, scholars may apply the CRISPR/Cas9 system of genetic modification. This provides breeders with a method to directly test the roles of the introduced genetic variants through editing of the genes implicated by multi-omics profiling and the evaluation of the phenotypic implications. Together with multiple omics and genome editing, the process of gene function identification and the translation of genomic information for canola improvement with increased oil content and stress/tolerance to diseases is facilitated.

Table 2. QTL mapping in Brassica napus

Trait	QTL Information	Chromosome	Additive Effect/ Variance Explained	Environmental Influence	References
Seed Oil Content (General)	Multiple QTLs identified across all nine C genome chromosomes.	C1–C9	Additive effect: 0.2–1.0% Phenotypic variance: 2–20%	Significant influence observed in several studies.	(Chao et al., 2017; Chen et al., 2010; Würschum et al., 2013).
Stable Oil Content QTL (C5)	High stability observed for QTL on C5 across nine field trials, likely introgressed from *B. oleracea.*	C5	QTL detected across 9 field trials, indicating high stability.	Minimal environmental effect in Alberta trials.	(Chao et al., 2017; Javed et al., 2016; Körber et al., 2016)
Protein Content QTL (C5)	QTL colocalized with oil content QTL on C5. Allele for high protein content derived from *B. oleracea.*	C5	N/A	Environmentally sensitive; contradicts reports of negative correlation between oil and protein QTLs.	(Chao et al., 2017; Zhao et al., 2008)
Independent Protein QTL	Independent QTL for protein content identified, highlighting the complexity of trait improvement.	Various	Additive effects varied widely.	Environmental influence varies depending on QTL location and study conditions.	(Zhao et al., 2008)

continued on following page

Table 2. Continued

Trait	QTL Information	Chromosome	Additive Effect/ Variance Explained	Environmental Influence	References
Glucosinolate Content QTL	Major QTL on C9 identified in multiple studies, influencing total GSL content.	C9	Major QTL on C9, with positions ranging from 1.1 Mb to 8.0 Mb.	Detected across multiple environments; suggests more than one QTL on C9.	(Körber et al., 2016; Olivos-Trujillo et al., 2015)
Consistent GSL QTL (C9)	QTL on C9 detected in *B. napus* and *B. oleracea* populations; aligns with findings from multiple studies.	C9	Position of QTL varies from 3.9 Mb to 16.9 Mb on C9.	Evidence of environmental influence, though stable QTL also reported.	(Olivos-Trujillo et al., 2015; Sotelo et al., 2014)
Oil Content QTL (C5)	Additional QTL on C5 identified; positions vary but suggest novel recombination or introgression from *B. oleracea*.	C5	Varies across studies, with some detecting QTL in all trials and others only in specific environments.	Influence of environment reported by some, but stability in others suggests robust QTL.	(Fu et al., 2017; Körber et al., 2016; Zou et al., 2019)
Oil Content Variation	Evidence of transgressive segregation for oil content in progeny, even when parents have similar oil content.	Various	Phenotypic variance explained ranges from 2–20%.	Environment plays a significant role in the expression of these QTLs.	(Chao et al., 2017; Zhao et al., 2008)
Protein Content Complexity	Protein content QTL sensitive to environmental conditions; challenges in simultaneous improvement of oil and protein due to negative correlation.	Various	Positive and negative alleles for protein and oil contents often co-localized, complicating breeding efforts.	Significant environment × QTL interaction reported.	(Chao et al., 2017)
Multiple GSL QTL (C9)	Second QTL on C9 identified through association mapping, located at 3–4 Mb, though not detected in all studies.	C9	QTLs at 7–8 Mb and 3–4 Mb positions on C9, indicating multiple QTLs affecting GSL content.	Variability in detection across different environments and studies.	(Körber et al., 2016) (Chao et al., 2017; Zhao et al., 2008)

5.3. Transcriptomics

Transcriptomics helps to identify new genes involved in the development of the traits as well as the regulatory processes by which they occur (Lowe et al., 2017). Transcriptome analysis of seeds of canola plants grown under different conditions of the environment with the help of a transcriptome it is possible to compare genes that react to abiotic and biotic stresses in the conditions when canola plants are grown (Zhang et al., 2019). For example, transcriptomic analyses have shown the heightened expressions of particular genes in reaction to drought stress, suggesting biomarkers in increasing drought resistance in canola (Ai et al., 2024). Several earlier investigations have indicated that transcriptome profiling is a crucial factor in elucidating the canola response to alkaline salt stress (Wang et al., 2021). Specifically, transcriptomics has been shown that under such stress conditions, canola roots up or down regulate several genes involved in amino acid and fatty acid synthesis processes (Wang et al., 2021). These studies found evidence that while carbohydrate metabolism declines, the levels of amino acids and fatty acids increases, and therefore these pathways become essential for the functioning of plant stress. This shows that the changes in the transcriptome level of these metabolic pathways help canola to tolerate and respond to alkaline salt stress.

Specifically, transcriptome studies have shown that under such stress conditions, canola roots undergo alterations in amino acid and fatty acid metabolism associated genes (Raza et al., 2021; Wang et al., 2021). These studies have revealed that whereas the rate of carbohydrate breakdown is lowered that of amino acid and fatty acids' build-up rises thus pointing at those pathways' importance for the plant's stress response. This suggests that under alkaline salt stress, alteration of these metabolic pathways may facilitate canola bears this stress (Gill et al., 2012). The study published earlier worked on the transcriptomic analysis of *Brassica napus* with different clubroot resistance (CR) genes under stress from *Plasmodiophora brassicae* pathotype X (L-G2). This work used RNA-seq to investigate the samples collected from root tissues at 14 dpi, when the pathogen had finished cortical infection, although there were no distinct symptoms to manifest the infection (Tonu et al., 2023). This time point was chosen because prior work revealed that regulatory loci associated with pathogen recognition, signal transduction and defense response pathways are highly expressed at this time point. Further, the gene expression patterns associated with the CR genes CRaM and Crr1rutb were found to be highly significantly different (Wen et al., 2024).

A combination of the transcriptomic information with the genomic data improves the accuracy of the breeding models. Surprisingly, currently published literature has demonstrated that by integrating transcriptomic data into the hybrid prediction models, there is a marked increase in the overall accuracy of estimating canola hybrid performance on complex traits. It also creates an opportunity for the breeders to select hybrids that should do well in specific conditions.

5.4. Proteomics and Metabolomics

A proteomics and metabolomics analysis provide hierarchical information regarding the biochemical changes occurring in canola plants. Proteomics is focused on protein identification and quantification (Brewis & Brennan, 2010) while on the other hand, metabolomics is more a study of small bio molecules (Liu & Locasale, 2017). These approaches are important especially in determining the flow of the biochemical processes that defines important traits like quality of the end product and the nutritional value of the oil. For instance, using metabolites to predict oil profiles in the canola seed has provided a method of understanding the enzymes that contribute to lipid deposition (lipidomics) (Woodfield et al., 2018). Thus, since breeders seek to alter traits associated with these pathways, it will be possible to create canola varieties with improved oil quality and nutrition. In addition, proteomic analysis can help to understand function of individual proteins involved in stress responses that will facilitate the identification of stress-tolerant canola varieties (Nesatyy & Suter, 2007). The most abundant proteins reported in the seeds of Brassica are the seed storage proteins cruciferin and napin, belonging to the 12S globulin and 2S albumin families of proteins, respectively (Rahman et al., 2021). Previous study compound current various research results to investigate the proteomic and metabolomics changes associated with the salt stress in *Brassica napus* (canola). Finally, proteomics evaluation indicated that proteins linked to the ABA signaling pathway, such as BnPP2Cs and BnABFs, were changed evidently, thus over again affirming the part of these proteins in energy conservation and stress acclimatization (Shu et al., 2022; Wei et al., 2021). It also revealed that jasmonic acid (JA) signaling plays a vital role in maintaining the energy homeostasis during salt stress as reported earlier (Wang et al., 2020). Glycine betaine (GB) and sn-glycero-3-phosphocholine were identified as important metabolites affecting salt tolerance by metabolomic studies (Shu et al., 2022). As demonstrated in prior investigations, GB is critical for the preservation of cellular osmotica and membrane stability; a condition evidenced by a 2-fold increase of GB at the 24 hours post-stress stage (Subbarao et al., 2001). Further, the surge in sn-glycero-3-phosphocholine (GPC), which is a precursor in the synthesis of phosphatidylcholine (PC), and converted into choline, that pointed out lipid changes are part of stress response (Subbarao et al., 2001).

Furthermore, a decrease in N-Acetylserotonin (NAS) may suggest its biosynthesis into melatonin in response to ROS scavenging consistent with prior studies pointing to melatonin's stress regulation role (Arnao & Hernández-Ruiz, 2015). A higher accumulation of L-cysteine and HSP proteins also points to a strong induction of antioxidant machinery in maize roots agreeing with (Wang et al., 2017) who stressed the antioxidant benefits of these compounds under salt stress conditions. Overall, these findings present a multiple-level integration of proteomic and metabolomics changes; while these respective data aggregate from multiple studies on how canola plant performs in saline condition.

5.5. Phenomics in Canola

Different imaging techniques are changing the face of plant phenomics by facilitating the measurement of numerous attributes so central to plant breeding (Kumar et al., 2015). Morphological features are analyzed using visible light imaging while functional properties of the photosynthetic machinery are assessed using fluorescence imaging (Rahman et al., 2015). In the thermal imagery, the status of water as well as the stomatal conductance is recorded, while the hyperspectral imagery enables the sharing of data related to biochemical content (Sobejano-Paz et al., 2020). Further, spectral reflectance sensors quantify biomass and nutrient content, LiDAR quantifies the distribution of plant structures in 3D space, and chlorophyll fluorescence sensors record the capacity of the plant's photosynthetic machinery (Haworth et al., 2023; Tardieu et al., 2017). Advanced technical tools employ Robotics and automation in handling the numerous samples, conveyor belts for high throughput imaging while other robotic arms for sample manipulation. The complexity of the traits, states of traits and their interconnections require efficient data management assisted by the machine learning algorithms, big data systems, statistical tools (Kehoe et al., 2015).

In canola, the integration of phenomics has proven invaluable for optimizing breeding programs and improving stress tolerance, nutrient use efficiency, and disease resistance (Yang et al., 2021). By leveraging technologies such as hyperspectral imaging and thermal imaging, researchers have identified critical traits and genetic markers that contribute to canola's adaptability under stress conditions (Dhakal, 2023). For instance, the application of chlorophyll fluorescence sensors has facilitated the assessment of photosynthetic efficiency, revealing variations among genotypes that can inform breeding strategies (Fu et al., 2022). Additionally, the use of spectral reflectance sensors has allowed for the precise estimation of biomass and nutrient status, aiding in the selection of genotypes with enhanced nutrient use efficiency (Banerjee et al., 2020). These advancements in canola phenomics not only contribute to higher yield and better stress management but also support the development of sustainable agricultural practices by enabling the cultivation of more resilient and resource-efficient varieties (Chaudhary et al., 2024). Figure 3 depicts cross-omics integration techniques.

Figure 3. Cross omics integration techniques

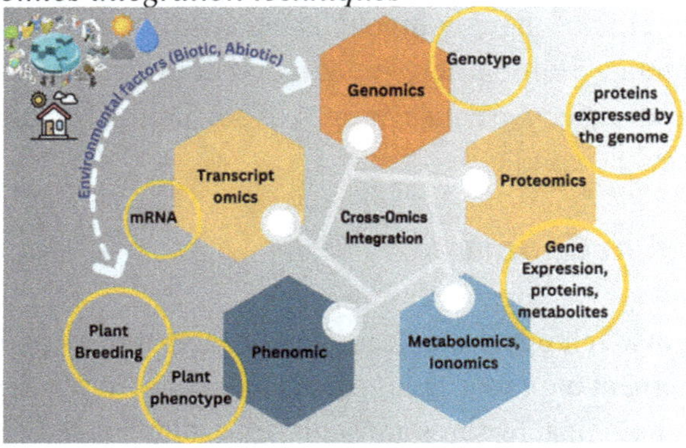

6. FUTURE CHALLENGES AND CONSIDERATIONS IN OMICS AND CANOLA BREEDING

While the potential of omics technologies in canola breeding is vast, several challenges (Figure 4) must be addressed:

6.1. Data Management and Analysis

Multi-omics approach produces massive amounts of data, which can only be analyzed using advanced computational methods. In order for these platforms to be of value to breeders, it is crucial that they are designed to be user friendly and can process these large chunks of data appropriately (Kaur et al., 2021).

Figure 4. Future challenges and trends in omics-based canola breeding

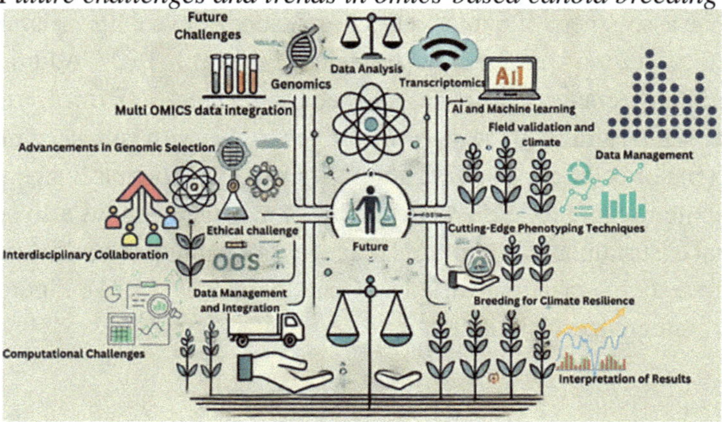

6.2. Field Validation

This is particularly important since many of the 'ideal' traits that are discovered with the help of omics technologies are benchtop realities and not actual field gains. Thus, the molecular biologists should work hand in hand with the agronomists (Rajczewski et al., 2022).

6.3. Ethical and Regulatory Considerations

With new methods in plant breeding, issues to do with genetic engineering and the use of new characteristics comes into view. It is clear that regulatory frameworks only partially help to address the importance of accelerated technology development in plant breeding (Acquaah, 2009).

6.4. Interdisciplinary Collaboration

Multi-omics approaches implementation is the integration of traditional plant science, data analysis proficiency, and plant breeding (Jaramillo-Botero et al., 2022). The multi-disciplinary associations will enable dissemination of information as well as the formulation of new strategies in breeding (Ortolani et al., 2017).

Future trends in omics-based canola breeding are expected to bring efficiency and effectiveness to a new level of breeding programs. Integration of the various omics methods including genomics, transcriptomics, proteomics, metabolomics, and phenomics is a well-coordinated and systematic approach toward improvement of canola traits (Razzaq et al., 2019).

7. TRENDS IN GENETIC IMPROVEMENT OF CANOLA: OMICS PERSPECTIVE

However, one of the most difficult issues in multi-omics integration is the question of how to combine multiple data sets originating from various omics data platforms (Figure 4). In the same manner, every omics layer has unique data structure, data size, and kind of information including genomics, transcriptomics, proteomics, metabolomics and others (Canzler et al., 2020). For example, genomic data may include SNPs and CNVs, where transcriptomes mainly deal with gene expression levels, and metabolomics mainly refers to many types of metabolites. There is, therefore, a seeming call for procedural replication for assimilation of such data sets into a more coherent framework for analysis and interpretation (Wortmann et al., 2022).

7.1. Integration of Multi-Omics Data

Its omics perspective genomics, transcriptomics, proteomics and metabolomics are fast becoming the game changers in the canola breeding program. Thus, this broad strategy enables the researchers to inform more about the genetics and biochemistry that underlie important phenotypes (Tahir ul Qamar et al., 2020). Thus, by detecting the behaviour of different material layers at once, the breeders make right prediction about the outcome of hybrid behaviour and the transfer of certain traits (Cooper et al., 2014). Over the last few years, investigations have shown that the incorporation of genetic marker data with transcriptomic and metabolomics profiles further improves the ability of specific agronomic traits such as yield and stress tolerance to be predicted. Such an approach should eventually result in the release of canola varieties capable of coping with the current and future problems affecting agriculture (Yandeau-Nelson et al., 2016).

Since this integrative approach is only relatively developed, its use in canola breeding can be expected to grow systematically, particularly within the framework of climate change (E. Wang et al., 2022). Advanced precision in predicting such multiple-characterized traits will be useful in refining the canola varieties capable of withstanding weathers. New developments in data preprocessing and analytical methods are expected to enhance the ability of breeders in maximizing the usage of multi-omics data for accurate and better breeding strategies (Mahmood et al., 2022).

Since the beginning of the commentary, precisely in the first decade of the modern millennium, the huge potential of Genomic Selection (GS) has been captivated by the development of new populace resources (Tagkouli, 2021).

Genomic selection commonly known as GS is poised for being an important tool in breeding for the future of canola (Merrick et al., 2022). Analyzing the genomic data, GS helps to estimate breeding based on the genetic information and to identify the most suitable genotypes much faster than using the traditional selection methods (Crossa et al., 2017). They found that as sequencing costs come down, and genetic data become more abundant, there's great potential for GS to shave time off the breeding cycle to rapidly generate new high yield varieties of canola. One such breakthrough that has caught the attention of animal breeders is the genomic estimated breeding values (GEBV) that can also help determine the alleles that are linked to superior traits such as yield, disease resistance as well as abiotic stress tolerance (Mir et al., 2012).

It is with the understanding of the above said factors of modern agriculture that the implementation of GS will be central towards facing these challenges such as climate change (Mensah et al., 2021). The breeders are continually improving those genomics tools, so in later selections they are closer to desirable characteristics of canola that improves flexibility and productivity of the crop; thus leading to the development of elite varieties designed for certain agriculture requirements (Mohd Saad et al., 2021). With systems biology approaches, it is possible to build the models that could help predict the combinations of the interactions of the omics layers to certain phenotypes (Sidak et al., 2022). For example, conducting an eQTL, pQTL and mQTL study would enhance the predictability of the traits because these would explain the connections between the expressions, functions and metabolic profiles of a gene (Lee, 2022).

7.2. Advancement in Phenotyping Technology

New high throughput phenotyping techniques are the game changer for the assessment of canola traits when it comes to breeding improvement (Kumar et al.). Systems like data acquisition, imaging and phenotyping platforms are fast helping to assess the traits such as plant height, leaf area and chlorophyll content remotely (Feng et al., 2021). Together with omics data these technologies provided more comprehensive knowledge about the relationship between genotype and phenotype that in turn helped to find relevant traits underlying yield potential and stress tolerance (Singh et al., 2021).

This interaction between phenotyping and omics technologies should improve breeding approaches by providing a better method for selecting on traits (Langridge & Fleury, 2011). For example, hyperspectral image can give information on the biochemical content of the plants; this can help in the identification of potential genotypes with better nutrient use efficiency or tolerances to stress (Banerjee et al., 2020). In the future, these technologies will become more important to the task of breeding new and improved, high-yielding, and canola types that are more resistant to biotic stresses (Singh et al., 2021).

7.3. Breeding for Climate Resilience

Another consideration with reference to global climate change is breeding of canola varieties that are favorable to abiotic stresses such as drought, heat or salinity (Wu et al., 2018). Omics technologies play an important role for detecting genetic elements and molecular pathways related to these stress responses (Razzaq et al., 2021). Thus transcriptomic and proteomic approaches provide breeders a basis for understanding the mechanisms of stress response and further selection of desirable Canola traits for growth in different growing conditions (Zhu et al., 2016).

That is why it will be important to concentrate on climate-smart breeding as the future of canola production sustainability heavily depends on it (Raman et al., 2019). Through deliberately breeding for water use efficiency, heat tolerance, salinity tolerance in canola crop, breeders can develop new exciting varieties to suit a wide range of environments commonly associated with drought, heat and salt stress. This approach also goes hand in hand with food security and acts as a precursor to sustainable agriculture (Beddington et al., 2012).

7.4. Exploring Wild Relatives and Underutilized Species

The genetic variation study of wild relatives and low erudite Brassica species creates new prospects for canola production (Farnia et al., 2015). Such species always harbor favorable alleles which offer diseases, improved contents and stress resistance among others (Dwivedi et al., 2008). The omics technologies facilitate the identification and introgression of these valuable traits into the improved cultigen types of canola, expanding the genetic resources base and improving the performance of the cultivated varieties.

The listing genetic resource of wild relatives and underexploited species will be critical in the development of new canola varieties to serve the dynamic needs of agriculture. Through the enhancement of the trait that would enhance yield and the nutritional quality breeders equally solve the ailment that affect the crop and the market preference score by maintaining a competitive value of the canola crop (Raboanatahiry et al., 2021).

7.5. AI and Machine Learning Integration

The advancement of AI and ML with the omics data is expected to bring the drastic change in canola breeding. These technologies provide algorithms as means to analyze and interpret big data derived from multi-omics experiments, detect patterns and make prediction, which would be nearly impossible by conventional methods. AI and ML can quicken the breeding plan, increase the possibilities of the primary character success, and support with decision-making (Mahmood et al., 2022).

In the future, the development of AI and ML technologies will increase and will be used in canola breeding. For example, active learning algorithms can be trained to for predicting the hybrid performance evaluation via integrating the multiOME data to allow the breeders to select the best performing candidates (Melchinger et al., 2023). This confluence of AI and ML with omics data will go a long way to push breeder efficiency as well as generate new traits that can be utilized to solve modern breeders' complex problems. Computer intensive is the nature of analysis of multi-omics data. The generation of OMICS data, especially through high through-put sequencing, creates large data volumes for storage, processing and subsequent analysis and calls for effective bioinformatics algorithms and computational infrastructure. Further, to increase the application of multi-omics data in breeding programs, it is imperative to build friendly interface and analyst software to process the data (Jimenez-Lopez et al., 2013).

7.6. Navigating Regulatory and Ethical Challenges

As omics technologies advance, navigating the regulatory and ethical landscape of canola breeding will become increasingly important. With the rise of new techniques such as gene editing, it is essential that breeding practices comply with safety and ethical standards (Jimenez-Lopez et al., 2013). Breeders must maintain transparency and public trust while developing new canola varieties, ensuring that the benefits of these innovations are clearly communicated to all stakeholders (Smith, 2019).

Engaging with farmers, consumers, and policymakers will be critical in addressing concerns and promoting the responsible use of omics-based technologies. As the field of canola breeding continues to evolve, clear and consistent regulatory frameworks will be necessary to support the safe and effective application of these powerful tools, ultimately contributing to the sustainability and success of global agriculture (Nasar et al., 2024).

8. CONCLUSION

The canola breeders should consider responding to these questions to achieve the benefits of multi-omics data integration in breeding. Thus, to improve the integration of multi-omics data, researchers can benefit from interdisciplinary collaboration, follow up with the construction of solid computational tools, and provide efficient field validation. When applying the omics technologies in the further development of sustainable agricultural practices and paradigms, these challenges will have to be met at some point. The emerging omics technology provides many opportunities in the canola breeding programs. Applying multi-omics approaches and combining them with conventional breeding, it is possible to create new canola varieties with high yield but at the same time with low vulnerability to biotic and abiotic stress factors as well as enhanced nutritional value. The change in the measures of farming will be important for the global need of foods and the adoption of these innovations for sustainable farming. Canola breeding of the future can help to feed the world and make agriculture more climate-resistant.

REFERENCES

Acquaah, G. (2009). *Principles of plant genetics and breeding*. John Wiley & Sons.

Ai, X., El-Badri, A. M., Batool, M., Lou, H., Hu, J., Wang, Z., Wang, C., Xiao, Y., Xiao, J., & Xiong, Y. (2024). Integrated germination related traits and transcriptomic analysis elucidate the potential mechanism of rapeseed under drought stress. *Plant Growth Regulation, 104*(2), 1–19. DOI: 10.1007/s10725-024-01196-0

Andrews, S. S., Peria, W. J., Richard, C. Y., Colman-Lerner, A., & Brent, R. (2016). Push-pull and feedback mechanisms can align signaling system outputs with inputs. *Cell systems, 3*(5), 444-455. e442.

Antoniewicz, M. R. (2021). A guide to metabolic flux analysis in metabolic engineering: Methods, tools and applications. *Metabolic Engineering, 63*, 2–12. DOI: 10.1016/j.ymben.2020.11.002 PMID: 33157225

Arnao, M. B., & Hernández-Ruiz, J. (2015). Functions of melatonin in plants: A review. *Journal of Pineal Research, 59*(2), 133–150. DOI: 10.1111/jpi.12253 PMID: 26094813

Ashraf, U., Mahmood, S., Shahid, N., Imran, M., Siddique, M., & Abrar, M. (2022). Multi-omics approaches for strategic improvements of crops under changing climatic conditions. In *Principles and practices of OMICS and genome editing for crop improvement* (pp. 57–92). Springer. DOI: 10.1007/978-3-030-96925-7_3

Babadi, A. A., Rahmati, S., Fakhlaei, R., Barati, B., Wang, S., Doherty, W., & Ostrikov, K. K. (2022). Emerging technologies for biodiesel production: Processes, challenges, and opportunities. *Biomass and Bioenergy, 163*, 106521. DOI: 10.1016/j.biombioe.2022.106521

Banerjee, B. P., Joshi, S., Thoday-Kennedy, E., Pasam, R. K., Tibbits, J., Hayden, M., Spangenberg, G., & Kant, S. (2020). High-throughput phenotyping using digital and hyperspectral imaging-derived biomarkers for genotypic nitrogen response. *Journal of Experimental Botany, 71*(15), 4604–4615. DOI: 10.1093/jxb/eraa143 PMID: 32185382

Bayer, P. E., Hurgobin, B., Golicz, A. A., Chan, C. K. K., Yuan, Y., Lee, H., Renton, M., Meng, J., Li, R., Long, Y., Zou, J., Bancroft, I., Chalhoub, B., King, G. J., Batley, J., & Edwards, D. (2017). Assembly and comparison of two closely related Brassica napus genomes. *Plant Biotechnology Journal, 15*(12), 1602–1610. DOI: 10.1111/pbi.12742 PMID: 28403535

Beddington, J. R., Asaduzzaman, M., Fernandez, A., Clark, M. E., Guillou, M., Jahn, M. M., Erda, L., Mamo, T., Bo, N. V., & Nobre, C. A. (2012). Achieving food security in the face of climate change: Final report from the Commission on Sustainable Agriculture and Climate Change.

Benevenuto, R. F., Zanatta, C. B., Waßmann, F., Eckerstorfer, M. F., & Agapito-Tenfen, S. Z. (2023). Integration of omics analyses into GMO risk assessment in Europe: A case study from soybean field trials. *Environmental Sciences Europe*, *35*(1), 14. DOI: 10.1186/s12302-023-00715-6

Berindean, I. V., Taoutaou, A., Rida, S., Ona, A. D., Stefan, M. F., Costin, A., Racz, I., & Muntean, L. (2024). Modern Breeding Strategies and Tools for Durable Late Blight Resistance in Potato. *Plants*, *13*(12), 1711. DOI: 10.3390/plants13121711 PMID: 38931143

Brewis, I. A., & Brennan, P. (2010). Proteomics technologies for the global identification and quantification of proteins. *Advances in Protein Chemistry and Structural Biology*, *80*, 1–44. DOI: 10.1016/B978-0-12-381264-3.00001-1 PMID: 21109216

Canzler, S., Schor, J., Busch, W., Schubert, K., Rolle-Kampczyk, U. E., Seitz, H., Kamp, H., von Bergen, M., Buesen, R., & Hackermüller, J. (2020). Prospects and challenges of multi-omics data integration in toxicology. *Archives of Toxicology*, *94*(2), 371–388. DOI: 10.1007/s00204-020-02656-y PMID: 32034435

Cao, P., Zhao, Y., Wu, F., Xin, D., Liu, C., Wu, X., Lv, J., Chen, Q., & Qi, Z. (2022). Multi-omics techniques for soybean molecular breeding. *International Journal of Molecular Sciences*, *23*(9), 4994. DOI: 10.3390/ijms23094994 PMID: 35563386

Chalhoub, B., Denoeud, F., Liu, S., Parkin, I. A., Tang, H., Wang, X., Chiquet, J., Belcram, H., Tong, C., Samans, B., Corréa, M., Da Silva, C., Just, J., Falentin, C., Koh, C. S., Le Clainche, I., Bernard, M., Bento, P., Noel, B., & Wincker, P. (2014). Early allopolyploid evolution in the post-Neolithic Brassica napus oilseed genome. *Science*, *345*(6199), 950–953. DOI: 10.1126/science.1253435 PMID: 25146293

Chao, H., Wang, H., Wang, X., Guo, L., Gu, J., Zhao, W., Li, B., Chen, D., Raboanatahiry, N., & Li, M. (2017). Genetic dissection of seed oil and protein content and identification of networks associated with oil content in Brassica napus. *Scientific Reports*, *7*(1), 46295. DOI: 10.1038/srep46295 PMID: 28393910

Chapman, S. C., Chakraborty, S., Dreccer, M. F., & Howden, S. M. (2012). Plant adaptation to climate change—Opportunities and priorities in breeding. *Crop & Pasture Science*, *63*(3), 251–268. DOI: 10.1071/CP11303

Chaudhary, D., Pal, N., Arora, A., Prashant, B. D., & Venadan, S. (2024). Plant Functional Traits in Crop Breeding: Advancement and Challenges. In *Plant Functional Traits for Improving Productivity* (pp. 169-202). Springer.

Chauhan, J., Singh, K., Singh, V., & Kumar, S. (2011). Hundred years of rapeseed-mustard breeding in India: Accomplishments and future strategies. *Indian Journal of Agricultural Sciences*, *81*(12), 1093–1109.

Chen, G., Geng, J., Rahman, M., Liu, X., Tu, J., Fu, T., Li, G., McVetty, P. B., & Tahir, M. (2010). Identification of QTL for oil content, seed yield, and flowering time in oilseed rape (Brassica napus). *Euphytica*, *175*(2), 161–174. DOI: 10.1007/s10681-010-0144-9

Chen, X., Tong, C., Zhang, X., Song, A., Hu, M., Dong, W., Chen, F., Wang, Y., Tu, J., Liu, S., Tang, H., & Zhang, L. (2021). A high-quality Brassica napus genome reveals expansion of transposable elements, subgenome evolution and disease resistance. *Plant Biotechnology Journal*, *19*(3), 615–630. DOI: 10.1111/pbi.13493 PMID: 33073445

Cooper, M., Messina, C. D., Podlich, D., Totir, L. R., Baumgarten, A., Hausmann, N. J., Wright, D., & Graham, G. (2014). Predicting the future of plant breeding: Complementing empirical evaluation with genetic prediction. *Crop & Pasture Science*, *65*(4), 311–336. DOI: 10.1071/CP14007

Crossa, J., Pérez-Rodríguez, P., Cuevas, J., Montesinos-López, O., Jarquín, D., De Los Campos, G., Burgueño, J., González-Camacho, J. M., Pérez-Elizalde, S., Beyene, Y., Dreisigacker, S., Singh, R., Zhang, X., Gowda, M., Roorkiwal, M., Rutkoski, J., & Varshney, R. K. (2017). Genomic selection in plant breeding: Methods, models, and perspectives. *Trends in Plant Science*, *22*(11), 961–975. DOI: 10.1016/j.tplants.2017.08.011 PMID: 28965742

Dhakal, K. (2023). Application of Machine Learning and Hyperspectral Imaging in Plant Phenomics Research.

Dwivedi, S. L., Upadhyaya, H. D., Stalker, H. T., Blair, M. W., Bertioli, D. J., Nielen, S., & Ortiz, R. (2008). Enhancing crop gene pools with beneficial traits using wild relatives. *Plant Breeding Reviews*, *30*, 179–230.

Estevez, R. L., Duarte, J. B., Chambo, A., & Cruz, M. d. (2014). The canola culture (Brassica napus var. oleifera).

Farnia, A., Mansouri, M., Farnia, A., & Branch, B. (2015). Study on morphological characteristics of maize (Zea mays L.) cultivars under different plant densities. *Indian Journal of Natural Sciences*, *5*(30), 34–56.

Feng, L., Chen, S., Zhang, C., Zhang, Y., & He, Y. (2021). A comprehensive review on recent applications of unmanned aerial vehicle remote sensing with various sensors for high-throughput plant phenotyping. *Computers and Electronics in Agriculture*, *182*, 106033. DOI: 10.1016/j.compag.2021.106033

Fu, P., Montes, C. M., Siebers, M. H., Gomez-Casanovas, N., McGrath, J. M., Ainsworth, E. A., & Bernacchi, C. J. (2022). Advances in field-based high-throughput photosynthetic phenotyping. *Journal of Experimental Botany*, *73*(10), 3157–3172. DOI: 10.1093/jxb/erac077 PMID: 35218184

Fu, Y., Zhang, D., Gleeson, M., Zhang, Y., Lin, B., Hua, S., Ding, H., Frauen, M., Li, J., Qian, W., & Yu, H. (2017). Analysis of QTL for seed oil content in Brassica napus by association mapping and QTL mapping. *Euphytica*, *213*(1), 1–15. DOI: 10.1007/s10681-016-1817-9

Gaber, M. A. F. M., Tujillo, F. J., Mansour, M. P., & Juliano, P. (2018). Improving oil extraction from canola seeds by conventional and advanced methods. *Food Engineering Reviews*, *10*(4), 198–210. DOI: 10.1007/s12393-018-9182-1

Gazave, E., Tassone, E. E., Ilut, D. C., Wingerson, M., Datema, E., Witsenboer, H. M., Davis, J. B., Grant, D., Dyer, J. M., Jenks, M. A., Brown, J., & Gore, M. A. (2016). Population genomic analysis reveals differential evolutionary histories and patterns of diversity across subgenomes and subpopulations of Brassica napus L. *Frontiers in Plant Science*, *7*, 525. DOI: 10.3389/fpls.2016.00525 PMID: 27148342

Gill, S. S., Gill, R., Kumar, G., Pareek, A., Sharma, P. C., Anjum, N. A., & Tuteja, N. (2012). Mustard: approaches for crop improvement and abiotic stress tolerance. *Improving crop resistance to abiotic stress*, 1351-1368.

Gu, J., Guan, Z., Jiao, Y., Liu, K., & Hong, D. (2024). The story of a decade: Genomics, functional genomics and molecular breeding in Brassica napus. *Plant Communications*, *5*(4), 100884. DOI: 10.1016/j.xplc.2024.100884 PMID: 38494786

Haworth, M., Marino, G., Atzori, G., Fabbri, A., Daccache, A., Killi, D., Carli, A., Montesano, V., Conte, A., Balestrini, R., & Centritto, M. (2023). Plant Physiological Analysis to Overcome Limitations to Plant Phenotyping. *Plants*, *12*(23), 4015. DOI: 10.3390/plants12234015 PMID: 38068650

Herselman, L., Thwaites, R., Kimmins, F., Courtois, B., Van Der Merwe, P., & Seal, S. E. (2004). Identification and mapping of AFLP markers linked to peanut (Arachis hypogaea L.) resistance to the aphid vector of groundnut rosette disease. *Theoretical and Applied Genetics*, *109*(7), 1426–1433. DOI: 10.1007/s00122-004-1756-z PMID: 15290049

Hussain, M., Khan, M. A., Ali, Y., Javaid, M. M., Iqbal, B., Nasir, M., Sabir, W., & Muhammad, F. (2017). Wheat breeding for durable rust resistance and high yield potential in historical prospective and current status. *Advances in Zoology and Botany*, *5*(4), 55–63. DOI: 10.13189/azb.2017.050404

Jamaloddin, M., Mahender, A., Gokulan, C. G., Balachiranjeevi, C., Maliha, A., Patel, H. K., & Ali, J. (2021). Molecular approaches for disease resistance in rice. In *Rice improvement: Physiological, molecular breeding and genetic perspectives* (pp. 315–378). Springer International Publishing Cham. DOI: 10.1007/978-3-030-66530-2_10

Jamaloddin, M., Maliha, A., Gokulan, C., Gaur, N., & Patel, H. K. (2021). Metabolomics-assisted breeding for crop improvement: An emerging approach. *Omics Technologies for Sustainable Agriculture and Global Food Security*, *1*, 241–279. DOI: 10.1007/978-981-16-0831-5_11

Jaramillo-Botero, A., Colorado, J., Quimbaya, M., Rebolledo, M. C., Lorieux, M., Ghneim-Herrera, T., Arango, C. A., Tobón, L. E., Finke, J., Rocha, C., Muñoz, F., Riascos, J. J., Silva, F., Chirinda, N., Caccamo, M., Vandepoele, K., & Goddard, W. A.III. (2022). The ÓMICAS alliance, an international research program on multi-omics for crop breeding optimization. *Frontiers in Plant Science*, *13*, 992663. DOI: 10.3389/fpls.2022.992663 PMID: 36311093

Javed, N., Geng, J., Tahir, M., McVetty, P., Li, G., & Duncan, R. W. (2016). Identification of QTL influencing seed oil content, fatty acid profile and days to flowering in Brassica napus L. *Euphytica*, *207*(1), 191–211. DOI: 10.1007/s10681-015-1565-2

Jiang, F., & Doudna, J. A. (2017). CRISPR–Cas9 structures and mechanisms. *Annual Review of Biophysics*, *46*(1), 505–529. DOI: 10.1146/annurev-biophys-062215-010822 PMID: 28375731

Jimenez-Lopez, J. C., Gachomo, E. W., Sharma, S., & Kotchoni, S. O. (2013). Genome sequencing and next-generation sequence data analysis: A comprehensive compilation of bioinformatics tools and databases.

Jovovic, Z., Andjelkovic, V., Przulj, N., & Mandic, D. (2020). Untapped genetic diversity of wild relatives for crop improvement. *Rediscovery of genetic and genomic resources for future food security*, 25-65.

Kaur, P., Singh, A., & Chana, I. (2021). Computational techniques and tools for omics data analysis: State-of-the-art, challenges, and future directions. *Archives of Computational Methods in Engineering*, *28*(7), 4595–4631. DOI: 10.1007/s11831-021-09547-0 PMID: 33519179

Kaur, S., Cogan, N., Ye, G., Baillie, R., Hand, M., Ling, A., Mcgearey, A., Kaur, J., Hopkins, C., Todorovic, M., Mountford, H., Edwards, D., Batley, J., Burton, W., Salisbury, P., Gororo, N., Marcroft, S., Kearney, G., Smith, K. F., & Spangenberg, G. C. (2009). Genetic map construction and QTL mapping of resistance to blackleg (Leptosphaeria maculans) disease in Australian canola (Brassica napus L.) cultivars. *Theoretical and Applied Genetics*, *120*(1), 71–83. DOI: 10.1007/s00122-009-1160-9 PMID: 19821065

Kehoe, B., Patil, S., Abbeel, P., & Goldberg, K. (2015). A survey of research on cloud robotics and automation. *IEEE Transactions on Automation Science and Engineering*, *12*(2), 398–409. DOI: 10.1109/TASE.2014.2376492

Kim, K.-S., Diers, B., Hyten, D., Rouf Mian, M., Shannon, J., & Nelson, R. L. (2012). Identification of positive yield QTL alleles from exotic soybean germplasm in two backcross populations. *Theoretical and Applied Genetics*, *125*(6), 1353–1369. DOI: 10.1007/s00122-012-1944-1 PMID: 22869284

Körber, N., Bus, A., Li, J., Parkin, I. A., Wittkop, B., Snowdon, R. J., & Stich, B. (2016). Agronomic and seed quality traits dissected by genome-wide association mapping in Brassica napus. *Frontiers in Plant Science*, *7*, 386. DOI: 10.3389/fpls.2016.00386 PMID: 27066036

Kumar, A., Singh, V., & Priyamedha. (2022). Botanical Descriptions of Brassica juncea: Taxonomy, Cytology, Cytogenetics and Phylogenetic Relationships. In *The Brassica juncea Genome* (pp. 15-28). Springer.

Kumar, B. S., Jeannie, S., Doppalapudi, V. R., & Kumar, A. A. REVOLUTIONIZING CROP PHENOTYPING: BREAKTHROUGHS IN HIGH-THROUGHPUT METHODS AND THEIR MULTIFACETED IMPLEMENTATIONS. *Advances in Agronomy Vol.*

Kumar, J., Pratap, A., & Kumar, S. (2015). Plant phenomics: an overview. *Phenomics in crop plants: trends, options and limitations*, 1-10.

Langridge, P., & Fleury, D. (2011). Making the most of 'omics' for crop breeding. *Trends in Biotechnology*, *29*(1), 33–40. DOI: 10.1016/j.tibtech.2010.09.006 PMID: 21030098

Lee, C. (2022). Towards the genetic architecture of complex gene expression traits: Challenges and prospects for eQTL mapping in humans. *Genes*, *13*(2), 235. DOI: 10.3390/genes13020235 PMID: 35205280

Lee, H., Chawla, H. S., Obermeier, C., Dreyer, F., Abbadi, A., & Snowdon, R. (2020). Chromosome-scale assembly of winter oilseed rape Brassica napus. *Frontiers in Plant Science*, *11*, 496. DOI: 10.3389/fpls.2020.00496 PMID: 32411167

Li, B., & Ritchie, M. D. (2021). From GWAS to gene: Transcriptome-wide association studies and other methods to functionally understand GWAS discoveries. *Frontiers in Genetics*, *12*, 713230. DOI: 10.3389/fgene.2021.713230 PMID: 34659337

Liu, X., & Locasale, J. W. (2017). Metabolomics: A primer. *Trends in Biochemical Sciences*, *42*(4), 274–284. DOI: 10.1016/j.tibs.2017.01.004 PMID: 28196646

Lowe, R., Shirley, N., Bleackley, M., Dolan, S., & Shafee, T. (2017). Transcriptomics technologies. *PLoS Computational Biology*, *13*(5), e1005457. DOI: 10.1371/journal.pcbi.1005457 PMID: 28545146

Mahalingam, R. (2017). Phenotypic, physiological and malt quality analyses of US barley varieties subjected to short periods of heat and drought stress. *Journal of Cereal Science*, *76*, 199–205. DOI: 10.1016/j.jcs.2017.06.007

Mahmood, U., Li, X., Fan, Y., Chang, W., Niu, Y., Li, J., Qu, C., & Lu, K. (2022). Multi-omics revolution to promote plant breeding efficiency. *Frontiers in Plant Science*, *13*, 1062952. DOI: 10.3389/fpls.2022.1062952 PMID: 36570904

Malik, M. M. (2023). *Genetic Divergence among Local and Exotic Brassica juncea Germplasm as Revealed by Using Simple Sequence Repeats (SSRs)*. CAPITAL UNIVERSITY.

Mansouripour, S. (2018). *Identification of Brassica napus L. sources of resistance against blackleg (Leptosphaeria maculans)*. North Dakota State University.

McVetty, P. B., Mietkiewska, E., Omonov, T., Curtis, J., Taylor, D. C., & Weselake, R. J. (2016). Brassica spp. oils. In *Industrial oil crops* (pp. 113–156). Elsevier. DOI: 10.1016/B978-1-893997-98-1.00005-1

Melchinger, A. E., Fernando, R., Stricker, C., Schön, C.-C., & Auinger, H.-J. (2023). Genomic prediction in hybrid breeding: I. Optimizing the training set design. *Theoretical and Applied Genetics*, *136*(8), 176. DOI: 10.1007/s00122-023-04413-y PMID: 37532821

Mensah, H., Ahadzie, D. K., Takyi, S. A., & Amponsah, O. (2021). Climate change resilience: Lessons from local climate-smart agricultural practices in Ghana. *Energy, Ecology & Environment*, *6*(3), 271–284. DOI: 10.1007/s40974-020-00181-3

Merrick, L. F., Herr, A. W., Sandhu, K. S., Lozada, D. N., & Carter, A. H. (2022). Optimizing plant breeding programs for genomic selection. *Agronomy (Basel), 12*(3), 714. DOI: 10.3390/agronomy12030714

Mir, R. R., Zaman-Allah, M., Sreenivasulu, N., Trethowan, R., & Varshney, R. K. (2012). Integrated genomics, physiology and breeding approaches for improving drought tolerance in crops. *Theoretical and Applied Genetics, 125*(4), 625–645. DOI: 10.1007/s00122-012-1904-9 PMID: 22696006

Mmbando, G. S. (2024). Omics: A new, promising technologies for boosting crop yield and stress resilience in African agriculture. *Plant Stress, 11*, 100366. DOI: 10.1016/j.stress.2024.100366

Mohd Saad, N. S., Severn-Ellis, A. A., Pradhan, A., Edwards, D., & Batley, J. (2021). Genomics armed with diversity leads the way in Brassica improvement in a changing global environment. *Frontiers in Genetics, 12*, 600789. DOI: 10.3389/fgene.2021.600789 PMID: 33679880

Mora-Oberlaender, J., Jiménez-Barreto, J., Rodríguez-Abril, Y., Estrada-Arteaga, M., & Chaparro-Giraldo, A. (2022). Opportunities for Generic Cisgenic Crops. In *Cisgenic Crops: Potential and Prospects* (pp. 89-119). Springer. DOI: 10.1007/978-3-031-06628-3_6

Nansamba, M., Sibiya, J., Tumuhimbise, R., Karamura, D., Kubiriba, J., & Karamura, E. (2020). Breeding banana (Musa spp.) for drought tolerance: A review. *Plant Breeding, 139*(4), 685–696. DOI: 10.1111/pbr.12812

Nasar, S., Muhammad, M., Siddiqi, E. H., Majeed, M., Ameen, N., Ullah, S., Kousar, S., & Abdullah, M. (2024). OMICS-Based Knowledge for Achieving Food and Nutritional Security. *OMICs-based Techniques for Global Food Security*, 67-90.

Nesatyy, V. J., & Suter, M. J.-F. (2007). Proteomics for the analysis of environmental stress responses in organisms. *Environmental Science & Technology, 41*(20), 6891–6900. DOI: 10.1021/es070561r PMID: 17993125

Ntui, V. O., Tripathi, J. N., Kariuki, S. M., & Tripathi, L. (2024). Cassava molecular genetics and genomics for enhanced resistance to diseases and pests. *Molecular Plant Pathology, 25*(1), e13402. DOI: 10.1111/mpp.13402 PMID: 37933591

Olivos-Trujillo, M., Gajardo, H. A., Salvo, S., González, A., & Muñoz, C. (2015). Assessing the stability of parameters estimation and prediction accuracy in regression methods for estimating seed oil content in Brassica napus L. using NIR spectroscopy. *2015 CHILEAN Conference on Electrical, Electronics Engineering, Information and Communication Technologies (CHILECON)* DOI: 10.1109/Chilecon.2015.7400347

Ortolani, L., Bocci, R., Bàrberi, P., Howlett, S., & Chable, V. (2017). Changes in knowledge management strategies can support emerging innovative actors in organic agriculture: The case of participatory plant breeding in Europe. *Organic Farming*, *3*(1), 20–33. DOI: 10.12924/of2017.03010020

Oyekunle, M., Adamu, R. S., Ndou, E., Beyene, Y., Abdulmalik, M., & Oikeh, S. O. (2023). Efficacy of drought-tolerant and insect-protected transgenic TELA® maize traits in Nigeria. *Transgenic Research*, *32*(3), 169–178. DOI: 10.1007/s11248-023-00345-x PMID: 37043164

Pereira-Lorenzo, S., Fischer, M., Ramos-Cabrer, A. M., & Castro, I. (2018). Apple (Malus spp.) breeding: Present and future. *Advances in Plant Breeding Strategies: Fruits*, *3*, 3–29. DOI: 10.1007/978-3-319-91944-7_1

Potter, T., Burton, W., Edwards, J., Wratten, N., Mailer, R., Salisbury, P., & Pearce, A. (2016). Assessing progress in breeding to improve grain yield, quality and blackleg (Leptosphaeria maculans) resistance in selected Australian canola cultivars (1978–2012). *Crop & Pasture Science*, *67*(4), 308–316. DOI: 10.1071/CP15290

Qu, C., Hasan, M., Lu, K., Liu, L., Liu, X., Xie, J., Wang, M., Lu, J., Odat, N., & Wang, R. (2012). Genetic diversity and relationship analysis of the Brassica napus germplasm using simple sequence repeat (SSR) markers. *African Journal of Biotechnology*, *11*(27), 6923–6933.

Raboanatahiry, N., Li, H., Yu, L., & Li, M. (2021). Rapeseed (Brassica napus): Processing, utilization, and genetic improvement. *Agronomy (Basel)*, *11*(9), 1776. DOI: 10.3390/agronomy11091776

Rahman, H., Bennett, R. A., & Kebede, B. (2017). Mapping of days to flower and seed yield in spring oilseed Brassica napus carrying genome content introgressed from Brassica oleracea. *Molecular Breeding*, *37*(1), 1–15. DOI: 10.1007/s11032-016-0608-2

Rahman, H., Ramanathan, V., Jagadeeshselvam, N., Ramasamy, S., Rajendran, S., Ramachandran, M., Sudheer, P. D., Chauhan, S., Natesan, S., & Muthurajan, R. (2015). Phenomics: technologies and applications in plant and agriculture. *PlantOmics: The Omics of Plant Science*, 385-411.

Rahman, M., Guo, Q., Baten, A., Mauleon, R., Khatun, A., Liu, L., & Barkla, B. J. (2021). Shotgun proteomics of Brassica rapa seed proteins identifies vicilin as a major seed storage protein in the mature seed. *PLoS One*, *16*(7), e0253384. DOI: 10.1371/journal.pone.0253384 PMID: 34242257

Rajczewski, A. T., Jagtap, P. D., & Griffin, T. J. (2022). An overview of technologies for MS-based proteomics-centric multi-omics. *Expert Review of Proteomics*, *19*(3), 165–181. DOI: 10.1080/14789450.2022.2070476 PMID: 35466851

Raman, H., Uppal, R. K., & Raman, R. (2019). Genetic solutions to improve resilience of canola to climate change. *Genomic designing of climate-smart oilseed crops*, 75-131.

Rauf, S., da Silva, J. T., Khan, A. A., & Naveed, A. (2010). Consequences of plant breeding on genetic diversity. *International Journal of Plant Breeding*, *4*(1), 1–21.

Raza, A., Hafeez, M. B., Zahra, N., Shaukat, K., Umbreen, S., Tabassum, J., Charagh, S., Khan, R. S. A., & Hasanuzzaman, M. (2020). The plant family Brassicaceae: Introduction, biology, and importance. *The Plant Family Brassicaceae: Biology and Physiological Responses to Environmental Stresses*, 1-43.

Raza, A., Su, W., Hussain, M. A., Mehmood, S. S., Zhang, X., Cheng, Y., Zou, X., & Lv, Y. (2021). Integrated analysis of metabolome and transcriptome reveals insights for cold tolerance in rapeseed (Brassica napus L.). *Frontiers in Plant Science*, *12*, 721681. DOI: 10.3389/fpls.2021.721681 PMID: 34691103

Razzaq, A., Sadia, B., Raza, A., Khalid Hameed, M., & Saleem, F. (2019). Metabolomics: A way forward for crop improvement. *Metabolites*, *9*(12), 303. DOI: 10.3390/metabo9120303 PMID: 31847393

Razzaq, M. K., Aleem, M., Mansoor, S., Khan, M. A., Rauf, S., Iqbal, S., & Siddique, K. H. (2021). Omics and CRISPR-Cas9 approaches for molecular insight, functional gene analysis, and stress tolerance development in crops. *International Journal of Molecular Sciences*, *22*(3), 1292. DOI: 10.3390/ijms22031292 PMID: 33525517

Riar, A., Gill, G., & McDonald, G. (2020). Different post-sowing nitrogen management approaches required to improve nitrogen and water use efficiency of canola and mustard. *Frontiers in Plant Science*, *11*, 1111. DOI: 10.3389/fpls.2020.01111 PMID: 32793266

Sarkar, A., Kisiala, A., Adhikary, D., Basu, U., Emery, R. N., Rahman, H., & Kav, N. N. (2023). Silicon ameliorates clubroot responses in canola (Brassica napus): A "multi-omics"-based investigation into possible mechanisms. *Physiologia Plantarum*, *175*(2), e13900. DOI: 10.1111/ppl.13900 PMID: 36992551

Shafighi, A., Ardakani, M. R., Rad, A. H. S., Alavifazel, M., & Rafiei, F. (2021). Grain yield and associated physiological traits of rapeseed (Brassica napus L.) cultivars under different planting dates and drought stress at the flowering stage. *Italian Journal of Agronomy*, *16*(1), 1648. DOI: 10.4081/ija.2020.1648

Shah, F., & Wu, W. (2019). Soil and crop management strategies to ensure higher crop productivity within sustainable environments. *Sustainability (Basel), 11*(5), 1485. DOI: 10.3390/su11051485

Shu, J., Ma, X., Ma, H., Huang, Q., Zhang, Y., Guan, M., & Guan, C. (2022). Transcriptomic, proteomic, metabolomic, and functional genomic approaches of Brassica napus L. during salt stress. *PLoS One, 17*(3), e0262587. DOI: 10.1371/journal.pone.0262587 PMID: 35271582

Sidak, D., Schwarzerová, J., Weckwerth, W., & Waldherr, S. (2022). Interpretable machine learning methods for predictions in systems biology from omics data. *Frontiers in Molecular Biosciences, 9*, 926623. DOI: 10.3389/fmolb.2022.926623 PMID: 36387282

Singh, R. K., Rai, N., Singh, A. K., Kumar, P., & Singh, B. (2019). A critical review on Tomato leaf curl virus resistance in tomato. *International Journal of Vegetable Science, 25*(4), 373–393. DOI: 10.1080/19315260.2018.1520379

Singh, R. K., Sood, P., Prasad, A., & Prasad, M. (2021). Advances in omics technology for improving crop yield and stress resilience. *Plant Breeding, 140*(5), 719–731. DOI: 10.1111/pbr.12963

Smith, S. (2019). The foundations, continuing evolution, and outcomes from the application of intellectual property protection in plant breeding and agriculture. *Plant Breeding Reviews, 43*, 121–213. DOI: 10.1002/9781119616801.ch5

So, K. K., & Duncan, R. W. (2021). Breeding canola (Brassica napus L.) for protein in feed and food. *Plants, 10*(10), 2220. DOI: 10.3390/plants10102220 PMID: 34686029

Sobejano-Paz, V., Mikkelsen, T. N., Baum, A., Mo, X., Liu, S., Köppl, C. J., Johnson, M. S., Gulyas, L., & García, M. (2020). Hyperspectral and thermal sensing of stomatal conductance, transpiration, and photosynthesis for soybean and maize under drought. *Remote Sensing (Basel), 12*(19), 3182. DOI: 10.3390/rs12193182

Song, J.-M., Guan, Z., Hu, J., Guo, C., Yang, Z., Wang, S., Liu, D., Wang, B., Lu, S., Zhou, R., Xie, W.-Z., Cheng, Y., Zhang, Y., Liu, K., Yang, Q.-Y., Chen, L.-L., & Guo, L. (2020). Eight high-quality genomes reveal pan-genome architecture and ecotype differentiation of Brassica napus. *Nature Plants, 6*(1), 34–45. DOI: 10.1038/s41477-019-0577-7 PMID: 31932676

Sotelo, T., Soengas, P., Velasco, P., Rodríguez, V. M., & Cartea, M. E. (2014). Identification of metabolic QTLs and candidate genes for glucosinolate synthesis in Brassica oleracea leaves, seeds and flower buds. *PLoS One, 9*(3), e91428. DOI: 10.1371/journal.pone.0091428 PMID: 24614913

Subbarao, G., Levine, L. H., Wheeler, R. M., & Stutte, G. W. (2001). Glycine betaine accumulation: its role in stress resistance in crop plants. In *Handbook of plant and crop physiology* (pp. 903–930). CRC Press. DOI: 10.1201/9780203908426.ch45

Sun, F., Fan, G., Hu, Q., Zhou, Y., Guan, M., Tong, C., Li, J., Du, D., Qi, C., Jiang, L., Liu, W., Huang, S., Chen, W., Yu, J., Mei, D., Meng, J., Zeng, P., Shi, J., Liu, K., & Wang, H. (2017). The high-quality genome of Brassica napus cultivar 'ZS 11'reveals the introgression history in semi-winter morphotype. *The Plant Journal*, *92*(3), 452–468. DOI: 10.1111/tpj.13669 PMID: 28849613

Tagkouli, V. (2021). *Evaluation of recurrent selection as a method to achieve rapid re-adaptation of faba bean to a niche agro-climate*. University of Reading.

Tahir ul Qamar, M., Faryad, A., Bari, A., Zahid, B., Zhu, X., & Chen, L.-L. (2020). Effectiveness of conventional crop improvement strategies vs. omics. *Environment, climate, plant and vegetation growth*, 253-284.

Tang, T., Chen, G., Bu, C., Liu, F., Liu, L., & Zhao, X. (2018). Transgene introgression from Brassica napus to different varieties of Brassica juncea. *Plant Breeding*, *137*(2), 171–180. DOI: 10.1111/pbr.12567

Tardieu, F., Cabrera-Bosquet, L., Pridmore, T., & Bennett, M. (2017). Plant phenomics, from sensors to knowledge. *Current Biology*, *27*(15), R770–R783. DOI: 10.1016/j.cub.2017.05.055 PMID: 28787611

Tavares, S., Ramos, A. P., Pires, A. S., Azinheira, H. G., Caldeirinha, P., Link, T., Abranches, R., Silva, M. C., Voegele, R. T., Loureiro, J., & Talhinhas, P. (2014). Genome size analyses of Pucciniales reveal the largest fungal genomes. *Frontiers in Plant Science*, *5*, 422. DOI: 10.3389/fpls.2014.00422 PMID: 25206357

Tocher, D. R., Betancor, M. B., Sprague, M., Olsen, R. E., & Napier, J. A. (2019). Omega-3 long-chain polyunsaturated fatty acids, EPA and DHA: Bridging the gap between supply and demand. *Nutrients*, *11*(1), 89. DOI: 10.3390/nu11010089 PMID: 30621155

Ton, L. B., Neik, T. X., & Batley, J. (2020). The use of genetic and gene technologies in shaping modern rapeseed cultivars (Brassica napus L.). *Genes*, *11*(10), 1161. DOI: 10.3390/genes11101161 PMID: 33008008

Tonu, N. N., Wen, R., Song, T., Guo, X., Murphy, L. A., Gossen, B. D., Yu, F., & Peng, G. (2023). Canola with stacked genes shows moderate resistance and resilience against a field population of Plasmodiophora brassicae (clubroot) pathotype X. *Plants*, *12*(4), 726. DOI: 10.3390/plants12040726 PMID: 36840074

Vanhercke, T., Dyer, J. M., Mullen, R. T., Kilaru, A., Rahman, M. M., Petrie, J. R., Green, A. G., Yurchenko, O., & Singh, S. P. (2019). Metabolic engineering for enhanced oil in biomass. *Progress in Lipid Research, 74*, 103–129. DOI: 10.1016/j.plipres.2019.02.002 PMID: 30822461

Verma, R. P. S., Lal, C., Malik, R., Kharub, A. S., Kumar, L., & Kumar, D. (2022). Barley improvement: current status and future prospects in changing scenario. *New Horizons in Wheat and Barley Research: Global Trends, Breeding and Quality Enhancement*, 93-134.

Wang, E., He, D., Wang, J., Lilley, J. M., Christy, B., Hoffmann, M. P., O'Leary, G., Hatfield, J. L., Ledda, L., Deligios, P. A., Grant, B., Jing, Q., Nendel, C., Kage, H., Qian, B., Eyshi Rezaei, E., Smith, W., Weymann, W., & Ewert, F. (2022). How reliable are current crop models for simulating growth and seed yield of canola across global sites and under future climate change? *Climatic Change, 172*(1), 20. DOI: 10.1007/s10584-022-03375-2

Wang, J., Kong, L., Zhang, L., Shi, X., Yu, B., Li, J., Zhang, B., Gao, M., Liu, X., Li, X., Gao, Y., Peng, D., & Liu, S. (2022). Breeding a Soybean Cultivar Heinong 531 with Peking-Type Cyst Nematode Resistance, Enhanced Yield, and High Seed-Oil Contents. *Phytopathology, 112*(6), 1345–1349. DOI: 10.1094/PHYTO-08-21-0347-R PMID: 34879718

Wang, J., Song, L., Gong, X., Xu, J., & Li, M. (2020). Functions of jasmonic acid in plant regulation and response to abiotic stress. *International Journal of Molecular Sciences, 21*(4), 1446. DOI: 10.3390/ijms21041446 PMID: 32093336

Wang, W., Pang, J., Zhang, F., Sun, L., Yang, L., Zhao, Y., Yang, Y., Wang, Y., & Siddique, K. H. (2021). Integrated transcriptomics and metabolomics analysis to characterize alkali stress responses in canola (Brassica napus L.). *Plant Physiology and Biochemistry, 166*, 605–620. DOI: 10.1016/j.plaphy.2021.06.021 PMID: 34186284

Wang, X. (2016). Broadening of genetic diversity in spring canola (Brassica napus L.) by use of the C genome of Brassica oleracea var. italica and Brassica oleracea var. capitata.

Wang, Y., Gu, W., Meng, Y., Xie, T., Li, L., Li, J., & Wei, S. (2017). γ-Aminobutyric acid imparts partial protection from salt stress injury to maize seedlings by improving photosynthesis and upregulating osmoprotectants and antioxidants. *Scientific Reports, 7*(1), 43609. DOI: 10.1038/srep43609 PMID: 28272438

Wei, J., Zheng, G., Yu, X., Liu, S., Dong, X., Cao, X., Fang, X., Li, H., Jin, J., Mi, W., & Liu, Z. (2021). Comparative transcriptomics and proteomics analyses of leaves reveals a freezing stress-responsive molecular network in winter rapeseed (Brassica rapa L.). *Frontiers in Plant Science*, *12*, 664311. DOI: 10.3389/fpls.2021.664311 PMID: 33995460

Wei, X., Eglinton, J., Piperidis, G., Atkin, F., Morgan, T., Parfitt, R., & Hu, F. (2022). Sugarcane breeding in Australia. *Sugar Tech*, *24*(1), 151–165. DOI: 10.1007/s12355-021-00969-5

Welter, K. C., Martins, C. M. M. R., de Palma, A. S. V., Martins, M. M., Dos Reis, B. R., Schmidt, B. L. U., & Saran Netto, A. (2016). Canola oil in lactating dairy cow diets reduces milk saturated fatty acids and improves its omega-3 and oleic fatty acid content. *PLoS One*, *11*(3), e0151876. DOI: 10.1371/journal.pone.0151876 PMID: 27015405

Wen, R., Song, T., Gossen, B. D., & Peng, G. (2024). Comparative transcriptome analysis of canola carrying a single vs stacked resistance genes against clubroot. *Frontiers in Plant Science*, *15*, 1358605. DOI: 10.3389/fpls.2024.1358605 PMID: 38835867

Woodfield, H. K., Cazenave-Gassiot, A., Haslam, R. P., Guschina, I. A., Wenk, M. R., & Harwood, J. L. (2018). Using lipidomics to reveal details of lipid accumulation in developing seeds from oilseed rape (Brassica napus L.). *Biochimica et Biophysica Acta (BBA)-. Biochimica et Biophysica Acta. Molecular and Cell Biology of Lipids*, *1863*(3), 339–348. DOI: 10.1016/j.bbalip.2017.12.010 PMID: 29275220

Wortmann, S. B., Oud, M. M., Alders, M., Coene, K. L., van der Crabben, S. N., Feichtinger, R. G., Garanto, A., Hoischen, A., Langeveld, M., Lefeber, D., Mayr, J. A., Ockeloen, C. W., Prokisch, H., Rodenburg, R., Waterham, H. R., Wevers, R. A., van de Warrenburg, B. P. C., Willemsen, M. A. A. P., Wolf, N. I., & van Karnebeek, C. D. M. (2022). How to proceed after "negative" exome: A review on genetic diagnostics, limitations, challenges, and emerging new multiomics techniques. *Journal of Inherited Metabolic Disease*, *45*(4), 663–681. DOI: 10.1002/jimd.12507 PMID: 35506430

Wu, D., Liang, Z., Yan, T., Xu, Y., Xuan, L., Tang, J., Zhou, G., Lohwasser, U., Hua, S., Wang, H., Chen, X., Wang, Q., Zhu, L., Maodzeka, A., Hussain, N., Li, Z., Li, X., Shamsi, I. H., Jilani, G., & Jiang, L. (2019). Whole-genome resequencing of a worldwide collection of rapeseed accessions reveals the genetic basis of ecotype divergence. *Molecular Plant*, *12*(1), 30–43. DOI: 10.1016/j.molp.2018.11.007 PMID: 30472326

Wu, W., Ma, B. L., & Whalen, J. K. (2018). Enhancing rapeseed tolerance to heat and drought stresses in a changing climate: Perspectives for stress adaptation from root system architecture. *Advances in Agronomy*, *151*, 87–157. DOI: 10.1016/bs.agron.2018.05.002

Würschum, T., Maurer, H. P., Dreyer, F., & Reif, J. C. (2013). Effect of inter-and intragenic epistasis on the heritability of oil content in rapeseed (Brassica napus L.). *Theoretical and Applied Genetics*, *126*(2), 435–441. DOI: 10.1007/s00122-012-1991-7 PMID: 23052025

Yadava, D., Vasudev, S., Singh, N., Mohapatra, T., & Prabhu, K. (2012). Breeding major oil crops: Present status and future research needs. *Technological innovations in major world oil crops, volume 1: breeding*, 17-51.

Yandeau-Nelson, M. D., Lauter, N., & Zabotina, O. A. (2016). Advances in metabolomic applications in plant genetics and breeding. *CABI Reviews*(2015), 1-15.

Yang, Y., Saand, M. A., Huang, L., Abdelaal, W. B., Zhang, J., Wu, Y., Li, J., Sirohi, M. H., & Wang, F. (2021). Applications of multi-omics technologies for crop improvement. *Frontiers in Plant Science*, *12*, 563953. DOI: 10.3389/fpls.2021.563953 PMID: 34539683

Zhang, X., & Fernando, W. D. (2017). Insights into fighting against blackleg disease of Brassica napus in Canada. *Crop & Pasture Science*, *69*(1), 40–47. DOI: 10.1071/CP16401

Zhang, Y., Ali, U., Zhang, G., Yu, L., Fang, S., Iqbal, S., Li, H., Lu, S., & Guo, L. (2019). Transcriptome analysis reveals genes commonly responding to multiple abiotic stresses in rapeseed. *Molecular Breeding*, *39*(10-11), 1–19. DOI: 10.1007/s11032-019-1052-x

Zhao, J., Dimov, Z., Becker, H. C., Ecke, W., & Möllers, C. (2008). Mapping QTL controlling fatty acid composition in a doubled haploid rapeseed population segregating for oil content. *Molecular Breeding*, *21*(1), 115–125. DOI: 10.1007/s11032-007-9113-y

Zhao, W., Wang, X., Wang, H., Tian, J., Li, B., Chen, L., Chao, H., Long, Y., Xiang, J., Gan, J., Liang, W., & Li, M. (2016). Genome-wide identification of QTL for seed yield and yield-related traits and construction of a high-density consensus map for QTL comparison in Brassica napus. *Frontiers in Plant Science*, *7*, 17. DOI: 10.3389/fpls.2016.00017 PMID: 26858737

Zhou, Q., Zhou, C., Zheng, W., Mason, A. S., Fan, S., Wu, C., Fu, D., & Huang, Y. (2017). Genome-wide SNP markers based on SLAF-seq uncover breeding traces in rapeseed (Brassica napus L.). *Frontiers in Plant Science*, *8*, 648. DOI: 10.3389/fpls.2017.00648 PMID: 28503182

Zhu, M., Monroe, J. G., Suhail, Y., Villiers, F., Mullen, J., Pater, D., Hauser, F., Jeon, B. W., Bader, J. S., Kwak, J. M., Schroeder, J. I., McKay, J. K., & Assmann, S. M. (2016). Molecular and systems approaches towards drought-tolerant canola crops. *The New Phytologist*, *210*(4), 1169–1189. DOI: 10.1111/nph.13866 PMID: 26879345

Zou, J., Mao, L., Qiu, J., Wang, M., Jia, L., Wu, D., He, Z., Chen, M., Shen, Y., Shen, E., Huang, Y., Li, R., Hu, D., Shi, L., Wang, K., Zhu, Q., Ye, C., Bancroft, I., King, G. J., & Fan, L. (2019). Genome-wide selection footprints and deleterious variations in young Asian allotetraploid rapeseed. *Plant Biotechnology Journal*, *17*(10), 1998–2010. DOI: 10.1111/pbi.13115 PMID: 30947395

Chapter 14
Proteomics of Canola Plant

Rooma Waqar
http://orcid.org/0009-0009-1654-2860
Quaid-i-Azam University, Islamabad, Pakistan

Javed Iqbal
http://orcid.org/0000-0002-9032-2622
Bacha Khan University, Charsadda, Pakistan

Abdul Samad Mumtaz
http://orcid.org/0000-0002-7034-9203
Quaid-i-Azam University, Islamabad, Pakistan

Lubna Anjum Minhas
Quaid-i-Azam University, Islamabad, Pakistan

Shumaila Ijaz
Shenzhen University, China

Zakir Ullah
Quaid-i-Azam University, Islamabad, Pakistan

Ghulam Murtaza
Yunnan University, Kunming, China

Rashid Iqbal
The Islamia University of Bahawalpur, Pakistan

ABSTRACT

Proteomics, or the large-scale study of proteins and their functions, is an effective method for improving our understanding of the canola plant (Brassica napus). This study focusses on the use of proteomics in canola research, namely in growth, development, and responses to biotic and abiotic challenges. Proteomics allows for the identification and measurement of proteins implicated in essential metabolic pathways, stress responses, and disease resistance, yielding insights that can lead to the development of more resilient and productive canola varieties. Furthermore, the study of post-translational modifications and comparative proteomics provide crucial insights into protein regulation and evolutionary links. These analyses rely

DOI: 10.4018/979-8-3693-8603-3.ch014

heavily on techniques like mass spectrometry, 2D gel electrophoresis, and LC-MS/MS. The combination of proteomics and bioinformatics technologies makes it easier to comprehend complex data, hence advancing marker-assisted breeding, genetic engineering, and nutritional enhancement.

INTRODUCTION

The *Brassica* genus includes a wide variety of plant species that have been adapted to be used as turnips, broccoli, Brussels sprouts, cabbage, and cauliflower (all under the *Brassica oleracea* species), as well as for seed oil production (*B. napus*, *B. juncea* and *B. rapa*) and condiments (*B. nigra*, *B. juncea* and *B. carinata*,). *B. rapa*, *B. juncea* and *B. napus* are the principal oilseed crops, whereas *B. nigra* is a prominent condiment within the *Brassica* genus (Fan et al., 2021). *B. juncea*, and *B. napus* which contain larger quantities of erucic acid, are commonly employed in oleochemical businesses because derivatives of erucic acid serve as vital renewable raw materials. These plants' oils and byproducts are used to make coating agents, printing inks, polymers, lubricants, soaps, and surfactants, among other things. Furthermore, *B. carinata* exhibits significant promise for usage as a crop for producing industrial oil and biodiesel (Seepaul et al., 2021).

B. juncea has great promise for phytoextraction of metals from soil among the *Brassica* species. This capacity has been used to bioaccumulate heavy and toxic metals, assisting in the phytoremediation process to reduce soil contaminants and make the area acceptable for farming (Szczygłowska, Piekarska, Konieczka, & Namieśnik, 2011). Increasing yield is the main goal of many *Brassica* breeding initiatives across the world, which have had considerable success using both traditional breeding techniques like hybridisation and selection as well as molecular marker-assisted breeding programs. Thanks to recent developments in high-throughput sequencing, the genomes of all six Brassica species have been sequenced. With a genome size of 238.8 Mb and 41,174 genes (Wang XiaoWu et al., 2011), *B. rapa* was the first species to be sequenced. It was followed by *B. oleracea*, whose genome size was 630 Mb and had 45,758 genes. *B. carinata*, which has a very large genome of 1.087 Gb with 97,149 predicted genes, was the last to be sequenced (X. Song et al., 2021). Through the identification and development of many markers made possible by these sequencing efforts, researchers have been able to piece together the genetic landscape of loci linked to a variety of agronomic parameters as well as phenotypes resulting from abiotic and biotic pressures. Plant breeders now have access to an unparalleled quantity of loci for genetic modification as a result. However, comprehending the geographical and temporal expression, roles, and interactions of gene products is necessary for the efficient use of this knowledge. Due to these requirements, the utility of genomics and transcriptomics achievements is limited, as transcript levels only offer approximations of the amounts of proteins that are available. Lower protein yields can be caused by either insufficient translation or rapid degradation of highly transcribed mRNAs (Feder & Walser, 2005). Moreover, unlike the genome, the proteome is not as homogeneous or stable. Genes are made up of four nucleotides that are sequenced the same in all cells, whereas proteins are made up of several amino acids that are frequently altered post-translationally by the addition of sugars, lipids, and other substances. Additionally, the set of proteins varies based on developmental stage, tissue type and various environmental signals. Proteins can also arise from a single transcript due to alternative splicing (A. Singh & Roychoudhury, 2021).

Focussing on the proteome is crucial to understanding the mechanisms governing the phenotypic of traits and processes since proteins are the functional products of genes and act as the main metabolic mediators that link transcript expression to cellular metabolism. Cataloguing and measuring every protein present in a cell, tissue, or organelle at a given time is known as a proteomic study. It's also essential to identify proteins and measure them in certain cells or tissues under particular circumstances. Proteomic studies rely heavily on PTMs as essential mechanisms for controlling and diversifying the cellular proteome according to developmental and environmental circumstances. When compared to other species such as tomato, tobacco, and rice, *Arabidopsis thaliana*'s proteome is the best characterised among model plant organisms (Y. S. Kwon et al., 2010). This chapter provides a comprehensive report on various proteomic investigations undertaken in different *Brassica* species, even if some prior reviews have examined proteomes in *Brassica* under biotic and abiotic stresses (Neik, Amas, Barbetti, Edwards, & Batley, 2020) (Raza et al., 2021).

METHODS FOR PROTEIN EXTRACTION AND ANALYSIS IN *BRASSICA*

In proteomic investigations, the extraction of proteins from plants is an essential step since the quality and integrity of the proteins have a big influence on the analyses that come after. Protein solubilisation is maximised, modifications are limited, and protein breakdown is minimised in an optimal extraction process. However, due to the presence of substances such phenolics, organic acids, polysaccharides, and pigments, extracting proteins from plants is more difficult than from other systems (Balotf, Wilson, Tegg, Nichols, & Wilson, 2022). These substances may cause problems for proteins during their extraction, purification, and identification processes. Trichloroacetic acid (TCA) and acetone are commonly used in *Brassica* proteomics to precipitate proteins, which is then followed by phenol extraction (Damerval, De Vienne, Zivy, & Thiellement, 1986). By successfully eliminating interfering chemicals and inhibiting plant proteases, this method improves the stability of the proteins that have been extracted. Although this is a simple procedure, it may make handling the dissolved proteins challenging (Sixue Chen & Harmon, 2006). Similarly, nucleic acids, polysaccharides, and ions are also eliminated during the protein extraction process that uses phenol and methanol. When opposed to the TCA/acetone approach, the method of phenol-methanol extraction typically produces samples of greater quality, however some proteins may be lost in the process. The phenol-based approach is thought to be superior in terms of resolution in two-dimensional gel electrophoresis (2-DE); however, because of its toxicity and lengthy processing time, the TCA/acetone method is frequently used for different proteomics investigations (Wu, Gong, & Wang, 2014). From 2-DGE to sophisticated large-scale methods like liquid chromatography-mass spectrometry (LC-MS) (Agrawal, Yonekura, Iwahashi, Iwahashi, & Rakwal, 2005) and matrix-assisted laser desorption/ionization time-of-flight mass spectrometry (MALDI-TOF/TOF MS), proteomic analysis has advanced dramatically. The emergence of MS-based proteomics in plants has mitigated the necessity of producing antibodies against particular proteins, which has been a significant obstacle in plant proteomics because of the scarcity of antibodies in contrast to animal systems (Mergner & Kuster, 2022). Multiplexing in proteome analysis is made easier by high-throughput techniques such as label-free proteomics and isobaric tagging for relative and absolute quantification (iTRAQ). Taken as a whole, these techniques have transformed protein identification, quantification, and functional analysis. **Figure 1** shows an overview of proteomics methods. Applications of proteomics techniques presented in **figure 2.**

Figure 1. An overview of proteomics methods

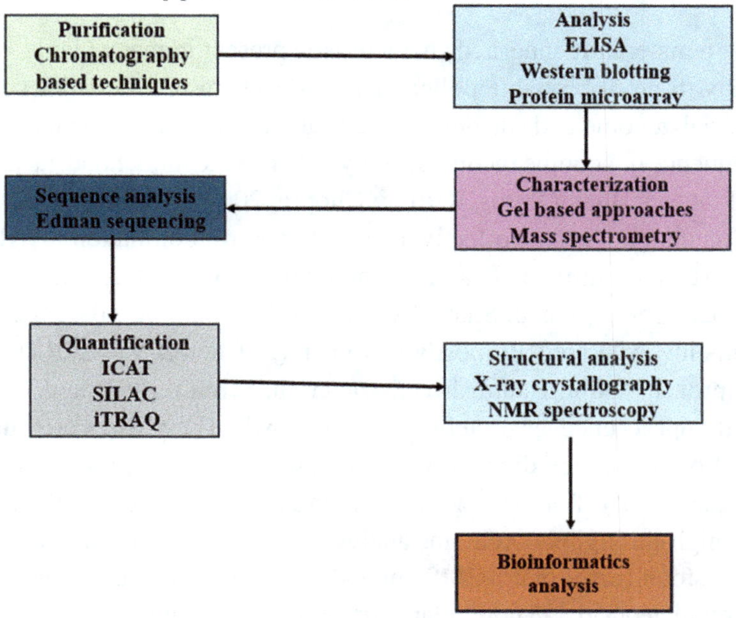

Figure 2. Applications of proteomics techniques encompass a wide range of fields

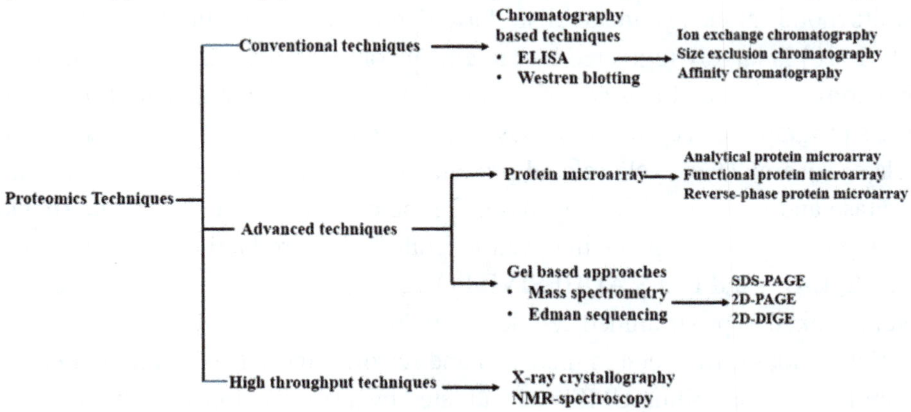

CYTOPLASMIC MALE STERILITY

CMS is a maternally inherited characteristic present in many plants, defined by the inability to produce viable pollen. The CMS characteristic is encoded by the mitochondrial genome and can develop naturally through mutational alterations or as a consequence of genome incompatibility between the mitochondrial and nuclear systems (Prakash, Bhat, Quiros, Kirti, & Chopra, 2009). Because the CMS trait is simple to maintain, it can be effectively used to control pollination in hybrid crops that are vital to agriculture. In Brassica, a number of CMS systems have been found, including ogu, pol, nap, hau, Shaan2A, and orf 220; however, the precise protein components involved are still mostly unknown (L. Chen & Liu, 2014). Using the 2-DGE approach, (Mihr, Baumgärtner, Dieterich, Schmitz, & Braun, 2001) made the first attempt to uncover proteins linked to CMS. They discovered unique mitochondrial proteins, including as ATP synthase, heat shock proteins, and NADH dehydrogenase, that differed between CMS and fertile lines. Using the iTRAQ approach for high-throughput proteome analysis, 833 differentially abundant proteins (DAPs) were found in the ogura CMS line and its restorer line in cabbage, comprising 295 downregulated and 538 upregulated proteins. Interestingly, proteins involved in sporopollenin production and lipid transfer were downregulated in the CMS line, but cysteine ABC transporters were increased in the restorer line. Key DAPs involved in protein processing in the endoplasmic reticulum (ER) and all identified heat shock proteins were upregulated in the CMS line (Xing et al., 2018). Subsequent studies on mitochondrial energy metabolism have demonstrated that the CMS trait (Han et al., 2018) is largely regulated by proteins involved in the tetrad wall and pollen exine construction and degradation, as well as those that are essential for anther tissue's programmed cell death. It was discovered that the CMS line had increased levels of several TCA cycle-related proteins, including citrate synthase, aconitate hydratase and pyruvate dehydrogenase. On the other hand, the CMS line showed downregulation of the genes involved in pollen wall production, CYP703A/CYP703A2, LAP6 and LESS ADHESIVE POLLEN 5 (LAP5). Different increase of proteins linked to programmed cell death, including as peroxidase, catalase-3, and glycolate oxidase, was seen in the CMS and restorer lines. The function of energy metabolism in controlling CMS was validated by a study comparing the proteome alterations of the Shaan2A CMS line to that of the near-isogenic restoration line Shaan2B in B. napus. DAPs are mostly linked to energy and carbohydrate metabolism (Ning et al., 2019). The LAP6 and CYP703A2 proteins were found in this study; these were expressed in anthers and connected to the formation of pollen but were downregulated in the CMS line. In *Arabidopsis*, knockout mutants for these genes displayed incomplete CMS along with faulty exine (Kim et al., 2010). These investigations showed that the transcription factors MYB80, which are essential for the development of anther, were expressed differently in the CMS and restorer lines. 186 DAPs were found between the maintainer line and the hau CMS line in a different iTRAQ-based proteomic study of hau type CMS in *B. napus* flower buds (Heng et al., 2019). Proteins that associated with stress (oxidative), resistance to death of cell, and protein processing were more prevalent in the maintenance line compared to the CMS line, which displayed a decreased accumulation of these proteins. Comparing the CMS line to the maintenance line, less proteins, including pyruvate kinase (Rippka, Deruelles, Waterbury, Herdman, & Stanier, 1979), malate dehydrogenase (MDH), and NADH dehydrogenase, were present. Additionally, proteins including adenine phosphoribosyl transferase and esterase known to be responsible for the CMS characteristic, were downregulated. Enzymes such as peroxidase, SOD, and monodehydroascorbate reductase (MDHCR) were also over-expressed in the CMS line, suggesting an overexpression of stress-related molecules and ROS. All of the proteomic investigations' findings point to the possibility that reduced energy output and ROS buildup in CMS lines could damage mitochondrial function and cause male sterility.

Proteomics of Floral Organs

A plant's life cycle significantly changes as it develops flowers, moving it from the juvenile to the reproductive stage. Many genes and proteins are involved in this complex and tightly controlled process. The roles of these proteins and their variations in various developmental stages and environments have been the subject of several investigations. Important for reproduction, the stigma acts as a receptor for pollen, collecting, sorting, and moistening it before directing suitable pollen tubes to the egg cells for fertilisation. Numerous proteins, such as calnexin, pectin methyltransferase and PIS 63, as well as a number of self-incompatibility-related proteins that are known to be expressed exclusively at the stigma, were found in the mature stigma of B. napus by (Nazemof, Couroux, Xing, & Robert, 2016). Auxin Response Factor 8 (ARF8), RPL27A and protein phosphatase 2A were among the stigma proteins classified functionally; these classifications indicated involvement in protein metabolism, cellular metabolism, and development. Angiosperm evolution and diversification are greatly aided by polyploidy, a heritable condition arising from genomic merging and doubling (Van de Peer, Ashman, Soltis, & Soltis, 2021). (Yang et al., 2019) examined the proteomes colchicine-induced autotetraploid *B. rapa* flower buds during meiosis in order to evaluate the impact of polyploidy on the reproductive cycle. By comparing the autotetraploid to the diploids, their LC-MS/MS analysis revealed 526 differentially abundant proteins (DAPs), of which 267 were upregulated and 295 downregulated (Zhou et al., 2019).

Proteomic Analysis Of Seeds And Siliques

Plants rely heavily on their seeds for economic purposes, as they offer evolutionary advantages for both survival and propagation (McKey, Elias, Pujol, & Duputié, 2010). Morphogenesis, seed filling, and seed desiccation are the three overlapping stages in the dynamic and controlled process of seed formation. During these stages, the tegmen and testa, the endosperm that supplies nutrients, and the embryo all develop in unison. To explore the involvement of proteins in seed growth, Hajduch et al. (2006) undertook a proteomic investigation on seed filling in *B. napus* at various intervals (2, 3, 4, 5, and 6 weeks after flowering). After two, four, and six weeks after flowering, their morphological characterisation showed that the majority of the seed mass was in the embryo. Five weeks after flowering, an increased protein content of roughly 10% of the fresh seed weight was observed. Four to six weeks following blooming, high-abundance seed storage proteins (SSPs) such as cruciferin and vicilin-like proteins were detected. Rubisco and phosphoribulokinase were among the DAPs linked to energy-related metabolic activities; most proteins were classified as being involved in metabolism (e.g. glutamine synthetase, and sucrose synthase) (Hajduch et al., 2006). (Li, Gao, Xu, Zhang, & Wang, 2012) further clarified the roles of proteins in seed development and identified 209 DAPs associated with primary metabolism, protein processing, cell structure, and signal transduction. Throughout the experiment, the relative quantity of metabolic proteins varied, revealing different metabolic activities during the maturation of the embryo. It is noteworthy that throughout the later phases of seed development, proteins associated with defence mechanisms increased, whilst proteins involved in synthesis were plentiful 25 days following pollination. The majority of protein groups showed increased abundance 20–25 days post-pollination, indicating considerable cellular activity during this time~45.

(Rahman et al., 2021) also found several DAPs in *B. rapa*. Numerous genes implicated in stress reactions, cellular homeostasis, and the metabolism of fats and carbohydrates were found by functional enrichment analysis. Among the nonredundant proteins, napins, cruciferins, and vicilins accounted for 4.72%. Notably, *B. rapa* (Rahman et al., 2021) was the first organism to identify five of the vicilins. Few studies have used proteomics in B. napus to examine the molecular regulation of seed germination, despite the fact that many have concentrated on seed growth and filling (Agrawal & Thelen, 2006)~47,48. In one such study, (Gu et al., 2016) found that during germination, there was a drop in oil content and the expression of proteins linked to the metabolism of sulphur amino acids. Since the main purpose of brassica crops is oil extraction, breeding strategies for species such as *B. napus*, *B. carinata* and *B. juncea* (Pal, Sandhu, & Bhatia, 2021) must prioritise increasing the seed oil content. In one of the first investigations into the proteins influencing oil content in *B. napus* cultivars, (Gan et al., 2013) found 119 DAPs separating lines with high oil content (HO) and low oil content (LO). Interestingly, the carbon fixation enzyme was found to be upregulated in the line (HO), indicating that greater carbon fixation is linked to increased oil accumulation. Additional DAPs comprised proteins associated with oil-related processes and storage activities. After 30, 40, and 50 days after pollination, respectively (Gan et al., 2013), recent transcriptome and proteome analyses have shown 112, 151, and 138 DAPs, illuminating the molecular mechanisms controlling oil content accumulation under dry circumstances in growing siliques of *B. napus*. LACS, FAD3, and ACCase were among the upregulated proteins involved in fatty acid synthesis and were primarily located in the plastids and cytosol. The fatty acid biosynthesis-related downregulated proteins suggest that drought stress lowers oil content by promoting fatty acid breakdown and decreasing fatty acid synthesis.

Proteomics and Biotic Stress Response

Plants, whether in the wild or in agricultural settings, encounter a variety of biotic stressors during their life cycles. Plant health and agricultural yield can be greatly impacted by interactions with bacteria, viruses, fungus, and other pests (Gan et al., 2013). Plants have responded by evolving defence mechanisms against these threats, such as triggering defence responses in response to pathogen identification, which triggers the transcription of genes involved in defence. Proteomics has become an important resource for figuring out which important proteins are involved in these resistance responses (Poveda, 2020). Particularly vulnerable to many viruses, bacteria, and fungus include the *Brassica* species, which causes significant crop losses globally (Poveda, 2020). Studies reveal that transcription factors and kinases, which are implicated in modulating responses to both abiotic and biotic stressors, show up as convergence points in stress signalling pathways. A multitude of proteome research involving several Brassica species have been carried out, contributing to our comprehension of their defence mechanisms against biotic stress.

Table 1. Summary of Proteomic Studies on Silique and Seed Development, as well as other tissues in various Brassica Species

Brassica sp.	Stress factor	Technique(s)used	References
B. rapa	Cold	RPLC,LC-MS/MS	(Yuan et al., 2019)
B. rapa	Cold	iTRAQ-LC-MS/MS	(Y. Xu et al., 2018)
B. rapa	Turnip mosaic virus	iTRAQ-LC-MS/MS	(Lyu, Gao, Zhang, Zhang, & Hou, 2020)
B. rapa	Freezing stress	LC with tandem MS	(Wei et al., 2021)
B. rapa	Seed proteome	Nano HPLC-MS/MS	(Rahman et al., 2021)
B. rapa	Plasmodiophora brassicae	UHPLC-MS/MS	(T. Song, Chu, Lahlali, Yu, & Peng, 2016)
B. rapa	Proteomic analyses of floral buds	LC-MS/MS	(Yang et al., 2019)
B. oleracea	X. campestris	2-DE, MALDI-TOF/TOFMS	(Andrade et al., 2008)
B. oleracea	Xanthomonas campestris	1D nano flow LC-MS/ MS(Shotgun)	(Santos et al., 2019)
B. oleracea	High temperature and water logging stress	MALDI-TOFMS 127	(Lin, Lin, Chen, Shen, & Lo, 2015)
B. napus	Arsenic toxicity	iTRAQ proteomics	(Farooq et al., 2021)134
B. napus	Salt stress	MALDI-TOF/TOFMS	(Bandehagh, Salekdeh, Toorchi, Mohammadi, & Komatsu, 2011)
B. napus	Salt stress	MALDI-TOF/TOFMS	(Jia et al., 2015)150
B. napus	Cold	Nano-HPLC-MS/MS	(Mi et al., 2021)120
B. napus	Copper	LC-MS/MS	(Ju et al., 2021)136
B. napus	Seed filling	MALDI-TOF,LC-MS/MS	(Hajduch et al., 2006)44
B. napus	Salt stress	2-DE,MALDITOF/TOFMS	(Yousuf, Ahmad, Ganie, & Iqbal, 2016)154
B. napus	Salt stress and Exogenous Lipoic Acid	2-DE,MALDITOF/TOFMS	(Yıldız, Akçalı, & Terzi, 2015)
B. juncea	Cadmium	2D-DIGE and iTRAQ	(Alvarez et al., 2009)
B. juncea	Salt stress	2-DE, MALDITOF/TOFMS	(Yousuf et al., 2016)

VIRUSES

Viruses in *Brassica* Species

Viruses are important plant diseases, particularly in *Brassica* species, where their parasitic nature and tiny genomes cause large yield losses. Aphids in particular are important insect vectors for the propagation of these viruses, as are other agricultural practices. When a virus and a suitable host plant come into contact, the viral genome replicates inside the host cells, creating new viral particles that travel throughout the plant and have negative consequences. Cauliflower mosaic virus (CaMV), beet western yellows virus (BWYV) and turnip mosaic virus (TuMV), are three prominent pathogenic viruses that infect *Brassica* crops. These viruses are spread by aphids and cause significant economic losses on a worldwide scale. TuMV particularly affects Brassica rapa, for which many resistance genes have been identified and cloned (Z.-X. Song et al., 2022). In *B. rapa*, Liu et al. (Liu et al., 2019) investigated the genes and complexes involved in the movement and dissemination of viral particles via plasmodesmata, xylem, and phloem. 6,043 unique proteins were found in their investigation, many of which are essential to the intricate regulatory networks controlling TuMV's long-distance migration. Heat shock proteins (HSP90s), upregulated patellin (PATL) family proteins, PEN2 and PEN3 proteins involved in callose deposition, and pathogenesis-related proteins (e.g., chitinases, thaumatin-like proteins) were identified as key proteins associated with plasmodesmata and calcium transporters. Proteins that scavenge ROS, such as APX1 and APX2, were also identified. Proteins that scavenge ROS, such as APX1 and APX2, were also found. Comparing TuMV-infected plants with healthy controls in a corresponding investigation, (Lyu et al., 2020) discovered that 75 proteins had differing abundances. The pathways linked to plant hormone signalling, endoplasmic reticulum protein processing, and plant–pathogen interactions were shown to be enriched in these proteins. A network of putative TuMV-infected genes was discovered by additional correlation analysis. These genes included those connected to nonspecific lipid transfer proteins, WRKY transcription factors, and calcium signalling. When taken as a whole, these findings improve our knowledge of the antiviral systems in affected *Brassica* plants and offer insightful information for developing plants resistant to disease.

Bacteria in *Brassica* Species

Pathogenic bacteria, in contrast to viruses, enter plant tissues through wounds or stomata, or other natural openings. Extracellular space is where most molecular interactions take place. The genomes of the pathogenic bacteria and the plant are required to comprehend the molecular mechanisms of host-pathogen interactions. Many plant pathogenic bacteria, such as *Ralstonia solanacearum, Pseudomonas syringae*, and *Xanthomonas campestris* (J. Xu & Wang, 2019), have had their entire genomes sequenced thanks to developments in genome sequencing. Particularly well-known for producing black rot in *Brassica* crops like broccoli, cauliflower, and cabbage is Xanthomonas campestris. With a susceptible cultivar of *B. oleracea*, Andrade et al. (Andrade et al., 2008) performed in vivo proteome analysis to study the relationships between *X. campestris* pv. *campestris* (Xcc) and host plants. It was also discovered that other proteins, like as ATP synthase and elongation factor Tu, are involved in metabolism. (Villeth et al., 2009) investigated the proteome alterations in *B. rapa* in response to interactions with *B. oleracea* cultivars that are sensitive and resistant. They discovered various isoforms of some proteins, indicating that the same protein may be involved in different activities depending on the sort of interaction with Xcc. They also identified proteins that were related to photosynthetic proteins. The pathogen's pathogenicity-related membrane proteins were also shown to be increased in this investigation.

According to (Ribeiro et al., 2018) there were no appreciable alterations to the plant protein composition in the early phases of infection. In contrast to controls, however, a greater number of proteins—particularly those involved in defence and antioxidant production—were found in the sensitive plants. Photosynthetic proteins were significantly reduced in susceptible plants, but only resistant plants exhibited the presence of the epithiospecifier protein (ESP), suggesting that this protein may have a function in preventing Xcc colonisation. (Santos et al., 2019) identified 30 DAPs by proteome analysis on a resistant cultivar of *B. oleracea*. A substantially elevated gene coding for endochitinase was found during validation using qRT-PCR, and it has been demonstrated that this gene confers resistance against X. campestris. They concluded that, even in young plants, photosynthesis is essential for the development of resistance.

By employing label-free quantitative proteomics to uncover DAPs implicated in proteolysis, photosynthesis, antioxidative systems, redox homeostasis, and innate immunity, Islam et al. (Islam et al., 2021) validated these results in B. napus. Major resistance responses in *B. napus* infected with Xcc have been proposed to include enhanced reducing potential, upregulation of redox signalling proteins, and programmed cell death. The mechanisms underpinning resistance to *X. campestris* are better understood as a result of these studies, and the candidate genes found in these studies present exciting opportunities for breeding initiatives targeted at creating resistant *Brassica* cultivars.

Fungi

While viruses and bacteria play a large role in the loss of agricultural production, the majority of plant illnesses are caused by fungus, which are accountable for around 70% of these diseases. Many investigations of the relationships between plants and fungus belonging to the biotrophic and necrotrophic (feeding on dead tissue) groups have made use of proteomics. Various fungal pathogens, such as, *Leptosphaeria maculans*, *Albugo candida*, *Sclerotinia sclerotiorum*, *Plasmodiophora brassicae* and *Verticillium longisporum*, have been examined in relation to diverse *Brassica* species, including *B. napus*, *B. juncea*, *B. oleracea*, *B. carinata*, and *B. rapa* (Amas, Anderson, Edwards, Cowling, & Batley, 2021; Cornelsen et al., 2021; Ding et al., 2021). Among these, the obligatory parasite *P. brassicae* is the cause of clubroot disease, a serious problem for canola crops in western Canada (Cornelsen et al., 2021).

Proteomics has been used in studies examining *P. brassica*'s interactions with Brassica species to evaluate changes in the proteome following infection. In order to identify proteins involved in lignin biosynthesis, calcium homeostasis, glycolysis, cytokinin metabolism and ROS detoxification at different times following inoculation, for example, examined the roots of a susceptible variety of *B. napus* (Hatakeyama et al., 2013). Adenosine kinase was less abundant, which corroborated earlier findings about the function of cytokinin in the early stages of infection. Decreased ROS metabolism indicated that cytokinin plays a critical role in disease susceptibility. Proteomic analysis have been employed in studies to further explore the intricate connections between *B. rapa* and *P. Brassicae* (W. Zhang et al., 2019). At the moment, a number of clubroot resistance (CR) genes, such as Crr3, CRa, CRb, and CRk, have been found and mapped (Diederichsen, Frauen, Linders, Hatakeyama, & Hirai, 2009). Transcriptomics and resistance mapping were used to identify "Rcr1," a well-characterized CR gene (Chang et al., 2012) (Chu et al., 2014). Following *P. brassicae* infection, Song et al. compared resistant (Rcrl present) and susceptible (Rcrl absent) *B. rapa* types using a proteomic study, which resulted in the identification of 527 differentially abundant proteins (DAPs). These proteins have been connected to a new signalling route that, following pathogen detection, triggers plant defence systems. In resistant plants, proteins associated with ROS homeostasis were upregulated, while several involved in redox regulation were downregulated. This study shed light on the molecular processes connected to Brassica's Rcrl gene (T. Song et al., 2016). (Ji et al., 2018)investigated the proteomics of B. rapa infected with P. brassicae and found proteins associated with the interplay between these pathways by identifying salicylic acid-induced SAR and jasmonic acid/ethylene-induced systemic resistance (ISR) (Ji et al., 2018). In resistant and susceptible lines of *B. rapa* L. ssp. pekinensis, (Lan et al., 2019) used iTRAQ analysis to find almost

5,000 DAPs, and the results showed substantial relationships with glutathione transferase activity and other defense-related pathways (Lan et al., 2019). These results emphasised important proteins, like glutamine synthetase and β-galactosidase, that are involved in resistance to disease.

In the most recent investigation by (J. Zhang, Ge, Chang, Xin, & Ji, 2021), a DAP discovered in susceptible *B. rapa* roots exhibited strong similarity to the Arabidopsis ATG6 gene, suggesting its potential function in resistance against *P. brassicae* (J. Zhang et al., 2021). (Moon et al., 2020) investigated the early proteomics of *B. oleracea* in relation to P. brassicae, highlighting the function of these proteins in modulating plant defences by identifying DAPs involved in hormone signalling pathways and the oxidative stress response (Moon et al., 2020). Proteomics has also been used to study *S. sclerotiorum*, a necrotrophic fungus that is known to cause stem rot in oil seed *Brassica*. The connection between *B. napus* and *S. sclerotiorum* was originally studied by (Liang, Srivastava, Rahman, Strelkov, & Kav, 2008), who found 32 proteins connected to several biological processes. Near-isogenic lines (NILs) were employed by (Wen et al., 2013) to identify 20 proteins linked to stem rot resistance that were involved in antioxidation and pathogen defence. These findings revealed that specific proteins in resistant lines promoted ROS removal and slowed pathogen spread (Cao, Xu, & Cai, 2016). In following investigations utilising TMT-based quantitative proteomics, researchers investigated interactions between *B. napus* and both wild-type and nonpathogenic strains of *S. sclerotiorum*, expanding our understanding of the host-pathogen dynamics (Cao et al., 2016). Using six generations from a hybrid between one resistant and two susceptible genotypes, (M. Singh et al., 2021) performed genetic and proteomic investigations of stem rot resistance in *B. juncea* (Indian mustard). Their genomic research showed that duplicate epistasis controls *B. juncea* resistance. An important step towards understanding the molecular mechanisms of *B. juncea* stem rot resistance was taken when a comparative proteomics approach revealed higher expression levels of peptidyl-prolyl cis-trans isomerase in the resistant variety during the early stages of infection compared to the susceptible lines.

B. juncea and other rapeseed types suffer from "white rust" disease, which is caused by the biotrophic oomycete *Albugo candida* (Sokhi & Khangura, 1992). In order to explore the molecular mechanisms behind the B. juncea–A. candida interaction, Kaur et al. (Kaur, Jost, Sivasithamparam, & Barbetti, 2011) conducted transcript analysis in conjunction with a comparative time-course protein profiling experiment. Nineteen proteins were found to exhibit consistent variations in expression between an A. candida-inoculated susceptible and resistant strain of *B. juncea*. A thaumatin-like protein (PR-5) was exclusively found in the resistant line, and its quantity changed dramatically during the course of the inoculation, according to Q-TOF MS/MS analysis. Furthermore, PPIase isoform CYP20-3, which exhibits elevated expression in response to the infection, was only found in the susceptible variety. This research not only identified potential key molecules involved in the defense response but also provided genetic sources for developing resistant lines through conventional breeding.

(Sharma et al., 2007) used two lines created by crossing *B. napus* and *B. carinata* to assess tolerance to Alternaria brassicae, which causes leaf spot disease. Among 48 proteins 41 upregulated and 7 downregulated with statistically different levels at various time intervals in the tolerant line were found by their 2-DE analysis. There were 23 proteins that showed substantial changes in the susceptible line (4 upregulated and 19 downregulated). The tolerant line of proteins included auxin signal transduction (peptidyl-prolyl isomerase and auxin-responsive protein), ROS-mediated signalling (peroxiredoxin, glutathione peroxidase, and putative dehydroascorbate reductase), and enzymes linked to the production of reactive oxygen species (ROS) and metabolic pathways.

The saprophytic fungus *Fusarium oxysporum* f. sp. conlutinans (Foc) causes Fusarium wilt in *B. oleracea* by invading roots and colonising xylem. (Pu et al., 2016) examined the proteome of xylem sap following in-solution protein digestion using liquid chromatography-tandem mass spectrometry (LC-MS/MS), comparing its composition in plants from resistant and susceptible cultivars as well as non-infected plants. (Subramanian, Bansal, & Kav, 2005) used tandem mass spectrometry and 2-DE to examine protein profiles from leaf samples of two lines in order to comprehend the molecular underpinnings of the defence mechanism. In addition to 28 proteins that showed differential expression in the resistant genotype 48 hours after immunisation, the study found five proteins that were exclusive to the susceptible genotype and seven proteins that were particular to the resistant genotype.

(Sharma et al., 2008) further studied the pathogen-stress response in *Brassica* species against *L. maculans* by proteomic analysis. Using tandem mass spectrometry, they were able to identify 51 protein locations in two species of *Brassica* that were extensively impacted by the disease (*B. napus* and *B. carinata*). In a proteomics-based investigation, (Marra et al., 2010) reported alterations in cotyledon protein patterns following inoculation with virulent and avirulent isolates of *L. maculans* in the *B. napus* cv. Surpass 400, which carries a resistance gene from *B. rapa* ssp. Sylvestris. Another important fungal pathogen of the Brassicaceae family is *Verticillium longisporum*, which spends the majority of its life cycle mostly in xylem. (Floerl et al., 2008) examined the effects of *V. longisporum* (isolate VL43) on disease symptoms, photosynthesis, and nutritional status by examining the xylem sap and leaf apoplast of *B. napus* var. napus. Of the 170 sites found by two-dimensional protein separation of the leaf apoplast, 12 were considerably elevated by VL43 infection. VL43-responsive proteins were shown to share similarities with endochitinase, peroxidase, PR-4 protein, and β-1,3-glucanase, according to LC-MS/MS analysis and database searches. The authors deduced that *V. longisporum's* ability to spread beyond the hypocotyl was probably inhibited by the activation of a unique array of extracellular PR-proteins, and that ethylene may have a role in upregulating these responsive proteins based on in silico analysis.

PROTEOMIC RESEARCH IN RESPONSE TO ABIOTIC STRESSORS

Cold

Of the six *Brassica* species, Brassica juncea is most common in the Indian subcontinent, but Brassica rapa and Brassica napus are widely grown in cold temperate climates. Ethiopian mustard, Brassica carinata, grows well in the highlands of Africa, while Brassica oleracea is a vegetable that is grown all over the world. Most of the world grows *Brassica nigra*, a condiment variety, in tropical Asia, Africa, and temperate Europe. Proteomics studies addressing cold stress in *Brassica* species other than *B. rapa* and *B. napus* are scarce, despite their importance. Interestingly, these two species are more resilient to cold than others; *B. rapa* is more resilient than *B. napus* (Aakanksha et al., 2021; Sun et al., 2007). Plants acquire osmolytes to increase their capacity to withstand cold stress (Dietz et al., 2006; Noctor, Mhamdi, & Foyer, 2014). Reactive oxygen species (ROS), which can cause cell death, such as hydrogen peroxide (H_2O) and superoxide anion (O_2), also build up as a result of cold stress (Asada, 2006; Banerjee, Tripathi, & Roychoudhury, 2018). Numerous chemicals, including myrosinase-associated proteins and nitric oxide, have been shown to lessen the effects of low temperature stress in *B. juncea* (Sehrawat & Deswal, 2014; Sehrawat, Sougrakpam, & Deswal, 2019).

Differentially abundant proteins (DAPs) in *B. rapa* were found through a quantitative proteome analysis to be responsible for the cultivar Longyou 7's resistance to cold by modifying metabolic pathways like the pentose phosphate pathway and the TCA cycle. Aconitate hydratase, glutamine synthetase (GS) (X. Wang et al., 2016; Y. Xu et al., 2018), and oxygen-evolving enhancer protein (OEE) are important proteins displaying variable expression. A total of 139 and 96 DAPs were found in the cold-tolerant (Longyou 7) and sensitive (Tianyou 4) cultivars, respectively, 114 in trials where plants were exposed to -4 °C to simulate cold stress. The proteins that were upregulated were connected to the synthesis of ribosomes and secondary metabolites, whereas the downregulated proteins were linked to the metabolism of carbohydrates and oxidative phosphorylation. Subsequent examination revealed important proteins including NADH dehydrogenase and peroxidase, highlighting their significance in cold stress reactions (Niu et al., 2021; Wei et al., 2021; Yuan et al., 2019).

Heat Stress

Despite the substantial effects that high temperatures have on plant morphology, there is a dearth of proteomics-based research on the consequences of heat stress in *Brassica* crops. Heat stress can reduce productivity, impede the development of pods and seeds, and cause pollen sterility (Sheng Chen, Stefanova, Siddique, & Cowling, 2021; Huang et al., 2019). Free radicals produced by this stress have a detrimental effect on macromolecular metabolism(Ismaili, Salavati, & Pour Mohammadi, 2015; Wang, Vinocur, & Altman, 2003). It can also impair the metabolism of ROS. Ascorbate peroxidase was one of the main upregulated proteins found in the first high-throughput proteomic analysis conducted on *B. napus*. The study examined the hypocotyls of seedlings that were two days old at increased temperatures and found 28 downregulated and 34 upregulated proteins. This shows that under heat stress, increased expression of some proteins may help optimise nutrition utilisation, leading to stunted growth. Proteins like SGT1 and HSP70 are essential for the heat stress response, according to similar discoveries in B. oleracea (Lin et al., 2015; Su et al., 2020). Additionally, under combined heat and flood conditions, distinct protein expressions related to photosynthesis and redox homeostasis were found in B. oleracea, according to a study that integrated physiological and proteomics methods.

Metal Stress

Significant environmental issues are being presented by the rising heavy metal contamination in agricultural lands as a result of industrialisation. The potential for metal buildup and tolerance is significant in *Brassica* species (Basu, Good, & Taylor, 2001; Meng et al., 2009). Proteomics-based studies on cadmium stress showed elevated pathways linked to redox homeostasis and sulphur compound production (Alvarez et al., 2009). 177 and 200 DAPs were found in various cultivars of *B. napus* under arsenic stress using an iTRAQ proteome analysis, indicating the presence of proteins involved in photosynthesis and stress response. There was a noticeable upregulation of proteins such catalase 2 and fructokinase-5 (Luo & Zhang, 2019). Analyses conducted under conditions of copper toxicity also discovered important upregulated proteins such as glutathione S-transferase and ascorbate peroxidase (Ju et al., 2021), as well as significant alterations in a number of metabolic pathways. All of these findings point to the possibility that *Brassica* plants can become resistant to heavy metals by means of improved detoxification, metabolic control, and ROS scavenging.

Drought Stress

Climate change has made drought an urgent problem in agriculture, impacting yield overall, growth, flowering time, and seed germination. The mechanisms by which *Brassica* plants tolerate drought are poorly known. Differential protein expression related to protein metabolism and photosynthesis under drought stress was found in *B. napus*, according to an iTRAQ proteomic analysis (Koh et al., 2015). A comparative analysis of many lines revealed that under dry conditions, proteins involved in signalling and light reaction control are downregulated. According to recent research, 138 DAPs and several phosphoproteins have been found, indicating that proteins controlling glutathione metabolism and photosynthesis play a major part in drought tolerance (S.-W. Kwon, Kim, Kim, & Lee, 2016; L. Wang et al., 2016). Specific proteins linked to signalling and amino acid pathways that were essential for drought adaptation were identified by proteomic profiling in *B. napus* (Shokri-Gharelo, Bandehagh, & Hossain, 2022).

Salinity Stress

Although salinisation of agricultural soils reduces crop productivity and yield, it poses a threat to food security. Elevated amounts of salt disrupt the osmotic and ionic equilibrium in plants, resulting in decreased absorption of water and oxidative damage as a result of the build-up of ROS(Rahnama, James, Poustini, & Munns, 2010). Studies evaluating *B. napus* proteome responses to salt found that 44 and 31 proteins, respectively, were differently expressed between genotypes that were tolerant and susceptible to salinity; these proteins included significant players in energy generation and oxidative stress (Bandehagh et al., 2011). Similar results showed that important proteins increased in response to salt stress (Jia et al., 2015)150 included RuBisCO activase and copper/zinc SOD. Additional proteome investigations of *B. oleracea* and *B. juncea* have revealed a variety of DAPs implicated in ion homeostasis and stress mitigation, suggesting that complicated metabolic pathways and ROS scavenging techniques are part of *Brassica* species' mechanisms for surviving salt stress (Banaei-Asl et al., 2015).

POST-TRANSLATIONAL MODIFICATIONS IN *BRASSICA*

Post-translational modifications (PTMs) introduce functional groups or modify molecules to change freshly synthesised proteins. These modifications, which include ubiquitination, glycosylation, and phosphorylation, improve the activity of proteins and allow plants to adapt to changes in their environment.(Millar et al., 2019). Proteomics has advanced to the point that research on Brassica has demonstrated the importance of PTMs in particular, phosphorylation in controlling plant responses to stress. For example, proteins involved in energy metabolism and signalling were found during seed development in rapeseed by phosphor proteomic analysis (Agrawal & Thelen, 2006). Furthermore, in *B. juncea*, PTMs such as S-nitrosylation have been connected to metabolic control and ROS detoxification during cold stress. Additionally, studies showed that different PTMs influence proteins linked to stress responses and lipid metabolism, highlighting the critical role PTMs play in *Brassica*'s ability to respond to environmental stressors (Arya, Prakash, Sougrakpam, & Deswal, 2021). Mechanisms of Abiotic Stress Induction and Cellular Responses in *Brassica* Species are presented in **Figure 3.**

Figure 3. Mechanisms of Abiotic Stress Induction and Cellular Responses in Brassica Species

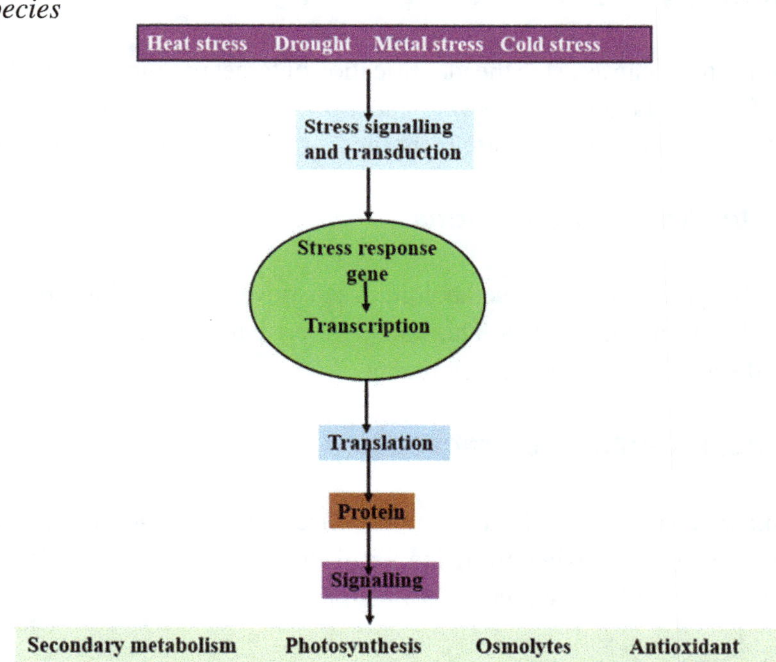

The future prospects of proteomics in canola (*Brassica napus)* research are promising and multifaceted. Here are some key areas where proteomics can make a significant impact.

Improved Crop Resilience

Proteomics can help identify proteins associated with stress responses (abiotic and biotic). Understanding these proteins can lead to the development of canola varieties that are more resilient to environmental challenges such as drought, salinity, and diseases.

Nutritional Quality Enhancement

By analyzing the proteome of canola, researchers can identify proteins that enhance the nutritional value of canola oil and seeds. This can support breeding programs aimed at increasing healthy fatty acid profiles and essential amino acids.

Biotechnology and Genetic Engineering

Proteomic tools can assist in the identification of target proteins for genetic modification. This can facilitate the development of transgenic canola with improved traits, such as higher yields, disease resistance, or enhanced nutritional content.

Metabolite-Protein Interactions

Integrating proteomics with metabolomics can provide insights into how proteins interact with various metabolites. This holistic approach can uncover pathways that regulate oil synthesis and seed development.

Sustainable Agriculture Practices

Proteomics can aid in understanding the mechanisms of nutrient uptake and utilization, which can help in optimizing fertilizer use and reducing environmental impacts. This is vital for sustainable farming practices.

Biomarker Discovery

Identifying biomarkers for growth, health, and yield can improve breeding strategies. Proteomic analyses can lead to the discovery of markers that predict performance in diverse environments.

Phytoremediation Applications

Canola can be used in phytoremediation to remove contaminants from soil. Proteomics can help identify the mechanisms by which canola plants tolerate and accumulate heavy metals or other pollutants.

Collaboration with Other Omics Technologies

Combining proteomics with genomics and transcriptomics can provide a comprehensive understanding of canola biology, enhancing breeding and research efforts.

CONCLUSION

In conclusion, the application of proteomics in *canola* research offers transformative potential for enhancing our understanding of the plant's growth, development, and resilience to environmental challenges. By leveraging advanced techniques such as mass spectrometry and bioinformatics, researchers can uncover critical protein interactions and regulatory mechanisms that underpin stress responses and disease resistance. These insights pave the way for the development of improved canola varieties through marker-assisted breeding and genetic engineering, ultimately contributing to sustainable agricultural practices and bolstering food security. As proteomics continues to evolve, it stands to play a pivotal role in addressing the pressing challenges facing global agriculture.

REFERENCES

Aakanksha, Y., Kumar, S., Yadav, B. G., Gupta, V., Mukhopadhyay, A., Pental, D., & Pradhan, A. K. (2021). Genetic analysis of heterosis for yield influencing traits in Brassica juncea using a doubled haploid population and its backcross progenies. *Frontiers in Plant Science*, *12*, 721631.

Agrawal, G. K., & Thelen, J. J. (2006). Large Scale Identification and Quantitative Profiling of Phosphoproteins Expressed during Seed Filling in Oilseed Rape* S. *Molecular & Cellular Proteomics*, *5*(11), 2044–2059.

Agrawal, G. K., Yonekura, M., Iwahashi, Y., Iwahashi, H., & Rakwal, R. (2005). System, trends and perspectives of proteomics in dicot plants: Part I: Technologies in proteome establishment. *Journal of Chromatography. B, Analytical Technologies in the Biomedical and Life Sciences*, *815*(1-2), 109–123.

Alvarez, S., Berla, B. M., Sheffield, J., Cahoon, R. E., Jez, J. M., & Hicks, L. M. (2009). Comprehensive analysis of the Brassica juncea root proteome in response to cadmium exposure by complementary proteomic approaches. *Proteomics*, *9*(9), 2419–2431.

Amas, J., Anderson, R., Edwards, D., Cowling, W., & Batley, J. (2021). Status and advances in mining for blackleg (Leptosphaeria maculans) quantitative resistance (QR) in oilseed rape (Brassica napus). *Theoretical and Applied Genetics*, *134*(10), 3123–3145.

Andrade, Aretusa E, Silva, Luciano P, Pereira, Jackeline L, Noronha, Eliane F, & Jr, Reis, Fabio B, Bloch Jr, Carlos, Mehta, Angela. (2008). In vivo proteome analysis of Xanthomonas campestris pv. campestris in the interaction with the host plant Brassica oleracea. *FEMS Microbiology Letters*, *281*(2), 167–174.

Arya, M., Prakash, S., Sougrakpam, Y., & Deswal, R. (2021). Brassica juncea leaf cuticle proteome analysis shows myrosinase protein, antifreeze activity, and post-translationally modified secretory proteins. *Plant Physiology and Biochemistry*, *161*, 234–247.

Asada, K. (2006). Production and scavenging of reactive oxygen species in chloroplasts and their functions. *Plant Physiology*, *141*(2), 391–396.

Balotf, S., Wilson, R., Tegg, R. S., Nichols, D. S., & Wilson, C. R. (2022). Shotgun proteomics as a powerful tool for the study of the proteomes of plants, their pathogens, and plant–pathogen interactions. *Proteomes*, *10*(1), 5.

Banaei-Asl, F., Bandehagh, A., Uliaei, E. D., Farajzadeh, D., Sakata, K., Mustafa, G., & Komatsu, S. (2015). Proteomic analysis of canola root inoculated with bacteria under salt stress. *Journal of Proteomics*, *124*, 88–111.

Bandehagh, A., Salekdeh, G. H., Toorchi, M., Mohammadi, A., & Komatsu, S. (2011). Comparative proteomic analysis of canola leaves under salinity stress. *Proteomics*, *11*(10), 1965–1975.

Banerjee, A., Tripathi, D. K., & Roychoudhury, A. (2018). Hydrogen sulphide trapeze: Environmental stress amelioration and phytohormone crosstalk. *Plant Physiology and Biochemistry*, *132*, 46–53.

Basu, U., Good, A. G., & Taylor, G. J. (2001). Transgenic Brassica napus plants overexpressing aluminium-induced mitochondrial manganese superoxide dismutase cDNA are resistant to aluminium. *Plant, Cell & Environment*, *24*(12), 1278–1269.

Cao, J.-Y., Xu, Y.-P., & Cai, X.-Z. (2016). TMT-based quantitative proteomics analyses reveal novel defense mechanisms of Brassica napus against the devastating necrotrophic pathogen Sclerotinia sclerotiorum. *Journal of Proteomics*, *143*, 265–277.

Chang, A., Peng, G., Chu, M., Yu, F., Falk, K. C., Liu, X., & Zhang, X. (2012). *Identification of the clubroot resistance gene Rpb1 and introgression of the resistance into canola breeding lines using a marker-assisted approach.* Paper presented at the VI International Symposium on Brassicas and XVIII Crucifer Genetics Workshop 1005.

Chen, L., & Liu, Y.-G. (2014). Male sterility and fertility restoration in crops. *Annual Review of Plant Biology*, *65*(1), 579–606.

Chen, Sheng, & Stefanova, Katia, Siddique, Kadambot HM, & Cowling, Wallace A. (2021). Transient daily heat stress during the early reproductive phase disrupts pod and seed development in Brassica napus L. *Food and Energy Security*, *10*(1), e262.

Chen, S., & Harmon, A. C. (2006). Advances in plant proteomics. *Proteomics*, *6*(20), 5504–5516.

Chu, M., Song, T., Falk, K. C., Zhang, X., Liu, X., Chang, A., & Yu, F. (2014). Fine mapping of Rcr1 and analyses of its effect on transcriptome patterns during infection by Plasmodiophora brassicae. *BMC Genomics*, *15*, 1–20.

Cornelsen, Justine, Zou, Zhongwei, Huang, Shuanglong, Parks, Paula, Lange, Ralph, & Peng, Gary, & Fernando, WG Dilantha. (2021). Validating the strategic deployment of blackleg resistance gene groups in commercial canola fields on the Canadian prairies. *Frontiers in Plant Science*, *12*, 669997.

Damerval, C., De Vienne, D., Zivy, M., & Thiellement, H. (1986). Technical improvements in two-dimensional electrophoresis increase the level of genetic variation detected in wheat-seedling proteins. *Electrophoresis*, *7*(1), 52–54.

Diederichsen, Elke, & Frauen, Martin, Linders, Enrico GA, Hatakeyama, Katsunori, & Hirai, Masashi. (2009). Status and perspectives of clubroot resistance breeding in crucifer crops. *Journal of Plant Growth Regulation*, *28*, 265–281.

Dietz, K.-J., Jacob, S., Oelze, M.-L., Laxa, M., Tognetti, V., de Miranda, S. M. N., & Finkemeier, I. (2006). The function of peroxiredoxins in plant organelle redox metabolism. *Journal of Experimental Botany*, *57*(8), 1697–1709.

Ding, L.-N., Li, T., Guo, X.-J., Li, M., Liu, X.-Y., Cao, J., & Tan, X.-L. (2021). Sclerotinia stem rot resistance in rapeseed: Recent progress and future prospects. *Journal of Agricultural and Food Chemistry*, *69*(10), 2965–2978.

Fan, Y., Niu, Y., Li, X., Li, S., Qu, C., Li, J., & Lu, K. (2021). *Population genomics of Brassica species Population Genomics: Crop Plants*. Springer.

Farooq, M. A., Hong, Z., Islam, F., Noor, Y., Hannan, F., Zhang, Y., & Song, W. (2021). Comprehensive proteomic analysis of arsenic induced toxicity reveals the mechanism of multilevel coordination of efficient defense and energy metabolism in two Brassica napus cultivars. *Ecotoxicology and Environmental Safety*, *208*, 111744.

Feder, M. E., & Walser, J.-C. (2005). The biological limitations of transcriptomics in elucidating stress and stress responses. *Journal of Evolutionary Biology*, *18*(4), 901–910.

Floerl, S., Druebert, C., Majcherczyk, A., Karlovsky, P., Kües, U., & Polle, A. (2008). Defence reactions in the apoplastic proteome of oilseed rape (Brassica napus var. napus) attenuate Verticillium longisporum growth but not disease symptoms. *BMC Plant Biology*, *8*, 1–15.

Gan, Lu, Zhang, Chun-yu, Wang, Xiao-dong, Wang, Hao, Long, Yan, Yin, Yong-tai, . . . Lin, Zhi-wei. (2013). Proteomic and comparative genomic analysis of two Brassica napus lines differing in oil content. *Journal of Proteome Research*, *12*(11), 4965–4978.

Gu, J., Chao, H., Gan, L., Guo, L., Zhang, K., Li, Y., & Li, M. (2016). Proteomic dissection of seed germination and seedling establishment in Brassica napus. *Frontiers in Plant Science*, *7*, 1482.

Hajduch, M., Casteel, J. E., Hurrelmeyer, K. E., Song, Z., Agrawal, G. K., & Thelen, J. J. (2006). Proteomic analysis of seed filling in Brassica napus. Developmental characterization of metabolic isozymes using high-resolution two-dimensional gel electrophoresis. *Plant Physiology*, *141*(1), 32–46.

Han, F., Zhang, X., Yang, L., Zhuang, M., Zhang, Y., Li, Z., & Lv, H. (2018). iTRAQ-based proteomic analysis of Ogura-CMS cabbage and its maintainer line. *International Journal of Molecular Sciences*, *19*(10), 3180.

Hatakeyama, K., Suwabe, K., Tomita, R. N., Kato, T., Nunome, T., Fukuoka, H., & Matsumoto, S. (2013). Identification and characterization of Crr1a, a gene for resistance to clubroot disease (Plasmodiophora brassicae Woronin) in Brassica rapa L. *PLoS One*, *8*(1), e54745.

Heng, S., Chen, F., Wei, C., Li, X., Yi, B., Ma, C., & Wen, J. (2019). Cytological and iTRAQ-based quantitative proteomic analyses of hau CMS in Brassica napus L. *Journal of Proteomics*, *193*, 230–238.

Huang, R., Liu, Z., Xing, M., Yang, Y., Wu, X., Liu, H., & Liang, W. (2019). Heat stress suppresses Brassica napus seed oil accumulation by inhibition of photosynthesis and BnWRI1 pathway. *Plant & Cell Physiology*, *60*(7), 1457–1470.

Islam, M. T., Lee, B.-R., La, V. H., Bae, D.-W., Jung, W.-J., & Kim, T.-H. (2021). Label-free quantitative proteomics analysis in susceptible and resistant Brassica napus cultivars infected with Xanthomonas campestris pv. campestris. *Microorganisms*, *9*(2), 253.

Ismaili, A., Salavati, A., & Pour Mohammadi, P. (2015). A comparative proteomic analysis of responses to high temperature stress in hypocotyl of canola (Brassica napus L.). *Protein and Peptide Letters*, *22*(3), 285–299.

Ji, R., Wang, Y., Wang, X., Liu, Y., Shen, X., & Feng, H. (2018). Proteomic analysis of the interaction between Plasmodiophora brassicae and Chinese cabbage (Brassica rapa L. ssp. Pekinensis) at the initial infection stage. *Scientia Horticulturae*, *233*, 386–393.

Jia, H., Shao, M., He, Y., Guan, R., Chu, P., & Jiang, H. (2015). Proteome dynamics and physiological responses to short-term salt stress in Brassica napus leaves. *PLoS One*, *10*(12), e0144808.

Ju, Y.-H., Roy, S. K., Choudhury, R., Aritra, K., Soo-Jeong, C., Ju-Young, R., Atikur, M., & Cho, K. (2021). Proteome changes reveal the protective roles of exogenous citric acid in alleviating Cu toxicity in Brassica napus L. *International Journal of Molecular Sciences*, *22*(11), 5879.

Kaur, P., Jost, R., Sivasithamparam, K., & Barbetti, M. J. (2011). Proteome analysis of the Albugo candida–Brassica juncea pathosystem reveals that the timing of the expression of defence-related genes is a crucial determinant of pathogenesis. *Journal of Experimental Botany*, *62*(3), 1285–1298.

Kim, Sung Soo, Grienenberger, Etienne, Lallemand, Benjamin, Colpitts, Che C, & Kim, Sun Young, Souza, Clarice de Azevedo, . . . Kaiser, Markus. (2010). LAP6/ POLYKETIDE SYNTHASE A and LAP5/POLYKETIDE SYNTHASE B encode hydroxyalkyl α-pyrone synthases required for pollen development and sporopollenin biosynthesis in Arabidopsis thaliana. *The Plant Cell*, *22*(12), 4045–4066.

Koh, J., Chen, G., Yoo, M.-J., Zhu, N., Dufresne, D., Erickson, J. E., & Chen, S. (2015). Comparative proteomic analysis of Brassica napus in response to drought stress. *Journal of Proteome Research*, *14*(8), 3068–3081.

Kwon, S.-W., Kim, M., Kim, H., & Lee, J. (2016). Shotgun quantitative proteomic analysis of proteins responding to drought stress in Brassica rapa L.(inbred line “Chiifu”). *International Journal of Genomics*, *2016*(1), 4235808.

Kwon, Y. S., Ryu, C.-M., Lee, S., Park, H. B., Han, K. S., Lee, J. H., & Kim, H. K. (2010). Proteome analysis of Arabidopsis seedlings exposed to bacterial volatiles. *Planta*, *232*, 1355–1370.

Lan, M., Li, G., Hu, J., Yang, H., Zhang, L., Xu, X., & Sun, R. (2019). iTRAQ-based quantitative analysis reveals proteomic changes in Chinese cabbage (Brassica rapa L.) in response to Plasmodiophora brassicae infection. *Scientific Reports*, *9*(1), 12058.

Li, W., Gao, Y., Xu, H., Zhang, Y., & Wang, J. (2012). A proteomic analysis of seed development in Brassica campestri L. *PLoS One*, *7*(11), e50290.

Liang, Yue, Srivastava, Sanjeeva, Rahman, Muhammad H, & Strelkov, Stephen E, & Kav, Nat NV. (2008). Proteome changes in leaves of Brassica napus L. as a result of Sclerotinia sclerotiorum challenge. *Journal of Agricultural and Food Chemistry*, *56*(6), 1963–1976.

Lin, H.-H., Lin, K.-H., Chen, S.-C., Shen, Y.-H., & Lo, H.-F. (2015). Proteomic analysis of broccoli (Brassica oleracea) under high temperature and waterlogging stresses. *Botanical Studies*, *56*, 1–11.

Liu, C., Sun, G. S., Zhang, R. J., Lv, S. W., Gao, L., Gao, L. W., & Zhang, C. W. (2019). Proteomic analysis provides integrated insight into mechanisms of Turnip mosaic virus long distance movement in Brassica rapa. *Biologia Plantarum*, *63*(1), 164–173.

Luo, J.-S., & Zhang, Z. (2019). Proteomic changes in the xylem sap of Brassica napus under cadmium stress and functional validation. *BMC Plant Biology*, *19*, 1–14.

Lyu, S., Gao, L., Zhang, R., Zhang, C., & Hou, X. (2020). Correlation analysis of expression profile and quantitative iTRAQ-LC-MS/MS proteomics reveals resistance mechanism against TuMV in Chinese cabbage (Brassica rapa ssp. pekinensis). *Frontiers in Genetics*, *11*, 963.

Marra, Roberta, & Li, Hua, Barbetti, MJ, Sivasithamparam, Krishnapillai, Vinale, Francesco, Cavallo, P, & Lorito, Matteo. (2010). Proteomic analysis of the interaction between Brassica napus cv. Surpass 400 and virulent or avirulent isolates of Leptosphaeria maculans. *Journal of Plant Pathology*, •••, 89–101.

McKey, D., Elias, M., Pujol, B., & Duputié, A. (2010). The evolutionary ecology of clonally propagated domesticated plants. *The New Phytologist*, *186*(2), 318–332.

Meng, H., Hua, S., Shamsi, I. H., Jilani, G., Li, Y., & Jiang, L. (2009). Cadmium-induced stress on the seed germination and seedling growth of Brassica napus L., and its alleviation through exogenous plant growth regulators. *Plant Growth Regulation*, *58*, 47–59.

Mergner, J., & Kuster, B. (2022). Plant proteome dynamics. *Annual Review of Plant Biology*, *73*(1), 67–92.

Mi, W., Liu, Z., Jin, J., Dong, X., Xu, C., Zou, Y., & Fang, X. (2021). Comparative proteomics analysis reveals the molecular mechanism of enhanced cold tolerance through ROS scavenging in winter rapeseed (Brassica napus L.). *PLoS One*, *16*(1), e0243292.

Mihr, C., Baumgärtner, M., Dieterich, J.-H., Schmitz, U. K., & Braun, H.-P. (2001). Proteomic approach for investigation of cytoplasmic male sterility (CMS) in Brassica. *Journal of Plant Physiology*, *158*(6), 787–794.

Millar, A Harvey, Heazlewood, Joshua L, Giglione, Carmela, Holdsworth, Michael J, Bachmair, Andreas, & Schulze, Waltraud X. (2019). The scope, functions, and dynamics of posttranslational protein modifications. *Annual Review of Plant Biology*, *70*(1), 119–151.

Moon, J. Y., Kim, S. T., Choi, G. J., Kwon, S.-Y., Cho, H. S., Kim, H.-S., & Park, J. M. (2020). Comparative proteomic analysis of host responses to Plasmodiophora brassicae infection in susceptible and resistant Brassica oleracea. *Plant Biotechnology Reports*, *14*, 263–274.

Nazemof, N., Couroux, P., Xing, T., & Robert, L. S. (2016). Proteomic analysis of the mature Brassica stigma reveals proteins with diverse roles in vegetative and reproductive development. *Plant Science, 250*, 51–58.

Neik, T. X., Amas, J., Barbetti, M., Edwards, D., & Batley, J. (2020). Understanding host–pathogen interactions in Brassica napus in the omics era. *Plants, 9*(10), 1336.

Ning, L., Wang, H., Li, D., Lin, Z., Li, Y., Zhao, W., & Li, M. (2019). Transcriptomic and proteomic analysis of Shaan2A cytoplasmic male sterility and its maintainer line in Brassica napus. *Frontiers in Plant Science, 10*, 252.

Niu, Z., Liu, L., Pu, Y., Ma, L., Wu, J., Hu, F., & Wang, W. (2021). iTRAQ-based quantitative proteome analysis insights into cold stress of Winter Rapeseed (Brassica rapa L.) grown in the field. *Scientific Reports, 11*(1), 23434.

Noctor, G., Mhamdi, A., & Foyer, C. H. (2014). The roles of reactive oxygen metabolism in drought: Not so cut and dried. *Plant Physiology, 164*(4), 1636–1648.

Pal, L., Sandhu, S. K., & Bhatia, D. (2021). Genome-wide association study and identification of candidate genes for seed oil content in Brassica napus. *Euphytica, 217*, 1–15.

Poveda, J. (2020). Trichoderma parareesei favors the tolerance of rapeseed (Brassica napus L.) to salinity and drought due to a chorismate mutase. *Agronomy (Basel), 10*(1), 118.

Prakash, Shyam, Bhat, SR, Quiros, CF, Kirti, PB, & Chopra, VL. (2009). 2 Brassica and its close allies: cytogenetics and evolution. *Plant breeding reviews, 31*(23), 21-187.

Pu, Z., Ino, Y., Kimura, Y., Tago, A., Shimizu, M., Natsume, S., & Shea, D. J. (2016). Changes in the proteome of xylem sap in Brassica oleracea in response to Fusarium oxysporum stress. *Frontiers in Plant Science, 7*, 31.

Rahman, M., Guo, Q., Baten, A., Mauleon, R., Khatun, A., Liu, L., & Barkla, B. J. (2021). Shotgun proteomics of Brassica rapa seed proteins identifies vicilin as a major seed storage protein in the mature seed. *PLoS One, 16*(7), e0253384.

Rahnama, A., James, R. A., Poustini, K., & Munns, R. (2010). Stomatal conductance as a screen for osmotic stress tolerance in durum wheat growing in saline soil. *Functional Plant Biology, 37*(3), 255–263.

Raza, A., Razzaq, A., Mehmood, S. S., Hussain, M. A., Wei, S., He, H., & Hasanuzzaman, M. (2021). Omics: The way forward to enhance abiotic stress tolerance in Brassica napus L. *GM Crops and Food: Biotechnology in Agriculture and the Food Chain, 12*(1), 251–281.

Ribeiro, D. G., da Cunha, G. C. R., dos Santos, C., Silva, L. P., de Oliveira, N., Brilhante, O., Labuto, L. B. D., & Mehta, A. (2018). Brassica oleracea resistance-related proteins identified at an early stage of black rot disease. *Physiological and Molecular Plant Pathology*, *104*, 9–14.

Rippka, R., Deruelles, J., Waterbury, J. B., Herdman, M., & Stanier, R. Y. (1979). Generic assignments, strain histories and properties of pure cultures of cyanobacteria. *Microbiology*, *111*(1), 1–61.

Santos, Cristiane, Nogueira, Fábio CS, Domont, Gilberto B, Fontes, Wagner, Prado, Guilherme S, Habibi, Peyman, . . . Jorrín-Novo, Jesus V. (2019). Proteomic analysis and functional validation of a Brassica oleracea endochitinase involved in resistance to Xanthomonas campestris. *Frontiers in Plant Science*, *10*, 414.

Seepaul, R., Kumar, S., Iboyi, J. E., Bashyal, M., Stansly, T. L., Bennett, R., & George, S. (2021). Brassica carinata: Biology and agronomy as a biofuel crop. *Global Change Biology. Bioenergy*, *13*(4), 582–599.

Sehrawat, A., & Deswal, R. (2014). S-nitrosylation analysis in Brassica juncea apoplast highlights the importance of nitric oxide in cold-stress signaling. *Journal of Proteome Research*, *13*(5), 2599–2619.

Sehrawat, A., Sougrakpam, Y., & Deswal, R. (2019). Cold modulated nuclear S-nitrosoproteome analysis indicates redox modulation of novel Brassicaceae specific, myrosinase and napin in Brassica juncea. *Environmental and Experimental Botany*, *161*, 312–333.

Sharma, Nidhi, Hotte, Naomi, Rahman, Muhammad H, Mohammadi, Mohsen, & Deyholos, Michael K, & Kav, Nat NV. (2008). Towards identifying Brassica proteins involved in mediating resistance to Leptosphaeria maculans: A proteomics-based approach. *Proteomics*, *8*(17), 3516–3535.

Sharma, Nidhi, Rahman, Muhammad Hafizur, Strelkov, Stephen, Thiagarajah, Mohan, & Bansal, Vipan K, & Kav, Nat NV. (2007). Proteome-level changes in two Brassica napus lines exhibiting differential responses to the fungal pathogen Alternaria brassicae. *Plant Science*, *172*(1), 95–110.

Shokri-Gharelo, Reza, Bandehagh, Ali, & Hossain, Mohammad Anwar. (2022). Proteomic Profiling and Protein-Protein Interaction Network Reveal the Molecular Mechanisms of Susceptibility to Drought Stress in Canola (Brassica napus L.). *Phyton (0031-9457)*, *91*(7).

Singh, A., & Roychoudhury, A. (2021). Gene regulation at transcriptional and post-transcriptional levels to combat salt stress in plants. *Physiologia Plantarum, 173*(4), 1556–1572.

Singh, M., Avtar, R., Lakra, N., Hooda, E., Singh, V. K., Bishnoi, M., & Choudhary, R. R. (2021). Genetic and proteomic basis of sclerotinia stem rot resistance in Indian mustard [Brassica juncea (L.) czern & coss.]. *Genes, 12*(11), 1784.

Sokhi, S. S., & Khangura, R. K. (1992). White rust of rapeseed and mustard. *Punjab Agricultural University, Ludhiana, DR/Publ, 92*(23), 1-16.

Song, T., Chu, M., Lahlali, R., Yu, F., & Peng, G. (2016). Shotgun label-free proteomic analysis of clubroot (Plasmodiophora brassicae) resistance conferred by the gene Rcr1 in Brassica rapa. *Frontiers in Plant Science, 7*, 1013.

Song, X., Wei, Y., Xiao, D., Gong, K., Sun, P., Ren, Y., & Li, X. (2021). Brassica carinata genome characterization clarifies U's triangle model of evolution and polyploidy in Brassica. *Plant Physiology, 186*(1), 388–406.

Song, Z.-X., Seo, E.-Y., Hu, W.-X., Jeong, J.-H., Moon, J. S., Kim, K.-H., & Lim, H.-S. (2022). Construction of full-length infectious cDNA clones of two Korean isolates of turnip mosaic virus breaking resistance in Brassica napus. *Archives of Virology, 167*(4), 1157–1162.

Su, H., Chen, G., Yang, L., Zhang, Y., Wang, Y., Fang, Z., & Lv, H. (2020). Proteomic variations after short-term heat shock treatment reveal differentially expressed proteins involved in early microspore embryogenesis in cabbage (Brassica oleracea). *PeerJ, 8*, e8897.

Subramanian, Babu, & Bansal, Vipan K, & Kav, Nat NV. (2005). Proteome-level investigation of Brassica carinata-derived resistance to Leptosphaeria maculans. *Journal of Agricultural and Food Chemistry, 53*(2), 313–324.

Sun, W. C., Wu, J. Y., Zeng, J., Zhu, H. X., Liu, Y. L., & Zhang, Y. H. (2007). Primary evaluation of cold tolerance among eight winter Brassica rapa. *J. Hunan Agric. Univ. Nature and Science, 33*, 151–155.

Szczygłowska, M., Piekarska, A., Konieczka, P., & Namieśnik, J. (2011). Use of brassica plants in the phytoremediation and biofumigation processes. *International Journal of Molecular Sciences, 12*(11), 7760–7771.

Van de Peer, Y., Ashman, T.-L., Soltis, P. S., & Soltis, D. E. (2021). Polyploidy: An evolutionary and ecological force in stressful times. *The Plant Cell, 33*(1), 11–26.

Villeth, Gabriela R, & Jr, Reis, Fabio B, Tonietto, Angela, Huergo, Luciano, De Souza, Emanuel M, Pedrosa, Fabio O, . . . Mehta, Angela. (2009). Comparative proteome analysis of Xanthomonas campestris pv. campestris in the interaction with the susceptible and the resistant cultivars of Brassica oleracea. *FEMS Microbiology Letters*, *298*(2), 260–266.

Wang, L., Jin, X., Li, Q., Wang, X., Li, Z., & Wu, X. (2016). Comparative proteomics reveals that phosphorylation of β carbonic anhydrase 1 might be important for adaptation to drought stress in Brassica napus. *Scientific Reports*, *6*(1), 39024.

Wang, W., Vinocur, B., & Altman, A. (2003). Plant responses to drought, salinity and extreme temperatures: Towards genetic engineering for stress tolerance. *Planta*, *218*, 1–14.

Wang XiaoWu, Wang XiaoWu, Wang HanZhong, Wang HanZhong, Wang Jun, Wang Jun, Sun RiFei, Sun RiFei, Wu Jian, Wu Jian, Liu ShengYi, Liu ShengYi, . . . Cheng Feng, Cheng Feng. (2011). The genome of the mesopolyploid crop species Brassica rapa.

Wang, X., Shan, X., Wu, Y., Su, S., Li, S., Liu, H., & Yuan, Y. (2016). iTRAQ-based quantitative proteomic analysis reveals new metabolic pathways responding to chilling stress in maize seedlings. *Journal of Proteomics*, *146*, 14–24.

Wei, J., Zheng, G., Yu, X., Liu, S., Dong, X., Cao, X., & Mi, W. (2021). Comparative transcriptomics and proteomics analyses of leaves reveals a freezing stress-responsive molecular network in winter rapeseed (Brassica rapa L.). *Frontiers in Plant Science*, *12*, 664311.

Wen, L., Tan, T.-L., Shu, J.-B., Chen, Y., Liu, Y., Yang, Z.-F., & Guan, C.-Y. (2013). Using proteomic analysis to find the proteins involved in resistance against Sclerotinia sclerotiorum in adult Brassica napus. *European Journal of Plant Pathology*, *137*, 505–523.

Wu, X., Gong, F., & Wang, W. (2014). Protein extraction from plant tissues for 2DE and its application in proteomic analysis. *Proteomics*, *14*(6), 645–658.

Xing, M., Sun, C., Li, H., Hu, S., Lei, L., & Kang, J. (2018). Integrated analysis of transcriptome and proteome changes related to the Ogura cytoplasmic male sterility in cabbage. *PLoS One*, *13*(3), e0193462.

Xu, J., & Wang, N. (2019). Where are we going with genomics in plant pathogenic bacteria? *Genomics*, *111*(4), 729–736.

Xu, Y., Zeng, X., Wu, J., Zhang, F., Li, C., Jiang, J., & Sun, W. (2018). iTRAQ-based quantitative proteome revealed metabolic changes in winter turnip rape (Brassica rapa L.) under cold stress. *International Journal of Molecular Sciences*, *19*(11), 3346.

Yang, Y., Wei, F., Braynen, J., Wei, X., Tian, B., Shi, G., & Zhang, X. (2019). Cytological and proteomic analyses of floral buds reveal an altered atlas of meiosis in autopolyploid Brassica rapa. *Cell & Bioscience*, *9*, 1–15.

Yıldız, M., Akçalı, N., & Terzi, H. (2015). Proteomic and biochemical responses of canola (Brassica napus L.) exposed to salinity stress and exogenous lipoic acid. *Journal of Plant Physiology*, *179*, 90–99.

Yousuf, P. Y., Ahmad, A., Ganie, A. H., & Iqbal, M. (2016). Salt stress-induced modulations in the shoot proteome of Brassica juncea genotypes. *Environmental Science and Pollution Research International*, *23*, 2391–2401.

Yuan, L., Wang, J., Xie, S., Zhao, M., Nie, L., Zheng, Y., & Wang, C. (2019). Comparative proteomics indicates that redox homeostasis is involved in high-and low-temperature stress tolerance in a novel Wucai (Brassica campestris L.) genotype. *International Journal of Molecular Sciences*, *20*(15), 3760.

Zhang, J., Ge, W., Chang, H., Xin, X., & Ji, R. (2021). Discovery of BrATG6 and its potential role in Brassica rapa L. resistance to infection by Plasmodiophora brassicae. *Gene*, *791*, 145711.

Zhang, W., Wang, S., Yu, F., Tang, J., Yu, L., Wang, H., & Li, J. (2019). Genome-wide identification and expression profiling of sugar transporter protein (STP) family genes in cabbage (Brassica oleracea var. capitata L.) reveals their involvement in clubroot disease responses. *Genes*, *10*(1), 71.

Zhou, X., Shi, F., Zhou, L., Zhou, Y., Liu, Z., Ji, R., & Feng, H. (2019). iTRAQ-based proteomic analysis of fertile and sterile flower buds from a genetic male sterile line 'AB01'in Chinese cabbage (*Brassica campestris* L. ssp. pekinensis). *Journal of Proteomics*, *204*, 103395.

Compilation of References

Aakanksha, Y., Kumar, S., Yadav, B. G., Gupta, V., Mukhopadhyay, A., Pental, D., & Pradhan, A. K. (2021). Genetic analysis of heterosis for yield influencing traits in Brassica juncea using a doubled haploid population and its backcross progenies. *Frontiers in Plant Science*, *12*, 721631.

Abbadi, A., & Leckband, G. (2011). Rapeseed breeding for oil content, quality, and sustainability. *European Journal of Lipid Science and Technology*, *113*(10), 1198–1206. DOI: 10.1002/ejlt.201100063

Abdullah, S., Majeed, S., & Nisar, M. (2019). RNA interference technology: A new approach in plant pest control. *Journal of Agricultural Biotechnology*, *25*(4), 672–682.

Abu, M., & Mengistu, B. (2023). *Correlation Coefficient and Path Coefficient Analysis for Seed Yield and Yield Related Traits of Rapeseed (Brassica Napus)*. Genotypes.

Acquaah, G. (2009). *Principles of plant genetics and breeding*. John Wiley & Sons.

Adetunji, A., Varghese, B., & Pammenter, N. (2021). Effects of exogenous application of five antioxidants on vigour, viability, oxidative metabolism and germination enzymes in aged cabbage and lettuce seeds. *South African Journal of Botany*, *137*, 85–97. DOI: 10.1016/j.sajb.2020.10.001

Adeyemi, K., Sabow, A., Ebrahimi, M., Samsudin, A., & Sazili, A. (2016). Fatty acid composition, cholesterol and antioxidant status of infraspinatus muscle, liver and kidney of goats fed blend of palm oil and canola oil. *Italian Journal of Animal Science*, *15*(2), 181–190. DOI: 10.1080/1828051X.2016.1158081

Adhikary, D., Kisiała, A., Sarkar, A., Basu, U., Rahman, H., Emery, R., & Kav, N. (2022). Early-stage responses toplasmodiophora brassicaeat the transcriptome and metabolome levels in clubroot resistant and susceptible oilseedbrassica napus. *Molecular Omics*, *18*(10), 991–1014. DOI: 10.1039/D2MO00251E PMID: 36382681

Adhikary, D., Mehta, D., Kisiala, A., Basu, U., Uhrig, R. G., Emery, R. J. N., Rahman, H., & Kav, N. N. V. (2024). Proteome- and metabolome-level changes during early stages of clubroot infection in brassica napus canola. *Molecular Omics*, *20*(4), 265–282. DOI: 10.1039/D3MO00210A PMID: 38334713

Adhikary, D., Mehta, D., Uhrig, R., Rahman, H., & Kav, N. (2022). A proteome-level investigation into plasmodiophora brassicae resistance in brassica napus canola. *Frontiers in Plant Science*, *13*, 860393. Advance online publication. DOI: 10.3389/fpls.2022.860393 PMID: 35401597

Afzal, M., Alghamdi, S., & Rahman, M. H. (2018). Current status and future possibilities of molecular genetics techniques in Brassica napus. *Biotechnology Letters*, *40*(3), 479–492. DOI: 10.1007/s10529-018-2510-y PMID: 29344848

Agarwal, M., Shrivastava, N., & Padh, H. (2008). Advances in molecular marker techniques and their applications in plant sciences. *Plant Cell Reports*, *27*(4), 617–631. DOI: 10.1007/s00299-008-0507-z PMID: 18246355

Agrawal, G. K., & Thelen, J. J. (2006). Large Scale Identification and Quantitative Profiling of Phosphoproteins Expressed during Seed Filling in Oilseed Rape* S. *Molecular & Cellular Proteomics*, *5*(11), 2044–2059.

Agrawal, G. K., Yonekura, M., Iwahashi, Y., Iwahashi, H., & Rakwal, R. (2005). System, trends and perspectives of proteomics in dicot plants: Part I: Technologies in proteome establishment. *Journal of Chromatography. B, Analytical Technologies in the Biomedical and Life Sciences*, *815*(1-2), 109–123.

Ahanger, M., Bhat, J., Siddiqui, M., Rinklebe, J., & Ahmad, P. (2020). Integration of silicon and secondary metabolites in plants: A significant association in stress tolerance. *Journal of Experimental Botany*, *71*(21), 6758–6774. DOI: 10.1093/jxb/eraa291 PMID: 32585681

Ahmad, M., Waraich, E. A., Zulfiqar, U., Yong, J. W. H., Ishfaq, M., din, K., Ullah, A., Abbas, A., Awan, M. I., Moussa, I. M., & Elshikh, M. S. (2024). Thiourea improves yield and quality traits of brassica napus l. by upregulating the antioxidant defense system under high temperature stress. *Scientific Reports*, *14*(1), 12195. Advance online publication. DOI: 10.1038/s41598-024-62257-y PMID: 38806561

Ahmad, Y., Haider, S., Iqbal, J., Abbasi, B. A., Yaseen, T., & Mahmood, T. (2022). The mechanisms of genome editing technologies in crop plants. In *Principles and Practices of OMICS and Genome Editing for Crop Improvement* (pp. 295–313). Springer International Publishing. DOI: 10.1007/978-3-030-96925-7_13

Ahmad, Y., Haider, S., Iqbal, J., Naseer, S., Attia, K. A., Mohammed, A. A., & Mahmood, T. (2024). In-silico analysis and transformation of OsMYB48 transcription factor driven by CaMV35S promoter in model plant–Nicotiana tabacum L. conferring abiotic stress tolerance. *GM Crops and Food: Biotechnology in Agriculture and the Food Chain*, *15*(1), 130–149. DOI: 10.1080/21645698.2024.2334476 PMID: 38551174

Ahmad, Z., Anjum, S., Skalický, M., Waraich, E., Tariq, R., Ayub, M., & Sabagh, A. (2021). Selenium alleviates the adverse effect of drought in oilseed crops camelina (camelina sativa l.) and canola (brassica napus l.). *Molecules (Basel, Switzerland)*, *26*(6), 1699. DOI: 10.3390/molecules26061699 PMID: 33803724

Aitken, K., Farmer, A., Berkman, P., Muller, C., Wei, X., Demano, E., & Kota, R. (2016). Generation of a 345K sugarcane SNP chip. *Proceedings of the Australian Society of Sugar Cane Technologists*, *29*, 1165–1172.

Ai, X., El-Badri, A. M., Batool, M., Lou, H., Hu, J., Wang, Z., Wang, C., Xiao, Y., Xiao, J., & Xiong, Y. (2024). Integrated germination related traits and transcriptomic analysis elucidate the potential mechanism of rapeseed under drought stress. *Plant Growth Regulation*, *104*(2), 1–19. DOI: 10.1007/s10725-024-01196-0

Akbar, W., Asif, M. U., Ismail, M., Bux, M., & Memon, R. M. (2017). Management of aphids on canola (*Brassica napus* L) through cultural practices. *Pakistan Entomologist*, *39*(2).

Aksouh-Harradj, N., Campbell, L., & Mailer, R. J. (2006). Flowering time plasticity and cold tolerance in Brassica napus. *Australian Journal of Agricultural Research*, *57*(5), 479–485.

Aksoy, E., Demirel, U., Öztürk, Z., Çalışkan, S., & Çalışkan, M. (2015). Recent advances in potato genomics, transcriptomics, and transgenicsunder drought and heat stresses: A review. *Turkish Journal of Botany*, *39*, 920–940. DOI: 10.3906/bot-1506-25

Albrechtsen, A., Nielsen, F. C., & Nielsen, R. (2010). Ascertainment biases in SNP chips affect measures of population divergence. *Molecular Biology and Evolution*, *27*(11), 2534–2547. DOI: 10.1093/molbev/msq148 PMID: 20558595

Aldridge, S., & Teichmann, S. (2020). Single cell transcriptomics comes of age. *Nature Communications*, *11*(1), 4307. Advance online publication. DOI: 10.1038/s41467-020-18158-5 PMID: 32855414

Ali, I., Salah, K. B. H., Sher, H., Ali, H., Ullah, Z., Ali, A., & Mahmood, T. (2022). Drought stress enhances the efficiency of floral dip method of Agrobacterium-mediated transformation in Arabidopsis thaliana. *Brazilian Journal of Biology*, *84*, e259326. DOI: 10.1590/1519-6984.259326 PMID: 35703626

Al-Lami, H. F. D., You, M. P., Banga, S. S., & Barbetti, M. J. (2023). Novel resistances provide new avenues to manage alternaria leaf spot (*Alternaria brassicae*) in canola (*Brassica napus*), mustard (B. *juncea*), and other brassicaceae crops. *Plant Disease*, *107*(2), 372–381. DOI: 10.1094/PDIS-05-22-1153-RE PMID: 35802017

Alvarez, S., Berla, B. M., Sheffield, J., Cahoon, R. E., Jez, J. M., & Hicks, L. M. (2009). Comprehensive analysis of the Brassica juncea root proteome in response to cadmium exposure by complementary proteomic approaches. *Proteomics*, *9*(9), 2419–2431.

Amas, J., Anderson, R., Edwards, D., Cowling, W., & Batley, J. (2021). Status and advances in mining for blackleg (Leptosphaeria maculans) quantitative resistance (QR) in oilseed rape (*Brassica napus*). *Theoretical and Applied Genetics*, *134*(10), 3123–3145. DOI: 10.1007/s00122-021-03877-0 PMID: 34104999

Anderson, K., Lee, S., & Johnson, M. (2018). Glycerol in skincare: A comprehensive review. *Journal of Cosmetic Science*, *69*(3), 1–15.

Andrade, Aretusa E, Silva, Luciano P, Pereira, Jackeline L, Noronha, Eliane F, & Jr, Reis, Fabio B, Bloch Jr, Carlos, Mehta, Angela. (2008). In vivo proteome analysis of Xanthomonas campestris pv. campestris in the interaction with the host plant Brassica oleracea. *FEMS Microbiology Letters*, *281*(2), 167–174.

Andrews, S. S., Peria, W. J., Richard, C. Y., Colman-Lerner, A., & Brent, R. (2016). Push-pull and feedback mechanisms can align signaling system outputs with inputs. *Cell systems, 3*(5), 444-455. e442.

Antoniewicz, M. R. (2021). A guide to metabolic flux analysis in metabolic engineering: Methods, tools and applications. *Metabolic Engineering*, *63*, 2–12. DOI: 10.1016/j.ymben.2020.11.002 PMID: 33157225

Aoki, Y., Okamura, Y., Tadaka, S., Kinoshita, K., & Obayashi, T. (2016). ATTED-II in 2016: A plant coexpression database towards lineagespecific coexpression. *Plant & Cell Physiology*, *57*(1), e5. DOI: 10.1093/pcp/pcv165 PMID: 26546318

Apel, K., & Hirt, H. (2004). REACTIVE OXYGEN SPECIES: Metabolism, Oxidative Stress, and Signal Transduction. *Annual Review of Plant Biology*, *55*(1), 373–399. DOI: 10.1146/annurev.arplant.55.031903.141701 PMID: 15377225

Arabidopsis Genome Initiative. (2000). Analysis of the genome sequence of the flowering plant Arabidopsis thaliana. *Nature*, *408*(6814), 796–815. DOI: 10.1038/35048692 PMID: 11130711

Arifuzzaman, M., Mamidi, S., McClean, P., & Rahman, M. (2016). QTL mapping for root vigor and days to flowering in Brassica napus L. *Canadian Journal of Plant Science*, CJPS-2016-0048. Advance online publication. DOI: 10.1139/CJPS-2016-0048

Arifuzzaman, M., Oladzadabbasabadi, A., McClean, P., & Rahman, M. (2019). Shovelomics for phenotyping root architectural traits of rapeseed/canola (Brassica napus L.) and genome-wide association mapping. *Molecular Genetics and Genomics*, *294*(4), 985–1000. DOI: 10.1007/s00438-019-01563-x PMID: 30968249

Arnao, M. B., & Hernández-Ruiz, J. (2015). Functions of melatonin in plants: A review. *Journal of Pineal Research*, *59*(2), 133–150. DOI: 10.1111/jpi.12253 PMID: 26094813

Arora, L., & Narula, A. (2017). Gene editing and crop improvement using CRISPR-Cas9 system. *Frontiers in Plant Science*, *8*, 1932. DOI: 10.3389/fpls.2017.01932 PMID: 29167680

Arya, M., Prakash, S., Sougrakpam, Y., & Deswal, R. (2021). Brassica juncea leaf cuticle proteome analysis shows myrosinase protein, antifreeze activity, and post-translationally modified secretory proteins. *Plant Physiology and Biochemistry*, *161*, 234–247.

Asada, K. (2006). Production and scavenging of reactive oxygen species in chloroplasts and their functions. *Plant Physiology*, *141*(2), 391–396.

Asaduzzaman, A. (2024). Genotype-by-environment interaction and stability of canola (<em>brassica napus </em>l) for weed suppression through improved interference. https://doi.org/DOI: 10.20944/preprints202408.0227.v1

Asaduzzaman, M., Pratley, J., An, M., Luckett, D., & Lemerle, D. (2015). Metabolomics differentiation of canola genotypes: Toward an understanding of canola allelochemicals. *Frontiers in Plant Science*, *5*. Advance online publication. DOI: 10.3389/fpls.2014.00765 PMID: 25620971

Ash, M. (2012). Soybeans & oil crops: Canola. United States Department of Agriculture, Economic Research Service. https://www.ers.usda.gov/topics/crops/soybeans-oil-crops/canola.aspx

Ashraf, U., Mahmood, S., Shahid, N., Imran, M., Siddique, M., & Abrar, M. (2022). Multi-omics approaches for strategic improvements of crops under changing climatic conditions. In *Principles and practices of OMICS and genome editing for crop improvement* (pp. 57–92). Springer. DOI: 10.1007/978-3-030-96925-7_3

Assis, R., Silva, F., Carneiro, L., Silva, G., Cruz, S., & Machado, C. (2020). Canola agronomic performance in three sowing dates in jataí-go. *Bioscience Journal, 36*, 36–47. DOI: 10.14393/BJ-v36n0a2020-48258

Assmann, G., & Gotto, A. M.Jr. (2004). HDL cholesterol and protective factors in atherosclerosis. *Circulation, 109*(23, Suppl. 1), III8–III14. DOI: 10.1161/01.CIR.0000131512.50667.46 PMID: 15198960

Atri, C., Akhatar, J., Gupta, M., Gupta, N., Goyal, A., Rana, K., Kaur, R., Mittal, M., Sharma, A., Singh, M. P., Sandhu, P. S., Barbetti, M. J., & Banga, S. S. (2019). Molecular and genetic analysis of defensive responses of Brassica juncea—B. Fruticulosa introgression lines to Sclerotinia infection. *Scientific Reports, 9*(1), 17089. DOI: 10.1038/s41598-019-53444-3 PMID: 31745129

Azadmard-Damirchi, S., Habibi-Nodeh, F., Hesari, J., Nemati, M., & Achachlouei, B. (2010). Effect of pretreatment with microwaves on oxidative stability and nutraceuticals content of oil from rapeseed. *Food Chemistry, 121*(4), 1211–1215. DOI: 10.1016/j.foodchem.2010.02.006

Babadi, A. A., Rahmati, S., Fakhlaei, R., Barati, B., Wang, S., Doherty, W., & Ostrikov, K. K. (2022). Emerging technologies for biodiesel production: Processes, challenges, and opportunities. *Biomass and Bioenergy, 163*, 106521. DOI: 10.1016/j.biombioe.2022.106521

Balotf, S., Wilson, R., Tegg, R. S., Nichols, D. S., & Wilson, C. R. (2022). Shotgun proteomics as a powerful tool for the study of the proteomes of plants, their pathogens, and plant–pathogen interactions. *Proteomes, 10*(1), 5.

Banaei-Asl, F., Bandehagh, A., Uliaei, E. D., Farajzadeh, D., Sakata, K., Mustafa, G., & Komatsu, S. (2015). Proteomic analysis of canola root inoculated with bacteria under salt stress. *Journal of Proteomics, 124*, 88–111.

Bancroft, I., Morgan, C., Fraser, F., Higgins, J., Wells, R., Clissold, L., Baker, D., Long, Y., Meng, J., Wang, X., Liu, S., & Trick, M. (2011). Dissecting the genome of the polyploid crop oilseed rape by transcriptome sequencing. *Nature Biotechnology, 29*(8), 762–766. DOI: 10.1038/nbt.1926 PMID: 21804563

Bandehagh, A., Dehghanian, Z., Henry, R., & Hossain, M. A. (2021). *Salinity tolerance in canola: Insights from proteomic studies Brassica Breeding and Biotechnology*. IntechOpen., DOI: 10.5772/intechopen.96649

Bandehagh, A., Salekdeh, G. H., Toorchi, M., Mohammadi, A., & Komatsu, S. (2011). Comparative proteomic analysis of canola leaves under salinity stress. *Proteomics*, *11*(10), 1965–1975.

Banerjee, A., Tripathi, D. K., & Roychoudhury, A. (2018). Hydrogen sulphide trapeze: Environmental stress amelioration and phytohormone crosstalk. *Plant Physiology and Biochemistry*, *132*, 46–53.

Banerjee, B. P., Joshi, S., Thoday-Kennedy, E., Pasam, R. K., Tibbits, J., Hayden, M., Spangenberg, G., & Kant, S. (2020). High-throughput phenotyping using digital and hyperspectral imaging-derived biomarkers for genotypic nitrogen response. *Journal of Experimental Botany*, *71*(15), 4604–4615. DOI: 10.1093/jxb/eraa143 PMID: 32185382

Bankevich, A., Nurk, S., Antipov, D., Gurevich, A., Dvorkin, M., Kulikov, A., Lesin, V. M., Nikolenko, S. I., Pham, S., Prjibelski, A. D., Pyshkin, A. V., Sirotkin, A. V., Vyahhi, N., Tesler, G., Alekseyev, M. A., & Pevzner, P. A. (2012). SPAdes: A new genome assembly algorithm and its applications to single-cell sequencing. *Journal of Computational Biology*, *19*(5), 455–477. DOI: 10.1089/cmb.2012.0021 PMID: 22506599

Baranyk, P., & Fábry, A. (Eds.). (1999). History of the rapeseed (Brassica napus L.) growing and breeding from Middle Age Europe to Canberra. International Rapeseed Congress. https://www.regional.org.au/au/gcirc/4/374.htm

Barrales-Cureño, H., Herrera-Cabrera, B., Montiel-Montoya, J., López-Valdez, L., Salgado-Garciglia, R., Ocaño-Higuera, V., Sánchez-Herrera, L. M., Lucho-Constantino, G. G., & Zaragoza-Martínez, F. (2022). Metabolomics studies of allelopathy: A review. *Revista Colombiana De Ciencias Químico Farmacéuticas*, *51*(1). Advance online publication. DOI: 10.15446/rcciquifa.v51n1.102693

Barupal, D., Fan, S., & Fiehn, O. (2018). Integrating bioinformatics approaches for a comprehensive interpretation of metabolomics datasets. *Current Opinion in Biotechnology*, *54*, 1–9. DOI: 10.1016/j.copbio.2018.01.010 PMID: 29413745

Bassil, N. V., Davis, T. M., Zhang, H., Ficklin, S., Mittmann, M., Webster, T., & Van de Weg, E. (2015). Development and preliminary evaluation of a 90 K Axiom® SNP array for the allo-octoploid cultivated strawberry Fragaria× ananassa. *BMC Genomics*, *16*(1), 1–30. DOI: 10.1186/s12864-015-1310-1 PMID: 25886969

Basunanda, P., Radoev, M., & Ecke, W. (2010). Comparative mapping of quantitative trait loci involved in oil content in Brassica napus. *Theoretical and Applied Genetics*, *120*(2), 271–281. DOI: 10.1007/s00122-009-1133-z PMID: 19707740

Basu, U., Good, A. G., & Taylor, G. J. (2001). Transgenic Brassica napus plants overexpressing aluminium-induced mitochondrial manganese superoxide dismutase cDNA are resistant to aluminium. *Plant, Cell & Environment*, *24*(12), 1278–1269.

Batista, C., Barros, L., Carvalho, A. M., & Ferreira, I. C. F. R. (2011). Nutritional and nutraceutical potential of rape (Brassica napus L. var. napus) and "tronchuda" cabbage (Brassica oleracea L. var. costata) inflorescences. *Food and Chemical Toxicology*, *49*(6), 1208–1214. DOI: 10.1016/j.fct.2011.02.023 PMID: 21376770

Bawa, A., & Anilakumar, K. (2012). Genetically modified foods: Safety, risks and public concerns—a review. *Journal of Food Science and Technology*, *50*(6), 1035–1046. DOI: 10.1007/s13197-012-0899-1 PMID: 24426015

Bayer, P. E., Hurgobin, B., Golicz, A. A., Chan, C. K. K., Yuan, Y., Lee, H., Renton, M., Meng, J., Li, R., Long, Y., Zou, J., Bancroft, I., Chalhoub, B., King, G. J., Batley, J., & Edwards, D. (2017). Assembly and comparison of two closely related Brassica napus genomes. *Plant Biotechnology Journal*, *15*(12), 1602–1610. DOI: 10.1111/pbi.12742 PMID: 28403535

Beale, D. J., Kouremenos, K. A., & Palombo, E. A. (2016). *Microbial metabolomics*. Springer International Publishing., DOI: 10.1007/978-3-319-46326-1

Becker, H. C., & Engqvist, G. M. (2011). Heterosis and hybrid breeding in Brassica napus. *Plant Breeding*, *130*(4), 421–431.

Becker, M., Zhang, X., Walker, P., Wan, J., Millar, J., Khan, D., Granger, M. J., Cavers, J. D., Chan, A. C., Fernando, D. W. G., & Belmonte, M. (2017). Transcriptome analysis of the brassica napus–leptosphaeria maculans pathosystem identifies receptor, signaling and structural genes underlying plant resistance. *The Plant Journal*, *90*(3), 573–586. DOI: 10.1111/tpj.13514 PMID: 28222234

Beckie, H. J., Harker, K. N., & Hall, L. M. (2006). Impact of herbicide-tolerant crops on weed management: A case study of canola. *Weed Science*, *54*(4), 793–800.

Beckie, H. J., Harker, K. N., Hall, L. M., Warwick, S. I., Légère, A., Sikkema, P. H., Clayton, G. W., Thomas, A. G., Leeson, J. Y., Séguin-Swartz, G., & Simard, M.-J. (2006). A decade of herbicide-resistant crops in Canada. *Canadian Journal of Plant Science*, *86*(4), 1243–1264. DOI: 10.4141/P05-193

Beckie, H. J., Harker, K. N., & Légère, A. (2011). Impact of herbicide-resistant canola on weed management and crop rotation. *Weed Technology*, *25*(3), 468–476.

Beddington, J. R., Asaduzzaman, M., Fernandez, A., Clark, M. E., Guillou, M., Jahn, M. M., Erda, L., Mamo, T., Bo, N. V., & Nobre, C. A. (2012). Achieving food security in the face of climate change: Final report from the Commission on Sustainable Agriculture and Climate Change.

Beger, R., Dunn, W., Schmidt, M., Gross, S., Kirwan, J., Cascante, M., Brennan, L., Wishart, D. S., Oresic, M., Hankemeier, T., Broadhurst, D. I., Lane, A. N., Suhre, K., Kastenmüller, G., Sumner, S. J., Thiele, I., Fiehn, O., & Kaddurah-Daouk, R. (2016). Metabolomics enables precision medicine: "a white paper, community perspective". *Metabolomics*, *12*(9), 149. Advance online publication. DOI: 10.1007/s11306-016-1094-6 PMID: 27642271

Behla, R., Hirani, A., Zelmer, C., Yu, F., Fernando, W., McVetty, P., & Li, G. (2017). Identification of common QTL for resistance to Sclerotinia sclerotiorum in three doubled haploid populations of Brassica napus (L.). *Euphytica*, *213*(11), 1–16. DOI: 10.1007/s10681-017-2047-5

Belamkar, V., Guttieri, M. J., Hussain, W., Jarquín, D., El-Basyoni, I., Poland, J., & Baenziger, P. S. (2018). Genomic selection in preliminary yield trials in a winter wheat breeding program. *G3 (Bethesda, Md.)*, *8*(8), 2735–2747. DOI: 10.1534/g3.118.200415 PMID: 29945967

Belhaj, K., Chaparro-Garcia, A., Kamoun, S., Patron, N., & Nekrasov, V. (2015). Editing plant genomes with CRISPR/Cas9. *Current Opinion in Biotechnology*, *32*, 76–84. DOI: 10.1016/j.copbio.2014.11.007 PMID: 25437637

Bellec, A., Courtial, A., Cauet, S., Rodde, N., Vautrin, S., & Salse, J. (2016). Long read sequencing technology to solve complex genomic regions assembly in plants. *Next Generation, Sequencing & Applications*, *3*(2), 1–13. DOI: 10.4172/2469-9853.1000128

Benevenuto, R. F., Zanatta, C. B., Waßmann, F., Eckerstorfer, M. F., & Agapito-Tenfen, S. Z. (2023). Integration of omics analyses into GMO risk assessment in Europe: A case study from soybean field trials. *Environmental Sciences Europe*, *35*(1), 14. DOI: 10.1186/s12302-023-00715-6

Beretta, S., Patterson, M., Zaccaria, S., Vedova, G., & Bonizzoni, P. (2017). HapCHAT: Adaptive haplotype assembly for efficiently leveraging high coverage in long reads. *bioRxiv*. DOI: 10.1101/170225

Berindean, I. V., Taoutaou, A., Rida, S., Ona, A. D., Stefan, M. F., Costin, A., Racz, I., & Muntean, L. (2024). Modern Breeding Strategies and Tools for Durable Late Blight Resistance in Potato. *Plants*, *13*(12), 1711. DOI: 10.3390/plants13121711 PMID: 38931143

Berry, P. M., Sterling, M., Spink, J. H., Baker, C. J., Sylvester-Bradley, R., Mooney, S. J., Tams, A. R., & Ennos, A. R. (2004). Understanding and reducing lodging in cereals. *Advances in Agronomy*, *84*(04), 215–269. DOI: 10.1016/S1006-8104(13)60030-6

Beszterda, M., & Nogala-Kałucka, M. (2019). Current research developments on the processing and improvement of the nutritional quality of rapeseed (brassica napus l.). *European Journal of Lipid Science and Technology*, *121*(5), 1800045. Advance online publication. DOI: 10.1002/ejlt.201800045

Bhardwaj, H. L., Hamama, A. A., & Vick, B. A. (2003). Genotypic and environmental effects on oxidative stability of canola oil. *Industrial Crops and Products*, *18*(3), 251–257. DOI: 10.1016/S0926-6690(03)00049-1

Bianco, L., Cestaro, A., Linsmith, G., Muranty, H., Denancé, C., Théron, A., & Troggio, M. (2016). Development and validation of the Axiom® Apple480K SNP genotyping array. *The Plant Journal*, *86*(1), 62–74. DOI: 10.1111/tpj.13145 PMID: 26919684

Bologa, A. M., Stoica, I., Ratiu, A. C., Constantin, N. D., & Ecovoiu, A. A. (2022). ONT-based alternative assemblies impact on the annotations of unique versus repetitive features in the genome of a Romanian strain of *Drosophila melanogaster. International Journal of Molecular Sciences*, *23*(23), 14892. DOI: 10.3390/ijms232314892 PMID: 36499217

Bongaerts, M., Bonte, R., Demirdas, S., Huidekoper, H., Langendonk, J., Wilke, M., de Valk, W., Blom, H. J., Reinders, M. J. T., & Ruijter, G. (2022). Integration of metabolomics with genomics: Metabolic gene prioritization using metabolomics data and genomic variant (cadd) scores. *Molecular Genetics and Metabolism*, *136*(3), 199–218. DOI: 10.1016/j.ymgme.2022.05.002 PMID: 35660124

Bonnardeaux, J. (2007). *Uses for Canola Meal*. Department of Agriculture and Food.

Breckell, G., & Silander, O. (2021). Do you want to build a genome? Benchmarking hybrid bacterial genome assembly methods. *bioRxiv*. DOI: 10.1101/2021.11.07.467652

Brewis, I. A., & Brennan, P. (2010). Proteomics technologies for the global identification and quantification of proteins. *Advances in Protein Chemistry and Structural Biology*, *80*, 1–44. DOI: 10.1016/B978-0-12-381264-3.00001-1 PMID: 21109216

Brookes, G., & Barfoot, P. (2014). Economic impact of GM crops: The global income and production effects 1996–2012. *GM Crops and Food: Biotechnology in Agriculture and the Food Chain*, *5*(1), 65–75. DOI: 10.4161/gmcr.28098 PMID: 24637520

Brown, A. K., Chen, X., & Davis, R. M. (2021). Genetic complexity in allotetraploid crop species: Insights from genomic studies in Brassica napus. *The Plant Genome*, *14*(2), 1–15.

Brown, A., Smith, C., & Davis, R. (2021). Valorization of canola oil by-products: Current trends and future prospects. *Bioresource Technology*, *315*, 123894.

Brzáková, M., Veselá, Z., Vařeka, J., & Bauer, J. (2023). Improving Breeding Value Reliability with Genomic Data in Breeding Groups of Charolais. *Genes*, *14*(12), 2139. DOI: 10.3390/genes14122139 PMID: 38136964

Bus, A., Hecht, J., Huettel, B., Reinhardt, R., & Stich, B. (2012). High-throughput polymorphism detection and genotyping in Brassica napus using next-generation RAD sequencing. *BMC Genomics*, *13*(1), 281. DOI: 10.1186/1471-2164-13-281 PMID: 22726880

Bus, A., Körber, N., Snowdon, R. J., & Stich, B. (2011). Patterns of molecular variation in a species-wide germplasm set of Brassica napus. *Theoretical and Applied Genetics*, *123*(8), 1413–1423. DOI: 10.1007/s00122-011-1676-7 PMID: 21847624

Butler, J. B., Vaillancourt, R. E., Potts, B. M., Lee, D. J., King, G. J., Baten, A., Shepherd, M., & Freeman, J. S. (2017). Comparative genomics of Eucalyptus and Corymbia reveals low rates of genome structural rearrangement. *BMC Genomics*, *18*(1), 397. DOI: 10.1186/s12864-017-3782-7 PMID: 28532390

Buzdin, A., & Патрушев, M. (2021). Will plant genome editing play a decisive role in "quantum-leap" improvements in crop yield to feed an increasing global human population? *Plants*, *10*(8), 1667. DOI: 10.3390/plants10081667 PMID: 34451712

Cai, D., Xiao, Y., Yang, W., Ye, W., Wang, B., Younas, M., Wu, J., & Liu, K. (2014). Association mapping of six yield-related traits in rapeseed (Brassica napus L.). *Theoretical and Applied Genetics*, *127*(1), 85–96. DOI: 10.1007/s00122-013-2203-9 PMID: 24121524

Cai, G., Wang, G., Kim, S., Li, J., Zhou, Y., & Wang, X. (2021). Increased expression of fatty acid and abc transporters enhances seed oil production in camelina. *Biotechnology for Biofuels*, *14*(1), 49. Advance online publication. DOI: 10.1186/s13068-021-01899-w PMID: 33640013

Cai, G., Yang, Q., Yang, Q., Zhao, Z., Chen, H., Wu, J., & Zhou, Y. (2020). Identification of candidate genes of QTLs for seed weight in Brassica napus through comparative mapping among Arabidopsis and Brassica species. *BMC Plant Biology*, *20*(1), 21. PMID: 31931712

Callinan, P. A., & Feinberg, A. P. (2006). The emerging science of epigenomics. *Human Molecular Genetics, 15*(suppl_1), R95–R101. DOI: 10.1093/hmg/ddl095 PMID: 16651376

Canola Council of Canada. (1990). *Canola oil and meal: Standards and regulations*. Canola Council of Canada.

Cantila, A., Saad, N., Amas, J., Edwards, D., & Batley, J. (2020). Recent findings unravel genes and genetic factors underlying leptosphaeria maculans resistance in brassica napus and its relatives. *International Journal of Molecular Sciences, 22*(1), 313. DOI: 10.3390/ijms22010313 PMID: 33396785

Canzler, S., Schor, J., Busch, W., Schubert, K., Rolle-Kampczyk, U. E., Seitz, H., Kamp, H., von Bergen, M., Buesen, R., & Hackermüller, J. (2020). Prospects and challenges of multi-omics data integration in toxicology. *Archives of Toxicology, 94*(2), 371–388. DOI: 10.1007/s00204-020-02656-y PMID: 32034435

Cao, J.-Y., Xu, Y.-P., & Cai, X.-Z. (2016). TMT-based quantitative proteomics analyses reveal novel defense mechanisms of Brassica napus against the devastating necrotrophic pathogen Sclerotinia sclerotiorum. *Journal of Proteomics, 143*, 265–277.

Cao, P., Zhao, Y., Wu, F., Xin, D., Liu, C., Wu, X., Lv, J., Chen, Q., & Qi, Z. (2022). Multi-omics techniques for soybean molecular breeding. *International Journal of Molecular Sciences, 23*(9), 4994. DOI: 10.3390/ijms23094994 PMID: 35563386

Cardoza, V. S., & Stewart, C. N. (2004). Agrobacterium-mediated transformation of canola. In *Transgenic Crops of the World: Essential Protocols* (pp. 379–387). Springer. DOI: 10.1007/978-1-4020-2333-0_28

Cardoza, V., & Stewart, C. N.Jr. (2004). Brassica biotechnology: Progress in cellular and molecular biology. *In Vitro Cellular & Developmental Biology. Plant, 40*(6), 542–551. DOI: 10.1079/IVP2004568

Cartea, M. E., & Velasco, P. (2008). Glucosinolates in Brassica foods: Bioavailability in food and significance for human health. *Phytochemistry Reviews, 7*(2), 213–229. DOI: 10.1007/s11101-007-9072-2

Cavanagh, C. R., Chao, S., Wang, S., Huang, B. E., Stephen, S., Kiani, S., & Akhunov, E. (2013). Genome-wide comparative diversity uncovers multiple targets of selection for improvement in hexaploid wheat landraces and cultivars. *Proceedings of the National Academy of Sciences of the United States of America, 110*(20), 8057–8062. DOI: 10.1073/pnas.1217133110 PMID: 23630259

Chalhoub, B., Denoeud, F., Liu, S., Parkin, I. A. P., Tang, H., Wang, X., Chiquet, J., Belcram, H., Tong, C., Samans, B., Corréa, M., Da Silva, C., Just, J., Falentin, C., Koh, C. S., Le Clainche, I., Bernard, M., Bento, P., Noel, B., & Wincker, P. (2014). Early allopolyploid evolution in the post-Neolithic Brassica napus oilseed genome. *Science*, *345*(6199), 950–953. DOI: 10.1126/science.1253435 PMID: 25146293

Challa, S., & Neelapu, N. R. (2018). Genome-wide association studies (GWAS) for abiotic stress tolerance in plants. In *Biochemical, physiological and molecular avenues for combating abiotic stress tolerance in plants* (pp. 135-150). Academic press.

Chang, A., Peng, G., Chu, M., Yu, F., Falk, K. C., Liu, X., & Zhang, X. (2012). *Identification of the clubroot resistance gene Rpb1 and introgression of the resistance into canola breeding lines using a marker-assisted approach.* Paper presented at the VI International Symposium on Brassicas and XVIII Crucifer Genetics Workshop 1005.

Chao, H., Wang, H., Wang, X., Guo, L., Gu, J., Zhao, W., Li, B., Chen, D., Raboanatahiry, N., & Li, M. (2017). Genetic dissection of seed oil and protein content and identification of networks associated with oil content in *Brassica napus*. *Scientific Reports*, *7*(1), 46295. Advance online publication. DOI: 10.1038/srep46295 PMID: 28393910

Chao, W. S., Horvath, D. P., Stamm, M. J., & Anderson, J. V. (2021). Genome-wide association mapping of freezing tolerance loci in canola (Brassica napus L.). *Agronomy (Basel)*, *11*(2), 233. DOI: 10.3390/agronomy11020233

Chapman, S. C., Chakraborty, S., Dreccer, M. F., & Howden, S. M. (2012). Plant adaptation to climate change—Opportunities and priorities in breeding. *Crop & Pasture Science*, *63*(3), 251–268. DOI: 10.1071/CP11303

Chat, V., Ferguson, R., Morales, L., & Kirchhoff, T. (2022). Ultra low-coverage whole-genome sequencing as an alternative to genotyping arrays in genome-wide association studies. *Frontiers in Genetics*, *12*, 790445. DOI: 10.3389/fgene.2021.790445 PMID: 35251117

Chaudhary, D., Pal, N., Arora, A., Prashant, B. D., & Venadan, S. (2024). Plant Functional Traits in Crop Breeding: Advancement and Challenges. In *Plant Functional Traits for Improving Productivity* (pp. 169-202). Springer.

Chauhan, J., Singh, K., Singh, V., & Kumar, S. (2011). Hundred years of rapeseed-mustard breeding in India: Accomplishments and future strategies. *Indian Journal of Agricultural Sciences*, *81*(12), 1093–1109.

Chen, C. L., Rodiger, J., Chung, V., Viswanatha, R., Mohr, S. E., Hu, Y., & Perrimon, N. (2020). SNP-CRISPR: A web tool for SNP-specific genome editing. *G3 (Bethesda, Md.)*, *10*(2), 489–494. DOI: 10.1534/g3.119.400904 PMID: 31822517

Cheng, H., Concepcion, G., Feng, X., Zhang, H., & Li, H. (2020). Haplotype-resolved de novo assembly with phased assembly graphs. *arXiv: Genomics*. https://doi.org/ DOI: 10.1101/2020.06.09.175539

Chen, G., Geng, J., Rahman, M., Liu, X., Tu, J., Fu, T., Li, G., McVetty, P. B., & Tahir, M. (2010). Identification of QTL for oil content, seed yield, and flowering time in oilseed rape (Brassica napus). *Euphytica*, *175*(2), 161–174. DOI: 10.1007/s10681-010-0144-9

Cheng, F., Liu, S., & Wu, J. (2011). BRAD, the Brassica database: A resource for Brassica genomics. *Nucleic Acids Research*, *39*(Database issue), D681–D687.

Cheng, F., Wu, J., & Wang, X. (2014). Genome Triplication Drove the Diversification of Brassica Plants. *Horticulture Research*, *1*(1), 14024. DOI: 10.1038/hortres.2014.24 PMID: 26504539

Chen, H., Wang, Y., Liu, J., Zhao, T., Yang, C., Ding, Q., & Wang, D. (2022). Identification of WRKY transcription factors responding to abiotic stresses in Brassica napus L. *Planta*, *255*(1), 1–17. DOI: 10.1007/s00425-021-03733-x PMID: 34837557

Chen, H., Xie, W., He, H., Yu, H., Chen, W., Li, J., & Zhang, Q. (2014). A high-density SNP genotyping array for rice biology and molecular breeding. *Molecular Plant*, *7*(3), 541–553. DOI: 10.1093/mp/sst135 PMID: 24121292

Chen, H., Yu, X., Zhang, X., Yang, L., Huang, X., Zhang, J., Pritchard, H. W., & Li, W. (2017). Phospholipase dα1-mediated phosphatidic acid change is a key determinant of desiccation-induced viability loss in seeds. *Plant, Cell & Environment*, *41*(1), 50–63. DOI: 10.1111/pce.12925 PMID: 28152567

Chen, J., Wang, B., Zhang, Y., Yue, X., Li, Z., & Liu, K. (2017). High-density ddRAD linkage and yield-related QTL mapping delimits a chromosomal region responsible for oil content in rapeseed (Brassica napus L.). *Breeding Science*, *67*(3), 296–306. DOI: 10.1270/jsbbs.16116 PMID: 28744183

Chen, K., Wang, Y., Zhang, R., Zhang, H., & Gao, C. (2019). CRISPR/Cas genome editing and precision plant breeding in agriculture. *Annual Review of Plant Biology*, *70*(1), 667–697. DOI: 10.1146/annurev-arplant-050718-100049 PMID: 30835493

Chen, L., & Liu, Y.-G. (2014). Male sterility and fertility restoration in crops. *Annual Review of Plant Biology*, *65*(1), 579–606.

Chen, S., & Harmon, A. C. (2006). Advances in plant proteomics. *Proteomics, 6*(20), 5504–5516.

Chen, Sheng, & Stefanova, Katia, Siddique, Kadambot HM, & Cowling, Wallace A. (2021). Transient daily heat stress during the early reproductive phase disrupts pod and seed development in Brassica napus L. *Food and Energy Security, 10*(1), e262.

Chen, W., Zhang, Y., Liu, X., Chen, B., Tu, J., & Tingdong, F. (2007). Detection of QTL for six yield-related traits in oilseed rape (Brassica napus) using DH and immortalized F2 populations. *Theoretical and Applied Genetics, 115*(6), 849–858. DOI: 10.1007/s00122-007-0613-2 PMID: 17665168

Chen, X., Li, H., & Pandey, M. K. (2018). Dissecting the genetic architecture of seed oil content in Brassica napus through genome-wide association studies. *Journal of Experimental Botany, 69*(20), 4917–4933.

Chen, X., Tong, C., Zhang, X., Song, A., Hu, M., Dong, W., Chen, F., Wang, Y., Tu, J., Liu, S., Tang, H., & Zhang, L. (2021). A high-quality Brassica napus genome reveals expansion of transposable elements, subgenome evolution and disease resistance. *Plant Biotechnology Journal, 19*(3), 615–630. DOI: 10.1111/pbi.13493 PMID: 33073445

Chen, Z. Y., Wang, R. F., & Wu, T. C. (1996). Antioxidant activity of natural flavonoids is governed by number and location of their aromatic hydroxyl groups. *Chemistry and Physics of Lipids, 79*(2), 157–163. DOI: 10.1016/0009-3084(96)02523-6 PMID: 8640902

Chin, C., Peluso, P., Sedlazeck, F. J., Nattestad, M., Concepcion, G., Clum, A., & Rank, D. (2016). Phased diploid genome assembly with single-molecule real-time sequencing. *Nature Methods, 13*(12), 1050–1054. DOI: 10.1038/nmeth.4035 PMID: 27749838

Choudhury, R., Beezley, J., Davis, B., Tomeck, J., Gratzl, S., Golzarri-Arroyo, L., & O'Connell, T. (2020). Viime: Visualization and integration of metabolomics experiments. *Journal of Open Source Software, 5*(54), 2410. DOI: 10.21105/joss.02410 PMID: 33768193

Chu, M., Song, T., Falk, K. C., Zhang, X., Liu, X., Chang, A., & Yu, F. (2014). Fine mapping of Rcr1 and analyses of its effect on transcriptome patterns during infection by Plasmodiophora brassicae. *BMC Genomics, 15*, 1–20.

Chu, U., Kumar, S., Sigmund, A., Johnson, K., Li, Y., Vongdeuane, P., & Jones, T. (2020). Genotype-independent transformation and genome editing of brassica napus using a novel explant material. *Frontiers in Plant Science*, *11*, 579524. Advance online publication. DOI: 10.3389/fpls.2020.579524 PMID: 33133118

Clarke, W. E., Higgins, E. E., Plieske, J., Wieseke, R., Sidebottom, C., Khedikar, Y., Batley, J., Edwards, D., Meng, J., Li, R., Lawley, C. T., Pauquet, J., Laga, B., Cheung, W., Iniguez-Luy, F., Dyrszka, E., Rae, S., Stich, B., Snowdon, R. J., & Parkin, I. A. P. (2016). A high-density SNP genotyping array for Brassica napus and its ancestral diploid species based on optimised selection of single-locus markers in the allotetraploid genome. *Theoretical and Applied Genetics*, *129*(10), 1887–1899. DOI: 10.1007/s00122-016-2746-7 PMID: 27364915

Clarke, W. E., Parkin, I. A., Gajardo, H. A., Gerhardt, D. J., Higgins, E., Sidebottom, C., & Iniguez-Luy, F. L. (2013). Genomic DNA enrichment using sequence capture microarrays: A novel approach to discover sequence nucleotide polymorphisms (SNP) in Brassica napus L. *PLoS One*, *8*(12), e81992. DOI: 10.1371/journal.pone.0081992 PMID: 24312619

Clevenger, J., Chavarro, C., Pearl, S. A., Ozias-Akins, P., & Jackson, S. A. (2015). Single nucleotide polymorphism identification in polyploids: A review, example, and recommendations. *Molecular Plant*, *8*(6), 831–846. DOI: 10.1016/j.molp.2015.02.002 PMID: 25676455

Clevenger, J., Chu, Y., Chavarro, C., Agarwal, G., Bertioli, D. J., Leal-Bertioli, S. C., & Ozias-Akins, P. (2017). Genome-wide SNP genotyping resolves signatures of selection and tetrasomic recombination in peanut. *Molecular Plant*, *10*(2), 309–322. DOI: 10.1016/j.molp.2016.11.015 PMID: 27993622

Cobb, J. N., Biswas, P. S., & Platten, J. D. (2019). Back to the future: Revisiting MAS as a tool for modern plant breeding. *Theoretical and Applied Genetics*, *132*(3), 647–667. DOI: 10.1007/s00122-018-3266-4 PMID: 30560465

Collard, B. C., & Mackill, D. J. (2008). Marker-assisted selection: An approach for precision plant breeding in the twenty-first century. *Philosophical Transactions of the Royal Society of London. Series B, Biological Sciences*, *363*(1491), 557–572. DOI: 10.1098/rstb.2007.2170 PMID: 17715053

Combs, G. F.Jr. (1992). *The vitamins: Fundamental aspects in nutrition and health*. Academic Press.

Cook, J., Douglas, G., Zhang, J., Glick, B., Langille, M., Liu, K., & Cheng, Z. (2020). Transcriptomic profiling of brassica napus responses to pseudomonas aeruginosa. *Innate Immunity*, *27*(2), 143–157. DOI: 10.1177/1753425920980512 PMID: 33353474

Cook, J., Hui, J., Zhang, J., Kember, M., Berrué, F., Zhang, J., & Cheng, Z. (2022). Production of quorum sensing-related metabolites and phytoalexins during pseudomonas aeruginosa–brassica napus interaction. *Microbiology, 168*(8). Advance online publication. DOI: 10.1099/mic.0.001212 PMID: 35980361

Cooper, M., Messina, C. D., Podlich, D., Totir, L. R., Baumgarten, A., Hausmann, N. J., Wright, D., & Graham, G. (2014). Predicting the future of plant breeding: Complementing empirical evaluation with genetic prediction. *Crop & Pasture Science, 65*(4), 311–336. DOI: 10.1071/CP14007

Cornelsen, J., Ort, N. W. W., Gabert, R. K., Epp, I., & Rempel, C. B. (2023). Current and potential pest threats for canola in the canadian prairies. *Pest Management Science, 80*(5), 2220–2234. DOI: 10.1002/ps.7858 PMID: 37899491

Cornelsen, Justine, Zou, Zhongwei, Huang, Shuanglong, Parks, Paula, Lange, Ralph, & Peng, Gary, & Fernando, WG Dilantha. (2021). Validating the strategic deployment of blackleg resistance gene groups in commercial canola fields on the Canadian prairies. *Frontiers in Plant Science, 12*, 669997.

Cowling, W. A. (2007). Genetic diversity in Australian canola and implications for crop breeding for changing future environments. *Field Crops Research, 104*(1-3), 103–111. DOI: 10.1016/j.fcr.2006.12.014

Cowling, W. A., Castro-Urrea, F. A., Stefanova, K. T., Li, L., Banks, R. G., Saradadevi, R., & Siddique, K. H. (2023). Optimal contribution selection improves the rate of genetic gain in grain yield and yield stability in spring canola in Australia and Canada. *Plants, 12*(2), 383. DOI: 10.3390/plants12020383 PMID: 36679096

Crossa, J., Pérez, P., & Hickey, J. (2017). Genomic selection in plant breeding: Methods, models, and perspectives. *Trends in Plant Science, 22*(11), 961–975. DOI: 10.1016/j.tplants.2017.08.011 PMID: 28965742

Cui, C., Zhang, K., Chai, L., Zheng, B., Zhang, J., Jiang, J., Tan, C., Li, H., Chen, D., & Jiang, L. (2024). Unraveling the mechanism of flower color variation in brassica napus by integrated metabolome and transcriptome analyses. *Frontiers in Plant Science, 15*, 1419508. Advance online publication. DOI: 10.3389/fpls.2024.1419508 PMID: 38933465

Cui, F., Sui, N., Duan, G., Liu, Y., Yan, H., Liu, S., & Li, G. (2018). Identification of metabolites and transcripts involved in salt stress and recovery in peanut. *Frontiers in Plant Science, 9*, 217. Advance online publication. DOI: 10.3389/fpls.2018.00217 PMID: 29520289

D'Hooghe, P., Escamez, S., Trouverie, J., & Avice, J. (2013). Sulphur limitation provokes physiological and leaf proteome changes in oilseed rape that lead to perturbation of sulphur, carbon and oxidative metabolisms. *BMC Plant Biology*, *13*(1), 23. Advance online publication. DOI: 10.1186/1471-2229-13-23 PMID: 23391283

Dalton-Morgan, J., Hayward, A., Alamery, S., Tollenaere, R., Mason, A. S., Campbell, E., & Batley, J. (2014). A high-throughput SNP array in the amphidiploid species Brassica napus shows diversity in resistance genes. *Functional & Integrative Genomics*, *14*(4), 643–655. DOI: 10.1007/s10142-014-0391-2 PMID: 25147024

Damerval, C., De Vienne, D., Zivy, M., & Thiellement, H. (1986). Technical improvements in two-dimensional electrophoresis increase the level of genetic variation detected in wheat-seedling proteins. *Electrophoresis*, *7*(1), 52–54.

Das, A., Rushton, P., & Rohila, J. (2017). Metabolomic profiling of soybeans (glycine max l.) reveals the importance of sugar and nitrogen metabolism under drought and heat stress. *Plants*, *6*(2), 21. DOI: 10.3390/plants6020021 PMID: 28587097

Das, R., Pradhan, S., & Parida, A. (2020). De-novo transcriptome analysis unveils differentially expressed genes regulating drought and salt stress response in panicum sumatrense. *Scientific Reports*, *10*(1), 21251. Advance online publication. DOI: 10.1038/s41598-020-78118-3 PMID: 33277539

Degen, B., & Müller, N. A. (2023). A simulation study comparing advanced marker-assisted selection with genomic selection in tree breeding programs. *G3 (Bethesda, Md.)*, *13*(10), jkad164. DOI: 10.1093/g3journal/jkad164 PMID: 37494068

Degirolamo, C., & Rudel, L. L. (2010). Dietary monounsaturated fatty acids appear not to provide cardioprotection. *Current Atherosclerosis Reports*, *12*(6), 391–396. DOI: 10.1007/s11883-010-0133-4 PMID: 20725810

Degrave, A., Wagner, M., George, P., Coudard, L., Pinochet, X., Ermel, M., Gay, E., Fudal, I., Moreno-Rico, O., Rouxel, T., & Balesdent, M. (2021). A new avirulence gene of Leptosphaeria maculans, AvrLm14, identifies a resistance source in American broccoli (Brassica oleracea) genotypes. *Molecular Plant Pathology*, *22*(12), 1599–1612. DOI: 10.1111/mpp.13131 PMID: 34467616

Delourme, R., Laperche, A., Bouchet, A., Jubault, M., Paillard, S., Manzanares-Dauleux, M., & Nesi, N. (2018). Genes and Quantitative Trait Loci Mapping for Major Agronomic Traits in Brassica napus L.., 41-85. https://doi.org/.DOI: 10.1007/978-3-319-43694-4_3

Delourme, R., Brun, H., & Renard, M. (2011). Genetics and molecular breeding in brassicas: From classical breeding to genomics. *Plant Breeding Reviews*, *35*, 123–149.

Delourme, R., Chevre, A. M., & Brun, H. (2013). Strategies for durable resistance to Leptosphaeria maculans in Brassica napus. *European Journal of Plant Pathology, 137*(1), 1–15.

Delourme, R., Falentin, C., & Fopa Fomeju, B. (2014). Genetic control of blackleg resistance in Brassica napus: Identification of QTLs and candidate genes. *Plant Pathology, 63*(5), 1035–1045.

Deng, C., Liu, H., Yao, Y., Guo, S., Xiao, L., Fu, Z., & Du, D. (2019). QTL analysis of four yield-related traits for Brassica napus L. in multiple environments. *Molecular Breeding, 39*(12), 166. Advance online publication. DOI: 10.1007/s11032-019-1067-3

Denti, L., Khorsand, P., Bonizzoni, P., Hormozdiari, F., & Chikhi, R. (2022). Improved structural variant discovery in hard-to-call regions using sample-specific string detection from accurate long reads. *bioRxiv*. DOI: 10.1101/2022.02.12.480198

Deschamps, S., Mudge, J., Cameron, C., Ramaraj, T., Anand, A., Fengler, K., Hayes, K., Llaca, V., Jones, T. J., & May, G. (2016). Characterization, correction and de novo assembly of an Oxford Nanopore genomic dataset from *Agrobacterium tumefaciens. Scientific Reports, 6*(1), 28625. DOI: 10.1038/srep28625 PMID: 27350167

Dhakal, K. (2023). Application of Machine Learning and Hyperspectral Imaging in Plant Phenomics Research.

Diederichsen, Elke, & Frauen, Martin, Linders, Enrico GA, Hatakeyama, Katsunori, & Hirai, Masashi. (2009). Status and perspectives of clubroot resistance breeding in crucifer crops. *Journal of Plant Growth Regulation, 28*, 265–281.

Diekstra, A., Bosgoed, E., Rikken, A., van Lier, B., Kamsteeg, E. J., Tychon, M., Derks, R. C., van Soest, R. A., Mensenkamp, A. R., Scheffer, H., Neveling, K., & Nelen, M. R. (2015). Translating Sanger-based routine DNA diagnostics into generic massive parallel ion semiconductor sequencing. *Clinical Chemistry, 61*(1), 154–162. DOI: 10.1373/clinchem.2014.225250 PMID: 25274553

Dietz, K.-J., Jacob, S., Oelze, M.-L., Laxa, M., Tognetti, V., de Miranda, S. M. N., & Finkemeier, I. (2006). The function of peroxiredoxins in plant organelle redox metabolism. *Journal of Experimental Botany, 57*(8), 1697–1709.

Ding, L.-N., Li, T., Guo, X.-J., Li, M., Liu, X.-Y., Cao, J., & Tan, X.-L. (2021). Sclerotinia stem rot resistance in rapeseed: Recent progress and future prospects. *Journal of Agricultural and Food Chemistry, 69*(10), 2965–2978.

Ding, Y., Shi, Y., & Yang, S. (2019). Advances and Challenges in Uncovering Cold Tolerance Regulatory Mechanisms in Plants. *The New Phytologist, 222*(4), 1690–1704. DOI: 10.1111/nph.15696 PMID: 30664232

Diouf, L., Pan, Z., He, S., Gong, W., Jia, Y., Magwanga, R., Romy, K., Rashid, H., Kirungu, J., & Du, X. (2017). High-Density Linkage Map Construction and Mapping of Salt-Tolerant QTLs at Seedling Stage in Upland Cotton Using Genotyping by Sequencing (GBS). *International Journal of Molecular Sciences*, *18*(12), 2622. Advance online publication. DOI: 10.3390/ijms18122622 PMID: 29206184

Dong, C., Dalton-Morgan, J., Vincent, K., & Sharp, P. (2009). A modified TILLING method for wheat breeding. *The Plant Genome*, *2*(1), plantgenome2008.10.0012. DOI: 10.3835/plantgenome2008.10.0012

Dong, Z., Alam, M. K., Xie, M., Yang, L., Liu, J., Helal, M. M. U., Huang, J., Cheng, X., Liu, Y., Tong, C., Zhao, C., & Liu, S. (2021). Mapping of a major QTL controlling plant height using a high-density genetic map and QTL-seq methods based on whole-genome resequencing in *Brassica napus. G3 (Bethesda, Md.)*, *11*(7), jkab118. Advance online publication. DOI: 10.1093/g3journal/jkab118 PMID: 33836054

Dowen, R. H., Pelizzola, M., Schmitz, R. J., Lister, R., Dowen, J. M., Nery, J. R., Dixon, J. E., & Ecker, J. R. (2012). Widespread dynamic DNA methylation in response to biotic stress. *Proceedings of the National Academy of Sciences of the United States of America*, *109*(32), E2183–E2191. DOI: 10.1073/pnas.1209329109 PMID: 22733782

Downey, R. (1990). Brassica oilseed breeding-achievements and opportunities.

Downey, R. K., & Röbbelen, G. (1989). Brassica species. In *Oil crops of the world* (pp. 339–362). McGraw-Hill.

Downing, W. L., Mauxion, F., Fauvarque, M. O., Reviron, M.-P., De Vienne, D., Vartanian, N., & Giraudat, J. (1992). A Brassica napus transcript encoding a protein related to the Künitz protease inhibitor family accumulates upon water stress in leaves, not in seeds. *The Plant Journal*, *2*(5), 685–693. DOI: 10.1111/j.1365-313X.1992.tb00137.x PMID: 1302628

Dwivedi, S. L., Upadhyaya, H. D., Stalker, H. T., Blair, M. W., Bertioli, D. J., Nielen, S., & Ortiz, R. (2008). Enhancing crop gene pools with beneficial traits using wild relatives. *Plant Breeding Reviews*, *30*, 179–230.

Ebersbach, J., Khan, N. A., McQuillan, I., Higgins, E. E., Horner, K., Bandi, V., Gutwin, C., Vail, S. L., Robinson, S. J., & Parkin, I. P. (2022). Exploiting High-Throughput Indoor Phenotyping to Characterize the Founders of a Structured B. napus Breeding Population. *Frontiers in Plant Science*, *12*, 780250. Advance online publication. DOI: 10.3389/fpls.2021.780250 PMID: 35069637

Ecke, W., Clemens, R., Honsdorf, N., & Becker, H. C. (2010). Extent and structure of linkage disequilibrium in canola quality winter rapeseed (Brassica napus L.). *Theoretical and Applied Genetics*, *120*(5), 921–931. DOI: 10.1007/s00122-009-1221-0 PMID: 19953219

Elbasyoni, I. S., Lorenz, A. J., Guttieri, M., Frels, K., Baenziger, P. S., Poland, J., & Akhunov, E. (2018). A comparison between genotyping-by-sequencing and array-based scoring of SNPs for genomic prediction accuracy in winter wheat. *Plant Science*, *270*, 123–130. DOI: 10.1016/j.plantsci.2018.02.019 PMID: 29576064

Eldin, S., Ziena, H., Khair, S., & Rozan, M. (2018). Canola seed meal as a potential source of natural antioxidant. *Alexandria Science Exchange Journal*, *39*(OCTOBER-DECEMBER), 615–619. DOI: 10.21608/asejaiqjsae.2018.19582

El-Metwally, S., Ouda, O. M., & Helmy, M. (2014). *Next generation sequencing technologies and challenges in sequence assembly* (Vol. 7). Springer Science & Business. DOI: 10.1007/978-1-4939-0715-1

El-Mezawy, A., Al-Forkan, M., Wu, L., Weselake, R. J., Shah, S., & Rahman, H. (2016). Down regulation of the IND gene causes male sterility in canola (Brassica napus L.). *Biocatalysis and Agricultural Biotechnology*, *6*, 9–18. DOI: 10.1016/j.bcab.2016.02.004

Elshire, R. J., Glaubitz, J. C., Sun, Q., Poland, J. A., Kawamoto, K., Buckler, E. S., & Mitchell, S. E. (2011). A robust, simple genotyping-by-sequencing (GBS) approach for high diversity species. *PLoS One*, *6*(5), e19379. DOI: 10.1371/journal.pone.0019379 PMID: 21573248

Emwas, A., Roy, R., McKay, R., Tenori, L., Saccenti, E., Gowda, G., Raftery, D., Alahmari, F., Jaremko, L., Jaremko, M., & Wishart, D. (2019). Nmr spectroscopy for metabolomics research. *Metabolites*, *9*(7), 123. DOI: 10.3390/metabo9070123 PMID: 31252628

Eskin, N. A. M., & McDonald, B. E. (1991). Canola oil. *Nutrition Bulletin*, *16*(3), 138–146. DOI: 10.1111/j.1467-3010.1991.tb01048.x

Eskin, N. A. M., & Przybylski, R. (2003). RAPE SEED OIL/CANOLA. In Caballero, B. (Ed.), *Encyclopedia of Food Sciences and Nutrition* (2nd ed., pp. 4911–4916). Academic Press. DOI: 10.1016/B0-12-227055-X/01349-3

Estevez, R. L., Duarte, J. B., Chambo, A., & Cruz, M. d. (2014). The canola culture (Brassica napus var. oleifera).

Faizan, M., Sultan, H., Alam, P., Karabulut, F., Cheng, S. H., Rajput, V. D., Minkina, T., Hayat, S., Khan, M. N., & Nie, L. (2024). Melatonin and its cross-talk with other signaling molecules under abiotic stress. *Plant Stress*, *11*, 100410. DOI: 10.1016/j.stress.2024.100410

Fan, Y., Niu, Y., Li, X., Li, S., Qu, C., Li, J., & Lu, K. (2021). *Population genomics of Brassica species Population Genomics: Crop Plants*. Springer.

Farag, M. A., Sharaf Eldin, M. G., Kassem, H., & Abou el Fetouh, M. (2013). Metabolome classification of Brassica napus L. organs via UPLC–QTOF–PDA–MS and their anti-oxidant potential. *Phytochemical Analysis*, *24*(3), 277–287. DOI: 10.1002/pca.2408 PMID: 23055344

Farid, M., Yang, R. C., Kebede, B., & Rahman, H. (2020). Evaluation of Brassica oleracea accessions for resistance to Plasmodiophora brassicae and identification of genomic regions associated with resistance. *Genome*, *63*(2), 91–101. DOI: 10.1139/gen-2019-0098 PMID: 31600449

Farjad, M., Rigault, M., Pateyron, S., Martin-Magniette, M., Krapp, A., Meyer, C., & Fagard, M. (2018). Nitrogen limitation alters the response of specific genes to biotic stress. *International Journal of Molecular Sciences*, *19*(11), 3364. DOI: 10.3390/ijms19113364 PMID: 30373239

Farnia, A., Mansouri, M., Farnia, A., & Branch, B. (2015). Study on morphological characteristics of maize (Zea mays L.) cultivars under different plant densities. *Indian Journal of Natural Sciences*, *5*(30), 34–56.

Farooq, M. A., Hong, Z., Islam, F., Noor, Y., Hannan, F., Zhang, Y., & Song, W. (2021). Comprehensive proteomic analysis of arsenic induced toxicity reveals the mechanism of multilevel coordination of efficient defense and energy metabolism in two Brassica napus cultivars. *Ecotoxicology and Environmental Safety*, *208*, 111744.

Fattahi, F., Fakheri, B., Solouki, M., Möllers, C., & Rezaizad, A. (2018). Mapping QTL controlling agronomic traits in a doubled haploid population of winter oilseed rape (Brassica napus L.). *Journal of Genetics*, *97*(5), 1389–1406. DOI: 10.1007/s12041-018-1044-3 PMID: 30555087

Feder, M. E., & Walser, J.-C. (2005). The biological limitations of transcriptomics in elucidating stress and stress responses. *Journal of Evolutionary Biology*, *18*(4), 901–910.

Feizabadi, A., Noormohammadi, G., & Fatehi, F. (2021). Changes in growth, physiology, and fatty acid profile of rapeseed cultivars treated with vermicompost under drought stress. *Journal of Soil Science and Plant Nutrition*, *21*(1), 200–208. DOI: 10.1007/s42729-020-00353-4

Felcher, K. J., Coombs, J. J., Massa, A. N., Hansey, C. N., Hamilton, J. P., Veilleux, R. E., & Douches, D. S. (2012). Integration of two diploid potato linkage maps with the potato genome sequence. *PLoS One*, *7*(4), e36347. DOI: 10.1371/journal.pone.0036347 PMID: 22558443

Feng, L., Chen, S., Zhang, C., Zhang, Y., & He, Y. (2021). A comprehensive review on recent applications of unmanned aerial vehicle remote sensing with various sensors for high-throughput plant phenotyping. *Computers and Electronics in Agriculture*, *182*, 106033. DOI: 10.1016/j.compag.2021.106033

Ferreira, M. E., Williams, P. H., & Osborn, T. C. (1994). RFLP mapping of Brassica napus using F1-derived doubled haploid lines. *Theoretical and Applied Genetics*, *89*(5), 615–621. DOI: 10.1007/BF00222456 PMID: 24177938

Fikere, M., Barbulescu, D. M., Malmberg, M. M., Spangenberg, G. C., Cogan, N. O., & Daetwyler, H. D. (2020). Meta-analysis of GWAS in canola blackleg (Leptosphaeria maculans) disease traits demonstrates increased power from imputed whole-genome sequence. *Scientific Reports*, *10*(1), 14300. DOI: 10.1038/s41598-020-71274-6 PMID: 32868838

Finnegan, E. J., Kovac, K. A., Jaligot, E., Sheldon, C. C., James, P. W., & Dennis, E. S. (2005). The downregulation of Flowering Locus C (FLC) expression in plants with low levels of DNA methylation and by vernalization occurs by distinct mechanisms. *The Plant Journal*, *44*(3), 420–432. DOI: 10.1111/j.1365-313X.2005.02541.x PMID: 16236152

Fletcher, R. S., Herrmann, D., Mullen, J. L., Li, Q., Schrider, D. R., Price, N., Lin, J., Grogan, K., Kern, A., & McKay, J. K. (2016). Identification of Polymorphisms Associated with Drought Adaptation QTL in *Brassica napus* by Resequencing. *G3 (Bethesda, Md.)*, *6*(4), 793–803. DOI: 10.1534/g3.115.021279 PMID: 26801646

Floerl, S., Druebert, C., Majcherczyk, A., Karlovsky, P., Kües, U., & Polle, A. (2008). Defence reactions in the apoplastic proteome of oilseed rape (Brassica napus var. napus) attenuate Verticillium longisporum growth but not disease symptoms. *BMC Plant Biology*, *8*, 1–15.

Food and Drug Administration. (2006). Unsaturated fatty acids from canola oil and reduced risk of coronary heart disease. U.S. Food and Drug Administration. https://www.fda.gov/food/ingredientspackaginglabeling/labelingnutrition/ucm073992.htm#cardio

Fordyce, S., Torrion, J., Lachowiec, J., Miller, P., Biddulph, B., & Carr, P. (2018). Field evidence for frost-induced cross-stress tolerance in canola. *Agricultural & Environmental Letters*, *3*(1), 180020. Advance online publication. DOI: 10.2134/ael2018.05.0020

Franzke, A., Lysak, M. A., & Al-Shehbaz, I. A. (2011). Phylogeny and evolutionary history of the Brassicaceae. *Plant Systematics and Evolution*, *291*(1-2), 75–87.

Friedt, W., & Snowdon, R. (2009). Oilseed rape: Genetics, genomics, and breeding. In *Biotechnology in Agriculture and Forestry* (pp. 123–156). Springer.

Fu, D. H., Jiang, L. Y., Mason, A. S., Xiao, M. L., Zhu, L. R., Li, L. Z., Zhou, Q., Shen, C., & Huang, C. (2016). Research progress and strategies for multifunctional rapeseed: A case study of China. *Journal of Integrative Agriculture*, *15*(8), 1673–1684. DOI: 10.1016/S2095-3119(16)61384-9

Fukai, E., Karim, M. M., Shea, D. J., Tonu, N. N., Falk, K. C., Funaki, T., & Okazaki, K. (2019). An LTR retrotransposon insertion was the cause of world's first low erucic acid *Brassica rapa* oilseed cultivar. *Molecular Breeding*, *39*(1), 1–12. DOI: 10.1007/s11032-018-0916-9

Fu, P., Montes, C. M., Siebers, M. H., Gomez-Casanovas, N., McGrath, J. M., Ainsworth, E. A., & Bernacchi, C. J. (2022). Advances in field-based high-throughput photosynthetic phenotyping. *Journal of Experimental Botany*, *73*(10), 3157–3172. DOI: 10.1093/jxb/erac077 PMID: 35218184

Fussell, G. E. (1955). History of cole (Brassica sp.). *Nature*, *176*(4471), 48–51. DOI: 10.1038/176048a0

Fu, Y., Lu, K., Qian, L., Mei, J., Wei, D., Peng, X., Xu, X., Li, J., Frauen, M., Dreyer, F., Snowdon, R. J., & Qian, W. (2015). Development of genic cleavage markers in association with seed glucosinolate content in canola. *Theoretical and Applied Genetics*, *128*(6), 1029–1037. DOI: 10.1007/s00122-015-2487-z PMID: 25748114

Fu, Y., Zhang, D., Gleeson, M., Zhang, Y., Lin, B., Hua, S., Ding, H., Frauen, M., Li, J., Qian, W., & Yu, H. (2017). Analysis of QTL for seed oil content in Brassica napus by association mapping and QTL mapping. *Euphytica*, *213*(1), 1–15. DOI: 10.1007/s10681-016-1817-9

Gaber, M. A. F. M., Tujillo, F. J., Mansour, M. P., & Juliano, P. (2018). Improving oil extraction from canola seeds by conventional and advanced methods. *Food Engineering Reviews, 10*(4), 198–210. DOI: 10.1007/s12393-018-9182-1

Gabur, I., Chawla, H., Liu, X., Kumar, V., Faure, S., Tiedemann, A., & Obermeier, C. (2018). Finding invisible quantitative trait loci with missing data. *Plant Biotechnology Journal, 16*(12), 2102–2112. DOI: 10.1111/pbi.12942 PMID: 29729219

Gacek, K., Bayer, P. E., Anderson, R., Severn-Ellis, A. A., Wolko, J., Łopatyńska, A., Matuszczak, M., Bocianowski, J., Edwards, D., & Batley, J. (2021). QTL Genetic Mapping Study for Traits Affecting Meal Quality in Winter Oilseed Rape (Brassica Napus L.). *Genes, 12*(8), 1235. DOI: 10.3390/genes12081235 PMID: 34440409

Galanty, A., Grudzińska, M., Paździora, W., & Paśko, P. (2023). Erucic acid—both sides of the story: A concise review on its beneficial and toxic properties. *Molecules (Basel, Switzerland), 28*(4), 1924. DOI: 10.3390/molecules28041924 PMID: 36838911

Ganal, M. W., Durstewitz, G., Polley, A., Bérard, A., Buckler, E. S., Charcosset, A., & Falque, M. (2011). A large maize (Zea mays L.) SNP genotyping array: Development and germplasm genotyping, and genetic mapping to compare with the B73 reference genome. *PLoS One, 6*(12), e28334. DOI: 10.1371/journal.pone.0028334 PMID: 22174790

Gan, Lu, Zhang, Chun-yu, Wang, Xiao-dong, Wang, Hao, Long, Yan, Yin, Yong-tai, . . . Lin, Zhi-wei. (2013). Proteomic and comparative genomic analysis of two Brassica napus lines differing in oil content. *Journal of Proteome Research, 12*(11), 4965–4978.

Gao, J., Thelen, K., Min, D., Smith, S., Hao, X., & Gehl, R. (2010). Effects of manure and fertilizer applications on canola oil content and fatty acid composition. *Agronomy Journal, 102*(2), 790–797. DOI: 10.2134/agronj2009.0368

Gao, L., Caldwell, C., & Jiang, Y. (2018). Photosynthesis and growth of camelina and canola in response to water deficit and applied nitrogen. *Crop Science, 58*(1), 393–401. DOI: 10.2135/cropsci2017.07.0406

Garcia, A., Martinez, S., & Lopez, R. (2021). Time and resource allocation in canola breeding programs: A cost-benefit analysis. *Crop Science, 61*(4), 2567–2580.

Garcia-Aloy, M., Hulshof, P., Estruel-Amades, S., Osté, M., Lankinen, M., Geleijnse, J., de Goede, J., Ulaszewska, M., Mattivi, F., Bakker, S. J. L., Schwab, U., & Andrés-Lacueva, C. (2019). Biomarkers of food intake for nuts and vegetable oils: An extensive literature search. *Genes & Nutrition*, *14*(1), 7. Advance online publication. DOI: 10.1186/s12263-019-0628-8 PMID: 30923582

Garcia, L., & Martinez, J. (2021). Formulation strategies for glycerin-based soaps: Effects on physicochemical properties. *Journal of Surfactants and Detergents*, *24*(2), 315–328.

Garg, S., Fungtammasan, A., Carroll, A., Chou, M., Schmitt, A., Zhou, X., & Sedlazeck, F. (2019). Efficient chromosome-scale haplotype-resolved assembly of human genomes. *bioRxiv*. DOI: 10.1101/810341

Garimella, K. V., Iqbal, Z., Krause, M. A., Campino, S., Kekre, M., Drury, E., & MacInnis, B. (2020). Detection of simple and complex de novo mutations with multiple reference sequences. *Genome Research*, *30*(8), 1154–1169. DOI: 10.1101/gr.255505.119 PMID: 32817236

Garvin, M. R., Saitoh, K., & Gharrett, A. J. (2010). Application of single nucleotide polymorphisms to non-model species: A technical review. *Molecular Ecology Resources*, *10*(6), 915–934. DOI: 10.1111/j.1755-0998.2010.02891.x PMID: 21565101

Gazave, E., Tassone, E. E., Ilut, D. C., Wingerson, M., Datema, E., Witsenboer, H. M., Davis, J. B., Grant, D., Dyer, J. M., Jenks, M. A., Brown, J., & Gore, M. A. (2016). Population genomic analysis reveals differential evolutionary histories and patterns of diversity across subgenomes and subpopulations of *Brassica napus* L. *Frontiers in Plant Science*, *7*, 525. DOI: 10.3389/fpls.2016.00525 PMID: 27148342

Geibel, J., Reimer, C., Weigend, S., Weigend, A., Pook, T., & Simianer, H. (2021). How array design creates SNP ascertainment bias. *PLoS One*, *16*(3), e0245178. DOI: 10.1371/journal.pone.0245178 PMID: 33784304

Gelpí, E., de la Paz, M. P., Terracini, B., Abaitua, I., de la Cámara, A. G., Kilbourne, E. M., Lahoz, C., Nemery, B., Philen, R. M., Soldevilla, L., & Tarkowski, S. (2002). The Spanish toxic oil syndrome 20 years after its onset: A multidisciplinary review of scientific knowledge. *Environmental Health Perspectives*, *110*(5), 457–464. DOI: 10.1289/ehp.02110457 PMID: 12003748

Gent, J. I., Ellis, N. A., Guo, L., Harkess, A. E., Yao, Y., Zhang, X., & Dawe, R. K. (2013). CHH islands: De novo DNA methylation in near-gene chromatin regulation in maize. *Genome Research*, *23*(4), 628–637. DOI: 10.1101/gr.146985.112 PMID: 23269663

Ghazani, S., García-Llatas, G., & Marangoni, A. (2014). Micronutrient content of cold-pressed, hot-pressed, solvent extracted and rbd canola oil: Implications for nutrition and quality. *European Journal of Lipid Science and Technology*, *116*(4), 380–387. DOI: 10.1002/ejlt.201300288

Gibrat, J. F. (2019). On the use of algebraic topology concepts to check the consistency of genome assembly. *Biophysics and Physicobiology*, *16*(0), 444–451. DOI: 10.2142/biophysico.16.0_444 PMID: 31984196

Gill, S. S., Gill, R., Kumar, G., Pareek, A., Sharma, P. C., Anjum, N. A., & Tuteja, N. (2012). Mustard: approaches for crop improvement and abiotic stress tolerance. *Improving crop resistance to abiotic stress*, 1351-1368.

Gilliham, M., Able, J. A., & Roy, S. J. (2017). Translating knowledge about abiotic stress tolerance to breeding programmes. *The Plant Journal*, *90*(5), 898–917. DOI: 10.1111/tpj.13456 PMID: 27987327

Gilroy, S., Białasek, M., Suzuki, N., Górecka, M., Devireddy, A. R., Karpiński, S., & Mittler, R. (2016). ROS, Calcium, and Electric Signals: Key Mediators of Rapid Systemic Signaling in Plants. *Plant Physiology*, *171*(3), 1606–1615. DOI: 10.1104/pp.16.00434 PMID: 27208294

Gomaa, W., Mosaad, G., & Yu, P. (2018). On a molecular basis, investigate association of molecular structure with bioactive compounds, anti-nutritional factors and chemical and nutrient profiles of canola seeds and co-products from canola processing: Comparison crusher plants within canada and within china as well as between canada and china. *Nutrients*, *10*(4), 519. DOI: 10.3390/nu10040519 PMID: 29690527

Gómez-Campo, C., & Prakash, S. (1999). 2 Origin and domestication. In Gómez-Campo, C. (Ed.), *Developments in Plant Genetics and Breeding* (Vol. 4, pp. 33–58). Elsevier.

Gowda, G., & Raftery, D. (2015). Can nmr solve some significant challenges in metabolomics? *Journal of Magnetic Resonance (San Diego, Calif.)*, *260*, 144–160. DOI: 10.1016/j.jmr.2015.07.014 PMID: 26476597

Gravel, S., Henn, B. M., Gutenkunst, R. N., Indap, A. R., Marth, G. T., Clark, A. G., & Wu, H. (2011). Demographic history and rare allele sharing among human populations. *Proceedings of the National Academy of Sciences of the United States of America*, *108*(29), 11983–11988. DOI: 10.1073/pnas.1019276108 PMID: 21730125

Green, A. G. (2007). Genetic modification of canola oil for improved nutritional profiles. *Journal of the American Oil Chemists' Society*, *84*(12), 1105–1112.

Green, J. M. (2018). The rise and future of glyphosate and glyphosate-resistant crops. *Pest Management Science, 74*(5), 1035–1039. DOI: 10.1002/ps.4462 PMID: 27758090

Greer, S., Hackenberg, D., Gegas, V., Mitrousia, G., Edwards, D., Batley, J., Teakle, G., Barker, G., & Walsh, J. (2021). Quantitative Trait Locus Mapping of Resistance to Turnip Yellows Virus in Brassica rapa and Brassica oleracea and Introgression of These Resistances by Resynthesis Into Allotetraploid Plants for Deployment in Brassica napus. *Frontiers in Plant Science, 12*, 781385. Advance online publication. DOI: 10.3389/fpls.2021.781385 PMID: 34956278

Grimberg, Å., Carlsson, A., Marttila, S., Bhalerao, R., & Hofvander, P. (2015). Transcriptional transitions in nicotiana benthamiana leaves upon induction of oil synthesis by wrinkled1 homologs from diverse species and tissues. *BMC Plant Biology, 15*(1), 192. Advance online publication. DOI: 10.1186/s12870-015-0579-1 PMID: 26253704

Guan, D., McCarthy, S. A., Wood, J., Howe, K., Wang, Y., & Durbin, R. (2020). Identifying and removing haplotypic duplication in primary genome assemblies. *Bioinformatics (Oxford, England), 36*(9), 2896–2898. Advance online publication. DOI: 10.1093/bioinformatics/btaa025 PMID: 31971576

Gu, J., Chao, H., Gan, L., Guo, L., Zhang, K., Li, Y., & Li, M. (2016). Proteomic dissection of seed germination and seedling establishment in Brassica napus. *Frontiers in Plant Science, 7*, 1482.

Gu, J., Guan, Z., Jiao, Y., Liu, K., & Hong, D. (2024). The story of a decade: Genomics, functional genomics and molecular breeding in Brassica napus. *Plant Communications, 5*(4), 100884. DOI: 10.1016/j.xplc.2024.100884 PMID: 38494786

Gulden, R. H., Warwick, S. I., & Thomas, A. G. (2008). The biology of Canadian weeds. 137. Brassica napus L. and B. rapa L. *Canadian Journal of Plant Science, 88*(5), 951–996. DOI: 10.4141/CJPS07203

Gul, H., Ulfat, M., Zafar, Z., Haider, W., Ali, Z., Manzoor, H., & Athar, H. (2022). Photosynthesis and salt exclusion are key physiological processes contributing to salt tolerance of canola (brassica napus l.): Evidence from physiology and transcriptome analysis. *Genes, 14*(1), 3. DOI: 10.3390/genes14010003 PMID: 36672744

Guo, L., Wang, X., & Zhao, Y. (2016). Genetic variation for drought tolerance in Brassica napus: Insights from root architecture and water use efficiency. *Frontiers in Plant Science, 7*, 1253.

Guo, X., Sarup, P., Jensen, J., Orabi, J., Kristensen, N., Mulder, F., Jahoor, A., & Jensen, J. (2020). Genetic variance of metabolomic features and their relationship with malting quality traits in spring barley. *Frontiers in Plant Science*, *11*, 575467. Advance online publication. DOI: 10.3389/fpls.2020.575467 PMID: 33193515

Guo, Y., Liu, J., & Zhang, L. (2020). Metabolomic analysis of oil biosynthesis pathways in Brassica napus seeds. *Metabolomics*, *16*(5), 62. PMID: 32335734

Guo, Z., Wang, H., Tao, J., Ren, Y., Xu, C., Wu, K., & Xu, Y. (2019). Development of multiple SNP marker panels affordable to breeders through genotyping by target sequencing (GBTS) in maize. *Molecular Breeding*, *39*(3), 1–12. DOI: 10.1007/s11032-019-0940-4

Guo, Z., Yang, W., Chang, Y., Ma, X., Tu, H., Xiong, F., & Xiong, L. (2018). Genome-wide association studies of image traits reveal genetic architecture of drought resistance in rice. *Molecular Plant*, *11*(6), 789–805. DOI: 10.1016/j.molp.2018.03.018 PMID: 29614319

Gurusinghe, S., Brooks, T., Barrow, R., Zhu, X., Thotagamuwa, A., Dennis, P., Gupta, V. V. S. R., Vanniasinkam, T., & Weston, L. (2019). Technologies for the selection, culture and metabolic profiling of unique rhizosphere microorganisms for natural product discovery. *Molecules (Basel, Switzerland)*, *24*(10), 1955. DOI: 10.3390/molecules24101955 PMID: 31117282

Haddadi, P., Larkan, N. J., Van deWouw, A., Zhang, Y., Neik, T. X., & Beynon, E.. (2022). *Brassica napus* genes Rlm4 and Rlm7, conferring resistance to *Leptosphaeria maculans*, are alleles of the Rlm9 wall-associated kinase-like resistance locus. *Plant Biotechnology Journal*, *20*(7), 1229–1231. DOI: 10.1111/pbi.13818 PMID: 35338565

Haider, S., Iqbal, J., Naseer, S., Shaukat, M., Abbasi, B. A., Yaseen, T., & Mahmood, T. (2022b). Unfolding molecular switches in plant heat stress resistance: A comprehensive review. *Plant Cell Reports*, *41*(3), 775–798. DOI: 10.1007/s00299-021-02754-w PMID: 34401950

Haider, S., Iqbal, J., Naseer, S., Yaseen, T., Shaukat, M., Bibi, H., & Mahmood, T. (2021). Molecular mechanisms of plant tolerance to heat stress: Current landscape and future perspectives. *Plant Cell Reports*, *40*(12), 1–25. DOI: 10.1007/s00299-021-02696-3 PMID: 33890138

Haider, S., Iqbal, J., Shaukat, M., Naseer, S., & Mahmood, T. (2021b). The epigenetic chromatin-based regulation of somatic heat stress memory in plants. *Plant Gene*, *27*, 100318. DOI: 10.1016/j.plgene.2021.100318

Haider, S., Raza, A., Iqbal, J., Shaukat, M., & Mahmood, T. (2022a). Analyzing the regulatory role of heat shock transcription factors in plant heat stress tolerance: A brief appraisal. *Molecular Biology Reports*, *49*(6), 5771–5785. DOI: 10.1007/s11033-022-07190-x PMID: 35182323

Hajduch, M., Casteel, J. E., Hurrelmeyer, K. E., Song, Z., Agrawal, G. K., & Thelen, J. J. (2006). Proteomic analysis of seed filling in Brassica napus. Developmental characterization of metabolic isozymes using high-resolution two-dimensional gel electrophoresis. *Plant Physiology*, *141*(1), 32–46.

Han, F., Zhang, X., Yang, L., Zhuang, M., Zhang, Y., Li, Z., & Lv, H. (2018). iTRAQ-based proteomic analysis of Ogura-CMS cabbage and its maintainer line. *International Journal of Molecular Sciences*, *19*(10), 3180.

Hanifah, A., Maharijaya, A., Putri, S., Laviña, W., & Sobir, . (2018). Untargeted metabolomics analysis of eggplant (solanum melongena l.) fruit and its correlation to fruit morphologies. *Metabolites*, *8*(3), 49. DOI: 10.3390/metabo8030049 PMID: 30200482

Han, L., Silvestre, S., Sayanova, O., Haslam, R., & Napier, J. (2022). Using field evaluation and systematic iteration to rationalize the accumulation of omega-3 long-chain polyunsaturated fatty acids in transgenic camelina sativa. *Plant Biotechnology Journal*, *20*(9), 1833–1852. DOI: 10.1111/pbi.13867 PMID: 35656640

Hanschen, F., Lamy, E., Schreiner, M., & Rohn, S. (2014). Reactivity and stability of glucosinolates and their breakdown products in foods. *Angewandte Chemie*, *53*(43), 11430–11450. DOI: 10.1002/anie.201402639 PMID: 25163974

Hansen, S. N., & Harris, W. S. (2007). New evidence for the cardiovascular benefits of long chain omega-3 fatty acids. *Current Atherosclerosis Reports*, *9*(6), 434–440. DOI: 10.1007/s11883-007-0058-8 PMID: 18377782

Harker, K. N., O'Donovan, J. T., & Blackshaw, R. E. (2012). Reduced herbicide inputs in herbicide-tolerant canola (Brassica napus) systems. *Weed Science*, *60*(2), 228–235.

Harper, A. L., Trick, M., Higgins, J., Fraser, F., Clissold, L., Wells, R., Hattori, C., Werner, P., & Bancroft, I. (2012). Associative transcriptomics of traits in the polyploid crop species Brassica napus. *Nature Biotechnology*, *30*(8), 798–802. DOI: 10.1038/nbt.2302 PMID: 22820317

Harris, W. S., Kris-Etherton, P. M., & Harris, K. A. (2008). Intakes of long-chain omega-3 fatty acids associated with reduced risk for death from coronary heart disease in healthy adults. *Current Atherosclerosis Reports, 10*(6), 503–509. DOI: 10.1007/s11883-008-0078-z PMID: 18937898

Hasan, J., Shaikh, R., Megha, S., Herrmann, D., Kebede, B., & Rahman, H. (2021). Mapping of flowering time, seed quality and clubroot resistance in rutabaga × spring canola populations and their association. Euphytica, 217. https://.oi.org/10.1007/s10681-021-02889-7

Hasan, J., Shaikh, R., Megha, S., Herrmann, D. T., Kebede, B., & Rahman, H. (2021). Mapping of flowering time, seed quality and clubroot resistance in rutabaga × spring canola populations and their association. *Euphytica, 217*(8), 160. Advance online publication. DOI: 10.1007/s10681-021-02889-7

Hashemi, A., Abdolzadeh, A., & Sadeghipour, H. (2010). Beneficial effects of silicon nutrition in alleviating salinity stress in hydroponically grown canola,brassica napusl., plants. *Soil Science and Plant Nutrition, 56*(2), 244–253. DOI: 10.1111/j.1747-0765.2009.00443.x

Hatakeyama, K., Suwabe, K., Tomita, R. N., Kato, T., Nunome, T., Fukuoka, H., & Matsumoto, S. (2013). Identification and characterization of Crr1a, a gene for resistance to clubroot disease (Plasmodiophora brassicae Woronin) in Brassica rapa L. *PLoS One, 8*(1), e54745.

Havlickova, L., He, Z., Wang, L., Langer, S., Harper, A. L., Kaur, H., Broadley, M. R., Gegas, V., & Bancroft, I. (2018). Validation of an updated Associative Transcriptomics platform for the polyploid crop species Brassica napus by dissection of the genetic architecture of erucic acid and tocopherol isoform variation in seeds. *The Plant Journal, 93*(1), 181–192. DOI: 10.1111/tpj.13767 PMID: 29124814

Haworth, M., Marino, G., Atzori, G., Fabbri, A., Daccache, A., Killi, D., Carli, A., Montesano, V., Conte, A., Balestrini, R., & Centritto, M. (2023). Plant Physiological Analysis to Overcome Limitations to Plant Phenotyping. *Plants, 12*(23), 4015. DOI: 10.3390/plants12234015 PMID: 38068650

Hayward, A. (2011). Introduction-Oilseed Brassicas. In Edwards, D., Batley, J., Parkin, I., & Kole, C. (Eds.), *Genetics, Genomics and Breeding of Oilseed Brassicas*. Taylor & Francis Group.

Heakel, R. (2019). Analysis of genetic diversity among a population of canola genotypes as revealed by ISSR-PCR and their associations to seed yield and oil content. *Annals of Agricultural Sciences, Moshtohor, 57*(2), 425–434. Advance online publication. DOI: 10.21608/assjm.2019.44917

Heffner, E. L., Sorrells, M. E., & Jannink, J. L. (2009). Genomic selection for crop improvement. *Crop Science*, *49*(1), 1–12. DOI: 10.2135/cropsci2008.08.0512

He, J., Zhao, X., Laroche, A., Lu, Z. X., Liu, H., & Li, Z. (2014). Genotyping-bysequencing (GBS), an ultimate marker-assisted selection (MAS) tool to accelerate plant breeding. *Frontiers in Plant Science*, *5*, 484. DOI: 10.3389/fpls.2014.00484 PMID: 25324846

Heng, S., Chen, F., Wei, C., Li, X., Yi, B., Ma, C., & Wen, J. (2019). Cytological and iTRAQ-based quantitative proteomic analyses of hau CMS in Brassica napus L. *Journal of Proteomics*, *193*, 230–238.

Hernández-Terán, A., Wegier, A., Benítez, M., Lira, R., & Escalante, A. (2017). Domesticated, genetically engineered, and wild plant relatives exhibit unintended phenotypic differences: A comparative meta-analysis profiling rice, canola, maize, sunflower, and pumpkin. *Frontiers in Plant Science*, *8*, 2030. Advance online publication. DOI: 10.3389/fpls.2017.02030 PMID: 29259610

Herselman, L., Thwaites, R., Kimmins, F., Courtois, B., Van Der Merwe, P., & Seal, S. E. (2004). Identification and mapping of AFLP markers linked to peanut (Arachis hypogaea L.) resistance to the aphid vector of groundnut rosette disease. *Theoretical and Applied Genetics*, *109*(7), 1426–1433. DOI: 10.1007/s00122-004-1756-z PMID: 15290049

Heslot, N., Rutkoski, J., Poland, J., Jannink, J. L., & Sorrells, M. E. (2013). Impact of marker ascertainment bias on genomic selection accuracy and estimates of genetic diversity. *PLoS One*, *8*(9), e74612. DOI: 10.1371/journal.pone.0074612 PMID: 24040295

Hickey, L. T., Hafeez, A. N., & Robinson, H. (2019). Breeding crops to feed 10 billion. *Nature Biotechnology*, *37*(9), 744–754. DOI: 10.1038/s41587-019-0152-9 PMID: 31209375

Hirsch, C. N., Foerster, J. M., Johnson, J. M., Sekhon, R. S., Muttoni, G., Vaillancourt, B., & Buell, C. R. (2014). Insights into the maize pan-genome and pan-transcriptome. *The Plant Cell*, *26*(1), 121–135. DOI: 10.1105/tpc.113.119982 PMID: 24488960

Hong, , Yang, L., Zhang, D., & Shi, J. (2016). Plant Metabolomics: An Indispensable System Biology Tool for Plant Science. *International Journal of Molecular Sciences*, *17*(6), 767. Advance online publication. DOI: 10.3390/ijms17060767 PMID: 27258266

Hooft, J., Mohimani, H., Bauermeister, A., Dorrestein, P., Duncan, K., & Medema, M. (2020). Linking genomics and metabolomics to chart specialized metabolic diversity. *Chemical Society Reviews*, *49*(11), 3297–3314. DOI: 10.1039/D0CS00162G PMID: 32393943

Hopkins, R. J., Damman, H., & Robinson, J. (2009). Glucosinolate-biosynthesis genes and herbivory resistance in Brassicaceae plants. *Journal of Chemical Ecology*, *35*(1), 1–10. PMID: 19052820

Hossain, S., Kadkol, G., Raman, R., Salisbury, P., & Raman, H. (2012). Breeding *Brassica napus* for shatter resistance. *Plant Breeding*, *14*(1), 313–332.

Howard, B. V., Robbins, D. C., Sievers, M. L., Lee, E. T., Rhoades, D., Devereux, R. B., Cowan, L. D., Gray, R. S., Welty, T. K., Go, O. T., & Howard, W. J. (2000). LDL cholesterol as a strong predictor of coronary heart disease in diabetic individuals with insulin resistance and low LDL: The Strong Heart Study. *Arteriosclerosis, Thrombosis, and Vascular Biology*, *20*(3), 830–835. DOI: 10.1161/01.ATV.20.3.830 PMID: 10712410

Howlett, B., Marcroft, S., & Wouw, A. (2015). Evolution of virulence in fungal plant pathogens: Exploiting fungal genomics to control plant disease. *Mycologia*, *107*(3), 441–451. DOI: 10.3852/14-317 PMID: 25725000

Huang, R., Liu, Z., Xing, M., Yang, Y., Wu, X., Liu, H., & Liang, W. (2019). Heat stress suppresses Brassica napus seed oil accumulation by inhibition of photosynthesis and BnWRI1 pathway. *Plant & Cell Physiology*, *60*(7), 1457–1470.

Huang, S., Deng, L., Guan, M., Li, J., Lu, K., Wang, H., Fu, D., Mason, A. S., Liu, S., & Hua, W. (2013). Identification of genome-wide single nucleotide polymorphisms in allopolyploid crop Brassica napus. *BMC Genomics*, *14*(1), 717. DOI: 10.1186/1471-2164-14-717 PMID: 24138473

Huang, S., Liu, Z., & Li, C. (2019). CRISPR-Cas9 mediated herbicide resistance in Brassica napus. *Plant Biotechnology Journal*, *17*(12), 2253–2264.

Huang, S., Xiang, Y., Xie, Y., Guo, Y., Zheng, L., Li, J., Wu, Z., Fu, D., Wang, H., Liu, S., & Hua, W. (2020). Exome capture-based target enrichment for genetic association discovery in Brassica napus. *The Plant Journal*, *104*(3), 805–819.

Huang, X., & Han, B. (2014). Natural variations and genome-wide association studies in crop plants. *Annual Review of Plant Biology*, *65*(1), 531–551. DOI: 10.1146/annurev-arplant-050213-035715 PMID: 24274033

Huang, X., Zhao, Y., Wei, X., Li, C., Wang, A., Zhao, Q., Li, W., Guo, Y., Deng, L., Zhu, C., Fan, D., Lu, Y., Weng, Q., Liu, K., Zhou, T., Jing, Y., Si, L., Dong, G., Huang, T., & Han, B. (2013). Genome-wide association study of flowering time and grain yield traits in a worldwide collection of rice germplasm. *Nature Genetics*, *44*(1), 32–39. DOI: 10.1038/ng.1018 PMID: 22138690

Huang, Y., Yan, X., Cai, Y., Yin, Z., Cai, R., Guo, W., & Qiu, L. (2019). Transcriptome analysis reveals genetic mechanisms of the fine regulation of the adaptation of Brassica napus to different environments. *The Plant Journal*, *100*(3), 629–643.

Huang, Y., Zhang, L., & Wang, X. (2023). Integrating genomics and phenomics for accelerated improvement of seed oil content in Brassica napus. *Nature Plants*, *9*(3), 265–279.

Huang, Z., Ban, Y. Y., Bao, R., Zhang, X. X., Xu, A. X., & Ding, J. (2014). Inheritance and gene mapping of the white flower in *Brassica napus* L. *New Zealand Journal of Crop and Horticultural Science*, *42*(2), 111–117. DOI: 10.1080/01140671.2013.863211

Hu, C., Shi, J., Quan, S., Cui, B., Kleeßen, S., Nikoloski, Z., & Zhang, D. (2014). Metabolic variation between japonica and indica rice cultivars as revealed by non-targeted metabolomics. *Scientific Reports*, *4*(1), 5067. Advance online publication. DOI: 10.1038/srep05067 PMID: 24861081

Hu, D., Jing, J., Snowdon, R. J., Mason, A. S., Shen, J., Meng, J., & Zou, J. (2021). Exploring the gene pool of Brassica napus by genomics-based approaches. *Plant Biotechnology Journal*, *19*(9), 1693–1712. DOI: 10.1111/pbi.13636 PMID: 34031989

Huiming, P., Jiefu, Z., Cunkou, Q., Jianqin, G., & Xinjun, C. (2003). Inheritance of herbicide resistance in *Brassica napus* and its utilization. *Jiangsu Journal of Agricultural Sciences*, *19*(2), 81–86. DOI: 10.3390/ijms21041345

Hu, J., Chen, B., Zhao, J., Zhang, F., Xie, T., Xu, K., Gao, G., Yan, G., Li, H., Li, L., Ji, G., An, H., Li, H., Huang, Q., Zhang, M., Wu, J., Song, W., Zhang, X., Luo, Y., & Wu, X. (2022). Genomic selection and genetic architecture of agronomic traits during modern rapeseed breeding. *Nature Genetics*, *54*(5), 694–704. DOI: 10.1038/s41588-022-01055-6 PMID: 35484301

Hu, L., Robert, C., Cadot, S., Zhang, X., Ye, M., Li, B., Manzo, D., Chervet, N., Steinger, T., van der Heijden, M. G. A., Schlaeppi, K., & Erb, M. (2018). Root exudate metabolites drive plant-soil feedback on growth and defense by shaping the rhizosphere microbiota. *Nature Communications*, *9*(1), 2738. Advance online publication. DOI: 10.1038/s41467-018-05122-7 PMID: 30013066

Hulse-Kemp, A. M., Lemm, J., Plieske, J., Ashrafi, H., Buyyarapu, R., Fang, D. D., & Stelly, D. M. (2015). Development of a 63K SNP array for cotton and high-density mapping of intraspecific and interspecific populations of Gossypium spp. *G3 (Bethesda, Md.)*, *5*(6), 1187–1209. DOI: 10.1534/g3.115.018416 PMID: 25908569

Hu, Q., Hua, W., Yin, Y., Zhang, X., Liu, L., Shi, J., Zhao, Y., Qin, L., Chen, C., & Wang, H. (2017). Rapeseed research and production in China. *The Crop Journal*, *5*(2), 127–135. DOI: 10.1016/j.cj.2016.06.005

Hurgobin, B., Golicz, A. A., Bayer, P. E., Chan, C. K. K., Tirnaz, S., Dolatabadian, A., & Edwards, D. (2018). Homoeologous exchange is a major cause of gene presence/absence variation in the amphidiploid Brassica napus. *Plant Biotechnology Journal*, *16*(7), 1265–1274. DOI: 10.1111/pbi.12867 PMID: 29205771

Hussain, M., Khan, M. A., Ali, Y., Javaid, M. M., Iqbal, B., Nasir, M., Sabir, W., & Muhammad, F. (2017). Wheat breeding for durable rust resistance and high yield potential in historical prospective and current status. *Advances in Zoology and Botany*, *5*(4), 55–63. DOI: 10.13189/azb.2017.050404

Hussain, S., Iqbal, M., & Cheema, N. M. (2019). Molecular and physiological mechanisms of salt tolerance in Brassica napus. *Plant Physiology and Biochemistry*, *142*, 401–411.

Hu, X., Li, W., & Chen, Q. (2015). Proteomic analysis of defense responses in Brassica napus under biotic and abiotic stress. *Journal of Proteomics*, *127*, 216–226.

Hu, X., Sullivan-Gilbert, M., Gupta, M., & Thompson, S. A. (2006). Mapping of the loci controlling oleic and linolenic acid contents and development of *fad2* and *fad3* allele-specific markers in canola (*Brassica napus* L.). *Theoretical and Applied Genetics*, *113*(3), 497–507. DOI: 10.1007/s00122-006-0315-1 PMID: 16767448

Huynh, B. L., Ehlers, J. D., Huang, B. E., Muñoz-Amatriaín, M., Lonardi, S., Santos, J., Ndeve, A., Batieno, B. J., Boukar, O., Cisse, N., Drabo, I., Fatokun, C., Kusi, F., Agyare, R. Y., Guo, Y.-N., Herniter, I., Lo, S., Wanamaker, S. I., Xu, S., & Roberts, P. A. (2018). A multi-parent advanced generation inter-cross (MAGIC) population for genetic analysis and improvement of cowpea (*Vigna unguiculata* L. Walp.). *The Plant Journal*, *93*(6), 1129–1142. DOI: 10.1111/tpj.13827 PMID: 29356213

Ijaz, S., Iqbal, J., Abbasi, B. A., Yaseen, T., Rehman, S., Kazi, M. & Mahmood, T. (2024b). Role of OMICS-Based Technologies in Plant Sciences. OMICs-based Techniques for Global Food Security, 45-66.

Ijaz, S., Iqbal, J., Abbasi, B. A., Yaseen, T., Rehman, S., Kazi, M., & Mahmood, T. (2024). Role of OMICS-Based Technologies in Plant Sciences. OMICs-based Techniques for Global Food Security, 45-66.

Ijaz, S., Iqbal, J., Abbasi, B. A., Yaseen, T., Rehman, S., Kazi, M., & Mahmood, T. (2024a). Role of OMICS-Based Technologies in Plant Sciences. OMICs-based Techniques for Global Food Security, 45-66.

Ijaz, S., Iqbal, J., Abbasi, B. A., Tufail, A., Ullah, Z., Sharifi-Rad, J., & Uddin, S. (2023). Biofortification: Lessons from the Past and Strategies for Future Food Security. In *Legumes Biofortification* (pp. 521–545). Springer International Publishing. DOI: 10.1007/978-3-031-33957-8_23

Ingvarsson, P. K., Hvidsten, T. R., & Street, N. R. (2016). Towards integration of population and comparative genomics in forest trees. *The New Phytologist*, *212*(2), 338–344. DOI: 10.1111/nph.14153 PMID: 27575589

Iniguez-Luy, F. L., & Federico, M. L. (2011). The genetics of *Brassica napus*. *Genetics and Genomics of the Brassicaceae*, 291-322. DOI: 10.1007/978-1-4419-7118-0_10

Islam, M. T., Lee, B.-R., La, V. H., Bae, D.-W., Jung, W.-J., & Kim, T.-H. (2021). Label-free quantitative proteomics analysis in susceptible and resistant Brassica napus cultivars infected with Xanthomonas campestris pv. campestris. *Microorganisms*, *9*(2), 253.

Ismaili, A., Salavati, A., & Pour Mohammadi, P. (2015). A comparative proteomic analysis of responses to high temperature stress in hypocotyl of canola (Brassica napus L.). *Protein and Peptide Letters*, *22*(3), 285–299.

Jagadeesan, B., Gerner-Smidt, P., Allard, M. W., Leuillet, S., Winkler, A., Xiao, Y., Chaffron, S., Van Der Vossen, J., Tang, S., Katase, M., McClure, P., Kimura, B., Ching Chai, L., Chapman, J., & Grant, K. (2019). The use of next-generation sequencing for improving food safety: Translation into practice. *Food Microbiology*, *79*, 96–115. DOI: 10.1016/j.fm.2018.11.005 PMID: 30621881

Jagadish, S., Pal, M., Sukumaran, S., Parani, M., & Siddique, K. (2020). Heat stress resilient crops for future hotter environments. *Plant Physiology Reports*, *25*(4), 529–532. DOI: 10.1007/s40502-020-00559-9

Jamaloddin, M., Mahender, A., Gokulan, C. G., Balachiranjeevi, C., Maliha, A., Patel, H. K., & Ali, J. (2021). Molecular approaches for disease resistance in rice. In *Rice improvement: Physiological, molecular breeding and genetic perspectives* (pp. 315–378). Springer International Publishing Cham. DOI: 10.1007/978-3-030-66530-2_10

Jamaloddin, M., Maliha, A., Gokulan, C., Gaur, N., & Patel, H. K. (2021). Metabolomics-assisted breeding for crop improvement: An emerging approach. *Omics Technologies for Sustainable Agriculture and Global Food Security, 1*, 241–279. DOI: 10.1007/978-981-16-0831-5_11

Jan, H., Abbadi, A., Lücke, S., Nichols, R., & Snowdon, R. (2016). Genomic prediction of testcross performance in canola (*Brassica napus*). *PLoS One, 11*(1), e0147769. Advance online publication. DOI: 10.1371/journal.pone.0147769 PMID: 26824924

Jaramillo-Botero, A., Colorado, J., Quimbaya, M., Rebolledo, M. C., Lorieux, M., Ghneim-Herrera, T., Arango, C. A., Tobón, L. E., Finke, J., Rocha, C., Muñoz, F., Riascos, J. J., Silva, F., Chirinda, N., Caccamo, M., Vandepoele, K., & Goddard, W. A.III. (2022). The ÓMICAS alliance, an international research program on multi-omics for crop breeding optimization. *Frontiers in Plant Science, 13*, 992663. DOI: 10.3389/fpls.2022.992663 PMID: 36311093

Jarquín, D., Crossa, J., & Lacaze, X. (2014). Genomic prediction of quantitative traits in plant breeding using GBLUP models. *BMC Genomics, 15*, 937.

Jarquin, D., Kocak, K., Posadas, L., Hyma, K., Jedlicka, J., Graef, G., & Lorenz, A. (2014). Genotyping by sequencing for genomic prediction in a soybean breeding population. *BMC Genomics, 15*(1), 740. DOI: 10.1186/1471-2164-15-740 PMID: 25174348

Jaspers, P., & Kangasjärvi, J. (2010). Reactive oxygen species in abiotic stress signaling. *Physiologia Plantarum, 138*(4), 405–413. DOI: 10.1111/j.1399-3054.2009.01321.x PMID: 20028478

Javed, N., Geng, J., Tahir, M., McVetty, P., Li, G., & Duncan, R. W. (2016). Identification of QTL influencing seed oil content, fatty acid profile and days to flowering in Brassica napus L. *Euphytica, 207*(1), 191–211. DOI: 10.1007/s10681-015-1565-2

Jean, M., Brown, G. G., & Landry, B. S. (1997). Genetic mapping of nuclear fertility restorer genes for the 'Polima' cytoplasmic male sterility in canola (Brassica napus L.) using DNA markers. *Theoretical and Applied Genetics, 95*(3-4), 321–328. DOI: 10.1007/s001220050566

Jendoubi, T. (2021). Approaches to integrating metabolomics and multi-omics data: A primer. *Metabolites, 11*(3), 184. DOI: 10.3390/metabo11030184 PMID: 33801081

Jenko, J., Gorjanc, G., Cleveland, M. A., Varshney, R. K., Whitelaw, C. B., Woolliams, J. A., & Hickey, J. M. (2015). Potential of promotion of alleles by genome editing to improve quantitative traits in livestock breeding programs. *Genetics, Selection, Evolution : GSE, 47*(1), 55. DOI: 10.1186/s12711-015-0135-3 PMID: 26133579

Jha, M., & Roy, S. (2020). Functional module extraction from gene expression data using data mining techniques. *ACM SIGBioinformatics Record*, *10*(1), 1–2. DOI: 10.1145/3383672.3383674

Jia, H., Shao, M., He, Y., Guan, R., Chu, P., & Jiang, H. (2015). Proteome dynamics and physiological responses to short-term salt stress in Brassica napus leaves. *PLoS One*, *10*(12), e0144808.

Jia, H., Shao, M., He, Y., Guan, R., Chu, P., & Jiang, H. (2015). Proteome dynamics and physiological responses to short-term salt stress in brassica napus leaves. *PLoS One*, *10*(12), e0144808. DOI: 10.1371/journal.pone.0144808 PMID: 26691228

Jiang, F., & Doudna, J. A. (2017). CRISPR–Cas9 structures and mechanisms. *Annual Review of Biophysics*, *46*(1), 505–529. DOI: 10.1146/annurev-biophys-062215-010822 PMID: 28375731

Jiang, N., Shi, S., Shi, H., Khanzada, H., Wassan, G., Zhu, C., Peng, X., Yu, Q., Chen, X., He, X., Fu, J., Hu, L., Xu, J., Ouyang, L., Sun, X., Zhou, D., He, H., & Bian, J. (2017). Mapping QTL for Seed Germinability under Low Temperature Using a New High-Density Genetic Map of Rice. *Frontiers in Plant Science*, *8*, 1223. Advance online publication. DOI: 10.3389/fpls.2017.01223 PMID: 28747923

Jiang, Y., Duan, Y., Yin, J., Ye, S., Zhu, J., Zhang, F., & Luo, K. (2014). Genome-wide identification and characterization of the populus wrky transcription factor family and analysis of their expression in response to biotic and abiotic stresses. *Journal of Experimental Botany*, *65*(22), 6629–6644. DOI: 10.1093/jxb/eru381 PMID: 25249073

Jian, H., Yang, B., Zhang, A., Zhang, L., Xu, X., Li, J., & Liu, L. (2017). Screening of Candidate Leaf Morphology Genes by Integration of QTL Mapping and RNA Sequencing Technologies in Oilseed Rape (*Brassica napus* L.). *PLoS One*, *12*(1), e0169641. DOI: 10.1371/journal.pone.0169641 PMID: 28068426

Jian-ping, W., Shu-fen, Z., Jia-cheng, Z., Jun-ping, H., Jin-hua, C., Yan-cheng, W., Kun, H., Lei, Z., Dong-guo, W., & Dong-fang, C. (2020). Genome-wide analysis of DNA methylation under drought in rapeseed (*Brassica napus* L.). *Zhongguo Youliao Zuowu Xuebao*, *42*(4).

Jiefu, Z., Cunkou, Q., Huiming, P., Song, C., Feng, C., Jianqin, G., Xinjun, C., Hui, G., & Shouzhong, F. (2010). Inheritance and QTL identification of oil content in rapeseed *Brassica napus* L.

Jimenez-Lopez, J. C., Gachomo, E. W., Sharma, S., & Kotchoni, S. O. (2013). Genome sequencing and next-generation sequence data analysis: A comprehensive compilation of bioinformatics tools and databases.

Jing, X. (2018). Peer review #1 of "molecular response of canola to salt stress: insights on tolerance mechanisms (v0.1)". https://doi.org/DOI: 10.7287/peerj.4822v0.1/reviews/1

Ji, R., Wang, Y., Wang, X., Liu, Y., Shen, X., & Feng, H. (2018). Proteomic analysis of the interaction between Plasmodiophora brassicae and Chinese cabbage (Brassica rapa L. ssp. Pekinensis) at the initial infection stage. *Scientia Horticulturae*, *233*, 386–393.

Johnson, I. T. (2002). Glucosinolates: Bioavailability and importance to health. *International Journal for Vitamin and Nutrition Research*, *72*(1), 26–31. DOI: 10.1024/0300-9831.72.1.26 PMID: 11887749

Johnson, L. M., & Lee, S. Y. (2022). Global trends in oilseed production and the role of canola in sustainable agriculture. *Journal of Crop Science*, *58*(4), 423–437.

Johnson, L., Wilson, P., & Taylor, R. (2019). Canola oil as a feedstock for biodiesel production: A review. *Renewable & Sustainable Energy Reviews*, *109*, 139–152.

Jones, P., Senanayake, V., Pu, S., Jenkins, D., Connelly, P., Lamarche, B., Couture, P., Charest, A., Baril-Gravel, L., West, S. G., Liu, X., Fleming, J. A., McCrea, C. E., & Kris-Etherton, P. (2014). Dha-enriched high–oleic acid canola oil improves lipid profile and lowers predicted cardiovascular disease risk in the canola oil multicenter randomized controlled trial. *The American Journal of Clinical Nutrition*, *100*(1), 88–97. DOI: 10.3945/ajcn.113.081133 PMID: 24829493

Jorge, T., Rodrigues, J., Caldana, C., Schmidt, R., Dongen, J., Thomas-Oates, J., & António, C. (2015). Mass spectrometry-based plant metabolomics: Metabolite responses to abiotic stress. *Mass Spectrometry Reviews*, *35*(5), 620–649. DOI: 10.1002/mas.21449 PMID: 25589422

Joughi, E. S. G., Hervan, E. M., Rad, A. S., & Noormohamadi, G. H. (2018). Fatty acid composition of oilseed rapeseed genotypes as affected by vermicompost application and different thermal regimes. *Agronomy Research (Tartu)*, *16*(1), 230–242.

Jovovic, Z., Andjelkovic, V., Przulj, N., & Mandic, D. (2020). Untapped genetic diversity of wild relatives for crop improvement. *Rediscovery of genetic and genomic resources for future food security*, 25-65.

Jung, H., Jeon, M., Hodgett, M., Waterhouse, P., & Eyun, S. (2020). A comparative evaluation of genome assemblers from long-read sequencing for plants and crops. *Journal of Agricultural and Food Chemistry*, *68*(29), 7670–7677. Advance online publication. DOI: 10.1021/acs.jafc.0c01647 PMID: 32530283

Jun, L., Hirani, A. H., & Zhe, L., ChunRen, W., McVetty, P., Daayf, F., & Genyi, L. (2016). QTL controlling glucosinolate content in seeds of "Brassica napus" L. *Australian Journal of Crop Science*, *10*, 152–160.

Ju, Y.-H., Roy, S. K., Choudhury, R., Aritra, K., Soo-Jeong, C., Ju-Young, R., Atikur, M., & Cho, K. (2021). Proteome changes reveal the protective roles of exogenous citric acid in alleviating Cu toxicity in Brassica napus L. *International Journal of Molecular Sciences*, *22*(11), 5879.

Kalantarahmadi, S., & Daneshian, J. (2023). Improving of canola (Brassica napus L.) yield and oil quality by foliar application of micro-nutrients under high-temperature stress. *Journal of Soil Science and Plant Nutrition*, *23*(1), 351–367. DOI: 10.1007/s42729-022-01016-2

Kang, L., Li, P., Wang, A., Ge, X., & Li, Z. (2017). A novel cytoplasmic male sterility in *Brassica napus* (inap CMS) with carpelloid stamens via protoplast fusion with Chinese woad. *Frontiers in Plant Science*, *8*, 529. DOI: 10.3389/fpls.2017.00529 PMID: 28428799

Kang, Y. J., Lee, T., Lee, J., Shim, S., Jeong, H., Satyawan, D., Kim, M. Y., & Lee, S. H. (2016). Translational genomics for plant breeding with the genome sequence explosion. *Plant Biotechnology Journal*, *14*(4), 1057–1069. DOI: 10.1111/pbi.12449 PMID: 26269219

Karabulut, F., & Khan, A. Abd_Allah, E. F., Parray, J. A. and Hashem, A. (2024). Omics and other biotechnological tools for biopesticide and microbial bioremediation. In Microbiome-Assisted Bioremediation (pp. 103-128). Academic Press.

Karim, M. M., Siddika, A., Tonu, N. N., Hossain, D. M., Meah, M. B., Kawanabe, T., Fujimoto, R., & Okazaki, K. (2014). Production of high yield short duration Brassica napus by interspecific hybridization between B. oleracea and B. rapa. *Breeding Science*, *63*(5), 495–502. DOI: 10.1270/jsbbs.63.495 PMID: 24757390

Katam, R., Lin, C., Grant, K., Katam, C., & Chen, S. (2022). Advances in plant metabolomics and its applications in stress and single-cell biology. *International Journal of Molecular Sciences*, *23*(13), 6985. DOI: 10.3390/ijms23136985 PMID: 35805979

Katche, E., Quezada-Martinez, D., Katche, E. I., Vasquez-Teuber, P., & Mason, A. S. (2019). Interspecific Hybridization for Brassica Crop Improvement. *Crop Breeding, Genetics and Genomics, 1*, e190007. DOI: 10.20900/cbgg20190007

Katz, A. M., & Messineo, F. C. (1982). Fatty acid effects on membranes: Possible role in the pathogenesis of ischemic myocardial damage. *Journal of Molecular and Cellular Cardiology, 14*(Suppl. 3), 119–122. DOI: 10.1016/0022-2828(82)90139-0 PMID: 7143451

Kaur, N., Alok, A., & Kumar, P. (2018). CRISPR-Cas9 mediated genome editing in Brassica napus for enhanced agronomic traits. *Plant Molecular Biology Reporter, 36*(5), 729–740.

Kaur, P., Jost, R., Sivasithamparam, K., & Barbetti, M. J. (2011). Proteome analysis of the Albugo candida–Brassica juncea pathosystem reveals that the timing of the expression of defence-related genes is a crucial determinant of pathogenesis. *Journal of Experimental Botany, 62*(3), 1285–1298.

Kaur, P., Singh, A., & Chana, I. (2021). Computational techniques and tools for omics data analysis: State-of-the-art, challenges, and future directions. *Archives of Computational Methods in Engineering, 28*(7), 4595–4631. DOI: 10.1007/s11831-021-09547-0 PMID: 33519179

Kaur, S., Cogan, N., Ye, G., Baillie, R., Hand, M., Ling, A., Mcgearey, A., Kaur, J., Hopkins, C., Todorovic, M., Mountford, H., Edwards, D., Batley, J., Burton, W., Salisbury, P., Gororo, N., Marcroft, S., Kearney, G., Smith, K. F., & Spangenberg, G. C. (2009). Genetic map construction and QTL mapping of resistance to blackleg (Leptosphaeria maculans) disease in Australian canola (Brassica napus L.) cultivars. *Theoretical and Applied Genetics, 120*(1), 71–83. DOI: 10.1007/s00122-009-1160-9 PMID: 19821065

Kebede, B., & Rahman, H. (2019). QTL mapping of root and aboveground biomass in the Brassica C genome using a B. napus population carrying genome content introgressed from B. oleracea. *Molecular Breeding, 39*(12), 169. Advance online publication. DOI: 10.1007/s11032-019-1064-6

Kehoe, B., Patil, S., Abbeel, P., & Goldberg, K. (2015). A survey of research on cloud robotics and automation. *IEEE Transactions on Automation Science and Engineering, 12*(2), 398–409. DOI: 10.1109/TASE.2014.2376492

Khodabin, G., Tahmasebi-Sarvestani, Z., Rad, A., & Modarres-Sanavy, S. (2020). Effect of drought stress on certain morphological and physiological characteristics of a resistant and a sensitive canola cultivar. *Chemistry & Biodiversity, 17*(2), e1900399. Advance online publication. DOI: 10.1002/cbdv.201900399 PMID: 31909554

Kim, H., Park, J., & Lee, Y. (2020). Effects of glycerin on hair moisturization and damage prevention. *International Journal of Cosmetic Science*, *42*(3), 267–276.

Kim, K.-S., Diers, B., Hyten, D., Rouf Mian, M., Shannon, J., & Nelson, R. L. (2012). Identification of positive yield QTL alleles from exotic soybean germplasm in two backcross populations. *Theoretical and Applied Genetics*, *125*(6), 1353–1369. DOI: 10.1007/s00122-012-1944-1 PMID: 22869284

Kim, S., Yamaoka, Y., Ono, H., Kim, H., Shim, D., Maeshima, M., Martinoia, E., Cahoon, E. B., Nishida, I., & Lee, Y. (2012). Atabca9 transporter supplies fatty acids for lipid synthesis to the endoplasmic reticulum. *Proceedings of the National Academy of Sciences of the United States of America*, *110*(2), 773–778. DOI: 10.1073/pnas.1214159110 PMID: 23269834

Kim, Sung Soo, Grienenberger, Etienne, Lallemand, Benjamin, Colpitts, Che C, & Kim, Sun Young, Souza, Clarice de Azevedo, . . . Kaiser, Markus. (2010). LAP6/POLYKETIDE SYNTHASE A and LAP5/POLYKETIDE SYNTHASE B encode hydroxyalkyl α-pyrone synthases required for pollen development and sporopollenin biosynthesis in Arabidopsis thaliana. *The Plant Cell*, *22*(12), 4045–4066.

Kiran, A., Wakeel, A., Snowdon, R., & Friedt, W. (2019). Genetic dissection of root architectural traits by qtl and genome-wide association mapping in rapeseed (brassica napus). *Plant Breeding*, *138*(2), 184–192. DOI: 10.1111/pbr.12665

Kittipol, V., He, Z., Wang, L., Doheny-Adams, T., Langer, S., & Bancroft, I. (2019). Genetic architecture of glucosinolate variation in brassica napus. *Journal of Plant Physiology*, *240*, 152988. DOI: 10.1016/j.jplph.2019.06.001 PMID: 31255878

Knoch, D., Abbadi, A., Grandke, F., Meyer, R. C., Samans, B., Werner, C. R., Snowdon, R. J., & Altmann, T. (2020). Strong temporal dynamics of QTL action on plant growth progression revealed through high-throughput phenotyping in canola. *Plant Biotechnology Journal*, *18*(1), 68–82. DOI: 10.1111/pbi.13171 PMID: 31125482

Knoch, D., Meyer, R. C., Heuermann, M. C., Riewe, D., Peleke, F. F., Szymański, J., & Altmann, T. (2024). Integrated multi-omics analyses and genome-wide association studies reveal prime candidate genes of metabolic and vegetative growth variation in canola. *The Plant Journal*, *117*(3), 713–728. DOI: 10.1111/tpj.16524 PMID: 37964699

Knoch, D., Werner, C., Meyer, R., Riewe, D., Abbadi, A., Lücke, S., Snowdon, R. J., & Altmann, T. (2021). Multi-omics-based prediction of hybrid performance in canola. *Theoretical and Applied Genetics*, *134*(4), 1147–1165. DOI: 10.1007/s00122-020-03759-x PMID: 33523261

Knoll, J. E., Ramos, M. L., Zeng, Y., Holbrook, C. C., Chow, M., Chen, S., & Ozias-Akins, P. (2011). TILLING for allergen reduction and improvement of quality traits in peanut (Arachis hypogaea L.). *BMC Plant Biology*, *11*(1), 1–13. DOI: 10.1186/1471-2229-11-81 PMID: 21569438

Koh, J., Barbulescu, D., Salisbury, P., & Slater, A. (2016). Pterostilbene is a potential candidate for control of blackleg in canola. *PLoS One*, *11*(5), e0156186. DOI: 10.1371/journal.pone.0156186 PMID: 27213274

Koh, J., Chen, G., Yoo, M.-J., Zhu, N., Dufresne, D., Erickson, J. E., & Chen, S. (2015). Comparative proteomic analysis of Brassica napus in response to drought stress. *Journal of Proteome Research*, *14*(8), 3068–3081.

Komar, A. (2009). *Single Nucleotide Polymorphisms*. Springer., DOI: 10.1007/978-1-60327-411-1

Körber, N., Bus, A., Li, J., Parkin, I. A., Wittkop, B., Snowdon, R. J., & Stich, B. (2016). Agronomic and seed quality traits dissected by genome-wide association mapping in Brassica napus. *Frontiers in Plant Science*, *7*, 386. DOI: 10.3389/fpls.2016.00386 PMID: 27066036

Körber, N., Wittkop, B., Bus, A., Friedt, W., Snowdon, R. J., & Stich, B. (2012). Seedling development in a Brassica napus diversity set and its relationship to agronomic performance. *Theoretical and Applied Genetics*, *125*(6), 1275–1287. DOI: 10.1007/s00122-012-1912-9 PMID: 22782254

Koscielny, C. B., Gardner, S. W., Technow, F., & Duncan, R. W. (2020). Linkage mapping and whole-genome predictions in canola (Brassica napus) subjected to differing temperature treatments. *Crop & Pasture Science*, *71*(3), 229–238. DOI: 10.1071/CP19387

Kráľová, K., Jampílek, J., & Ostrovský, I. (2012). Metabolomics - useful tool for study of plant responses to abiotic stresses. *Ecological Chemistry and Engineering. S*, *19*(2), 133–161. DOI: 10.2478/v10216-011-0012-0

Kris-Etherton, P. M., Krummel, D., Russell, M. E., Dreon, D., Mackey, S., Borchers, J., & Wood, P. D. (1988). The effect of diet on plasma lipids, lipoproteins, and coronary heart disease. *Journal of the American Dietetic Association*, *88*(11), 1373–1400. DOI: 10.1016/S0002-8223(21)08023-8 PMID: 2846672

Kruger, D., Fourie, J., & Malan, A. (2016). The effect of cover crops and their management on plant-parasitic nematodes in vineyards. *South African Journal of Enology and Viticulture*, *36*(2). Advance online publication. DOI: 10.21548/36-2-953

Kumar, A., Singh, V., & Priyamedha. (2022). Botanical Descriptions of Brassica juncea: Taxonomy, Cytology, Cytogenetics and Phylogenetic Relationships. In *The Brassica juncea Genome* (pp. 15-28). Springer.

Kumar, B. S., Jeannie, S., Doppalapudi, V. R., & Kumar, A. A. REVOLUTIONIZING CROP PHENOTYPING: BREAKTHROUGHS IN HIGH-THROUGHPUT METHODS AND THEIR MULTIFACETED IMPLEMENTATIONS. *Advances in Agronomy Vol.*

Kumar, J., Pratap, A., & Kumar, S. (2015). Plant phenomics: an overview. *Phenomics in crop plants: trends, options and limitations*, 1-10.

Kumari, M., Yagnik, K. N., Gupta, V., Singh, I. K., Gupta, R., Verma, P. K., & Singh, A. (2024). Metabolomics-driven investigation of plant defense response against pest and pathogen attack. *Physiologia Plantarum*, *176*(2), e14270. Advance online publication. DOI: 10.1111/ppl.14270 PMID: 38566280

Kumari, R., Kumar, P., Sharma, V., & Kumar, H. (2018). Allele Mining for Crop Improvement. *International Journal of Pure & Applied Bioscience*, *6*(1), 1456–1465. DOI: 10.18782/2320-7051.6073

Kumar, R., Bohra, A., Pandey, A., Pandey, M., & Kumar, A. (2017). Metabolomics for plant improvement: Status and prospects. *Frontiers in Plant Science*, *8*, 1302. Advance online publication. DOI: 10.3389/fpls.2017.01302 PMID: 28824660

Kupiec, M., Zbikowska, A., Marciniak-Lukasiak, K., & Kowalska, M. (2020). Rapeseed oil in new application: Assessment of structure of oleogels based on their physicochemical properties and microscopic observations. *Agriculture*, *10*(6), 211. DOI: 10.3390/agriculture10060211

Kurowska, M., Daszkowska-Golec, A., Gruszka, D., Marzec, M., Szurman, M., Szarejko, I., & Maluszynski, M. (2011). TILLING-a shortcut in functional genomics. *Journal of Applied Genetics*, *52*(4), 371–390. DOI: 10.1007/s13353-011-0061-1 PMID: 21912935

Kutcher, H. R., Warland, J. S., & Brandt, S. A. (2010). Temperature and precipitation effects on canola yields in Saskatchewan, Canada. *Agricultural and Forest Meteorology*, *150*(2), 161–165. DOI: 10.1016/j.agrformet.2009.09.011

Kwon, S.-W., Kim, M., Kim, H., & Lee, J. (2016). Shotgun quantitative proteomic analysis of proteins responding to drought stress in Brassica rapa L.(inbred line "Chiifu"). *International Journal of Genomics*, *2016*(1), 4235808.

Kwon, Y. S., Ryu, C.-M., Lee, S., Park, H. B., Han, K. S., Lee, J. H., & Kim, H. K. (2010). Proteome analysis of Arabidopsis seedlings exposed to bacterial volatiles. *Planta*, *232*, 1355–1370.

Kyriakidou, M., Tai, H. H., Anglin, N. L., Ellis, D., & Strömvik, M. V. (2018). Current strategies of polyploid plant genome sequence assembly. *Frontiers in Plant Science*, *9*, 1660. DOI: 10.3389/fpls.2018.01660 PMID: 30519250

Lahlali, R., Song, T., Chu, M., Yu, F., Kumar, S., Karunakaran, C., & Peng, G. (2017). Evaluating changes in cell-wall components associated with clubroot resistance using fourier transform infrared spectroscopy and rt-pcr. *International Journal of Molecular Sciences*, *18*(10), 2058. DOI: 10.3390/ijms18102058 PMID: 28954397

Lamaoui, M., Jemo, M., Datla, R., & Bekkaoui, F. (2018). Heat and drought stresses in crops and approaches for their mitigation. *Frontiers in Chemistry*, *6*, 26. Advance online publication. DOI: 10.3389/fchem.2018.00026 PMID: 29520357

Lammerink, J. (1967). The inheritance of clubroot resistance in *Brassica napus* L. *New Zealand Journal of Agricultural Research*, *10*(1), 109–115. DOI: 10.1080/00288233.1967.10423081

Landry, B. S., Hubert, N., Etoh, T., Harada, J., & Lincoln, S. (1991). A genetic map for Brassica napus based on restriction fragment length polymorphisms detected with expressed sequences. *Genome*, *34*(4), 543–552. DOI: 10.1139/g91-084

Langridge, P., & Fleury, D. (2011). Making the most of 'omics' for crop breeding. *Trends in Biotechnology*, *29*(1), 33–40. DOI: 10.1016/j.tibtech.2010.09.006 PMID: 21030098

Lan, M., Li, G., Hu, J., Yang, H., Zhang, L., Xu, X., & Sun, R. (2019). iTRAQ-based quantitative analysis reveals proteomic changes in Chinese cabbage (Brassica rapa L.) in response to Plasmodiophora brassicae infection. *Scientific Reports*, *9*(1), 12058.

Larkan, N. J., Lydiate, D. J., Yu, F., Rimmer, S. R., & Borhan, M. H. (2014). Co-localisation of the blackleg resistance genes Rlm2 and LepR3 on *Brassica napus* chromosome A10. *BMC Plant Biology*, *14*(1), 387. DOI: 10.1186/s12870-014-0387-z PMID: 25551287

Larkan, N., Lydiate, D., Parkin, I., Nelson, M., Epp, D., Cowling, W., Rimmer, S. R., & Borhan, M. (2012). The brassica napus blackleg resistance gene lepr3 encodes a receptor-like protein triggered by the leptosphaeria maculans effector avrlm1. *The New Phytologist*, *197*(2), 595–605. DOI: 10.1111/nph.12043 PMID: 23206118

Laver, T., Harrison, J., O'Neill, P. A., Moore, K., Farbos, A., Paszkiewicz, K., & Studholme, D. J. (2015). Assessing the performance of the Oxford Nanopore Technologies MinION. *Biomolecular Detection and Quantification*, *3*, 1–8. DOI: 10.1016/j.bdq.2015.02.001 PMID: 26753127

Lay, C., Bell, T., Hamel, C., Harker, K., Mohr, R., Greer, C., Yergeau, É., & St-Arnaud, M. (2018). Canola root–associated microbiomes in the canadian prairies. *Frontiers in Microbiology*, *9*, 1188. Advance online publication. DOI: 10.3389/fmicb.2018.01188 PMID: 29937756

Lee, C. (2022). Towards the genetic architecture of complex gene expression traits: Challenges and prospects for eQTL mapping in humans. *Genes*, *13*(2), 235. DOI: 10.3390/genes13020235 PMID: 35205280

Lee, H., Chawla, H. S., Obermeier, C., Dreyer, F., Abbadi, A., & Snowdon, R. (2020). Chromosome-scale assembly of winter oilseed rape Brassica napus. *Frontiers in Plant Science*, *11*, 496. DOI: 10.3389/fpls.2020.00496 PMID: 32411167

Lee, J. S., & Adams, K. L. (2020). Global insights into duplicated gene expression and alternative splicing in polyploid *Brassica napus* under heat, cold, and drought stress. *The Plant Genome*, *13*(3), e20057. Advance online publication. DOI: 10.1002/tpg2.20057 PMID: 33043636

Lee, J., Izzah, N. K., Jayakodi, M., Perumal, S., Joh, H. J., Lee, H. J., Lee, S.-C., Park, J. Y., Yang, K.-W., Nou, I.-S., Seo, J., Yoo, J., Suh, Y., Ahn, K., Lee, J. H., Choi, G. J., Yu, Y., Kim, H., & Yang, T.-J. (2015). Genome-wide SNP identification and QTL mapping for black rot resistance in cabbage. *BMC Plant Biology*, *15*(1), 32. DOI: 10.1186/s12870-015-0424-6 PMID: 25644124

Lee, J., & Park, S. (2022). Exploring physiological limits of seed oil accumulation in Brassica species. *Plant Physiology*, *188*(2), 1023–1037. PMID: 34747475

Lee, M., Zhang, Q., & Wang, H. (2021). Optimization of biodiesel production from canola oil: A response surface methodology approach. *Fuel*, *285*, 119255.

Lee, T., Yang, S., Kim, E., Ko, Y., Hwang, S., Shin, J., Shim, J. E., Shim, H., Kim, H., Kim, C., & Lee, I. (2015). AraNet v2: An improved database of co-functional gene networks for the study of Arabidopsis thaliana and 27 other nonmodel plant species. *Nucleic Acids Research*, *43*(Database issue), D996–D1002. DOI: 10.1093/nar/gku1053 PMID: 25355510

Lewin, H. A., Larkin, D. M., Pontius, J., & O'Brien, S. J. (2009). Every genome sequence needs a good map. *Genome Research*, *19*(11), 1925–1928. DOI: 10.1101/gr.094557.109 PMID: 19596977

Liang, Yue, Srivastava, Sanjeeva, Rahman, Muhammad H, & Strelkov, Stephen E, & Kav, Nat NV. (2008). Proteome changes in leaves of Brassica napus L. as a result of Sclerotinia sclerotiorum challenge. *Journal of Agricultural and Food Chemistry, 56*(6), 1963–1976.

Li, B., & Ritchie, M. D. (2021). From GWAS to gene: Transcriptome-wide association studies and other methods to functionally understand GWAS discoveries. *Frontiers in Genetics, 12*, 713230. DOI: 10.3389/fgene.2021.713230 PMID: 34659337

Li, F., Chen, B., Xu, K., Wu, J., Song, W., Bancroft, I., & Wu, X. (2018). Genome-wide association study dissects the genetic architecture of seed weight and seed quality in rapeseed (Brassica napus L.). *DNA Research : an International Journal for Rapid Publication of Reports on Genes and Genomes, 25*(5), 485–499. PMID: 29947776

Li, F.-W., & Harkess, A. (2018). A guide to sequence your favorite plant genomes. *Applications in Plant Sciences, 6*(3), e1030. DOI: 10.1002/aps3.1030 PMID: 29732260

Li, H., Feng, H., Guo, C., Yang, S., Huang, W., Xiong, X., & Yang, W. (2020). High-throughput phenotyping accelerates the dissection of the dynamic genetic architecture of plant growth and yield improvement in rapeseed. *Plant Biotechnology Journal, 18*(11), 2345–2353. DOI: 10.1111/pbi.13396 PMID: 32367649

Li, H., Liu, S., & Fan, Y. (2021). Heat stress responses in Brassica napus: Identification of heat shock protein genes and antioxidant pathways. *Plant Science, 302*, 110687.

Li, J., Li, Y., Wang, R., Fu, J., Zhou, X., Fang, Y., Wang, Y., & Liu, Y. (2022). Multiple Functions of MiRNAs in *Brassica napus* L. *Life (Chicago, Ill.), 12*(11), 1811. DOI: 10.3390/life12111811 PMID: 36362967

Li, K., Wang, D., Gong, L., Lyu, Y., Guo, H., Chen, W., & Luo, J. (2019). Comparative analysis of metabolome of rice seeds at three developmental stages using a recombinant inbred line population. *The Plant Journal, 100*(5), 908–922. DOI: 10.1111/tpj.14482 PMID: 31355982

Li, L., Hur, M., Lee, J., Zhou, W., Song, Z., Ransom, N., Demirkale, C. Y., Nettleton, D., Westgate, M., Arendsee, Z., Iyer, V., Shanks, J., Nikolau, B., & Wurtele, E. (2015). A systems biology approach toward understanding seed composition in soybean. *BMC Genomics, 16*(S3), S9. Advance online publication. DOI: 10.1186/1471-2164-16-S3-S9 PMID: 25708381

Li, N., Xu, C., & Li, Y. (2019). Improvement of seed oil content in canola: From quantitative genetics to molecular breeding. *Journal of Integrative Plant Biology*, *61*(3), 336–351.

Lin, H.-H., Lin, K.-H., Chen, S.-C., Shen, Y.-H., & Lo, H.-F. (2015). Proteomic analysis of broccoli (Brassica oleracea) under high temperature and waterlogging stresses. *Botanical Studies*, *56*, 1–11.

Lin, L., Allemekinders, H., Dansby, A., Campbell, L., Durance-Tod, S., Berger, A., & Jones, P. (2013). Evidence of health benefits of canola oil. *Nutrition Reviews*, *71*(6), 370–385. DOI: 10.1111/nure.12033 PMID: 23731447

Li, R., Jeong, K., Davis, J., Kim, S., Lee, S., Michelmore, R., Kim, S., & Maloof, J. (2018). Integrated QTL and eQTL Mapping Provides Insights and Candidate Genes for Fatty Acid Composition, Flowering Time, and Growth Traits in a F2 Population of a Novel Synthetic Allopolyploid Brassica napus. *Frontiers in Plant Science*, *9*, 1632. Advance online publication. DOI: 10.3389/fpls.2018.01632 PMID: 30483289

Li, S., Zhang, C., Li, J., Yan, L., Wang, N., & Xia, L. (2021). Present and future prospects for wheat improvement through genome editing and advanced technologies. *Plant Communications*, *2*(4), 100211. DOI: 10.1016/j.xplc.2021.100211 PMID: 34327324

Liu, C., Sun, G. S., Zhang, R. J., Lv, S. W., Gao, L., Gao, L. W., & Zhang, C. W. (2019). Proteomic analysis provides integrated insight into mechanisms of Turnip mosaic virus long distance movement in Brassica rapa. *Biologia Plantarum*, *63*(1), 164–173.

Liu, L., Li, Y., Li, S., Hu, N., He, Y., Pong, R., Lin, D., Lu, L., & Law, M. (2012). Comparison of next-generation sequencing systems. *BioMed Research International*, *251364*, 1–11. Advance online publication. DOI: 10.1155/2012/251364 PMID: 22829749

Liu, Q., Guo, Y., Li, J., Long, J., Zhang, B., & Shyr, Y. (2012). Steps to ensure accuracy in genotype and SNP calling from Illumina sequencing data. *BMC Genomics*, *13*(S8), 1–8. DOI: 10.1186/1471-2164-13-S8-S8 PMID: 23281772

Liu, S., Fan, C., Li, J., Cai, G., Yang, Q., Wu, J., Yi, X., Zhang, C., & Zhou, Y. (2016). A genome-wide association study reveals novel elite allelic variations in seed oil content of Brassica napus. *Theoretical and Applied Genetics*, *129*(6), 1203–1215. DOI: 10.1007/s00122-016-2697-z PMID: 26912143

Liu, X., & Locasale, J. W. (2017). Metabolomics: A primer. *Trends in Biochemical Sciences*, *42*(4), 274–284. DOI: 10.1016/j.tibs.2017.01.004 PMID: 28196646

Liu, Y., Zhou, X., Yan, M., Wang, P., Wang, H., Xin, Q., Yang, L., Hong, D., & Yang, G. (2019). Fine mapping and candidate gene analysis of a seed glucosinolate content QTL, qGSL-C2, in rapeseed (Brassica napus L.). *Theoretical and Applied Genetics, 133*(2), 479–490. DOI: 10.1007/s00122-019-03479-x PMID: 31832742

Li, W., Gao, Y., Xu, H., Zhang, Y., & Wang, J. (2012). A proteomic analysis of seed development in Brassica campestri L. *PLoS One, 7*(11), e50290.

Li, X., Wei, Y., Acharya, A., Jiang, Q., Kang, J., & Brummer, E. C. (2014). A saturated genetic linkage map of autotetraploid alfalfa (Medicago sativa L.) developed using genotyping-by-sequencing is highly syntenous with the Medicago truncatula genome. *G3 (Bethesda, Md.), 4*(10), 1971–1979. DOI: 10.1534/g3.114.012245 PMID: 25147192

Li, Z., Chen, Y., Mu, D., Yuan, J., Shi, Y., Zhang, H., Gan, J., Li, N., Hu, X., Liu, B., Yang, B., & Fan, W. (2012). Comparison of the two major classes of assembly algorithms: Overlap–layout–consensus and de Bruijn graph. *Briefings in Functional Genomics, 11*(1), 25–37. DOI: 10.1093/bfgp/elr035 PMID: 22184334

Li, Z., Song, Z., Zhang, Y., Hao, Q., Song, A., Liu, L., & Liang, Y. (2018). Silicon enhancement of estimated plant biomass carbon accumulation under abiotic and biotic stresses. a meta-analysis. *Agronomy for Sustainable Development, 38*(3), 26. Advance online publication. DOI: 10.1007/s13593-018-0496-4

Lohani, N., Golicz, A., Singh, M., & Bhalla, P. (2019). Genome-wide analysis of the Hsf gene family in Brassica oleracea and a comparative analysis of the Hsf gene family in B. oleracea, B. rapa and B. napus. *Functional & Integrative Genomics, 19*(3), 515–531. DOI: 10.1007/s10142-018-0649-1 PMID: 30618014

Lohani, N., Jain, D., Singh, M. B., & Bhalla, P. L. (2020). Engineering multiple abiotic stress tolerance in canola, *Brassica napus. Frontiers in Plant Science, 11*, 3. DOI: 10.3389/fpls.2020.00003 PMID: 32161602

Long, W., Hu, M., Gao, J., Chen, S., Zhang, J., Cheng, L., & Pu, H. (2018). Identification and Functional Analysis of Two New Mutant BnFAD2 Alleles That Confer Elevated Oleic Acid Content in Rapeseed. *Frontiers in Genetics, 9*, 399. Advance online publication. DOI: 10.3389/fgene.2018.00399 PMID: 30294343

Lopez, A., & Fernandez, B. (2022). Glycerol derivatives in cosmetic formulations: A systematic review. *International Journal of Cosmetic Science, 44*(1), 32–45.

Lowe, R., Shirley, N., Bleackley, M., Dolan, S., & Shafee, T. (2017). Transcriptomics technologies. *PLoS Computational Biology, 13*(5), e1005457. DOI: 10.1371/journal.pcbi.1005457 PMID: 28545146

Lu, K., Wei, L., Li, X., Wang, Y., Wu, J., Liu, M., Zhang, C., Chen, Z., Xiao, Z., Jian, H., Cheng, F., Zhang, K., Du, H., Cheng, X., Qu, C., Qian, W., Liu, L., Wang, R., Zou, Q., & Li, J. (2019). Whole-genome resequencing reveals Brassica napus origin and genetic loci involved in its improvement. *Nature Communications*, *10*(1), 1154. DOI: 10.1038/s41467-019-09134-9 PMID: 30858362

Lu, M., & Zhan, X. (2018). The crucial role of multiomic approach in cancer research and clinically relevant outcomes. *The EPMA Journal*, *9*(1), 77–102. DOI: 10.1007/s13167-018-0128-8 PMID: 29515689

Luo, J.-S., & Zhang, Z. (2019). Proteomic changes in the xylem sap of Brassica napus under cadmium stress and functional validation. *BMC Plant Biology*, *19*, 1–14.

Luo, R., Liu, B., Xie, Y., Li, Z., Huang, W., Yuan, J., He, G., Chen, Y., Pan, Q., Liu, Y., Tang, J., Wu, G., Zhang, H., Shi, Y., Liu, Y., Yu, C., Wang, B., Lu, Y., Han, C., & Wang, J. (2012). SOAPdenovo2: An empirically improved memory-efficient short-read de novo assembler. *GigaScience*, *1*(1), 18. DOI: 10.1186/2047-217X-1-18 PMID: 23587118

Luo, X., Ma, C., Yue, Y., Hu, K., Li, Y., Duan, Z., & Zhou, Y. (2017). Unravelling the complex trait of harvest index in rapeseed (Brassica napus L.) with association mapping. *BMC Genomics*, *18*(1), 238. PMID: 28320314

Luo, Y. X., Luo, C., Du, D., Fu, Z., Yao, Y. M., Xu, C. C., & Zhang, H. (2014). Quantitative trait analysis of flowering time in spring rapeseed (B. napus L.). *Euphytica*, *200*(3), 321–335. DOI: 10.1007/s10681-014-1140-2

Lu, S., Sturtevant, D., Aziz, M., Jin, C., Li, Q., Chapman, K., & Guo, L. (2018). Spatial analysis of lipid metabolites and expressed genes reveals tissue-specific heterogeneity of lipid metabolism in high- and low-oil brassica napus l. seeds. *The Plant Journal*, *94*(6), 915–932. DOI: 10.1111/tpj.13959 PMID: 29752761

Lyu, S., Gao, L., Zhang, R., Zhang, C., & Hou, X. (2020). Correlation analysis of expression profile and quantitative iTRAQ-LC-MS/MS proteomics reveals resistance mechanism against TuMV in Chinese cabbage (Brassica rapa ssp. pekinensis). *Frontiers in Genetics*, *11*, 963.

MacIntosh, S., Shaw, M., Connelly, M., & Yao, Z. (2021). Food and feed safety of ns-b5øø27-4 omega-3 canola (brassica napus): A new source of long-chain omega-3 fatty acids. *Frontiers in Nutrition*, *8*, 716659. Advance online publication. DOI: 10.3389/fnut.2021.716659 PMID: 34660659

Madgule, M., Vinayaka, N., Sonkhaskar, Y. M., Patel, D. R., Karthikeyan, R., & Selvakumar, P. (2023). Mechanical properties and microstructure of activated TIG welded similar joints of Inconel alloys by desirability approaches. *Materials Today: Proceedings*, *77*, 528–533. DOI: 10.1016/j.matpr.2022.12.250

Madoui, M. A., Engelen, S., Cruaud, C., Belser, C., Bertrand, L., Alberti, A., Lemainque, A., Wincker, P., & Aury, J.-M. (2015). Genome assembly using Nanopore-guided long and error-free DNA reads. *BMC Genomics*, *16*(1), 1. DOI: 10.1186/s12864-015-1519-z PMID: 25927464

Mahalingam, R. (2017). Phenotypic, physiological and malt quality analyses of US barley varieties subjected to short periods of heat and drought stress. *Journal of Cereal Science*, *76*, 199–205. DOI: 10.1016/j.jcs.2017.06.007

Maher, L. B. W., Salisbury, P., Debonte, L., & Deng, X. (2007).(Organising Committee of: Wuhan). High Oleic, low linolenic (HOLL) specialty canola development in Australia; *Proceedings of the 12th International Rapeseed Congress*; Wuhan, China. 23–30 March; pp. 22–24.

Mahmood, U., Li, X., Fan, Y., Chang, W., Niu, Y., Li, J., Qu, C., & Lu, K. (2022). Multi-omics revolution to promote plant breeding efficiency. *Frontiers in Plant Science*, *13*, 1062952. DOI: 10.3389/fpls.2022.1062952 PMID: 36570904

Malik, M. M. (2023). *Genetic Divergence among Local and Exotic Brassica juncea Germplasm as Revealed by Using Simple Sequence Repeats (SSRs)*. CAPITAL UNIVERSITY.

Malmberg, M. M., Barbulescu, D. M., Drayton, M. C., Shinozuka, M., Thakur, P., Ogaji, Y. O., Spangenberg, G. C., Daetwyler, H. D., & Cogan, N. O. I. (2018). Cogan N.O.I. Evaluation and Recommendations for Routine Genotyping Using Skim Whole Genome Re-sequencing in Canola. *Frontiers in Plant Science*, *9*, 1809. Advance online publication. DOI: 10.3389/fpls.2018.01809

Malmberg, M. M., Spangenberg, G. C., Daetwyler, H. D., & Cogan, N. O. I. (2019). Assessment of low-coverage nanopore long read sequencing for SNP genotyping in doubled haploid canola (Brassica napus L.). *Scientific Reports*, *9*(1), 8688. DOI: 10.1038/s41598-019-45131-0 PMID: 31213642

Malmberg, M., Pembleton, L., Baillie, R., Drayton, M., Sudheesh, S., Kaur, S., & Cogan, N. (2017). Genotyping-by-sequencing through transcriptomics: Implementation in a range of crop species with varying reproductive habits and ploidy levels. *Plant Biotechnology Journal*, *16*(4), 877–889. DOI: 10.1111/pbi.12835 PMID: 28913899

Mamnabi, S., Nasrollahzadeh, S., Ghassemi-Golezani, K., & Raei, Y. (2020). Improving yield-related physiological characteristics of spring rapeseed by integrated fertilizer management under water deficit conditions. *Saudi Journal of Biological Sciences, 27*(3), 797–804. DOI: 10.1016/j.sjbs.2020.01.008 PMID: 32127754

Manchanda, P., & Suneja, Y. (2018). Genome editing for crop improvement: Status and prospects. *Biotechnology for Crop Improvement, 3*, 75–104. DOI: 10.1007/978-3-319-94746-4_5

Mangin, B., Casadebaig, P., Cadic, E., Blanchet, N., Boniface, M. C., Carrère, S., Gouzy, J., Legrand, L., Mayjonade, B., Pouilly, N., André, T., Coque, M., Piquemal, J., Laporte, M., Vincourt, P., Munos, S., & Langlade, N. B. (2017). Genetic control of plasticity of oil yield for combined abiotic stresses using a joint approach of crop modelling and genome-wide association. *Plant, Cell & Environment, 40*(10), 2276–2291. DOI: 10.1111/pce.12961 PMID: 28418069

Mann, A., Kumari, J., Kumar, R., Kumar, P., Pradhan, A., Pental, D., & Bisht, N. (2023). Targeted editing of multiple homologues of GTR1 and GTR2 genes provides the ideal low-seed, high-leaf glucosinolate oilseed mustard with uncompromised defence and yield. *Plant Biotechnology Journal, 21*(11), 2182–2195. DOI: 10.1111/pbi.14121 PMID: 37539488

Mansouripour, S. (2018). *Identification of Brassica napus L. sources of resistance against blackleg (Leptosphaeria maculans).* North Dakota State University.

Marra, Roberta, & Li, Hua, Barbetti, MJ, Sivasithamparam, Krishnapillai, Vinale, Francesco, Cavallo, P, & Lorito, Matteo. (2010). Proteomic analysis of the interaction between Brassica napus cv. Surpass 400 and virulent or avirulent isolates of Leptosphaeria maculans. *Journal of Plant Pathology*, •••, 89–101.

Mason, A. S. (2015). Challenges of genotyping polyploid species. In *Plant Genotyping: Methods and Protocols* (pp. 161–168). Springer. DOI: 10.1007/978-1-4939-1966-6_12

Mason, A. S., & Batley, J. (2015). Creating new interspecific hybrid and polyploid crops. *Trends in Biotechnology, 33*(8), 436–441. DOI: 10.1016/j.tibtech.2015.06.004 PMID: 26164645

Mason, A. S., Higgins, E. E., Snowdon, R. J., Batley, J., Stein, A., Werner, C., & Parkin, I. A. (2017). A user guide to the Brassica 60K Illumina Infinium™ SNP genotyping array. *Theoretical and Applied Genetics, 130*(4), 621–633. DOI: 10.1007/s00122-016-2849-1 PMID: 28220206

Mason, A., Chauhan, P., Banga, S., Banga, S., Salisbury, P., Barbetti, M., & Batley, J. (2018). Agricultural selection and presence–absence variation in spring-type canola germplasm. *Crop & Pasture Science*, *69*(1), 55–64. DOI: 10.1071/CP17161

Mattson, F. H., & Grundy, S. M. (1985). Comparison of effects of dietary saturated, monounsaturated, and polyunsaturated fatty acids on plasma lipids and lipoproteins in man. *Journal of Lipid Research*, *26*(2), 194–202. https://www.jlr.org/content/26/2/194. DOI: 10.1016/S0022-2275(20)34389-3 PMID: 3989378

McBeath, T. M., Meier, E. A., Ware, A., Kirkegaard, J., Moodie, M., Davoren, B., & Hunt, E. (2020). Agronomic management combining early-sowing on establishment opportunities, cultivar options and adequate nitrogen is critical for canola (Brassica napus) productivity and profit in low-rainfall environments. *Crop & Pasture Science*, *71*(9), 807–821. DOI: 10.1071/CP20226

McCallum, C. M., Comai, L., Greene, E. A., & Henikoff, S. (2000). T argeting I nduced L ocal L esions IN G enomes (TILLING) for plant functional genomics. *Plant Physiology*, *123*(2), 439–442. DOI: 10.1104/pp.123.2.439 PMID: 10859174

McClinchey, S., Gillespie, J., Fisher, T., Taylor, K., Challender, M., & Schmidt, D. (2016). Quality characteristics of glyphosate-tolerant canola cultivars containing a glyphosate acetyltransferase transgene (event dp-ø73496-4). *Crop Science*, *56*(4), 1736–1746. DOI: 10.2135/cropsci2015.12.0744

McCouch, S. R., Wright, M. H., Tung, C. W., Maron, L. G., McNally, K. L., Fitzgerald, M., & Mezey, J. (2016). Open access resources for genome-wide association mapping in rice. *Nature Communications*, *7*(1), 10532. DOI: 10.1038/ncomms10532 PMID: 26842267

McKey, D., Elias, M., Pujol, B., & Duputié, A. (2010). The evolutionary ecology of clonally propagated domesticated plants. *The New Phytologist*, *186*(2), 318–332.

McNaughton, H. I. (1976). *Turpic and relatives* (Simmonds, N. W., Ed.). Longman.

McVetty, P. B. E., & Duncan, R. W. (2016).Reference Module in Food Science. Elsevier; Amsterdam, The Netherlands. Canola/Rapeseed: Genetics and Breeding.

McVetty, B. E., & Scarth, R. (2012). Breeding oil quality for edible and industrial applications in oilseed rape and linseed. *Technological Innovations in Major World Oil Crops, Volume 2. Perspectives*, •••, 105–122.

McVetty, P. B. E., & Duncan, R. W. (2015). *Canola, rapeseed, and mustard: for biofuels and bioproducts. Industrial crops. Handbook of plant breeding*. Springer.

McVetty, P. B. E., & Scarth, R. (2002). Breeding for improved oil quality in Brassica oilseed species. *Journal of Crop Production*, *5*(1-2), 345–369. DOI: 10.1300/J144v05n01_14

McVetty, P. B., Mietkiewska, E., Omonov, T., Curtis, J., Taylor, D. C., & Weselake, R. J. (2016). Brassica spp. oils. In *Industrial oil crops* (pp. 113–156). Elsevier. DOI: 10.1016/B978-1-893997-98-1.00005-1

Mei, J., Liu, Y., Wei, D., Wittkop, B., Ding, Y., Li, Q., Li, J., Wan, H., Li, Z., Ge, X., Frauen, M., Snowdon, R. J., Qian, W., & Friedt, W. (2015). Transfer of sclerotinia resistance from wild relative of *Brassica oleracea* into *Brassica napus* using a hexaploidy step. *Theoretical and Applied Genetics*, *128*(4), 639–644. DOI: 10.1007/s00122-015-2459-3 PMID: 25628163

Melchinger, A. E., Fernando, R., Stricker, C., Schön, C.-C., & Auinger, H.-J. (2023). Genomic prediction in hybrid breeding: I. Optimizing the training set design. *Theoretical and Applied Genetics*, *136*(8), 176. DOI: 10.1007/s00122-023-04413-y PMID: 37532821

Menendez, Y., Sánchez, D., Snowdon, R., Rondanini, D., & Botto, J. (2021). Unraveling the impact of plant-density genetic architecture on agronomic traits in canola. *Journal of Experimental Botany*. Advance online publication. DOI: 10.1093/jxb/erab191 PMID: 33940608

Meng, H., Hua, S., Shamsi, I. H., Jilani, G., Li, Y., & Jiang, L. (2009). Cadmium-induced stress on the seed germination and seedling growth of Brassica napus L., and its alleviation through exogenous plant growth regulators. *Plant Growth Regulation*, *58*, 47–59.

Meng, X., & Zhang, S. (2013). Mapk cascades in plant disease resistance signaling. *Annual Review of Phytopathology*, *51*(1), 245–266. DOI: 10.1146/annurev-phyto-082712-102314 PMID: 23663002

Mensah, H., Ahadzie, D. K., Takyi, S. A., & Amponsah, O. (2021). Climate change resilience: Lessons from local climate-smart agricultural practices in Ghana. *Energy, Ecology & Environment*, *6*(3), 271–284. DOI: 10.1007/s40974-020-00181-3

Mente, A., de Koning, L., Shannon, H. S., & Anand, S. S. (2009). A systematic review of the evidence supporting a causal link between dietary factors and coronary heart disease. *Archives of Internal Medicine*, *169*(7), 659–669. DOI: 10.1001/archinternmed.2009.38 PMID: 19364995

Mergner, J., & Kuster, B. (2022). Plant proteome dynamics. *Annual Review of Plant Biology*, *73*(1), 67–92.

Merkosky, K., Leibovitz, A., & Wong, S. (2024). Cabbage Seedpod Weevil (*Ceutorhynchus obstrictus*) Management in Canola (*Brassica napus*) across western Canada. *Plant Health Cases*(2024). DOI: 10.1079/planthealthcases.2024.0014

Merrick, L. F., Herr, A. W., Sandhu, K. S., Lozada, D. N., & Carter, A. H. (2022). Optimizing plant breeding programs for genomic selection. *Agronomy (Basel), 12*(3), 714. DOI: 10.3390/agronomy12030714

Miao, H., Zeng, W., Wang, J., Zhang, F., Sun, B., & Wang, Q. (2021). Improvement of glucosinolates by metabolic engineering in brassica crops. *aBIOTECH, 2*(3), 314–329. DOI: 10.1007/s42994-021-00057-y PMID: 36303883

Mibei, E., Owino, W., Ambuko, J., Giovannoni, J., & Onyango, A. (2017). Metabolomic analyses to evaluate the effect of drought stress on selected african eggplant accessions. *Journal of the Science of Food and Agriculture, 98*(1), 205–216. DOI: 10.1002/jsfa.8458 PMID: 28573744

Mihr, C., Baumgärtner, M., Dieterich, J.-H., Schmitz, U. K., & Braun, H.-P. (2001). Proteomic approach for investigation of cytoplasmic male sterility (CMS) in Brassica. *Journal of Plant Physiology, 158*(6), 787–794.

Mikheyev, A. S., & Tin, M. M. (2014). A first look at the Oxford Nanopore MinION sequencer. *Molecular Ecology Resources, 14*(6), 1097–1102. DOI: 10.1111/1755-0998.12324 PMID: 25187008

Millar, A Harvey, Heazlewood, Joshua L, Giglione, Carmela, Holdsworth, Michael J, Bachmair, Andreas, & Schulze, Waltraud X. (2019). The scope, functions, and dynamics of posttranslational protein modifications. *Annual Review of Plant Biology, 70*(1), 119–151.

Miller, J. R., Koren, S., & Sutton, G. (2010). Assembly algorithms for next-generation sequencing data. *Genomics, 95*(6), 315–327. DOI: 10.1016/j.ygeno.2010.03.001 PMID: 20211242

Mir, R. R., Zaman-Allah, M., Sreenivasulu, N., Trethowan, R., & Varshney, R. K. (2012). Integrated genomics, physiology and breeding approaches for improving drought tolerance in crops. *Theoretical and Applied Genetics, 125*(4), 625–645. DOI: 10.1007/s00122-012-1904-9 PMID: 22696006

Mishra, R., Joshi, R. K., & Zhao, K. (2020). Genome editing for abiotic stress tolerance in Brassica napus using CRISPR-Cas9. *Frontiers in Plant Science, 11*, 563.

Mi, W., Liu, Z., Jin, J., Dong, X., Xu, C., Zou, Y., & Fang, X. (2021). Comparative proteomics analysis reveals the molecular mechanism of enhanced cold tolerance through ROS scavenging in winter rapeseed (Brassica napus L.). *PLoS One*, *16*(1), e0243292.

Mmbando, G. S. (2024). Omics: A new, promising technologies for boosting crop yield and stress resilience in African agriculture. *Plant Stress*, *11*, 100366. DOI: 10.1016/j.stress.2024.100366

Mochizuki, T., Sakamoto, M., Tanizawa, Y., Nakayama, T., Tanifuji, G., Kamikawa, R., & Shirahige, K. (2023). A practical assembly guideline for genomes with various levels of heterozygosity. *Briefings in Bioinformatics*, *24*(6), bbad337. Advance online publication. DOI: 10.1093/bib/bbad337 PMID: 37798248

Moghaieb, R. E. A., Mohammed, E. H. K., & Youssief, S. S. (2014). Genetic diversity among some canola cultivars as revealed by RAPD, SSR and AFLP analyses. *3 Biotech, 4*, 403-410.

Mohd Saad, N. S., Severn-Ellis, A. A., Pradhan, A., Edwards, D., & Batley, J. (2021). Genomics armed with diversity leads the way in Brassica improvement in a changing global environment. *Frontiers in Genetics*, *12*, 600789. DOI: 10.3389/fgene.2021.600789 PMID: 33679880

Moon, J. Y., Kim, S. T., Choi, G. J., Kwon, S.-Y., Cho, H. S., Kim, H.-S., & Park, J. M. (2020). Comparative proteomic analysis of host responses to Plasmodiophora brassicae infection in susceptible and resistant Brassica oleracea. *Plant Biotechnology Reports*, *14*, 263–274.

Mora-Oberlaender, J., Jiménez-Barreto, J., Rodríguez-Abril, Y., Estrada-Arteaga, M., & Chaparro-Giraldo, A. (2022). Opportunities for Generic Cisgenic Crops. In *Cisgenic Crops: Potential and Prospects* (pp. 89-119). Springer. DOI: 10.1007/978-3-031-06628-3_6

Morsi, N., Hashem, O., El-Hady, M., Abd-Elkrem, Y., El-temsah, M., Galal, E., & Abdelkader, M. (2023). Assessing drought tolerance of newly developed tissue-cultured canola genotypes under varying irrigation regimes. *Agronomy (Basel)*, *13*(3), 836. DOI: 10.3390/agronomy13030836

Müller, M., & Munné-Bosch, S. (2015). Ethylene response factors: A key regulatory hub in hormone and stress signaling. *Plant Physiology*, *169*(1), 32–41. DOI: 10.1104/pp.15.00677 PMID: 26103991

Mulusew, F., Denise, M., Barbulescu, M., Malmberg, P., Maharjan, P., & Salisbury, S.. (2020). Genomic prediction and genetic correlation of agronomic, blackleg disease, and seed quality traits in canola (*Brassica napus* L.). *Plants*, *9*(6), 719. DOI: 10.3390/plants9060719 PMID: 32517116

Muthamilarasan, M., Singh, N. K., & Prasad, M. (2019). Multi-omics approaches for strategic improvement of stress tolerance in underutilized crop species: A climate change perspective. *Advances in Genetics*, *103*, 1–38. DOI: 10.1016/bs.adgen.2019.01.001 PMID: 30904092

Nadeem, M. A., Nawaz, M. A., Shahid, M. Q., Doğan, Y., Comertpay, G., Yıldız, M., Hatipoğlu, R., Ahmad, F., Alsaleh, A., Labhane, N., Özkan, H., Chung, G., & Baloch, F. S. (2018). DNA molecular markers in plant breeding: Current status and recent advancements in genomic selection and genome editing. *Biotechnology, Biotechnological Equipment*, *32*(2), 261–285. DOI: 10.1080/13102818.2017.1400401

Nagaharu, U., & Nagaharu, N. (1935). Genome analysis in Brassica with special reference to the experimental formation of B. napus and peculiar mode of fertilization. *Japanese Journal of Botany*, *7*(7), 389–452.

Naheed, R., Aslam, H., Kanwal, H., Farhat, F., Gamar, M., Al-Mushhin, A., & Hessini, K. (2021). Growth attributes, biochemical modulations, antioxidant enzymatic metabolism and yield in brassica napus varieties for salinity tolerance. *Saudi Journal of Biological Sciences*, *28*(10), 5469–5479. DOI: 10.1016/j.sjbs.2021.08.021 PMID: 34588857

Nandula, V. K. (2019). Herbicide resistance: Mechanisms and management in crops. *Weed Science*, *67*(6), 559–572.

Nansamba, M., Sibiya, J., Tumuhimbise, R., Karamura, D., Kubiriba, J., & Karamura, E. (2020). Breeding banana (Musa spp.) for drought tolerance: A review. *Plant Breeding*, *139*(4), 685–696. DOI: 10.1111/pbr.12812

Nasar, S., Muhammad, M., Siddiqi, E. H., Majeed, M., Ameen, N., Ullah, S., Kousar, S., & Abdullah, M. (2024). OMICS-Based Knowledge for Achieving Food and Nutritional Security. *OMICs-based Techniques for Global Food Security*, 67-90.

Nath, U. K., Kim, H.-T., Khatun, K., Park, J.-I., Kang, K.-K., & Ill-Sup, N. (2016). Modification of Fatty Acid Profiles of Rapeseed (Brassica napus L.) Oil for Using as Food, Industrial Feed-Stock and Biodiesel. *Plant Breeding and Biotechnology*, *4*(2), 123–134. DOI: 10.9787/PBB.2016.4.2.123

Nath, U. K., Wilmer, J. A., Wallington, E. J., Becker, H. C., & Möllers, C. (2009). Increasing erucic acid content through combination of endogenous low polyunsaturated fatty acids alleles with Ld-LPAAT + Bn-fae1 transgenes in rapeseed (Brassica napus L.). *Theoretical and Applied Genetics*, *118*(4), 765–773. Advance online publication. DOI: 10.1007/s00122-008-0936-7 PMID: 19050848

Nazemof, N., Couroux, P., Xing, T., & Robert, L. S. (2016). Proteomic analysis of the mature Brassica stigma reveals proteins with diverse roles in vegetative and reproductive development. *Plant Science*, *250*, 51–58.

Neik, T. X., Amas, J., Barbetti, M., Edwards, D., & Batley, J. (2020). Understanding Host–Pathogen Interactions in Brassica napus in the Omics Era. *Plants*, *9*(10), 10. Advance online publication. DOI: 10.3390/plants9101336 PMID: 33050509

Neik, T. X., Amas, J., Barbetti, M., Edwards, D., & Batley, J. (2020). Understanding host–pathogen interactions in Brassica napus in the omics era. *Plants*, *9*(10), 1336.

Nejat, N., & Mantri, N. (2017). Plant immune system: Crosstalk between responses to biotic and abiotic stresses the missing link in understanding plant defense. *Current Issues in Molecular Biology*, •••, 1–16. DOI: 10.21775/cimb.023.001 PMID: 28154243

Nekrasov, V., Wang, C., Win, J., Lanz, C., Weigel, D., & Kamoun, S. (2017). Rapid generation of a transgene-free powdery mildew resistant tomato by genome deletion. *Scientific Reports*, *7*(1), 482. DOI: 10.1038/s41598-017-00578-x PMID: 28352080

Nelson, R. J., Naylor, R. L., & Jahn, M. M. (2004). The role of genomics research in improvement of" orphan" crops. *Crop Science*, *44*(6), 1901–1904. DOI: 10.2135/cropsci2004.1901

Nesatyy, V. J., & Suter, M. J.-F. (2007). Proteomics for the analysis of environmental stress responses in organisms. *Environmental Science & Technology*, *41*(20), 6891–6900. DOI: 10.1021/es070561r PMID: 17993125

Nguyen, D., Le, V., Nguy n, Q., Malau-Aduli, B., Nichols, P., & Malau-Aduli, A. (2017). Omega–3 long-chain fatty acids in the heart, kidney, liver and plasma metabolite profiles of australian prime lambs supplemented with pelleted canola and flaxseed oils. *Nutrients*, *9*(8), 893. DOI: 10.3390/nu9080893 PMID: 28817082

Nichol, J., Ribano, A. K. B., Hickerson, N. M. N., Ali, N., Jamois, F., & Samuel, M. A. (2023). Plant growth regulator extracts from seaweeds promote plant growth and confer drought tolerance in canola (brassica napus). *Plant Signaling & Behavior*, *18*(1), 2267222. Advance online publication. DOI: 10.1080/15592324.2023.2267222 PMID: 37903454

Nikvash, N., Kraft, R., Kharazipour, A., & Euring, M. (2010). Comparative properties of bagasse, canola and hemp particle boards. *Holz als Roh- und Werkstoff, 68*(3), 323–327. DOI: 10.1007/s00107-010-0465-3

Ning, L., Wang, H., Li, D., Lin, Z., Li, Y., Zhao, W., & Li, M. (2019). Transcriptomic and proteomic analysis of Shaan2A cytoplasmic male sterility and its maintainer line in Brassica napus. *Frontiers in Plant Science, 10*, 252.

Niu, F., Wang, B., Wu, F., Yan, J., Li, L., Wang, C., & Jiang, Y. Q. (2014). Canola (Brassica napus L.) NAC103 transcription factor gene is a novel player inducing reactive oxygen species accumulation and cell death in plants. *Biochemical and Biophysical Research Communications, 454*(1), 30–35. DOI: 10.1016/j.bbrc.2014.10.057 PMID: 25450358

Niu, Z., Liu, L., Pu, Y., Ma, L., Wu, J., Hu, F., & Wang, W. (2021). iTRAQ-based quantitative proteome analysis insights into cold stress of Winter Rapeseed (Brassica rapa L.) grown in the field. *Scientific Reports, 11*(1), 23434.

Noctor, G., Mhamdi, A., & Foyer, C. H. (2014). The roles of reactive oxygen metabolism in drought: Not so cut and dried. *Plant Physiology, 164*(4), 1636–1648.

Nokhrina, K., Ray, H., Bock, C., & Georges, F. (2014). Metabolomic shifts in brassica napus lines with enhancedbnplc2expression impact their response to low temperature stress and plant pathogens. *GM Crops and Food: Biotechnology in Agriculture and the Food Chain, 5*(2), 120–131. DOI: 10.4161/gmcr.28942 PMID: 24787279

Norton, G., & Harris, J. F. (1975). Compositional changes in developing rape seed (Brassica napus L.). *Planta, 123*(2), 163–174. DOI: 10.1007/BF00383865 PMID: 24435083

Ntui, V. O., Tripathi, J. N., Kariuki, S. M., & Tripathi, L. (2024). Cassava molecular genetics and genomics for enhanced resistance to diseases and pests. *Molecular Plant Pathology, 25*(1), e13402. DOI: 10.1111/mpp.13402 PMID: 37933591

O'Brien, D. (2016). Canola: Good protein source for dairy cattle. *Agricultural Research, 64*(2), 1.

Obata, T., & Fernie, A. (2012). The use of metabolomics to dissect plant responses to abiotic stresses. *Cellular and Molecular Life Sciences : CMLS, 69*(19), 3225–3243. DOI: 10.1007/s00018-012-1091-5 PMID: 22885821

Oh, S., Imran, M., Kim, E., Park, S., Lee, S., Park, H., & Ryu, T. (2023). Approach strategies and application of metabolomics to biotechnology in plants. *Frontiers in Plant Science, 14*, 1192235. Advance online publication. DOI: 10.3389/fpls.2023.1192235 PMID: 37636096

Oikonomopoulos, S., Bayega, A., Fahiminiya, S., Djambazian, H., Bérubé, P., & Ragoussis, J. (2020). Methodologies for transcript profiling using long-read technologies. *Frontiers in Genetics*, *11*, 606. Advance online publication. DOI: 10.3389/fgene.2020.00606 PMID: 32733532

Oliver, R. E., Tinker, N. A., Lazo, G. R., Chao, S., Jellen, E. N., Carson, M. L., & Jackson, E. W. (2013). SNP discovery and chromosome anchoring provide the first physically-anchored hexaploid oat map and reveal synteny with model species. *PLoS One*, *8*(3), e58068. DOI: 10.1371/journal.pone.0058068 PMID: 23533580

Olivos-Trujillo, M., Gajardo, H. A., Salvo, S., González, A., & Muñoz, C. (2015). Assessing the stability of parameters estimation and prediction accuracy in regression methods for estimating seed oil content in Brassica napus L. using NIR spectroscopy. *2015 CHILEAN Conference on Electrical, Electronics Engineering, Information and Communication Technologies (CHILECON)* DOI: 10.1109/Chilecon.2015.7400347

Oluwadare, O., Highsmith, M., & Cheng, J. (2019). An Overview of Methods for Reconstructing 3-D Chromosome and Genome Structures from Hi-C Data. *Biological Procedures Online*, *21*(1), 7. DOI: 10.1186/s12575-019-0094-0 PMID: 31049033

Ortolani, L., Bocci, R., Bàrberi, P., Howlett, S., & Chable, V. (2017). Changes in knowledge management strategies can support emerging innovative actors in organic agriculture: The case of participatory plant breeding in Europe. *Organic Farming*, *3*(1), 20–33. DOI: 10.12924/of2017.03010020

Osakabe, Y., Arinaga, N., Umezawa, T., Katsura, S., Nagamachi, K., Tanaka, H., & Yamaguchi-Shinozaki, K. (2013). Osmotic stress responses and plant growth controlled by potassium transporters inarabidopsis. *The Plant Cell*, *25*(2), 609–624. DOI: 10.1105/tpc.112.105700 PMID: 23396830

Oyekunle, M., Adamu, R. S., Ndou, E., Beyene, Y., Abdulmalik, M., & Oikeh, S. O. (2023). Efficacy of drought-tolerant and insect-protected transgenic TELA® maize traits in Nigeria. *Transgenic Research*, *32*(3), 169–178. DOI: 10.1007/s11248-023-00345-x PMID: 37043164

Ozhuner, E., Eldem, V., Ipek, A., Okay, S., Sakcali, S., Zhang, B., & Unver, T. (2013). Boron stress responsive microRNAs and their targets in barley. *PLoS One*, *8*(3), e59543. DOI: 10.1371/journal.pone.0059543 PMID: 23555702

Öztürk, M., Ünal, B., García-Caparrós, P., Khursheed, A., Gul, A., & Hasanuzzaman, M. (2020). Osmoregulation and its actions during the drought stress in plants. *Physiologia Plantarum*, *172*(2), 1321–1335. DOI: 10.1111/ppl.13297 PMID: 33280137

Pal, L., Sandhu, S. K., & Bhatia, D. (2021). Genome-wide association study and identification of candidate genes for seed oil content in Brassica napus. *Euphytica*, *217*, 1–15.

Pandey, M. K., Agarwal, G., Kale, S. M., Clevenger, J., Nayak, S. N., Sriswathi, M., & Varshney, R. K. (2017). Development and evaluation of a high density genotyping 'Axiom_ Arachis' array with 58 K SNPs for accelerating genetics and breeding in groundnut. *Scientific Reports*, *7*(1), 40577. DOI: 10.1038/srep40577 PMID: 28091575

Pan, W., Maaz, T., Madsen, I., Reese, M., Hammac, W., Wysocki, D., & Brown, J. (2017). 4r nitrogen management when integrating canola into semi-arid wheat. *Crops & Soils*, *50*(3), 16–66. DOI: 10.2134/cs2017.50.0309

Patel, J., Khandwal, D., Choudhary, B., Ardeshana, D., Jha, R., Tanna, B., Yadav, S., Mishra, A., Varshney, R. K., & Siddique, K. (2022). Differential physio-biochemical and metabolic responses of peanut (arachis hypogaea l.) under multiple abiotic stress conditions. *International Journal of Molecular Sciences*, *23*(2), 660. DOI: 10.3390/ijms23020660 PMID: 35054846

Patel, R. V., Nahal, H. K., Breit, R., & Provart, N. J. (2012). BAR expressolog identification: Expression profile similarity ranking of homologous genes in plant species. *The Plant Journal*, *71*(6), 1038–1050. DOI: 10.1111/j.1365-313X.2012.05055.x PMID: 22607031

Patterson, M., Marschall, T., Pisanti, N., Van Iersel, L., Stougie, L., Klau, G., & Schönhuth, A. (2015). WhatsHap: Weighted haplotype assembly for future-generation sequencing reads. *Journal of Computational Biology*, *22*(6), 498–509. DOI: 10.1089/cmb.2014.0157 PMID: 25658651

Pedras, M., Yaya, E., & Glawischnig, E. (2011). The phytoalexins from cultivated and wild crucifers: Chemistry and biology. *Natural Product Reports*, *28*(8), 1381. DOI: 10.1039/c1np00020a PMID: 21681321

Peng, Z., Fan, W., Wang, L., Paudel, D., Leventini, D., Tillman, B. L., & Wang, J. (2017). Target enrichment sequencing in cultivated peanut (Arachis hypogaea L.) using probes designed from transcript sequences. *Molecular Genetics and Genomics*, *292*(5), 955–965. DOI: 10.1007/s00438-017-1327-z PMID: 28492983

Percopo, C. M., Dyer, K. D., Ochkur, S. I., Luo, J. L., Fischer, E. R., Lee, J. J., Lee, N. A., & Domachowske, J. B. (2010). Isolation of human eosinophils: Microbead method has no impact on IL-5 sustained viability. *Experimental Dermatology*, *19*(5), 467–469. DOI: 10.1111/j.1600-0625.2009.00974.x PMID: 19758339

Pereira-Lorenzo, S., Fischer, M., Ramos-Cabrer, A. M., & Castro, I. (2018). Apple (Malus spp.) breeding: Present and future. *Advances in Plant Breeding Strategies: Fruits*, *3*, 3–29. DOI: 10.1007/978-3-319-91944-7_1

Pérez-Llorca, M., Pollmann, S., & Müller, M. (2023). Ethylene and jasmonates signaling network mediating secondary metabolites under abiotic stress. *International Journal of Molecular Sciences*, *24*(6), 5990. DOI: 10.3390/ijms24065990 PMID: 36983071

Perumal, S., Koh, C. S., Jin, L., Buchwaldt, M., Higgins, E. E., Zheng, C., Sanko, D., Robinson, S. J., Kagale, S., & Navabi, Z.-K.. (2020). A high-contiguity Brassica nigra genome localizes active centromeres and defines the ancestral Brassica genome. *Nature Plants*, *6*(8), 929–941. DOI: 10.1038/s41477-020-0735-y PMID: 32782408

Peterson, R., & Johnson, M. (2021). Environmental factors affecting seed oil content in canola: Implications for breeding strategies. *Field Crops Research*, *263*, 108071.

Poland, J. A., Brown, P. J., Sorrells, M. E., & Jannink, J. L. (2012). Development of high-density genetic maps for barley and wheat using a novel two-enzyme genotyping-by-sequencing approach. *PLoS One*, *7*(2), e32253. DOI: 10.1371/journal.pone.0032253 PMID: 22389690

Popovici, V., Bucur, L., Gîrd, C., Rambu, D., Calcan, S., Cucolea, E., Costache, T., Ungureanu-Iuga, M., Oroian, M., Mironeasa, S., Schröder, V., Ozon, E.-A., Lupuliasa, D., Caraiane, A., & Badea, V. (2022). Antioxidant, cytotoxic, and rheological properties of canola oil extract of usnea barbata (l.) weber ex f.h. wigg from călimani mountains, romania. *Plants*, *11*(7), 854. DOI: 10.3390/plants11070854 PMID: 35406834

Porter, P. M., Hegstad, J., Luker, K., & Ochsner, T. (2020). Soil health and sustainability of canola cropping systems in the Northern Great Plains. *Agronomy Journal*, *112*(5), 3589–3601. DOI: 10.1002/agj2.20363

Potter, T., Burton, W., Edwards, J., Wratten, N., Mailer, R., Salisbury, P., & Pearce, A. (2016). Assessing progress in breeding to improve grain yield, quality and blackleg (Leptosphaeria maculans) resistance in selected Australian canola cultivars (1978–2012). *Crop & Pasture Science*, *67*(4), 308–316. DOI: 10.1071/CP15290

Poveda, J. (2020). Trichoderma parareesei favors the tolerance of rapeseed (Brassica napus L.) to salinity and drought due to a chorismate mutase. *Agronomy (Basel)*, *10*(1), 118.

Prakash, Shyam, Bhat, SR, Quiros, CF, Kirti, PB, & Chopra, VL. (2009). 2 Brassica and its close allies: cytogenetics and evolution. *Plant breeding reviews*, *31*(23), 21-187.

Priest, H., Fox, S., Rowley, E., Murray, J., Michael, T., & Mockler, T. (2014). Analysis of global gene expression in brachypodium distachyon reveals extensive network plasticity in response to abiotic stress. *PLoS One*, *9*(1), e87499. DOI: 10.1371/journal.pone.0087499 PMID: 24489928

Prober, J. M., Trainor, G. L., Dam, R. J., Hobbs, F. W., Robertson, C. W., Zagursky, R. J., Cocuzza, A. J., Jensen, M. A., & Baumeister, K. (1987). A system for rapid DNA sequencing with fluorescent chain-terminating dideoxynucleotides. *Science*, *238*(4825), 336–341. DOI: 10.1126/science.2443975 PMID: 2443975

Pu, Z., Ino, Y., Kimura, Y., Tago, A., Shimizu, M., Natsume, S., & Shea, D. J. (2016). Changes in the proteome of xylem sap in Brassica oleracea in response to Fusarium oxysporum stress. *Frontiers in Plant Science*, *7*, 31.

Qian, Y., Ren, Q., Zhang, J., & Chen, L. (2019). Transcriptomic analysis of the maize (Zea mays L.) inbred line B73 response to heat stress at the seedling stage. *Gene*, *692*, 68–78. DOI: 10.1016/j.gene.2018.12.062 PMID: 30641208

Qu, C., Hasan, M., Lu, K., Liu, L., Liu, X., Xie, J., Wang, M., Lu, J., Odat, N., & Wang, R. (2012). Genetic diversity and relationship analysis of the Brassica napus germplasm using simple sequence repeat (SSR) markers. *African Journal of Biotechnology*, *11*(27), 6923–6933.

Qu, C., Jia, L., Fu, F., Zhao, H., Lu, K., Wei, L., Xu, X., Liang, Y., Li, S., Wang, R., & Li, J. (2017). Genome-wide association mapping and Identification of candidate genes for fatty acid composition in Brassica napus L. using SNP markers. *BMC Genomics*, *18*(1), 232. DOI: 10.1186/s12864-017-3607-8 PMID: 28292259

Quijada, P. A., Udall, J. A., Lambert, B., & Osborn, T. C. (2006). Quantitative trait analysis of seed yield and other complex traits in hybrid spring rapeseed (Brassica napus L.): 1. Identification of genomic regions from winter germplasm. *Theoretical and Applied Genetics*, *113*(3), 549–561. DOI: 10.1007/s00122-006-0323-1 PMID: 16767447

Rabbani, B., Tekin, M., & Mahdieh, N. (2014). The promise of whole-exome sequencing in medical genetics. *Journal of Human Genetics*, *59*(1), 5–15. DOI: 10.1038/jhg.2013.114 PMID: 24196381

Raboanatahiry, N., Chao, H., Dalin, H., Pu, S., Yan, W., Yu, L., Wang, B., & Li, M. (2018). QTL Alignment for Seed Yield and Yield Related Traits in Brassica napus. *Frontiers in Plant Science*, *9*, 1127. Advance online publication. DOI: 10.3389/fpls.2018.01127 PMID: 30116254

Raboanatahiry, N., Chao, H., Guo, L., Gan, J., Xiang, J., Yan, M., Zhang, L., Yu, L., & Li, M. (2017). Synteny analysis of genes and distribution of loci controlling oil content and fatty acid profile based on QTL alignment map in Brassica napus. *BMC Genomics*, *18*(1), 776. Advance online publication. DOI: 10.1186/s12864-017-4176-6 PMID: 29025408

Raboanatahiry, N., Li, H., Yu, L., & Li, M. (2021). Rapeseed (Brassica napus): Processing, utilization, and genetic improvement. *Agronomy (Basel)*, *11*(9), 1776. DOI: 10.3390/agronomy11091776

Rafique, N., Ilyas, N., Aqeel, M., Raja, N., Shabbir, G., Ajaib, M., & Ansari, M. (2023). Interactive effects of melatonin and salicylic acid on brassica napus under drought condition. *Plant and Soil*. Advance online publication. DOI: 10.1007/s11104-023-05942-7

Rahaman, M., Mamidi, S., & Rahman, M. (2017). Association mapping of agronomic traits of canola (Brassica napus L.) subject to heat stress under field conditions. *Australian Journal of Crop Science*, *11*(09), 1094–1105. DOI: 10.21475/ajcs.17.11.09.pne512

Rahman, H., Peng, G., Yu, F., Falk, K. C., Kulkarni, M., & Selvaraj, G. (2014). Genetics and breeding for clubroot resistance in Canadian spring canola (Brassica napus L.). *Canadian Journal of Plant Pathology, 36*(sup1), 122–134.

Rahman, H., Peng, G., Yu, F., Falk, K. C., Kulkarni, M., & Selvaraj, G. (2014). Genetics and breeding for clubroot resistance in Canadian spring canola (*Brassica napus* L.). *Canadian Journal of Plant Pathology, 36*(sup1), 122-134. DOI: 10.1080/07060661.2013.862571

Rahman, H., Ramanathan, V., Jagadeeshselvam, N., Ramasamy, S., Rajendran, S., Ramachandran, M., Sudheer, P. D., Chauhan, S., Natesan, S., & Muthurajan, R. (2015). Phenomics: technologies and applications in plant and agriculture. *PlantOmics: The Omics of Plant Science*, 385-411.

Rahman, H. (2013). Review: Breeding spring canola (brassica napusl.) by the use of exotic germplasm. *Canadian Journal of Plant Science*, *93*(3), 363–373. DOI: 10.4141/cjps2012-074

Rahman, H., Bennett, R. A., & Kebede, B. (2017). Mapping of days to flower and seed yield in spring oilseed Brassica napus carrying genome content introgressed from Brassica oleracea. *Molecular Breeding*, *37*(1), 1–15. DOI: 10.1007/s11032-016-0608-2

Rahman, H., Bennett, R. A., & Kebede, B. (2018). Molecular mapping of QTL alleles of Brassica oleracea affecting days to flowering and photosensitivity in spring *Brassica napus*. *PLoS One*, *13*(1), e0189723. DOI: 10.1371/journal.pone.0189723 PMID: 29320498

Rahman, H., & Kebede, B. (2021). Mapping of seed quality traits in the C genome of Brassica napus by using a population carrying genome content of B. oleracea and their effect on other traits. *The Plant Genome*, *14*(2), e20078. Advance online publication. DOI: 10.1002/tpg2.20078 PMID: 33818008

Rahman, H., Kebede, B., & Snowdon, R. (2014). Gene pyramiding in Canola: Strategies for resistance breeding and genomics. *Frontiers in Plant Science*, *5*, 219.

Rahman, M., Guo, Q., Baten, A., Mauleon, R., Khatun, A., Liu, L., & Barkla, B. J. (2021). Shotgun proteomics of Brassica rapa seed proteins identifies vicilin as a major seed storage protein in the mature seed. *PLoS One*, *16*(7), e0253384. DOI: 10.1371/journal.pone.0253384 PMID: 34242257

Rahman, M., Hoque, A., & Sheikh, M. (2018). Genetic control of glucosinolate content in Brassica napus: QTL mapping and marker-assisted selection. *Journal of Agricultural Science*, *156*(6), 759–768.

Rahman, M., & McVetty, P. (2011). A review of Brassica seed color. *Canadian Journal of Plant Science*, *91*(3), 437–446. DOI: 10.4141/cjps10124

Rahman, M., Sun, Z., McVetty, P. B., & Li, G. (2008). High throughput genome-specific and gene-specific molecular markers for erucic acid genes in Brassica napus (L.) for marker-assisted selection in plant breeding. *Theoretical and Applied Genetics*, *117*(6), 895–904. DOI: 10.1007/s00122-008-0829-9 PMID: 18633592

Rahnama, A., James, R. A., Poustini, K., & Munns, R. (2010). Stomatal conductance as a screen for osmotic stress tolerance in durum wheat growing in saline soil. *Functional Plant Biology*, *37*(3), 255–263.

Raimi, A., & Adeleke, R. (2023). 16s rrna gene-based identification and plant growth-promoting potential of cultivable endophytic bacteria. *Agronomy Journal*, *115*(3), 1447–1462. DOI: 10.1002/agj2.21241

Rajczewski, A. T., Jagtap, P. D., & Griffin, T. J. (2022). An overview of technologies for MS-based proteomics-centric multi-omics. *Expert Review of Proteomics*, *19*(3), 165–181. DOI: 10.1080/14789450.2022.2070476 PMID: 35466851

Rakow, G., & Woods, D. L. (1987). Outcrossing in rape and mustard under Saskatchewan prairie conditions. *Canadian Journal of Plant Science*, *67*(1), 147–151. DOI: 10.4141/cjps87-017

Ramakrishnan, T., Gift, M. D. M., Chitradevi, S., Jegan, R., & Jose, P. S. H. (2022). Nagaraja, Sharma, H. N., Selvakumar, P., & Hailegiorgis, S. M. (2022). Study of numerous resins used in polymer matrix composite materials. *Advances in Materials Science and Engineering*, *2022*, 1088926. Advance online publication. DOI: 10.1155/2022/1088926

Raman, H., Uppal, R. K., & Raman, R. (2019). Genetic solutions to improve resilience of canola to climate change. *Genomic designing of climate-smart oilseed crops*, 75-131.

Raman, H., Uppal, R. K., & Raman, R. (2019). Genetic solutions to improve resilience of canola to climate change. *Genomic designing of climate-smart oilseed crops*, 75-131. DOI: 10.1007/978-3-319-93536-2_2

Raman, H., Raman, R., Coombes, N., Song, J., Prangnell, R., Bandaranayake, C., & Edwards, D. (2016). Genome-wide association analyses reveal complex genetic architecture underlying natural variation for flowering time in canola. *Plant, Cell & Environment*, *39*(6), 1228–1239. DOI: 10.1111/pce.12644 PMID: 26428711

Raman, H., Raman, R., Diffey, S., Qiu, Y., McVittie, B., Barbulescu, D., Salisbury, P., Marcroft, S., & Delourme, R. (2018). Stable Quantitative Resistance Loci to Blackleg Disease in Canola (Brassica napus L.) Over Continents. *Frontiers in Plant Science*, *9*, 1622. Advance online publication. DOI: 10.3389/fpls.2018.01622 PMID: 30532758

Raman, H., Raman, R., Eckermann, P., Coombes, N., Manoli, S., Zou, X., Edwards, D., Meng, J., Prangnell, R., Stiller, J., Batley, J., Luckett, D., Wratten, N., & Dennis, E. (2012). Genetic and physical mapping of flowering time loci in canola (Brassica napus L.). *Theoretical and Applied Genetics*, *126*(1), 119–132. DOI: 10.1007/s00122-012-1966-8 PMID: 22955939

Raman, H., Raman, R., Mathews, K., Diffey, S., & Salisbury, P. (2020). QTL mapping reveals genomic regions for yield based on an incremental tolerance index to drought stress and related agronomic traits in canola. *Crop & Pasture Science*, *71*(6), 562–577. DOI: 10.1071/CP20046

Raman, H., Raman, R., McVittie, B., Borg, L., Diffey, S., Yadav, A., & Farquhar, G. (2020). Genetic and physiological bases for variation in water use efficiency in canola. *Food and Energy Security*, *9*(4), e237. Advance online publication. DOI: 10.1002/fes3.237

Raman, H., Raman, R., Pirathiban, R., McVittie, B., Sharma, N., Liu, S., Qiu, Y., Zhu, A., Killian, A., Cullis, B., Farquhar, G., Stuart-Williams, H., White, R., Tabah, D., Easton, A., & Zhang, Y. (2022). Multienvironment QTL analysis delineates a major locus associated with homoeologous exchanges for water-use efficiency and seed yield in canola. *Plant, Cell & Environment*, *45*(7), 2019–2036. DOI: 10.1111/pce.14337 PMID: 35445756

Raman, H., Raman, R., Qiu, Y., Yadav, A., Sureshkumar, S., Borg, L., Rohan, M., Wheeler, D., Owen, O., Menz, I., & Balasubramanian, S. (2019). GWAS hints at pleiotropic roles for FLOWERING LOCUS T in flowering time and yield-related traits in canola. *BMC Genomics*, *20*(1), 636. Advance online publication. DOI: 10.1186/s12864-019-5964-y PMID: 31387521

Raman, H., Shamaya, N., Pirathiban, R., McVittie, B., Raman, R., Cullis, B., & Easton, A. (2023). Quantitative trait loci for genotype and genotype by environment interaction effects for seed yield plasticity to terminal water-deficit conditions in canola (brassica napus l.). *Plants*, *12*(4), 720. DOI: 10.3390/plants12040720 PMID: 36840067

Raman, R., Diffey, S., Barbulescu, D., Coombes, N., Luckett, D., Salisbury, P., Cowley, R., Marcroft, S., & Raman, H. (2020). Genetic and physical mapping of loci for resistance to blackleg disease in canola (*Brassica napus* L.). *Scientific Reports*, *10*(1), 1–14. DOI: 10.1038/s41598-020-61211-y PMID: 32157120

Raman, R., Taylor, B., Lindbeck, K., Coombes, N., Barbulescu, D., Salisbury, P., & Raman, H. (2012). Molecular mapping and validation of Rlm1 gene for resistance to *Leptosphaeria maculans* in canola (*Brassica napus* L.). *Crop & Pasture Science*, *63*(10), 1007–1017. DOI: 10.1071/CP12255

Raman, R., Taylor, B., Marcroft, S., Stiller, J., Eckermann, P., Coombes, N., Rehman, A., Lindbeck, K., Luckett, D., Wratten, N., Batley, J., Edwards, D., Wang, X., & Raman, H. (2012). Molecular mapping of qualitative and quantitative loci for resistance to Leptosphaeria maculans causing blackleg disease in canola (*Brassica napus* L.). *Theoretical and Applied Genetics*, *125*(2), 405–418. DOI: 10.1007/s00122-012-1842-6 PMID: 22454144

Ram, T., Majumder, N. D., Krishnaveni, D., & Ansari, M. M. (2007). Rice variety Dhanrasi, an example of improving yield potential and disease resistance by introgressing gene (s) from wild species (Oryza rufipogon). *Current Science*, •••, 987–992.

Rattray, N., Deziel, N., Wallach, J., Khan, S., Vasiliou, V., Ioannidis, J., & Johnson, C. (2018). Beyond genomics: Understanding exposotypes through metabolomics. *Human Genomics, 12*(1), 4. Advance online publication. DOI: 10.1186/s40246-018-0134-x PMID: 29373992

Rauf, S., da Silva, J. T., Khan, A. A., & Naveed, A. (2010). Consequences of plant breeding on genetic diversity. *International Journal of Plant Breeding, 4*(1), 1–21.

Rauf, S., Fatima, S., & Ortíz, R. (2023). Modification of fatty acid profile and oil contents using gene editing in oilseed crops for a changing climate. *GM Crops and Food: Biotechnology in Agriculture and the Food Chain, 14*(1), 1–12. DOI: 10.1080/21645698.2023.2243041 PMID: 37551783

Ray, D., Gerber, J., MacDonald, G., & West, P. (2015). Climate variation explains a third of global crop yield variability. *Nature Communications, 6*(1), 5989. Advance online publication. DOI: 10.1038/ncomms6989 PMID: 25609225

Raza, A., Hafeez, M. B., Zahra, N., Shaukat, K., Umbreen, S., Tabassum, J., Charagh, S., Khan, R. S. A., & Hasanuzzaman, M. (2020). The plant family Brassicaceae: Introduction, biology, and importance. *The Plant Family Brassicaceae: Biology and Physiological Responses to Environmental Stresses*, 1-43.

Raza, W., Wang, J., Jousset, A., Friman, V., Mei, X., Wang, S., . . . Shen, Q. (2020). Bacterial community richness shifts the balance between volatile organic compound-mediated microbe–pathogen and microbe–plant interactions. Proceedings of the Royal Society B Biological Sciences, 287(1925), 20200403. https://doi.org/DOI: 10.1098/rspb.2020.0403

Raza, A., Ashraf, F., Zou, X., Zhang, X., & Tosif, H. (2020). Plant adaptation and tolerance to environmental stresses: mechanisms and perspectives. In *Plant eco-physiology and adaptation under climate change: mechanisms and perspectives I* (pp. 117–145). Springer. DOI: 10.1007/978-981-15-2156-0_5

Raza, A., Charagh, S., Razzaq, A., Javed, R., Khan, R. S. A., & Hasanuzzaman, M. (2020). Brassicaceae plants response and tolerance to drought stress: physiological and molecular interventions. In *The plant family Brassicaceae* (pp. 229–261). Springer. DOI: 10.1007/978-981-15-6345-4_7

Raza, A., Razzaq, A., Mehmood, S. S., Hussain, M. A., Wei, S., He, H., & Hasanuzzaman, M. (2021). Omics: The way forward to enhance abiotic stress tolerance in Brassica napus L. *GM Crops and Food: Biotechnology in Agriculture and the Food Chain, 12*(1), 251–281.

Raza, A., Razzaq, A., Mehmood, S. S., Zou, X., Zhang, X., Lv, Y., & Xu, J. (2019). Impact of climate change on crops adaptation and strategies to tackle its outcome: A review. *Plants*, *8*(2), 34. DOI: 10.3390/plants8020034 PMID: 30704089

Raza, A., Su, W., Hussain, M. A., Mehmood, S. S., Zhang, X., Cheng, Y., Zou, X., & Lv, Y. (2021). Integrated analysis of metabolome and transcriptome reveals insights for cold tolerance in rapeseed (Brassica napus L.). *Frontiers in Plant Science*, *12*, 721681. DOI: 10.3389/fpls.2021.721681 PMID: 34691103

Razavizadeh, R. (2015). Protein pattern of canola (brassica napus l.) changes in response to salt and salicylic acid in vitro. *Biological Letters*, *52*(1-2), 19–36. DOI: 10.1515/biolet-2015-0012

Razzaq, A., Sadia, B., Raza, A., Hameed, M., & Saleem, F. (2019). Metabolomics: A way forward for crop improvement. *Metabolites*, *9*(12), 303. DOI: 10.3390/metabo9120303 PMID: 31847393

Razzaq, M. K., Aleem, M., Mansoor, S., Khan, M. A., Rauf, S., Iqbal, S., & Siddique, K. H. (2021). Omics and CRISPR-Cas9 approaches for molecular insight, functional gene analysis, and stress tolerance development in crops. *International Journal of Molecular Sciences*, *22*(3), 1292. DOI: 10.3390/ijms22031292 PMID: 33525517

Rezayian, M., & Ebrahimzadeh, H. (2019). Different effects of calcium and penconazole on primary and secondary metabolites of brassica napus under drought. *Physiology and Molecular Biology of Plants : an International Journal of Functional Plant Biology*, *25*(2), 497–509. DOI: 10.1007/s12298-018-00634-4 PMID: 30956431

Rezayian, M., Niknam, V., & Arabloo, M. (2023). Iron nanoparticle regulate succinate dehydrogenase activity in canola plants under drought stress. *Scientific Reports*, *13*(1), 9628. Advance online publication. DOI: 10.1038/s41598-023-36105-4 PMID: 37316501

Rhee, S. Y., & Mutwil, M. (2014). Towards revealing the functions of all genes in plants. *Trends in Plant Science*, *19*(4), 213–221. DOI: 10.1016/j.tplants.2013.10.006 PMID: 24231067

Riar, A., Gill, G., & McDonald, G. (2020). Different post-sowing nitrogen management approaches required to improve nitrogen and water use efficiency of canola and mustard. *Frontiers in Plant Science*, *11*, 1111. DOI: 10.3389/fpls.2020.01111 PMID: 32793266

Ribeiro, D. G., da Cunha, G. C. R., dos Santos, C., Silva, L. P., de Oliveira, N., Brilhante, O., Labuto, L. B. D., & Mehta, A. (2018). Brassica oleracea resistance-related proteins identified at an early stage of black rot disease. *Physiological and Molecular Plant Pathology*, *104*, 9–14.

Richardson, C., Hennebelle, M., Otoki, Y., Zamora, D., Yang, J., Hammock, B., & Taha, A. (2017). Lipidomic analysis of oxidized fatty acids in plant and algae oils. *Journal of Agricultural and Food Chemistry*, *65*(9), 1941–1951. DOI: 10.1021/acs.jafc.6b05559 PMID: 28157307

Rincent, R., Charpentier, J. P., Faivre-Rampant, P., Paux, E., Le Gouis, J., Bastien, C., & Segura, V. (2018). Phenomic selection is a low-cost and high-throughput method based on indirect predictions: Proof of concept on wheat and poplar. *G3 (Bethesda, Md.)*, *8*(12), 3961–3972. DOI: 10.1534/g3.118.200760 PMID: 30373914

Rippka, R., Deruelles, J., Waterbury, J. B., Herdman, M., & Stanier, R. Y. (1979). Generic assignments, strain histories and properties of pure cultures of cyanobacteria. *Microbiology*, *111*(1), 1–61.

Ritchie, M., Holzinger, E., Li, R., Pendergrass, S., & Kim, D. (2015). Methods of integrating data to uncover genotype–phenotype interactions. *Nature Reviews. Genetics*, *16*(2), 85–97. DOI: 10.1038/nrg3868 PMID: 25582081

Rohart, F., Gautier, B., Singh, A., & Cao, K. (2017). Mixomics: An r package for 'omics feature selection and multiple data integration. *PLoS Computational Biology*, *13*(11), e1005752. DOI: 10.1371/journal.pcbi.1005752 PMID: 29099853

Rose, T. J., Damon, P., & Rengel, Z. (2008). Phosphorus-efficient canola: Root morphology and nutrient acquisition. *Plant and Soil*, *380*(1-2), 115–125.

Rossi, L., Zhang, W., Lombardini, L., & Ma, X. (2016). The impact of cerium oxide nanoparticles on the salt stress responses of brassica napus l. *Environmental Pollution*, *219*, 28–36. DOI: 10.1016/j.envpol.2016.09.060 PMID: 27661725

Roy, J., Shaikh, T., Mendoza, L., Hosain, S., Chapara, V., & Rahman, M. (2021). Genome-wide association mapping and genomic prediction for adult stage sclerotinia stem rot resistance in *Brassica napus* (L) under field environments. *Scientific Reports*, *11*(1), 21773. Advance online publication. DOI: 10.1038/s41598-021-01272-9 PMID: 34741104

Sabbahi, R., Azzaoui, K., Rhazi, L., Ayerdi-Gotor, A., Aussenac, T., Depeint, F., Taleb, M., & Hammouti, B. (2023). Factors affecting the quality of canola grains and their implications for grain-based foods. *Foods*, *12*(11), 2219. DOI: 10.3390/foods12112219 PMID: 37297464

Saharan, G. S., Mehta, N. K., Meena, P. D., Saharan, G. S., Mehta, N. K., & Meena, P. D. (2021). Inheritance of Disease Resistance. *Genomics of Crucifer's Host-Resistance*, 195-264.

Samans, B., Chalhoub, B., & Snowdon, R. J. (2017). Surviving a genome collision: Genomic signatures of allopolyploidization in the recent crop species Brassica napus. *The Plant Genome*, *10*(3), plantgenome2017.02.0013. DOI: 10.3835/plantgenome2017.02.0013 PMID: 29293818

Sanger, F., Nicklen, S., & Coulson, A. R. (1977). DNA sequencing with chain-terminating inhibitors. *Proceedings of the National Academy of Sciences of the United States of America*, *74*(12), 5463–5467. DOI: 10.1073/pnas.74.12.5463 PMID: 271968

Santos, Cristiane, Nogueira, Fábio CS, Domont, Gilberto B, Fontes, Wagner, Prado, Guilherme S, Habibi, Peyman, . . . Jorrín-Novo, Jesus V. (2019). Proteomic analysis and functional validation of a Brassica oleracea endochitinase involved in resistance to Xanthomonas campestris. *Frontiers in Plant Science*, *10*, 414.

Sargent, D., Conaty, W., Tissue, D., & Sharwood, R. (2022). Synthetic biology and opportunities within agricultural crops. *Journal of Sustainable Agriculture and Environment*, *1*(2), 89–107. DOI: 10.1002/sae2.12014

Sarkar, A., Kisiała, A., Adhikary, D., Basu, U., Emery, R., Rahman, H., & Kav, N. (2023). Silicon ameliorates clubroot responses in canola (brassica napus): A "multi-omics"-based investigation into possible mechanisms. *Physiologia Plantarum*, *175*(2), e13900. Advance online publication. DOI: 10.1111/ppl.13900 PMID: 36992551

Sarkar, N. K., Kim, Y. K., & Grover, A. (2014). Coexpression network analysis associated with call of rice seedlings for encountering heat stress. *Plant Molecular Biology*, *84*(1-2), 125–143. DOI: 10.1007/s11103-013-0123-3 PMID: 23975147

Sarwar, M., Ahmad, N., & Tofique, M. (2003). Efficiency of different chemicals on canola strain Rainbow (Brassica napus L.) for aphids control. *Asian Journal of Plant Sciences*, *2*(11), 831–833. DOI: 10.3923/ajps.2003.831.833

Sarwar, M., Ahmad, N., & Tofique, M. (2004). Response of various Brassica genotypes against aphids infestation under natural conditions. *Pakistan Journal of Zoology*, *26*(4), 69–74.

Sato, Y., Murakami, T., Funatsuki, H., Matsuba, S., Saruyama, H., & Tanida, M. (2001). Heat Shock-mediated APX Gene Expression and protection against Chilling Injury in rice Seedlings. *Journal of Experimental Botany*, *52*(354), 145–151. DOI: 10.1093/jexbot/52.354.145 PMID: 11181723

Scarth, R., McVetty, P. B. E., Rimmer, S. R., & Stefansson, B. R. (1988). STELLAR LOW LINOLENIC-HIGH LINOLEIC ACID SUMMER RAPE. *Canadian Journal of Plant Science*, *68*(2), 509–511. DOI: 10.4141/cjps88-061

Scarth, R., & Tang, J. (2006). Modification of Brassica oil using conventional and transgenic approaches. *Crop Science*, *46*(3), 1225–1236. DOI: 10.2135/cropsci2005.08-0245

Schaefer, R. J., Schubert, M., Bailey, E., Bannasch, D. L., Barrey, E., Bar-Gal, G. K., & McCue, M. E. (2017). Developing a 670k genotyping array to tag~ 2M SNPs across 24 horse breeds. *BMC Genomics*, *18*(1), 1–18. DOI: 10.1186/s12864-017-3943-8 PMID: 28750625

Scheben, A., Batley, J., & Edwards, D. (2017). Genotyping-by-sequencing approaches to characterize crop genomes: Choosing the right tool for the right application. *Plant Biotechnology Journal*, *15*(2), 149–161. DOI: 10.1111/pbi.12645 PMID: 27696619

Schiessl, S. V., Quezada-Martinez, D., Tebartz, E., Snowdon, R. J., & Qian, L. (2019). The vernalisation regulator FLOWERING LOCUS C is differentially expressed in biennial and annual Brassica napus. *Scientific Reports*, *9*(1), 14911. DOI: 10.1038/s41598-019-51212-x PMID: 31624282

Schiessl, S., Huettel, B., Kuehn, D., Reinhardt, R., & Snowdon, R. (2017). Post-polyploidisation morphotype diversification associates with gene copy number variation. *Scientific Reports*, *7*(1), 41845. DOI: 10.1038/srep41845 PMID: 28165502

Schmutzer, T., Samans, B., Dyrszka, E., Ulpinnis, C., Weise, S., Stengel, D., Colmsee, C., Lespinasse, D., Micic, Z., Abel, S., Duchscherer, P., Breuer, F., Abbadi, A., Leckband, G., Snowdon, R., & Scholz, U. (2015). Species-wide genome sequence and nucleotide polymorphisms from the model allopolyploid plant Brassica napus. *Scientific Data*, *2*(1), 150072. DOI: 10.1038/sdata.2015.72 PMID: 26647166

Schrag, T. A., Westhues, M., & Schipprack, W. (2018). Genomic selection in hybrid breeding: Insights from canola and maize. *Theoretical and Applied Genetics*, *131*(6), 1285–1299.

Schranz, M. E., Lysak, M. A., & Mitchell-Olds, T. (2006). The ABC's of comparative genomics in the Brassicaceae: Building blocks of crucifer genomes. *Trends in Plant Science*, *11*(11), 535–542. DOI: 10.1016/j.tplants.2006.09.002 PMID: 17029932

Secchi, M. A., Fernandez, J. A., Stamm, M. J., Durrett, T., Prasad, P. V. V., Messina, C. D., & Ciampitti, I. A. (2023). Effects of heat and drought on canola (*Brassica napus* L.) yield, oil, and protein: A meta-analysis. *Field Crops Research*, *293*, 108848. DOI: 10.1016/j.fcr.2023.108848

Sedeek, K. E. M., Mahas, A., & Mahfouz, M. (2019). CRISPR-based genome editing in plants: A new era for crop improvement. *Plant Biotechnology Journal, 17*(8), 1583–1597.

Seepaul, R., Kumar, S., Iboyi, J. E., Bashyal, M., Stansly, T. L., Bennett, R., & George, S. (2021). Brassica carinata: Biology and agronomy as a biofuel crop. *Global Change Biology. Bioenergy, 13*(4), 582–599.

Sehrawat, A., & Deswal, R. (2014). S-nitrosylation analysis in Brassica juncea apoplast highlights the importance of nitric oxide in cold-stress signaling. *Journal of Proteome Research, 13*(5), 2599–2619.

Sehrawat, A., Sougrakpam, Y., & Deswal, R. (2019). Cold modulated nuclear S-nitrosoproteome analysis indicates redox modulation of novel Brassicaceae specific, myrosinase and napin in Brassica juncea. *Environmental and Experimental Botany, 161*, 312–333.

Seker, M., Gul, A., & Yildirim, M. (2008). Screening and comparing tocopherols in the rapeseed (Brassica napus L.) and olive (Olea europaea L.) varieties using high-performance liquid chromatography. *International Journal of Food Sciences and Nutrition, 59*(6), 483–490. DOI: 10.1080/09637480701539484 PMID: 19086241

Selvakumar, P. (2022). Phytochemical and pharmacological profile review of Bryophyllum pinnatum. *Biomedical and Biotechnology Research Journal, 6*(3), 295–301. DOI: 10.4103/bbrj.bbrj_126_22

Selvakumar, P. (2023). Novel Drug Target with Diverse Therapeutic Potential in Cancer Therapy. *Pharmacy Times, 55*(1), 11–14.

Selvakumar, P. (2023a). Biomedical applications of polymers in biosensors, cancer vaccines and drug delivery systems. *Biopolimery i Kletka, 39*(4), 245–256. DOI: 10.7124/bc.000AA0

Selvakumar, P. (2023b). Fiber reinforced polymer composites: Classification, chemical treatment, mechanical and tribological properties. *Research Journal of Chemistry and Environment, 27*(1), 111–119. DOI: 10.25303/2701rjce1110119

Selvakumar, P. (2023c). Novel drug target with diverse therapeutic potential in cancer therapy. *Pharmacy Times, 55*(1), 11–14.

Selvakumar, P. (2024a). Reasons for biodiversity loss. In Rathoure, A. K. (Ed.), *Impact of societal development and infrastructure on biodiversity decline* (pp. 37–49). IGI Global., DOI: 10.4018/979-8-3693-6950-0.ch004

Selvakumar, P. (2024b). Emerging engineering technologies: Design, development, and industrial applications of IoT-based aquaculture monitoring system. In El Kacimi, Y., & Alaoui, K. (Eds.), *Emerging engineering technologies and industrial applications* (pp. 332–341). IGI Global., DOI: 10.4018/979-8-3693-1335-0.ch016

Selvakumar, P., & Loganathan, V. (2012). Phytochemical screening and antioxidant activity of red flowered Mirabilis jalapa leaf in different solvents. *International Journal of Pharma and Bio Sciences*, *3*(4), 440–446.

Selvakumar, P., Mariappan, P., Petchiammal, S., & Sivagami, S. (2024a). Removal of chemicals from effluent using photobioreactor technology to improve environmental and health impacts. *Zeitschrift für Physikalische Chemie*, *238*(6), 1085–1102. DOI: 10.1515/zpch-2023-0450

Selvakumar, P., & Palanisamy, S. K., Cinthaikinian, Selvamurugan, Palanisamy, V., Mariappan, R., & Selvakumar, P. K. (. (2024b). Biosensors and its diverse applications in healthcare systems. *Zeitschrift für Physikalische Chemie*, *238*(6), 1103–1123. DOI: 10.1515/zpch-2023-0406

Sexton, A. C., & Howlett, B. J. (2004). Microsatellite markers reveal genetic differentiation among populations of *Sclerotinia sclerotiorum* from Australian canola fields. *Current Genetics*, *46*(6), 357–365. DOI: 10.1007/s00294-004-0543-3 PMID: 15549318

Shabala, S., & Cuin, T. A. (2008). Potassium transport and plant salt tolerance. *Physiologia Plantarum*, *133*(4), 651–669. DOI: 10.1111/j.1399-3054.2007.01008.x PMID: 18724408

Shafighi, A., Ardakani, M. R., Rad, A. H. S., Alavifazel, M., & Rafiei, F. (2021). Grain yield and associated physiological traits of rapeseed (Brassica napus L.) cultivars under different planting dates and drought stress at the flowering stage. *Italian Journal of Agronomy*, *16*(1), 1648. DOI: 10.4081/ija.2020.1648

Shahbaz, M., Noreen, N., & Perveen, S. (2013). Triacontanol modulates photosynthesis and osmoprotectants in canola (brassica napusl.) under saline stress. *Journal of Plant Interactions*, *8*(4), 350–359. DOI: 10.1080/17429145.2013.764469

Shah, F., & Wu, W. (2019). Soil and crop management strategies to ensure higher crop productivity within sustainable environments. *Sustainability (Basel)*, *11*(5), 1485. DOI: 10.3390/su11051485

Shahoveisi, F., Oladzad, A., Mendoza, L., Hosseinirad, S., Ruud, S., & Rissato, B. (2021). *Assessing the Effect of Phenotyping Scoring Systems and SNP Calling and Filtering Parameters on Detection of QTL Associated with Reaction of Brassica napus to Sclerotinia sclerotiorum*. PhytoFrontiers., DOI: 10.1094/PHYTOFR-10-20-0029-R

Sharma, A., & Kumar, S. (2021). Economic impact of increased seed oil content in canola: A global perspective. *Agricultural Economics, 52*(3), 405–417.

Sharma, Nidhi, Hotte, Naomi, Rahman, Muhammad H, Mohammadi, Mohsen, & Deyholos, Michael K, & Kav, Nat NV. (2008). Towards identifying Brassica proteins involved in mediating resistance to Leptosphaeria maculans: A proteomics-based approach. *Proteomics, 8*(17), 3516–3535.

Sharma, Nidhi, Rahman, Muhammad Hafizur, Strelkov, Stephen, Thiagarajah, Mohan, & Bansal, Vipan K, & Kav, Nat NV. (2007). Proteome-level changes in two Brassica napus lines exhibiting differential responses to the fungal pathogen Alternaria brassicae. *Plant Science, 172*(1), 95–110.

Sharon, B., Hulyalkar, N., Nguyen, V., Zimmern, P., Palmer, K., & Nisco, N. (2021). Hybrid de novo genome assembly for the generation of complete genomes of urinary bacteria using short- and long-read sequencing technologies. *Journal of Visualized Experiments, 174*(174). Advance online publication. DOI: 10.3791/62872 PMID: 34487123

Shaw, G. (2013). Polymorphism and single nucleotide polymorphisms (SNP s). *BJU International, 112*(5), 664–665. DOI: 10.1111/bju.12298 PMID: 23924425

Shen, W., Li, H., Teng, R., Wang, Y., Wang, W., & Zhuang, J. (2019). Genomic and transcriptomic analyses of HD-Zip family transcription factors and their responses to abiotic stress in tea plant (*Camellia sinensis*). *Genomics, 111*(5), 1142–1151. DOI: 10.1016/j.ygeno.2018.07.009 PMID: 30031053

Shen, Y., Xiang, Y., Xu, E., Ge, X., & Li, Z. (2018). Major Co-localized QTL for Plant Height, Branch Initiation Height, Stem Diameter, and Flowering Time in an Alien Introgression Derived Brassica napus DH Population. *Frontiers in Plant Science, 9*, 390. Advance online publication. DOI: 10.3389/fpls.2018.00390 PMID: 29643859

Shen, Y., Yang, Y., Xu, E., Ge, X., Xiang, Y., & Li, Z. (2017). Novel and major QTL for branch angle detected by using DH population from an exotic introgression in rapeseed (Brassica napus L.). *Theoretical and Applied Genetics, 131*(1), 67–78. DOI: 10.1007/s00122-017-2986-1 PMID: 28942459

Shokri-Gharelo, Reza, Bandehagh, Ali, & Hossain, Mohammad Anwar. (2022). Proteomic Profiling and Protein-Protein Interaction Network Reveal the Molecular Mechanisms of Susceptibility to Drought Stress in Canola (Brassica napus L.). *Phyton (0031-9457), 91*(7).

Shokri-Gharelo, R., & Motie-Noparvar, P. (2018). Molecular response of canola to salt stress: insights on tolerance mechanisms. https://doi.org/DOI: 10.7287/peerj.preprints.26716v1

Shokri-Gharelo, R., & Noparvar, P. (2018). Molecular response of canola to salt stress: Insights on tolerance mechanisms. *PeerJ*, *6*, e4822. DOI: 10.7717/peerj.4822 PMID: 29844974

Shu, J., Ma, X., Ma, H., Huang, Q., Zhang, Y., Guan, M., & Guan, C. (2022). Transcriptomic, proteomic, metabolomic, and functional genomic approaches of Brassica napus L. during salt stress. *PLoS One*, *17*(3), e0262587. DOI: 10.1371/journal.pone.0262587 PMID: 35271582

Sidak, D., Schwarzerová, J., Weckwerth, W., & Waldherr, S. (2022). Interpretable machine learning methods for predictions in systems biology from omics data. *Frontiers in Molecular Biosciences*, *9*, 926623. DOI: 10.3389/fmolb.2022.926623 PMID: 36387282

Singh, A., & Roychoudhury, A. (2021). Gene regulation at transcriptional and post-transcriptional levels to combat salt stress in plants. *Physiologia Plantarum*, *173*(4), 1556–1572.

Singh, M., Avtar, R., Lakra, N., Hooda, E., Singh, V. K., Bishnoi, M., & Choudhary, R. R. (2021). Genetic and proteomic basis of sclerotinia stem rot resistance in Indian mustard [Brassica juncea (L.) czern & coss.]. *Genes*, *12*(11), 1784.

Singh, R. K., Prasad, A., Muthamilarasan, M., Parida, S. K., & Prasad, M. (2020). Breeding and biotechnological interventions for trait improvement: Status and prospects. *Planta*, *252*(4), 54. DOI: 10.1007/s00425-020-03465-4 PMID: 32948920

Singh, R. K., Rai, N., Singh, A. K., Kumar, P., & Singh, B. (2019). A critical review on Tomato leaf curl virus resistance in tomato. *International Journal of Vegetable Science*, *25*(4), 373–393. DOI: 10.1080/19315260.2018.1520379

Singh, R. K., Sahu, P. P., Muthamilarasan, M., Dhaka, A., & Prasad, M. (2017). Genomics-assisted breeding for improving stress tolerance of graminaceous crops to biotic and abiotic stresses: Progress and prospects. In Senthil-Kumar, M. (Ed.), *Plant tolerance to individual and concurrent stresses* (pp. 59–81). Springer., DOI: 10.1007/978-81-322-3706-8_5

Singh, R. K., Sood, P., Prasad, A., & Prasad, M. (2021). Advances in omics technology for improving crop yield and stress resilience. *Plant Breeding, 140*(5), 719–731. DOI: 10.1111/pbr.12963

Slafer, G. A., Savin, R., & Sadras, V. O. (2020). Coarse and fine regulation of wheat yield components in response to genotype and environment. *Field Crops Research, 249*, 107716.

Slatko, B. E., Albright, L. M., Tabor, S., & Ju, J. (1999). DNA sequencing by the dideoxy method. *Current Protocols in Molecular Biology, 47*(1), 7.4.1–7.4.8.

Smith, J. R., Anderson, K. L., & Wilson, P. T. (2020). Canola: An overview of its agronomic importance and genetic improvement. *Annual Review of Plant Biology, 71*, 539–562.

Smith, J., & Brown, K. (2020). Advances in catalysis for efficient biodiesel production from canola oil. *Catalysis Reviews, 62*(4), 509–565.

Smith, S. (2019). The foundations, continuing evolution, and outcomes from the application of intellectual property protection in plant breeding and agriculture. *Plant Breeding Reviews, 43*, 121–213. DOI: 10.1002/9781119616801.ch5

Smith, S., & De Smet, I. (2012). Root system architecture: Insights from Arabidopsis and cereal crops. *Philosophical Transactions of the Royal Society of London. Series B, Biological Sciences, 367*(1595), 1441–1452. DOI: 10.1098/rstb.2011.0234 PMID: 22527386

Snowdon, R. J., Abbadi, A., Kox, T., Schmutzer, T., & Leckband, G. (2015). Heterotic haplotype capture: Precision breeding for hybrid performance. *Trends in Plant Science, 20*(7), 410–413. DOI: 10.1016/j.tplants.2015.04.013 PMID: 26027461

Snowdon, R. J., & Iniguez Luy, F. L. (2012). Potential to improve oilseed rape and canola breeding in the genomics era. *Plant Breeding, 131*(3), 351–360. DOI: 10.1111/j.1439-0523.2012.01976.x

Snowdon, R. J., Lühs, W., & Friedt, W. (2007). Oilseed rape genomics. *Plant Breeding Reviews, 27*, 19–54.

Sobejano-Paz, V., Mikkelsen, T. N., Baum, A., Mo, X., Liu, S., Köppl, C. J., Johnson, M. S., Gulyas, L., & García, M. (2020). Hyperspectral and thermal sensing of stomatal conductance, transpiration, and photosynthesis for soybean and maize under drought. *Remote Sensing (Basel), 12*(19), 3182. DOI: 10.3390/rs12193182

So, K. K., & Duncan, R. W. (2021). Breeding canola (Brassica napus L.) for protein in feed and food. *Plants, 10*(10), 2220. DOI: 10.3390/plants10102220 PMID: 34686029

Sokhi, S. S., & Khangura, R. K. (1992). White rust of rapeseed and mustard. *Punjab Agricultural University, Ludhiana, DR/Publ, 92*(23), 1-16.

Song, G., Li, X., & Wang, H. (2016). Identification of drought and heat tolerance loci in Brassica napus using GWAS. *Plant Biotechnology Journal, 14*(8), 1788–1799.

Song, J.-M., Guan, Z., Hu, J., Guo, C., Yang, Z., Wang, S., Liu, D., Wang, B., Lu, S., Zhou, R., Xie, W.-Z., Cheng, Y., Zhang, Y., Liu, K., Yang, Q.-Y., Chen, L.-L., & Guo, L. (2020). Eight high-quality genomes reveal pan-genome architecture and ecotype differentiation of Brassica napus. *Nature Plants, 6*(1), 34–45. DOI: 10.1038/s41477-019-0577-7 PMID: 31932676

Song, Q., Li, Q., Liu, Y., Zhang, F., Ma, B., Zhang, W., Man, W.-Q., Du, W.-G., Wang, G.-D., Chen, S.-Y., & Zhang, J. (2013). Soybean gmbzip123 gene enhances lipid content in the seeds of transgenic arabidopsis plants. *Journal of Experimental Botany, 64*(14), 4329–4341. DOI: 10.1093/jxb/ert238 PMID: 23963672

Song, T., Chu, M., Lahlali, R., Yu, F., & Peng, G. (2016). Shotgun label-free proteomic analysis of clubroot (Plasmodiophora brassicae) resistance conferred by the gene Rcr1 in Brassica rapa. *Frontiers in Plant Science, 7*, 1013.

Song, X., Wei, Y., Xiao, D., Gong, K., Sun, P., Ren, Y., & Li, X. (2021). Brassica carinata genome characterization clarifies U's triangle model of evolution and polyploidy in Brassica. *Plant Physiology, 186*(1), 388–406.

Song, Z.-X., Seo, E.-Y., Hu, W.-X., Jeong, J.-H., Moon, J. S., Kim, K.-H., & Lim, H.-S. (2022). Construction of full-length infectious cDNA clones of two Korean isolates of turnip mosaic virus breaking resistance in Brassica napus. *Archives of Virology, 167*(4), 1157–1162.

Sotelo, T., Soengas, P., Velasco, P., Rodríguez, V. M., & Cartea, M. E. (2014). Identification of metabolic QTLs and candidate genes for glucosinolate synthesis in Brassica oleracea leaves, seeds and flower buds. *PLoS One, 9*(3), e91428. DOI: 10.1371/journal.pone.0091428 PMID: 24614913

Spasibionek, S., Mikołajczyk, K., Ćwiek-Kupczyńska, H., Piętka, T., Krótka, K., Matuszczak, M., Nowakowska, J., Michalski, K., & Bartkowiak-Broda, I. (2020). Bartkowiak-Broda Marker assisted selection of new high oleic and low linolenic winter oilseed rape (Brassica napus L.) inbred lines revealing good agricultural value. *PLoS One, 15*(6), e0233959. DOI: 10.1371/journal.pone.0233959 PMID: 32497146

Spindel, J. E., Dahlberg, J., Colgan, M., Hollingsworth, J., Sievert, J., Staggenborg, S. H., & Vogel, J. P. (2018). Association mapping by aerial drone reveals 213 genetic associations for Sorghum bicolor biomass traits under drought. *BMC Genomics*, *19*(1), 1–18. DOI: 10.1186/s12864-018-5055-5 PMID: 30223789

Sreenivasulu, A., Manimozhi, T., Satheesh Kumar, P., Selvakumar, P., & Singh, P. (2022). The model of smart sensing device for sensitive nanoclusters modification in sensing properties. In 2022 2nd International Conference on Advance Computing and Innovative Technologies in Engineering (ICACITE) (pp. 1043–1047). IEEE. https://doi.org/DOI: 10.1109/ICACITE53722.2022.9823846

Stafen, C., Pedrolo, A., Schroeder, L., Viana, V., Busanello, C., Rosa, M., & Pegoraro, C. (2018). Seed germination and seedling growth of canola treated with abiotic factors that can affect growth and development. *International Journal of Vegetable Science*, *24*(6), 574–582. DOI: 10.1080/19315260.2018.1453909

Stanic, M., Hickerson, N., Arunraj, R., & Samuel, M. (2021). Gene-editing of the strigolactone receptor bnd14 confers promising shoot architectural changes in brassica napus (canola). *Plant Biotechnology Journal*, *19*(4), 639–641. DOI: 10.1111/pbi.13513 PMID: 33219600

STATISTA. Worldwide Oilseed Production since 2008. Available online: https://www.statista.com/statistics/267271/worldwide-oilseed-production-since-2008 (accessed on 8 July 2020).

Stefansson, B. R., & Hougen, F. W. (1964). Selection of rape plants (Brassica napus) with seed oil practically free from erucic acid. *Canadian Journal of Plant Science*, *44*(4), 359–364. DOI: 10.4141/cjps64-069

Stefansson, B. R., & Hougen, F. W. (1975). Selection for low erucic acid content in rapeseed (Brassica napus L.). *Canadian Journal of Plant Science*, *55*(2), 443–447.

Stefansson, B. R., Hougen, F. W., & Downey, R. K. (1961). Note on the isolation of rape plants with seed oil free from erucic acid. *Canadian Journal of Plant Science*, *41*(1), 218–219. DOI: 10.4141/cjps61-028

Strahl, B. D., & Allis, C. D. (2000). The language of covalent histonemodifications. *Nature*, *403*(6765), 41–45. DOI: 10.1038/47412 PMID: 10638745

Subbarao, G., Levine, L. H., Wheeler, R. M., & Stutte, G. W. (2001). Glycine betaine accumulation: its role in stress resistance in crop plants. In *Handbook of plant and crop physiology* (pp. 903–930). CRC Press. DOI: 10.1201/9780203908426.ch45

Subramanian, Babu, & Bansal, Vipan K, & Kav, Nat NV. (2005). Proteome-level investigation of Brassica carinata-derived resistance to Leptosphaeria maculans. *Journal of Agricultural and Food Chemistry*, *53*(2), 313–324.

Suchan, D., Bergsveinson, J., Manzon, L., Pierce, A., Kryachko, Y., Korber, D., & Yost, C. (2020). Transcriptomics reveal core activities of the plant growth-promoting bacterium delftia acidovorans ray209 during interaction with canola and soybean roots. *Microbial Genomics*, *6*(11). Advance online publication. DOI: 10.1099/mgen.0.000462 PMID: 33151138

Sugumaran, K. M., Pandiyarajan, V., Selvakumar, P., & Manoharan, S. (2024). Analysis of electromyography signals for control of mechanical prosthesis using machine learning techniques. In 2024 2nd International Conference on Intelligent Data Communication Technologies and Internet of Things (IDCIoT) (pp. 1059–1067). IEEE. DOI: 10.1109/IDCIoT59759.2024.10467707

Su, H., Chen, G., Yang, L., Zhang, Y., Wang, Y., Fang, Z., & Lv, H. (2020). Proteomic variations after short-term heat shock treatment reveal differentially expressed proteins involved in early microspore embryogenesis in cabbage (Brassica oleracea). *PeerJ*, *8*, e8897.

Sumitomo, K., Shirasawa, K., Isobe, S., Hirakawa, H., Hisamatsu, T., Nakano, Y., Yagi, M., & Ohmiya, A. (2019). Genome-wide association study overcomes the genome complexity in autohexaploid chrysanthemum and tags SNP markers onto the flower color genes. *Scientific Reports*, *9*(1), 13947. DOI: 10.1038/s41598-019-50028-z PMID: 31558738

Sumner, L., Lei, Z., Nikolau, B., & Saito, K. (2015). Modern plant metabolomics: Advanced natural product gene discoveries, improved technologies, and future prospects. *Natural Product Reports*, *32*(2), 212–229. DOI: 10.1039/C4NP00072B PMID: 25342293

Sun, C., Dong, Z., Zhao, L., Ren, Y., Zhang, N., & Chen, F. (2020). The Wheat 660K SNP array demonstrates great potential for marker-assisted selection in polyploid wheat. *Plant Biotechnology Journal*, *18*(6), 1354–1360. DOI: 10.1111/pbi.13361 PMID: 32065714

Sun, C., Wang, B., Wang, X., Hu, K., Li, K., Li, Z., & Li, Y. (2016). Genome-wide association study dissecting the genetic architecture underlying the branch angle trait in rapeseed (Brassica napus L.). *Scientific Reports*, *6*(1), 33673. DOI: 10.1038/srep33673 PMID: 27646167

Sun, D., Wang, C., Zhang, X., Zhang, W., Jiang, H., Yao, X., Liu, L., Wen, Z., Niu, G., & Shan, X. (2019). Draft genome sequence of cauliflower (*Brassica oleracea* L. var. botrytis) provides new insights into the C genome in Brassica species. *Horticulture Research*, *6*(1), 82. DOI: 10.1038/s41438-019-0164-0 PMID: 31645943

Sun, F., Fan, G., Hu, Q., Zhou, Y., Guan, M., Tong, C., Li, J., Du, D., Qi, C., Jiang, L., Liu, W., Huang, S., Chen, W., Yu, J., Mei, D., Meng, J., Zeng, P., Shi, J., Liu, K., & Wang, H. (2017). The high-quality genome of Brassica napus cultivar 'ZS 11'reveals the introgression history in semi-winter morphotype. *The Plant Journal*, *92*(3), 452–468. DOI: 10.1111/tpj.13669 PMID: 28849613

Sun, W. C., Wu, J. Y., Zeng, J., Zhu, H. X., Liu, Y. L., & Zhang, Y. H. (2007). Primary evaluation of cold tolerance among eight winter Brassica rapa. *J. Hunan Agric. Univ. Nature and Science*, *33*, 151–155.

Sun, W., Ma, W., Lei, J., Liu, Q., Yang, R., & Wu, J.. (2007). Study on Adaptation and Introduction Possibility of Winter Rapeseed to Dry and Cold Areas in Northwest China. *Zhongguo Nong Ye Ke Xue*, *40*, 2716–2726. DOI: 10.3321/j.issn:0578-1752.2007.12.008

Su, Z., Wang, T., Shrivastava, N., Chen, Y., Liu, X., Sun, C., & Lou, B. (2017). Piriformospora indica promotes growth, seed yield and quality of brassica napus l. *Microbiological Research*, *199*, 29–39. DOI: 10.1016/j.micres.2017.02.006 PMID: 28454707

Suzuki, N., Rivero, R., Shulaev, V., Blumwald, E., & Mittler, R. (2014). Abiotic and biotic stress combinations. *The New Phytologist*, *203*(1), 32–43. DOI: 10.1111/nph.12797 PMID: 24720847

Suzuki, S., Ono, N., Furusawa, C., Ying, B. W., & Yomo, T. (2011). Comparison of sequence reads obtained from three next-generation sequencing platforms. *PLoS One*, *6*(5), e20017. DOI: 10.1371/journal.pone.0019534 PMID: 21611185

Syed, A. (2016). *Oxidative stability and shelf life of vegetable oils Oxidative stability and shelf life of foods containing oils and fats*. Elsevier., DOI: 10.1016/B978-1-63067-056-6.00004-5

Szczygłowska, M., Piekarska, A., Konieczka, P., & Namieśnik, J. (2011). Use of brassica plants in the phytoremediation and biofumigation processes. *International Journal of Molecular Sciences*, *12*(11), 7760–7771.

Tagkouli, V. (2021). *Evaluation of recurrent selection as a method to achieve rapid re-adaptation of faba bean to a niche agro-climate*. University of Reading.

Tahir ul Qamar, M., Faryad, A., Bari, A., Zahid, B., Zhu, X., & Chen, L.-L. (2020). Effectiveness of conventional crop improvement strategies vs. omics. *Environment, climate, plant and vegetation growth*, 253-284.

Tan, C., Wei, X., & Zhang, Y. (2020). Trade-offs between seed oil content and other agronomic traits in Brassica napus: A meta-analysis. *Crop Science*, *60*(5), 2442–2458.

Tang, T., Chen, G., Bu, C., Liu, F., Liu, L., & Zhao, X. (2018). Transgene introgression from Brassica napus to different varieties of Brassica juncea. *Plant Breeding*, *137*(2), 171–180. DOI: 10.1111/pbr.12567

Tan, H., Xiang, X., Tang, J., & Wang, X. (2016). Nutritional functions of the funiculus in brassica napus seed maturation revealed by transcriptome and dynamic metabolite profile analyses. *Plant Molecular Biology*, *92*(4-5), 539–553. DOI: 10.1007/s11103-016-0530-3 PMID: 27539000

Tan, H., Xie, Q., Xiang, X., Li, J., Zheng, S., Xu, X., & Ye, W. (2015). Dynamic metabolic profiles and tissue-specific source effects on the metabolome of developing seeds of brassica napus. *PLoS One*, *10*(4), e0124794. DOI: 10.1371/journal.pone.0124794 PMID: 25919591

Tan, S. H., Morrison, W. R., & Karasov, W. H. (2011). Canola proteins for human consumption: Extraction, profile, and functional properties. *Journal of Food Science*, *76*(1), R16–R28. DOI: 10.1111/j.1750-3841.2010.01930.x PMID: 21535703

Tardieu, F., Cabrera-Bosquet, L., Pridmore, T., & Bennett, M. (2017). Plant phenomics, from sensors to knowledge. *Current Biology*, *27*(15), R770–R783. DOI: 10.1016/j.cub.2017.05.055 PMID: 28787611

Tardieu, F., & Tuberosa, R. (2010). Dissection and modeling of abiotic stress tolerance in plants. *Current Opinion in Plant Biology*, *13*(2), 206–212. DOI: 10.1016/j.pbi.2009.12.012 PMID: 20097596

Tavares, S., Ramos, A. P., Pires, A. S., Azinheira, H. G., Caldeirinha, P., Link, T., Abranches, R., Silva, M. C., Voegele, R. T., Loureiro, J., & Talhinhas, P. (2014). Genome size analyses of Pucciniales reveal the largest fungal genomes. *Frontiers in Plant Science*, *5*, 422. DOI: 10.3389/fpls.2014.00422 PMID: 25206357

Tettelin, H., Masignani, V., Cieslewicz, M. J., Donati, C., Medini, D., Ward, N. L., & Fraser, C. M. (2005). Genome analysis of multiple pathogenic isolates of Streptococcus agalactiae: Implications for the microbial "pan-genome". *Proceedings of the National Academy of Sciences of the United States of America*, *102*(39), 13950–13955. DOI: 10.1073/pnas.0506758102 PMID: 16172379

Theodoridou, K., & Yu, P. (2013). Effects of heat processing on the nutrient availability, protein molecular structure, and metabolic characteristics of canola meal. *Journal of Dairy Science*, *96*(6), 3991–4003. DOI: 10.3168/jds.2012-6512

Thomashow, M. F. (2016). The Role of cold-stress proteins in plant tolerance to freezing and subfreezing temperatures. *Plant Physiology*, *141*(4), 793–803.

Thompson, D. (2019). The role of humectants in oral care products: Focus on glycerol. *Journal of Dental Hygiene*, *93*(3), 42–49.

Tian, H., Lam, S., & Shui, G. (2016). Metabolomics, a powerful tool for agricultural research. *International Journal of Molecular Sciences*, *17*(11), 1871. DOI: 10.3390/ijms17111871 PMID: 27869667

Tian, Q., Li, B., Feng, Y., Zhao, W., Huang, J., & Chao, H. (2022). Application of CRISPR/Cas9 in rapeseed for gene function research and genetic improvement. *Agronomy (Basel)*, *12*(4), 824. DOI: 10.3390/agronomy12040824

Tocher, D. R., Betancor, M. B., Sprague, M., Olsen, R. E., & Napier, J. A. (2019). Omega-3 long-chain polyunsaturated fatty acids, EPA and DHA: Bridging the gap between supply and demand. *Nutrients*, *11*(1), 89. DOI: 10.3390/nu11010089 PMID: 30621155

Ton, L. B., Neik, T. X., & Batley, J. (2020). The use of genetic and gene technologies in shaping modern rapeseed cultivars (*Brassica napus* L.). *Genes*, *11*(10), 1161. DOI: 10.3390/genes11101161 PMID: 33008008

Tonu, N. N., Wen, R., Song, T., Guo, X., Murphy, L. A., Gossen, B. D., Yu, F., & Peng, G. (2023). Canola with stacked genes shows moderate resistance and resilience against a field population of Plasmodiophora brassicae (clubroot) pathotype X. *Plants*, *12*(4), 726. DOI: 10.3390/plants12040726 PMID: 36840074

Toriyama, K. (2021). Molecular basis of cytoplasmic male sterility and fertility restoration in rice. *Plant Biotechnology (Sheffield, England)*, *38*(3), 285–295. DOI: 10.5511/plantbiotechnology.21.0607a PMID: 34782814

Torkamaneh, D., Laroche, J., Bastien, M., Abed, A., & Belzile, F. (2018). Fast-GBS: A new pipeline for the efficient and highly accurate calling of SNPs from genotyping-by-sequencing data. *BMC Bioinformatics*, *19*(1), 5. DOI: 10.1186/s12859-016-1431-9 PMID: 28049422

Torkamaneh, D., Laroche, J., & Belzile, F. (2016). Genome-wide SNP calling from genotyping by sequencing (GBS) data: A comparison of seven pipelines and two sequencing technologies. *PLoS One*, *11*(8), e0161333. DOI: 10.1371/journal.pone.0161333 PMID: 27547936

Tripathi, P., Rabara, R., Reese, R., Miller, M., Rohila, J., Subramanian, S., & Rushton, P. (2016). A toolbox of genes, proteins, metabolites and promoters for improving drought tolerance in soybean includes the metabolite coumestrol and stomatal development genes. *BMC Genomics*, *17*(1), 102. Advance online publication. DOI: 10.1186/s12864-016-2420-0 PMID: 26861168

Ulfat, M., Athar, H., Khan, Z., & Kalaji, H. (2020). Rnaseq analysis reveals altered expression of key ion transporters causing differential uptake of selective ions in canola (brassica napus l.) grown under nacl stress. *Plants*, *9*(7), 891. DOI: 10.3390/plants9070891 PMID: 32674475

Umesh, M. (2023). Growth and physiological response of canola (brassica napus l.) tointeractive effect of temperature, moisture and nitrogen stressesunder controlled environment. Journal of Oilseeds Research, 40 (Specialissue). https://doi.org/DOI: 10.56739/5fxsvn54

United Soybean Board. (2008). 15th annual consumer attitudes about nutrition report. United Soybean Board. https://www.soyconnection.com/sites/default/files/ConsumerAttitudes2008.pdf

USDA Foreign Agricultural Service. (2024). *Rapeseed Explorer - World Production.* Retrieved from https://ipad.fas.usda.gov/cropexplorer/cropview/commodityView.aspx?cropid=2226000

Van de Peer, Y., Ashman, T.-L., Soltis, P. S., & Soltis, D. E. (2021). Polyploidy: An evolutionary and ecological force in stressful times. *The Plant Cell*, *33*(1), 11–26.

Vanhercke, T., Dyer, J. M., Mullen, R. T., Kilaru, A., Rahman, M. M., Petrie, J. R., Green, A. G., Yurchenko, O., & Singh, S. P. (2019). Metabolic engineering for enhanced oil in biomass. *Progress in Lipid Research*, *74*, 103–129. DOI: 10.1016/j.plipres.2019.02.002 PMID: 30822461

Vanhercke, T., Tahchy, A., Liu, Q., Zhou, X., Shrestha, P., Divi, U., & Petrie, J. (2013). Metabolic engineering of biomass for high energy density: Oilseed-like triacylglycerol yields from plant leaves. *Plant Biotechnology Journal*, *12*(2), 231–239. DOI: 10.1111/pbi.12131 PMID: 24151938

Varshney, R. K., Pandey, M. K., Bohra, A., Singh, V. K., Thudi, M., & Saxena, R. K. (2019). Toward the sequence-based breeding in legumes in the post-genome sequencing era. *Theoretical and Applied Genetics*, *132*(3), 797–816. DOI: 10.1007/s00122-018-3252-x PMID: 30560464

Varshney, R. K., Roorkiwal, M., & Sorrells, M. E. (2017). *Genomic selection for crop improvement: new molecular breeding strategies for crop improvement.* Springer International Publishing. DOI: 10.1007/978-3-319-63170-7

Varshney, R., (2007). Extending the repertoire of microsatellite markers for genetic linkage mapping and germplasm screening in chickpea. *Journal of SAT Agricultural Research, 5*, 1–3.

Verma, R. P. S., Lal, C., Malik, R., Kharub, A. S., Kumar, L., & Kumar, D. (2022). Barley improvement: current status and future prospects in changing scenario. *New Horizons in Wheat and Barley Research: Global Trends, Breeding and Quality Enhancement*, 93-134.

Vijayakumar, G., Muniyandi, R., Chandar, N. R., Selvakumar, P., & Duraisamy, R. (2024). Environmentally friendly TDS removal from wastewater by electrochemical ion exchange batch-type recirculation (EIR) technique. *Environmental Science. Water Research & Technology, 10*(4), 826–835. DOI: 10.1039/D3EW00793F

Villeth, Gabriela R, & Jr, Reis, Fabio B, Tonietto, Angela, Huergo, Luciano, De Souza, Emanuel M, Pedrosa, Fabio O, . . . Mehta, Angela. (2009). Comparative proteome analysis of Xanthomonas campestris pv. campestris in the interaction with the susceptible and the resistant cultivars of Brassica oleracea. *FEMS Microbiology Letters, 298*(2), 260–266.

Vo, K., Rahman, M., Rahman, M., Trinh, K., Kim, S., & Jeon, J. (2021). Proteomics and metabolomics studies on the biotic stress responses of rice: An update. *Rice (New York, N.Y.), 14*(1), 30. Advance online publication. DOI: 10.1186/s12284-021-00461-4 PMID: 33721115

Vos, P. G., Uitdewilligen, J. G., Voorrips, R. E., Visser, R. G., & van Eck, H. J. (2015). Development and analysis of a 20K SNP array for potato (Solanum tuberosum): An insight into the breeding history. *Theoretical and Applied Genetics, 128*(12), 2387–2401. DOI: 10.1007/s00122-015-2593-y PMID: 26263902

Waminal, N. E., Perumal, S., Lee, J., Kim, H. H., & Yang, T. J. (2016). Repeat evolution in *Brassica rapa* (AA), *B. oleracea* (CC), and *B. napus* (AACC) genomes. *Plant Breeding and Biotechnology, 4*(2), 107–122. DOI: 10.9787/PBB.2016.4.2.107

Wanasundara, J. P., McIntosh, T. C., Perera, S. P., Withana-Gamage, T. S., & Mitra, P. (2016). Canola/rapeseed protein-functionality and nutrition. *Oilseeds and Fats, Crops and Lipids, 23*(4), D407. Advance online publication. DOI: 10.1051/ocl/2016028

Wang XiaoWu, Wang XiaoWu, Wang HanZhong, Wang HanZhong, Wang Jun, Wang Jun, Sun RiFei, Sun RiFei, Wu Jian, Wu Jian, Liu ShengYi, Liu ShengYi, . . . Cheng Feng, Cheng Feng. (2011). The genome of the mesopolyploid crop species Brassica rapa.

Wang, J., Huang, Y., Li, X. L., & Li, H. Z. (2011). Research progress in clubroot of crucifers. *Plant prot, 37*, 153-158.

Wang, W., Das, A., Kainer, D., Schalamun, M., Morales-Suárez, A., Schwessinger, B., ... & Raley, M. (2020). The draft nuclear genome assembly of *Eucalyptus pauciflora*: A pipeline for comparing de novo assemblies. *GigaScience, 9*(giz160). DOI: 10.1093/gigascience/giz160

Wang, X. (2016). Broadening of genetic diversity in spring canola (Brassica napus L.) by use of the C genome of Brassica oleracea var. italica and Brassica oleracea var. capitata.

Wang, X., Liu, G., & Yang, Q. (2022). Global trends in oilseed production: Focus on canola's expanding role. Oilseed & fat crops and lipids, 29, 15.

Wang, B., Kumar, V., Olson, A., & Ware, D. (2019). Reviving the transcriptome studies: An insight into the emergence of single-molecule transcriptome sequencing. *Frontiers in Genetics, 10*, 384. Advance online publication. DOI: 10.3389/fgene.2019.00384 PMID: 31105749

Wang, E., He, D., Wang, J., Lilley, J. M., Christy, B., Hoffmann, M. P., O'Leary, G., Hatfield, J. L., Ledda, L., Deligios, P. A., Grant, B., Jing, Q., Nendel, C., Kage, H., Qian, B., Eyshi Rezaei, E., Smith, W., Weymann, W., & Ewert, F. (2022). How reliable are current crop models for simulating growth and seed yield of canola across global sites and under future climate change? *Climatic Change, 172*(1), 20. DOI: 10.1007/s10584-022-03375-2

Wang, F., Guo, Z., & Li, H. (2018). Cold acclimation and freezing tolerance in Brassica napus: Insights from transcriptomic analysis. *Frontiers in Plant Science, 9*, 1834.

Wang, H., Li, Y., & Zhang, Q. (2021). Advances in genome-wide association studies of complex traits in major crop species. *Nature Reviews. Genetics, 22*(7), 485–503.

Wang, J., Chu, S., Zhang, H., Zhu, Y., Cheng, H., & Yu, D. (2016). Development and application of a novel genome-wide SNP array reveals domestication history in soybean. *Scientific Reports, 6*(1), 20728. DOI: 10.1038/srep20728 PMID: 26856884

Wang, J., Jian, H., Wang, T., Wei, L., Li, J., Li, C., & Liu, L. (2016). Identification of microRNAs actively involved in fatty acid biosynthesis in developing *Brassica napus* seeds using high-throughput sequencing. *Frontiers in Plant Science*, *7*, 1570. DOI: 10.3389/fpls.2016.01570 PMID: 27822220

Wang, J., Kong, L., Zhang, L., Shi, X., Yu, B., Li, J., Zhang, B., Gao, M., Liu, X., Li, X., Gao, Y., Peng, D., & Liu, S. (2022). Breeding a Soybean Cultivar Heinong 531 with Peking-Type Cyst Nematode Resistance, Enhanced Yield, and High Seed-Oil Contents. *Phytopathology*, *112*(6), 1345–1349. DOI: 10.1094/PHYTO-08-21-0347-R PMID: 34879718

Wang, J., Song, L., Gong, X., Xu, J., & Li, M. (2020). Functions of jasmonic acid in plant regulation and response to abiotic stress. *International Journal of Molecular Sciences*, *21*(4), 1446. DOI: 10.3390/ijms21041446 PMID: 32093336

Wang, L., Jin, X., Li, Q., Wang, X., Li, Z., & Wu, X. (2016). Comparative proteomics reveals that phosphorylation of β carbonic anhydrase 1 might be important for adaptation to drought stress in Brassica napus. *Scientific Reports*, *6*(1), 39024.

Wang, P., Su, L., Gao, H., Jiang, X., Wu, X., Li, Y., Zhang, Q., Wang, Y., & Ren, F. (2018). Genome-wide characterization of bHLH genes in grape and analysis of their potential relevance to abiotic stress tolerance and secondary metabolite biosynthesis. *Frontiers in Plant Science*, *9*, 64. DOI: 10.3389/fpls.2018.00064 PMID: 29449854

Wang, P., Xiong, X., Zhang, X., Wu, G., & Liu, F. (2022). A review of erucic acid production in Brassicaceae oilseeds: Progress and prospects for the genetic engineering of high and low-erucic acid rapeseeds (*Brassica napus*). *Frontiers in Plant Science*, *13*, 899076. DOI: 10.3389/fpls.2022.899076 PMID: 35645989

Wang, P., Yang, C., Chen, H., Song, C., Zhang, X., & Wang, D. (2017). Transcriptomic basis for drought-resistance in Brassica napus L. *Scientific Reports*, *7*(1), 40532. DOI: 10.1038/srep40532 PMID: 28091614

Wang, S., Wong, D., Forrest, K., Allen, A., Chao, S., Huang, B. E., & Akhunov, E. (2014). Characterization of polyploid wheat genomic diversity using a high-density 90 000 single nucleotide polymorphism array. *Plant Biotechnology Journal*, *12*(6), 787–796. DOI: 10.1111/pbi.12183 PMID: 24646323

Wang, W., Pang, J., Zhang, F., Sun, L., Yang, L., Fu, T., & Siddique, K. (2022). Salt-responsive transcriptome analysis of canola roots reveals candidate genes involved in the key metabolic pathway in response to salt stress. *Scientific Reports*, *12*(1), 1666. Advance online publication. DOI: 10.1038/s41598-022-05700-2 PMID: 35102232

Wang, W., Pang, J., Zhang, F., Sun, L., Yang, L., Zhao, Y., Yang, Y., Wang, Y., & Siddique, K. H. (2021). Integrated transcriptomics and metabolomics analysis to characterize alkali stress responses in canola (Brassica napus L.). *Plant Physiology and Biochemistry*, *166*, 605–620. DOI: 10.1016/j.plaphy.2021.06.021 PMID: 34186284

Wang, W., Vinocur, B., & Altman, A. (2003). Plant responses to drought, salinity and extreme temperatures: Towards genetic engineering for stress tolerance. *Planta*, *218*, 1–14.

Wang, X., Chen, L., Wang, A., Wang, H., Tian, J., Zhao, X., Chao, H., Zhao, Y., Zhao, W., Xiang, J., Zhao, J., Li, M., & Hu, J. (2018). Quantitative trait loci analysis and genome-wide comparison for silique related traits in Brassica napus. *BMC Plant Biology*, *18*(1), 228. PMID: 30309330

Wang, X., Chen, Y., Thomas, C., Ding, G., Xu, P., Shi, D., & Shi, L. (2017). Genetic variants associated with the root system architecture of oilseed rape (brassica napus l.) under contrasting phosphate supply. *DNA Research : an International Journal for Rapid Publication of Reports on Genes and Genomes*, *24*(4), 407–417. DOI: 10.1093/dnares/dsx013 PMID: 28430897

Wang, X., Li, Y., Wang, X., Li, X., & Dong, S. (2022). Physiology and metabonomics reveal differences in drought resistance among soybean varieties. *Botanical Studies*, *63*(1), 8. Advance online publication. DOI: 10.1186/s40529-022-00339-8 PMID: 35332430

Wang, X., Shan, X., Wu, Y., Su, S., Li, S., Liu, H., & Yuan, Y. (2016). iTRAQ-based quantitative proteomic analysis reveals new metabolic pathways responding to chilling stress in maize seedlings. *Journal of Proteomics*, *146*, 14–24.

Wang, Y., Cheng, X., Shan, Q., Zhang, Y., Liu, J., Gao, C., & Qiu, J. L. (2014). Simultaneous editing of three homoeoalleles in hexaploid bread wheat confers heritable resistance to powdery mildew. *Nature Biotechnology*, *32*(9), 947–951. DOI: 10.1038/nbt.2969 PMID: 25038773

Wang, Y., Gu, W., Meng, Y., Xie, T., Li, L., Li, J., & Wei, S. (2017). γ-Aminobutyric acid imparts partial protection from salt stress injury to maize seedlings by improving photosynthesis and upregulating osmoprotectants and antioxidants. *Scientific Reports*, *7*(1), 43609. DOI: 10.1038/srep43609 PMID: 28272438

Wang, Y., Hou, Y. P., Chen, C. J., & Zhou, M. G. (2014). Detection of resistance in Sclerotinia sclerotiorum to carbendazim and dimethachlon in Jiangsu Province of China. *Australasian Plant Pathology*, *43*(3), 307–312. DOI: 10.1007/s13313-014-0271-1

Wang, Y., Liu, Q., Du, L., Hallerman, E., & Li, Y. (2020). Transcriptomic and metabolomic responses of rice plants to cnaphalocrocis medinalis caterpillar infestation. *Insects, 11*(10), 705. DOI: 10.3390/insects11100705 PMID: 33076419

Wang, Y., Strelkov, S. E., & Hwang, S. F. (2020). Yield losses in canola in response to blackleg disease. *Canadian Journal of Plant Science, 100*(5), 488–494. DOI: 10.1139/cjps-2019-0259

Wang, Z., Megha, S., Kebede, B., Kav, N., & Rahman, H. (2022). Genetic and molecular analysis reveals that two major loci and their interaction confer clubroot resistance in canola introgressed from rutabaga. *The Plant Genome, 15*(3), e20241. Advance online publication. DOI: 10.1002/tpg2.20241 PMID: 35818693

Wan, H., Chen, L., Guo, J., Li, Q., Wen, J., Yi, B., Ma, C., Tu, J., Fu, T., & Shen, J. (2017). Genome-wide association study reveals the genetic architecture underlying salt tolerance-related traits in rapeseed (Brassica napus L.). *Frontiers in Plant Science, 8*, 593. DOI: 10.3389/fpls.2017.00593 PMID: 28491067

Wan, H., Cui, Y., Ding, Y., Mei, J., Dong, H., Zhang, W., & Qian, W. (2017). Time-series analyses of transcriptomes and proteomes reveal molecular networks underlying oil accumulation in canola. *Frontiers in Plant Science, 7*, 2007. DOI: 10.3389/fpls.2016.02007 PMID: 28119706

Warner, K., Orr, P., & Glynn, M. (1997). Effects of fatty acid composition of oils on flavor and stability of fried foods. *Journal of the American Oil Chemists' Society, 74*(3), 347–356. DOI: 10.1007/s11746-997-0090-4

Weber, S. E., Chawla, H. S., Ehrig, L., Hickey, L. T., Frisch, M., & Snowdon, R. J. (2023). Accurate prediction of quantitative traits with failed SNP calls in canola and maize. *Frontiers in Plant Science, 14*, 1221750. DOI: 10.3389/fpls.2023.1221750 PMID: 37936929

Weeks, D. P. (2017). Gene editing in polyploid crops: Wheat, camelina, canola, potato, cotton, peanut, sugar cane, and citrus. In Weeks, D. P., & Yang, B. (Eds.), *Progress in Molecular Biology and Translational Science* (Vol. 149, pp. 65–80). Elsevier.

Wei, D., Cui, Y., He, Y., Xiong, Q., Qian, L., Tong, C., Lu, G., Ding, Y., Li, J., Jung, C., & Qian, W. (2017). A genome-wide survey with different rapeseed ecotypes uncovers footprints of domestication and breeding. *Journal of Experimental Botany, 68*(17), 4791–4801. DOI: 10.1093/jxb/erx311 PMID: 28992309

Wei, J., Zheng, G., Yu, X., Liu, S., Dong, X., Cao, X., Fang, X., Li, H., Jin, J., Mi, W., & Liu, Z. (2021). Comparative transcriptomics and proteomics analyses of leaves reveals a freezing stress-responsive molecular network in winter rapeseed (Brassica rapa L.). *Frontiers in Plant Science, 12*, 664311. DOI: 10.3389/fpls.2021.664311 PMID: 33995460

Wei, L., Li, Y., & Wang, H. (2014). Identification of quantitative trait loci for resistance to Sclerotinia sclerotiorum in Brassica napus. *Crop Science, 54*(4), 1442–1450.

Wei, X., Eglinton, J., Piperidis, G., Atkin, F., Morgan, T., Parfitt, R., & Hu, F. (2022). Sugarcane breeding in Australia. *Sugar Tech, 24*(1), 151–165. DOI: 10.1007/s12355-021-00969-5

Weldon, A., Davis, D. A., Rhodes, M., Morey, A., Iassonova, D., & Roy, L. A. (2023). Use of genetically modified canola oil as a replacement for fish oil in practical diets for whiteleg shrimp litopeneaus vannamei reared in green water conditions. *Aquaculture Research, 2023*, 1–12. DOI: 10.1155/2023/2999827

Welter, K. C., Martins, C. M. M. R., de Palma, A. S. V., Martins, M. M., Dos Reis, B. R., Schmidt, B. L. U., & Saran Netto, A. (2016). Canola oil in lactating dairy cow diets reduces milk saturated fatty acids and improves its omega-3 and oleic fatty acid content. *PLoS One, 11*(3), e0151876. DOI: 10.1371/journal.pone.0151876 PMID: 27015405

Wenger, A. M., Peluso, P., Rowell, W. J., Chang, P. C., Hall, R. J., Concepcion, G. T., Ebler, J., Fungtammasan, A., Kolesnikov, A., Olson, N. D., Töpfer, A., Alonge, M., Mahmoud, M., Qian, Y., Chin, C.-S., Phillippy, A. M., Schatz, M. C., Myers, G., DePristo, M. A., & Hunkapiller, M. W. (2019). Accurate circular consensus long-read sequencing improves variant detection and assembly of a human genome. *Nature Biotechnology, 37*(10), 1155–1162. DOI: 10.1038/s41587-019-0217-9 PMID: 31406327

Wen, L., Tan, T.-L., Shu, J.-B., Chen, Y., Liu, Y., Yang, Z.-F., & Guan, C.-Y. (2013). Using proteomic analysis to find the proteins involved in resistance against Sclerotinia sclerotiorum in adult Brassica napus. *European Journal of Plant Pathology, 137*, 505–523.

Wen, R., Song, T., Gossen, B. D., & Peng, G. (2024). Comparative transcriptome analysis of canola carrying a single vs stacked resistance genes against clubroot. *Frontiers in Plant Science, 15*, 1358605. DOI: 10.3389/fpls.2024.1358605 PMID: 38835867

Werner, C. R., Voss-Fels, K. P., Miller, C. N., Qian, W., Hua, W., Guan, C. Y., Snowdon, R. J., & Qian, L. (2018). Effective genomic selection in a narrow-genepool crop with low-density markers: Asian rapeseed as an example. *The Plant Genome, 11*(2), 170084. DOI: 10.3835/plantgenome2017.09.0084 PMID: 30025015

Wick, R. R., Judd, L. M., Gorrie, C. L., & Holt, K. E. (2017). Unicycler: Resolving bacterial genome assemblies from short and long sequencing reads. *PLoS Computational Biology, 13*(6), e1005595. DOI: 10.1371/journal.pcbi.1005595 PMID: 28594827

Wiebach, J., Nagel, M., Börner, A., Altmann, T., & Riewe, D. (2019). Age-dependent loss of seed viability is associated with increased lipid oxidation and hydrolysis. *Plant, Cell & Environment, 43*(2), 303–314. DOI: 10.1111/pce.13651 PMID: 31472094

Williams, J. G., Kubelik, A. R., Livak, K. J., Rafalski, J. A., & Tingey, S. V. (1990). DNA polymorphisms amplified by arbitrary primers are useful as genetic markers. *Nucleic Acids Research, 18*(22), 6531–6535. DOI: 10.1093/nar/18.22.6531 PMID: 1979162

Williams, S., Vail, S., & Arcand, M. M. (2021). Nitrogen use efficiency in parent vs. hybrid canola under varying nitrogen availabilities. *Plants, 10*(11), 2364. DOI: 10.3390/plants10112364 PMID: 34834725

Wilson, R., & Brown, J. (2023). Genotype by environment interactions in canola: Implications for breeding high oil content varieties. *Theoretical and Applied Genetics, 136*(2), 421–435.

Wilson, T., & Taylor, S. (2020). Glycerol from biodiesel production: The new corn syrup? *Renewable & Sustainable Energy Reviews, 124*, 109799.

Winfield, M. O., Allen, A. M., Burridge, A. J., Barker, G. L., Benbow, H. R., Wilkinson, P. A., & Edwards, K. J. (2016). High-density SNP genotyping array for hexaploid wheat and its secondary and tertiary gene pool. *Plant Biotechnology Journal, 14*(5), 1195–1206. DOI: 10.1111/pbi.12485 PMID: 26466852

Wishart, D., Cheng, L., Copié, V., Edison, A., Eghbalnia, H., Hoch, J., Gouveia, G. J., Pathmasiri, W., Powers, R., Schock, T. B., Sumner, L. W., & Uchimiya, M. (2022). Nmr and metabolomics—A roadmap for the future. *Metabolites, 12*(8), 678. DOI: 10.3390/metabo12080678 PMID: 35893244

Wittkop, B., Snowdon, R. J., & Friedt, W. (2009). Status and perspectives of breeding for enhanced yield and quality of oilseed crops for Europe. *Euphytica, 170*(1–2), 131–140. DOI: 10.1007/s10681-009-9940-5

Woodfield, H. K., Cazenave-Gassiot, A., Haslam, R. P., Guschina, I. A., Wenk, M. R., & Harwood, J. L. (2018). Using lipidomics to reveal details of lipid accumulation in developing seeds from oilseed rape (Brassica napus L.). *Biochimica et Biophysica Acta (BBA)-. Biochimica et Biophysica Acta. Molecular and Cell Biology of Lipids*, *1863*(3), 339–348. DOI: 10.1016/j.bbalip.2017.12.010 PMID: 29275220

Wortmann, S. B., Oud, M. M., Alders, M., Coene, K. L., van der Crabben, S. N., Feichtinger, R. G., Garanto, A., Hoischen, A., Langeveld, M., Lefeber, D., Mayr, J. A., Ockeloen, C. W., Prokisch, H., Rodenburg, R., Waterham, H. R., Wevers, R. A., van de Warrenburg, B. P. C., Willemsen, M. A. A. P., Wolf, N. I., & van Karnebeek, C. D. M. (2022). How to proceed after "negative" exome: A review on genetic diagnostics, limitations, challenges, and emerging new multiomics techniques. *Journal of Inherited Metabolic Disease*, *45*(4), 663–681. DOI: 10.1002/jimd.12507 PMID: 35506430

Wu, D., Liang, Z., Yan, T., Xu, Y., Xuan, L., Tang, J., & Huang, J. (2019). Whole-genome resequencing of a worldwide collection of rapeseed accessions reveals the genetic basis of ecotype divergence. *Molecular Plant*, *12*(1), 30–43. DOI: 10.1016/j.molp.2018.11.007 PMID: 30472326

Wu, J., Zhao, Q., & Liu, S. (2016). Genome-wide association study for seed oil content in Brassica napus. *BMC Plant Biology*, *16*, 144.

Wu, J., Zhao, Q., Liu, S., Shahid, M., Lan, L., Cai, G., & Zhou, Y. (2016). Genome-wide association study identifies new loci for resistance to Sclerotinia stem rot in Brassica napus. *Frontiers in Plant Science*, *7*, 1418. DOI: 10.3389/fpls.2016.01418 PMID: 27703464

Wu, J., Zhao, Q., Yang, Q., Liu, H., Li, Q., Yi, X., & Zhou, Y. (2021). Comparative transcriptome profiling uncovers a Lilium-derived allele in Brassica napus providing resistance to blackleg disease. *Plant Biotechnology Journal*, *19*(4), 693–708.

Wu, L., El-Mezawy, A., & Shah, S. (2013). Isolation and evaluation of three novel native promoters in Brassica napus. *Botany*, *91*(6), 414–419. DOI: 10.1139/cjb-2012-0245

Würschum, T., Maurer, H. P., Dreyer, F., & Reif, J. C. (2013). Effect of inter-and intragenic epistasis on the heritability of oil content in rapeseed (Brassica napus L.). *Theoretical and Applied Genetics*, *126*(2), 435–441. DOI: 10.1007/s00122-012-1991-7 PMID: 23052025

Würschum, T., Maurer, H. P., Dreyer, F., & Reif, J. C. (2014). Effect of inter- and intragenic epistasis on the heritability of oil content in rapeseed (Brassica napus L.). *Theoretical and Applied Genetics*, *127*(3), 615–622. PMID: 23052025

Wu, W., Duncan, R., & Ma, B. (2020). The stage sensitivity of short-term heat stress to lodging-resistant traits and yield determination in canola (brassica napus l.). *Journal Agronomy & Crop Science*, *207*(1), 74–87. DOI: 10.1111/jac.12464

Wu, W., Ma, B. L., & Whalen, J. K. (2018). Enhancing rapeseed tolerance to heat and drought stresses in a changing climate: Perspectives for stress adaptation from root system architecture. *Advances in Agronomy*, *151*, 87–157. DOI: 10.1016/bs.agron.2018.05.002

Wu, X., Gong, F., & Wang, W. (2014). Protein extraction from plant tissues for 2DE and its application in proteomic analysis. *Proteomics*, *14*(6), 645–658.

Xia, J., Sinelnikov, I., Han, B., & Wishart, D. (2015). Metaboanalyst 3.0—Making metabolomics more meaningful. *Nucleic Acids Research*, *43*(W1), W251–W257. DOI: 10.1093/nar/gkv380 PMID: 25897128

Xiao, Y., Chen, L., Zou, J., Tian, E., Xia, W., & Meng, J. (2010). Development of a population for substantial new type Brassica napus diversified at both A/C genomes. *Theoretical and Applied Genetics*, *121*(6), 1141–1150. DOI: 10.1007/s00122-010-1378-6 PMID: 20556596

Xiaoyu, D., Xu, J., He, H., Xing, Q., Shen, M., Cheng, Y., & Zhang, X. (2020). Unraveling waterlogging tolerance-related traits with QTL analysis in reciprocal intervarietal introgression lines using genotyping by sequencing in rapeseed (Brassica napus L.). *Journal of Integrative Agriculture*, *19*(8), 1974–1983. DOI: 10.1016/S2095-3119(19)62783-8

Xiao, Z., Zhang, C., Qu, C., Wei, L., Zhang, L., Yang, B., Lu, K., & Li, J. (2022). Identification of candidate genes regulating seed oil content by QTL mapping and transcriptome sequencing in Brassica napus. *Frontiers in Plant Science*, *13*, 1067121. Advance online publication. DOI: 10.3389/fpls.2022.1067121 PMID: 36570918

Xie, H., Konate, M., Sai, N., Tesfamicael, K. G., Cavagnaro, T., Gilliham, M., Breen, J., Metcalfe, A., Stephen, J. R., DeBei, R., Collins, C., & Lopez, C. M. R. (2017). Global DNA methylation patterns can play a role in defining terroir in grapevine (Vitis vinifera cv. Shiraz). *Frontiers in Plant Science*, *8*, 1860. DOI: 10.3389/fpls.2017.01860 PMID: 29163587

Xing, M., Sun, C., Li, H., Hu, S., Lei, L., & Kang, J. (2018). Integrated analysis of transcriptome and proteome changes related to the Ogura cytoplasmic male sterility in cabbage. *PLoS One*, *13*(3), e0193462.

Xin, S., Dong, H., Yang, L., Huang, D., Zheng, F., Cui, Y., Wu, S., Liao, J., He, Y., Wan, H., Liu, Z., Li, X., & Qian, W. (2021). Both overlapping and independent loci underlie seed number per pod and seed weight in Brassica napus by comparative quantitative trait loci analysis. *Molecular Breeding*, *41*(6), 41. Advance online publication. DOI: 10.1007/s11032-021-01232-1 PMID: 37309442

Xu, C., Ren, Y., Jian, Y., Guo, Z., Zhang, Y., Xie, C., & Zou, C. (2017). Development of a maize 55 K SNP array with improved genome coverage for molecular breeding. *Molecular Breeding*, *37*(3), 1–12. DOI: 10.1007/s11032-017-0622-z PMID: 28255264

Xu, J., & Wang, N. (2019). Where are we going with genomics in plant pathogenic bacteria? *Genomics*, *111*(4), 729–736.

Xu, Y., Zeng, X., Wu, J., Zhang, F., Li, C., Jiang, J., & Sun, W. (2018). iTRAQ-based quantitative proteome revealed metabolic changes in winter turnip rape (Brassica rapa L.) under cold stress. *International Journal of Molecular Sciences*, *19*(11), 3346.

Yadava, D., Vasudev, S., Singh, N., Mohapatra, T., & Prabhu, K. (2012). Breeding major oil crops: Present status and future research needs. *Technological innovations in major world oil crops, volume 1: breeding*, 17-51.

Yanagawa, A., & Fujimaki, H. (2013). Tolerance of canola to drought and salinity stresses in terms of root water uptake model parameters. *Journal of Hydrology and Hydromechanics*, *61*(1), 73–80. DOI: 10.2478/johh-2013-0009

Yandeau-Nelson, M. D., Lauter, N., & Zabotina, O. A. (2016). Advances in metabolomic applications in plant genetics and breeding. *CABI Reviews*(2015), 1-15.

Yang, Q., Liu, J., & Zhang, C. (2018). Metabolomic profiling of seed oil biosynthesis in Brassica napus. *Journal of Agricultural and Food Chemistry*, *66*(10), 2482–2490.

Yang, Y., Saand, M. A., Huang, L., Abdelaal, W. B., Zhang, J., Wu, Y., Li, J., Sirohi, M. H., & Wang, F. (2021). Applications of multi-omics technologies for crop improvement. *Frontiers in Plant Science*, *12*, 563953. DOI: 10.3389/fpls.2021.563953 PMID: 34539683

Yang, Y., Shen, Y., Li, S., Ge, X., & Li, Z. (2017). High Density Linkage Map Construction and QTL Detection for Three Silique-Related Traits in Orychophragmus violaceus Derived Brassica napus Population. *Frontiers in Plant Science*, *8*, 1512. Advance online publication. DOI: 10.3389/fpls.2017.01512 PMID: 28932230

Yang, Y., Wei, F., Braynen, J., Wei, X., Tian, B., Shi, G., & Zhang, X. (2019). Cytological and proteomic analyses of floral buds reveal an altered atlas of meiosis in autopolyploid Brassica rapa. *Cell & Bioscience*, *9*, 1–15.

Yao, S., Liang, F., Gill, R., Huang, J., Cheng, X., Liu, Y., Tong, C., & Liu, S. (2020). A global survey of the transcriptome of the allopolyploid Brassica napus based on single molecule long-read isoform sequencing and Illumina-based RNA-seq data.. *The Plant journal: for cell and molecular biology*. https://doi.org/.DOI: 10.1111/tpj.14754

Yasin, J. K., Khan, M., Wani, S. H., Pillai, M. A., Verma, N., Pandey, P., & Singh, K. (2022). Genomics in Crop Improvement: Potential Applications, Challenges and Future Prospects. *Agricultural Biotechnology: Latest Research and Trends*, 329-349.

Yıldız, M., Akçalı, N., & Terzi, H. (2015). Proteomic and biochemical responses of canola (Brassica napus L.) exposed to salinity stress and exogenous lipoic acid. *Journal of Plant Physiology*, *179*, 90–99.

Yim, W. C., Swain, M. L., Ma, D., An, H., Bird, K. A., Curdie, D. D., Wang, S., Ham, H. D., Luzuriaga-Neira, A., Kirkwood, J. S., Hur, M., Solomon, J. K. Q., Harper, J. F., Kosma, D. K., Alvarez-Ponce, D., Cushman, J. C., Edger, P. P., Mason, A. S., Pires, J. C., & Zhang, X. (2022). The final piece of the Triangle of U: Evolution of the tetraploid Brassica carinata genome. *The Plant Cell*, *34*(11), 4143–4172. DOI: 10.1093/plcell/koac249 PMID: 35961044

You, Q., Yang, X., Peng, Z., Xu, L., & Wang, J. (2018). Development and applications of a high throughput genotyping tool for polyploid crops: Single nucleotide polymorphism (SNP) array. *Frontiers in Plant Science*, *9*, 104. DOI: 10.3389/fpls.2018.00104 PMID: 29467780

Yousuf, P. Y., Ahmad, A., Ganie, A. H., & Iqbal, M. (2016). Salt stress-induced modulations in the shoot proteome of Brassica juncea genotypes. *Environmental Science and Pollution Research International*, *23*, 2391–2401.

Yuan, L., Wang, J., Xie, S., Zhao, M., Nie, L., Zheng, Y., & Wang, C. (2019). Comparative proteomics indicates that redox homeostasis is involved in high-and low-temperature stress tolerance in a novel Wucai (Brassica campestris L.) genotype. *International Journal of Molecular Sciences*, *20*(15), 3760.

Yu, B., Boyle, K., Zhang, W., Robinson, S. J., Higgins, E., Ehman, L., Relf-Eckstein, J., Rakow, G., Parkin, I. P., Sharpe, A. G., & Fobert, P. R. (2016). Multi-trait and multi-environment QTL analysis reveals the impact of seed colour on seed composition traits in Brassica napus. *Molecular Breeding*, *36*(8), 111. Advance online publication. DOI: 10.1007/s11032-016-0521-8

Yu, F., Lydiate, D. J., & Rimmer, S. R. (2005). Identification of two novel genes for blackleg resistance in *Brassica napus. Theoretical and Applied Genetics*, *110*(5), 969–979. DOI: 10.1007/s00122-004-1919-y PMID: 15798929

Yu, H., Xie, W., Li, J., Zhou, F., & Zhang, Q. (2014). A whole-genome SNP array (RICE 6 K) for genomic breeding in rice. *Plant Biotechnology Journal, 12*(1), 28–37. DOI: 10.1111/pbi.12113 PMID: 24034357

Zalatan, J., Lee, M., Almeida, R., Gilbert, L., Whitehead, E., Russa, M., & Lim, W. (2015). Engineering complex synthetic transcriptional programs with crispr rna scaffolds. *Cell, 160*(1-2), 339–350. DOI: 10.1016/j.cell.2014.11.052 PMID: 25533786

Zaman, M. A., Ullah, F., & Khan, A. (2019). Prospects of drought and heat tolerance in Canola (Brassica napus L.). *Advances in Agronomy, 153*, 121–178.

Zeki, Ö., Eylem, C., Reçber, T., Kır, S., & Nemutlu, E. (2020). Integration of gc–ms and lc–ms for untargeted metabolomics profiling. *Journal of Pharmaceutical and Biomedical Analysis, 190*, 113509. DOI: 10.1016/j.jpba.2020.113509 PMID: 32814263

Zhang, F., Liu, X., & Huang, G. (2022). Recent progress in reactor design for continuous biodiesel production from canola oil. *Chemical Engineering Journal, 430*, 132707. DOI: 10.1016/j.cej.2021.132707

Zhang, F., Liu, X., & Huang, G. (2023). Applications of GWAS-derived markers in genomic selection for yield improvement in Brassica napus. *Theoretical and Applied Genetics, 136*(5), 1089–1104. PMID: 37027012

Zhang, G., Zhou, J., Peng, Y., Tan, Z., Zhang, Y., Zhao, H., Liu, D., Liu, X., Li, L., Yu, L., Jin, C., Fang, S., Shi, J., Geng, Z., Yang, S., Chen, G., Liu, K., Yang, Q., Feng, H., & Yang, W. (2021). High-throughput phenotyping-based QTL mapping reveals the genetic architecture of the salt stress tolerance of *Brassica napus. Authorea (Authorea).* DOI: 10.22541/au.163256646.65671725/v1

Zhang, H., Liu, W., Zhang, Y., Deng, M., Niu, F., Yang, B., & Jiang, Y. (2014). Identification, expression and interaction analyses of calcium-dependent protein kinase (cpk) genes in canola (brassica napusl.). *BMC Genomics, 15*(1), 211. Advance online publication. DOI: 10.1186/1471-2164-15-211 PMID: 24646378

Zhang, H., Li, Y., & Wang, Z. (2017). Transcriptomic analysis of drought stress responses in Brassica napus. *Plant Molecular Biology, 95*(3), 287–300.

Zhang, H., Yang, B., Liu, W., Li, H., Wang, L., Wang, B., & Jiang, Y. (2014). Identification and characterization of cbl and cipk gene families in canola (brassica napus l.). *BMC Plant Biology, 14*(1), 8. Advance online publication. DOI: 10.1186/1471-2229-14-8 PMID: 24397480

Zhang, H., Yin, L., Wang, M., Yuan, X., & Liu, X. (2019). Factors affecting the accuracy of genomic selection for agricultural economic traits in maize, cattle, and pig populations. *Frontiers in Genetics*, *10*, 189. DOI: 10.3389/fgene.2019.00189 PMID: 30923535

Zhang, J., Ge, W., Chang, H., Xin, X., & Ji, R. (2021). Discovery of BrATG6 and its potential role in Brassica rapa L. resistance to infection by Plasmodiophora brassicae. *Gene*, *791*, 145711.

Zhang, J., Li, G., Li, H., Pu, X., Jiang, J., Chai, L., Zheng, B., Cui, C., Yang, Z., Zhu, Y., & Jiang, L. (2015). Transcriptome Analysis of Interspecific Hybrid between Brassica napusand B. Rapa Reveals Heterosis for Oil Rape Improvement. *International Journal of Genomics*, *2015*, 1–11. DOI: 10.1155/2015/230985 PMID: 26448924

Zhang, J., Ruhlman, T. A., Mower, J. P., & Jansen, R. K. (2013). Comparative analyses of two Geraniaceae transcriptomes using next-generation sequencing. *BMC Plant Biology*, *13*(1), 228. DOI: 10.1186/1471-2229-13-228 PMID: 24373163

Zhang, L., Cai, X., Wu, J., Liu, M., Grob, S., Cheng, F., Liang, J., Cai, C., Liu, Z., Liu, B., Wang, F., Li, S., Liu, F., Li, X., Cheng, L., Yang, W., Li, M., Grossniklaus, U., Zheng, H., & Wang, X. (2018). Improved Brassica rapa reference genome by single-molecule sequencing and chromosome conformation capture technologies. *Horticulture Research*, *5*(1), 50. DOI: 10.1038/s41438-018-0071-9 PMID: 30131865

Zhang, R., Zhang, C., Yu, C., Dong, J., & Hu, J. (2022). Integration of multi-omics technologies for crop improvement: Status and prospects. *Frontiers in Bioinformatics*, *2*, 1027457. Advance online publication. DOI: 10.3389/fbinf.2022.1027457 PMID: 36438626

Zhang, T., Vail, S., Duddu, H., Parkin, I., Guo, X., Johnson, E., & Shirtliffe, S. (2021). Phenotyping Flowering in Canola (Brassica napus L.) and Estimating Seed Yield Using an Unmanned Aerial Vehicle-Based Imagery. *Frontiers in Plant Science*, *12*, 686332. Advance online publication. DOI: 10.3389/fpls.2021.686332 PMID: 34220907

Zhang, W., Hu, D., Raman, R., Guo, S., Wei, Z., Shen, X., Meng, J., Raman, H., & Zou, J. (2017). Investigation of the Genetic Diversity and Quantitative Trait Loci Accounting for Important Agronomic and Seed Quality Traits in Brassica carinata. *Frontiers in Plant Science*, *8*, 615. Advance online publication. DOI: 10.3389/fpls.2017.00615 PMID: 28484482

Zhang, W., Wang, S., Yu, F., Tang, J., Yu, L., Wang, H., & Li, J. (2019). Genome-wide identification and expression profiling of sugar transporter protein (STP) family genes in cabbage (Brassica oleracea var. capitata L.) reveals their involvement in clubroot disease responses. *Genes*, *10*(1), 71.

Zhang, X., & Fernando, W. D. (2017). Insights into fighting against blackleg disease of Brassica napus in Canada. *Crop & Pasture Science*, *69*(1), 40–47. DOI: 10.1071/CP16401

Zhang, Y., Ali, U., Zhang, G., Yu, L., Fang, S., Iqbal, S., Li, H., Lu, S., & Guo, L. (2019). Transcriptome analysis reveals genes commonly responding to multiple abiotic stresses in rapeseed. *Molecular Breeding*, *39*(10-11), 1–19. DOI: 10.1007/s11032-019-1052-x

Zhang, Y., & Liu, J. (2020). Canola production and trade: A comprehensive review of global patterns and emerging trends. *Agronomy (Basel)*, *10*(12), 1956.

Zhang, Y., Wang, Y., Wu, D., Qu, D., Sun, X., & Zhang, X. (2022). SNP and haplotype variability in the *BnP5CR2* gene and association with resistance and susceptible cultivars for *Sclerotinia sclerotiorum* in *Brassica napus. Agronomy (Basel)*, *12*(2956), 2956. Advance online publication. DOI: 10.3390/agronomy12122956

Zhan, Z., Shah, N., Jia, R., Li, X., Zhang, C., & Piao, Z. (2022). Transferring of clubroot-resistant locus CRd from Chinese cabbage (Brassica rapa) to canola (Brassica napus) through interspecific hybridization. *Breeding Science*, *72*(3), 189–197. DOI: 10.1270/jsbbs.21052 PMID: 36408323

Zhao, C., Zhang, Z., & Xie, S. (2018). Genetic basis of cold tolerance in Brassica napus: Identification of cold-responsive genes. *Plant Biotechnology Journal*, *16*(9), 1692–1703.

Zhao, J., Dimov, Z., Becker, H. C., Ecke, W., & Möllers, C. (2008). Mapping QTL controlling fatty acid composition in a doubled haploid rapeseed population segregating for oil content. *Molecular Breeding*, *21*(1), 115–125. DOI: 10.1007/s11032-007-9113-y

Zhao, J., Paulo, M. J., Jamar, D., Lou, P., van Eeuwijk, F., Bonnema, G., Vreugdenhil, D., & Koornneef, M. (2007). Association mapping of leaf traits, flowering time, and phytate content in Brassica rapa. *Genome*, *50*(10), 963–973. DOI: 10.1139/G07-078 PMID: 18059559

Zhao, Q., Wu, J., Cai, G., Yang, Q., Shahid, M., Fan, C., Zhang, C., & Zhou, Y. (2019). A novel quantitative trait locus on chromosome A9 controlling oleic acid content in Brassica napus. *Plant Biotechnology Journal, 17*(12), 2313–2324. DOI: 10.1111/pbi.13142 PMID: 31037811

Zhao, W., Wang, X., Wang, H., Tian, J., Li, B., Chen, L., Chao, H., Long, Y., Xiang, J., Gan, J., Liang, W., & Li, M. (2016). Genome-Wide Identification of QTL for Seed Yield and Yield-Related Traits and Construction of a High-Density Consensus Map for QTL Comparison in Brassica napus. *Frontiers in Plant Science, 7*. Advance online publication. DOI: 10.3389/fpls.2016.00017 PMID: 26858737

Zhao, Y., Wang, H., Chen, W., & Li, Y. (2021). Genetic structure, linkage disequilibrium and association mapping of verticillium wilt resistance in elite cotton (Gossypium hirsutum L.) germplasm population. *BMC Genomics, 22*(1), 60. PMID: 33468056

Zheng, M., Peng, C., Liu, H., Tang, M., Yang, H., Li, X., & Cheng, X. (2020). Genome-wide association study reveals candidate genes for control of plant height, branch initiation height and branch number in rapeseed (Brassica napus L.). *Frontiers in Plant Science, 11*, 1875. PMID: 28769955

Zhou, G., & Bradbury, P. J. (2018). Genomic selection with machine learning. *PLOS Genetics, 14*(11), e1007735. PMID: 30399141

Zhou, Q., Galindo-González, L., Hwang, S. F., & Strelkov, S. E. (2021). Application of genomics and transcriptomics to accelerate development of clubroot resistant canola. *Canadian Journal of Plant Pathology, 43*(2), 189–208. DOI: 10.1080/07060661.2020.1794541

Zhou, Q., Zhou, C., Zheng, W., Mason, A. S., Fan, S., Wu, C., Fu, D., & Huang, Y. (2017). Genome-wide SNP markers based on SLAF-seq uncover breeding traces in rapeseed (Brassica napus L.). *Frontiers in Plant Science, 8*, 648. DOI: 10.3389/fpls.2017.00648 PMID: 28503182

Zhou, R., Yu, X., & Kjær, K. H. (2018). Genomic insights into temperature stress responses in Brassica napus. *Plant Physiology and Biochemistry, 131*, 110–119.

Zhou, X., Shi, F., Zhou, L., Zhou, Y., Liu, Z., Ji, R., & Feng, H. (2019). iTRAQ-based proteomic analysis of fertile and sterile flower buds from a genetic male sterile line 'AB01'in Chinese cabbage (*Brassica campestris* L. ssp. pekinensis). *Journal of Proteomics, 204*, 103395.

Zhou, Y. J., Chang, Y. N., You, J. Q., Li, S. Z., Zhuang, W., & Cao, C. J. (2019). Cold-pressed canola oil reduces hepatic steatosis by modulating oxidative stress and lipid metabolism in KM mice compared with refined bleached deodorized canola oil. *Journal of Food Science*, *84*(7), 1900–1908. DOI: 10.1111/1750-3841.14504 PMID: 31183867

Zhu, J., Zhang, J., Jiang, M., Wang, W., Jiang, J., Li, Y., Yang, L., & Zhou, X. (2021). Development of genome-wide SSR markers in rapeseed by next generation sequencing. *Gene*, *798*, 145798. DOI: 10.1016/j.gene.2021.145798 PMID: 34175391

Zhu, M., Monroe, J., Suhail, Y., Villiers, F., Mullen, J., Pater, D., & Assmann, S. (2016). Molecular and systems approaches towards drought-tolerant canola crops. *The New Phytologist*, *210*(4), 1169–1189. DOI: 10.1111/nph.13866 PMID: 26879345

Zhu, Q., King, G. J., Liu, X., Shan, N., Borpatragohain, P., Baten, A., Wang, P., Luo, S., & Zhou, Q. (2019). Identification of SNP loci and candidate genes related to four important fatty acid compositions in *Brassica napus* using genome-wide association study. *PLoS One*, *14*(8), e0221578. Advance online publication. DOI: 10.1371/journal.pone.0221578

Zimin, A. V., Puiu, D., Luo, M. C., Zhu, T., Koren, S., Marçais, G., Yorke, J. A., Dvořák, J., & Salzberg, S. L. (2017). Hybrid assembly of the large and highly repetitive genome of *Aegilops tauschii*, a progenitor of bread wheat, with the MaSuRCA mega-reads algorithm. *Genome Research*, *27*(5), 787–792. DOI: 10.1101/gr.213405.116 PMID: 28130360

Zong, W., Zhong, X., You, J., & Xiong, L. (2013). Genome-wide profiling of histone H3K4-tri-methylation and gene expression in rice under drought stress. *Plant Molecular Biology*, *81*(1-2), 175–188. DOI: 10.1007/s11103-012-9990-2 PMID: 23192746

Zou, J., Hu, D., & Liu, P. (2016). Genetic mapping of root architecture and drought tolerance traits in Brassica napus using GWAS. *Plant Breeding*, *135*(4), 466–474.

Zou, J., Hu, D., Mason, A. S., Shen, X., Wang, X., Wang, N., & Meng, J. (2018). Genetic changes in a novel breeding population of Brassica napus synthesized from hundreds of crosses between B. rapa and B. carinata. *Plant Biotechnology Journal*, *16*(2), 507–519. DOI: 10.1111/pbi.12791 PMID: 28703467

Zou, J., Mao, L., Qiu, J., Wang, M., Jia, L., Wu, D., He, Z., Chen, M., Shen, Y., Shen, E., Huang, Y., Li, R., Hu, D., Shi, L., Wang, K., Zhu, B., Zeng, P., Ma, B., Wang, H., & Fu, D. (2019). Genome-wide selection footprints and deleterious variations in young Asian allotetraploid rapeseed. *Plant Biotechnology Journal*, *17*(10), 1998–2010. DOI: 10.1111/pbi.13115 PMID: 30947395

Zou, J., Zhao, Y., Liu, P., Shi, L., Wang, X., Wang, M., Meng, J., & Reif, J. C. (2016). Seed Quality Traits Can Be Predicted with High Accuracy in Brassica napus Using Genomic Data. *PLoS One*, *11*(11), e0166624. DOI: 10.1371/journal.pone.0166624 PMID: 27880793

Zou, M., Shi, T., Wang, W., Ding, G., Xu, F., & Shi, L. (2021). Identification of QTLs for Seed Yield and Yield-Related Traits in Brassica Napus Grown with Contrasting Nitrogen Supplies. https://doi.org/DOI: 10.21203/rs.3.rs-674683/v1

Zou, Z., & Fernando, W. (2023). Overexpression of bnnac19 in brassica napus enhances resistance to leptosphaeria maculans, the blackleg pathogen of canola. *Plant Pathology*, *73*(1), 104–114. DOI: 10.1111/ppa.13791

Zou, Z., Liu, F., & Fernando, D. (2021). Genome-wide identification and analysis of the valine-glutamine motif-containing gene family inbrassica napusand functional characterization ofbnmks1in response toleptosphaeria maculans. *Phytopathology*, *111*(2), 281–292. DOI: 10.1094/PHYTO-04-20-0134-R PMID: 32804045

About the Contributors

Muhammad Waseem is an Associate Professor at College of Tropical Agriculture and Forestry, Hainan University, Haikou, Hainan, China. He is also serving as Assistant Professor of Botany at the Department of Botany, University of Narowal, Pakistan. He got a doctoral degree (Ph.D., specialization in Plant Molecular Biology) from Prof. Zhengguo Li's lab at the School of Life Sciences, Chongqing University, Chongqing, PR China. He then joined as also have four years of postdoctoral research. His interests include transcriptional regulation of flowering time regulation, fruit ripening and development, and abiotic stresses such as salinity and drought. In addition, he has hands-on experience in next-generation sequencing data analysis, transcriptomics, and data visualization. Currently, Dr. Waseem is actively in investigating the flowering time regulation in Brassica crops. He has been awarded with a National Science Foundation Project, RFIS worth 200000 RMB. In brief, he has published more than 40 research, 5 review, 9 book chapters research in various high-impact factor journals including New Phytologist, International Journal of Macromolecules, Journal of Biotechnology, Planta, Scientific reports, Pakistan Journal of Botany, GM Crops & Food, Environmental and Experimental Botany, International Journal of Molecular Biology, Genes, Plant Physiology, and Frontiers in Plant Science. He has also edited various book chapters in springer-nature, CRC press, and others. Moreover, currently, he is editing a few special issues as a leading guest editor in Frontiers in Plant Science, International Journal of Molecular Biology, GENES, and Functional Plant Biology.

***

Banzeer Ahsan Abbasi received her BS from Govt. Post Graduate Collage, 6th Road Satellite Town Rawalpindi and M.Phil and Ph.D degree in Botany from Quaid-i-Azam University, Islamabad, Pakistan. Dr. Banzeer is currently working as an Assistant Professor at Department of Botany, Rawalpindi Women University,

Rawalpindi, Pakistan. She has published more than 60 research articles in various international journals and 10 chapters in reputed publishers (ELSIEVER, RSC and SPRINGER). Currently, she is supervising 2 Ph.D, 6 MPhil and 6 BS students. She has expertise in the synthesis of novel nanomaterials from medicinal plants and their multiple biological and envirnemental applications. Her research is also focusing on phytochemistry, natural products of medicinal plants and Phylogenetics. She has a very decent research background with research published in well-reputed journals with H-index 27, i-10 index 44 and overall citations 2900s.

Ujjwal Agrawal, undergraduate physics student passionate about physics, astrophysics, biology, and machine learning, seeking opportunities to apply theoretical knowledge and practical skills in a dynamic and challenging environment. I am eager to contribute to interdisciplinary projects integrating expertise from diverse scientific domains and bridging the gap between physics, biology, and artificial intelligence.

Muhammad Aijaz Ahmed completed his PhD from the Institute of Botany, University of the Punjab. His field of expertise is mycology, where he has conducted extensive research on fungal biology and its applications. With significant contributions to the study of fungal interactions and their roles in various ecosystems, he has published numerous papers in reputed journals. His work is well-cited and respected within the scientific community. Muhammad Aijaz Ahmed continues to engage in advanced research and remains active as a contributor and reviewer for scholarly publications.

Asma Ajmal is currently pursuing an MPhil in Plant Sciences at Quaid-i-Azam University, Islamabad, with a focus on algal research. She completed her BS Honors degree in Botany from the University of Sargodha's Ex-sub campus, now known as Thal University Bhakkar. Her research specializes in algal cultivation, molecular and biochemical analyses, and data analysis. Asma is actively engaged in synthesizing nanoparticles and quantum dots from Arthrospira (Spirulina), exploring their applications in plant sciences and environmental sustainability. She is dedicated to advancing our understanding of algae's potential in biotechnology and contributing to innovative solutions in plant research. She has contributed a book chapter and worked as an author for research paper.

Zeeshan Arshed is a Research Associate in the Environmental Biotechnology Lab and a PhD candidate in Botany at the Institute of Botany, University of Punjab. He earned his MPhil degree from the same department. His research interests focus on abiotic stress responses, encompassing various morpho-physio-biochemical analyses in cereal crops. Zeeshan serves as a potential reviewer for several journals and has published numerous research papers that have garnered significant citations in reputable scientific publications.

Amin Fathi, Results-oriented researcher with a PhD in Agronomy and extensive interest in exploring plant responses to environmental stress, climate change, and plant-microbe interactions. Proven expertise in tillage, nanotechnology, biochar applications, and strategic management approaches to improve plant growth and resilience. Demonstrates leadership skills as advisor assistant and editor, coupled with a solid foundation in laboratory techniques, statistical analysis, and manuscript publishing.

Farhat Gul earned her M.Sc. and M.Phil. degrees in Plant Sciences from Quaid-i-Azam University, Islamabad, Pakistan, with distinction. She is currently pursuing a PhD in Plant Sciences, specializing in plant biochemistry and molecular biology. Her doctoral research focuses on the bio sustainable synthesis of metal oxide nanoparticles for biomedical and environmental applications. Additionally, her work encompasses phytochemistry, biotic stress responses, and the expression analysis of plants to address biotic and abiotic stresses. Farhat Gul is also proficient in cell line culturing. She has published several research articles, review papers, and book chapters in internationally recognized journals

Shumaila Ijaz has obtained her M.Phil degree in Botany (Plant Sciences) from Quaid-i-Azam University, Islamabad, Pakistan. He I currently pursuing her Ph.D from Department of Burn and Plastic Surgery, Shenzhen Institute of Translational Medicine, The First Affiliated Hospital of Shenzhen University, School of Biomedical Engineering, Shenzhen University Medical School, Shenzhen University, Shenzhen 518035, China. Her research group is using eco-friendly & sustainable avenues for the synthesis of novel functional nanomaterials from medicinal plants for biomedical and environmental applications. Further, she is also working on phytochemistry of medicinal plants, transcriptional regulation, and abiotic stresses and transformation of plants to combat various biotic and abiotic stresses. She has published more than 50 impact factor research and review articles in various journals of international repute and is a co-author of several book chapters.

Javed Iqbal has obtained his M.Sc. M.Phil and Ph.D degree in Botany (Plant Sciences) from Quaid-i-Azam University, Islamabad, Pakistan with a distinction in M.Phil (double Gold Medalist). Dr. Iqbal is currently working as an Assistant Professor at Department of Botany, Bacha Khan University, Charsadda, Pakistan. His research group is using eco-friendly & sustainable avenues for the synthesis of novel functional nanomaterials from medicinal plants for biomedical and environmental applications. Further, our research group is also working on phytochemistry of medicinal plants, transcriptional regulation, and abiotic stresses and transformation of plants to combat various biotic and abiotic stresses. He has published several research and review articles in various journals of international repute and is a co-author of several book chapters which are cited more than 4000 times. He is also the editorial board member and reviewer of many international journals of high repute.

Rashid Iqbal earned his PhD degree in Agronomy/Stress Physiology from the Islamia University of Bahawalpur. He also availed HEC IRSIP scholarship for Aarhus University Denmark. Currently, he is serving as an Assistant Professor at the Department of Agronomy. Dr. Rashid has published numerous research papers in renowned scientific journals with good impact factor. He is also a regular reviewer of the renowned research journals such as Remote Sensing, Agriculture, Sustainability and other MDPI Journals also Elsevier and Springer Journals. He has also competed one research project as Research Associate funded by NRPU (HEC) Pakistan. Dr. Rashid is also a HEC approved Supervisor to supervise M.phil and PhD students in the field of Agricultural Sciences.

Najeeba Parre Paker, a distinguished scholar in Plant Sciences, completed her M.Sc., M.Phil., and Ph.D. from Quaid-i-Azam University, Islamabad, Pakistan, earning distinction in her M.Phil. research. Currently, she is a Commonwealth Split-Site Scholar and Visiting Research Scholar at the University of York, UK, while continuing her Ph.D. at Quaid-i-Azam University. Paker's research bridges plant-microbe interactions and soil science, focusing on the role of Plant Growth-Promoting Rhizobacteria (PGPR) like Pseudomonas and Bacillus spp. in mitigating abiotic and biotic stresses. Her work innovates in areas such as pesticide assessment, organophosphate degradation, and plant health enhancement, advancing knowledge on nitrogen cycling through biological nitrogen fixation and nitrification. Passionate about microbial ecology, biocontrol strategies, and sustainable agriculture, Paker is dedicated to developing science-driven solutions for global agricultural and environmental challenges.

Amber Jabeen has obtained her MSc and MPhil degree in Botany (Plant Sciences) from Quaid-i- Azam University, Islamabad, Pakistan with a distinction in MSc. Amber is currently enrolled in PhD Plant Sciences at department of Plant Sciences, Quaid I Azam University, Islamabad. Her research emphasizes on the green synthesis of nanoparticles mediated by algae and their chemical and structural characterization. Her work spans across nanomedicine, and therapeutic applications of nanoparticles, contributing to advancements in drug delivery systems and disease treatments. She has actively contributed to research in molecular biology. Throughout her career, she has co-authored peer reviewed publications, including studies on the biomedical applications of nanoparticles, bioremediations, and stress responses of plants.

Munazza Jabeen has obtained her M.Sc. and M.Phil. degree in Plant Sciences from Quaid-i-Azam University, Islamabad, Pakistan. She has served as an Entomologist in the Department of Plant Protection, Pakistan. Her research group is using eco-friendly & sustainable avenues for the synthesis of novel functional nanomaterials from non-edible plants for biofuel production and environmental applications. Further, the research group is also working on the phytochemistry of medicinal plants, anatomical and palynological studies, and biodiversity conservation to get insights into the plants' interaction with other organisms and their environment, exploring the genetic diversity among plant species, their conservation, and sustainable utilization for future research studies. She has published several research articles in various journals of international repute.

Muhammad Kaleem obtained his M.Sc., M.Phil., and Ph.D. degrees in Botany (Plant Sciences) from Quaid-i-Azam University, Islamabad, Pakistan. His research focuses on the genetic characterization of microalgae, cyanobacteria, and plants, with additional emphasis on the environmental applications of microalgae and cyanobacteria. Dr. Kaleem is also involved in protein expression profiling in cyanobacteria and the development of cyanobacteria-mediated nanoparticles. He has published over 20 impact factor research articles, along with a review article and a book chapter in internationally recognized journals. Additionally, Dr. Kaleem serves as a reviewer for three prestigious international springer journals.

Ilham Khan holds an M.Sc. and M.Phil. in Plant Sciences from Quaid-i-Azam University, Islamabad. He is currently pursuing a Ph.D. in Plant Proteomics at the same institution, focusing on the therapeutic potential of medicinal plants. His research integrates metabolomics, transcriptomics, and proteomics, utilizing various bioinformatics tools to explore the antidiabetic properties of medicinal plants through in vitro, in silico, and in vivo studies. In addition, Mr. Khan has expertise in nanoparticle synthesis, investigating their biological activities and their role in mitigating biotic and abiotic stress in crops. He has contributed to the scientific community through several research articles and book chapters published in international journals.

T.C. Manjunath was born on 6th February 1967. He did his B.E. from RV College of Engg. in 1989 (B'lore Univ), M.E. in Robotics & Automation from Govt.'s LD College of Engg., (Gujarat Univ., Ahmedabad) in 1992 with FCD-First Rank & PhD in Systems & Control Engg. from IIT Bombay in the year 2007, got best research scholar award for the entire IITB. He is currently working as the Dean

Research (R & D) and as Professor in Dept. of Computer Science & Engg. with specialization in IoT, Cyber Security, Blockchain Technology in Rajarajeswari College of Engineering, Bengaluru, Karnataka. He has got a teaching (academic), research & administrative experience of more than 35+ years (10 years as Principal & 8 years as HoD, 1 year as Dean Research & rest as faculty) and has published more than 1500+ research papers in various national & international journals & conferences, with 175+ patents, 100+ book-chapters, 25 awards, 100s of Keynote talks and also published 25 text books in various field of science, engineering & technology, completed projects worth Rs. 9 Crores till date & guided 10 PhD scholars in VTU. He is a member of 30 Professional Society Chapters and is a fellow of IE, IETE, AMIEE, Sr. Member of IEEE and a chartered engineer.

Lubna Anjum Minhas obtained her B.S and M.Phil degrees in Botany (Plant Sciences) from Quaid-i-Azam University, Islamabad, Pakistan. Her M.Phil. research centered on cyanobacteria-mediated metal oxide nanoparticles, exploring their biomedical and environmental applications. She is the first author of four research articles and a review article, and she has contributed as a co-author to many other research articles. Additionally, Lubna serves as a reviewer for several prestigious international journals.

Abdul Samad Mumtaz has earned his M.Sc. and M.Phil. degrees from Quaid-i-Azam University and completed his Ph.D. at the University of Birmingham, UK. He is currently a professor at Quaid-i-Azam University in Islamabad. Dr. Mumtaz's extensive education has provided him with a profound understanding of molecular biology, genetics, climate change, mycology, phycology, and nanoparticles, along with their various applications. He leads a research group that focuses on eco-friendly and sustainable practices in the fields of phycology, bioremediation, and transcriptomics. The group's work in phycology investigates the ecological roles and potential applications of algae, particularly in addressing pressing environmental challenges. Through bioremediation, they explore innovative methods to utilize living organisms for detoxifying polluted environments, thereby promoting sustainable ecological practices. Additionally, the team's research on transcriptomics examines gene expression in response to environmental stimuli, which can inform strategies aimed at enhancing plant resilience and productivity. Dr. Mumtaz has contributed significantly to the field, publishing numerous articles and book chapters in reputable international journals. He also serves as an editorial board member and reviewer for several high-impact journals, further demonstrating his commitment to advancing scientific knowledge.

Ghulam Murtaza is a postdoc scholar at School of Agriculture, Yunnan University, Kunming, Yunnan 650504, China. He is currently working on the effect of biochar aging on physiochemical properties and adsorption behavior of biochar, antibiotic (Cpirofloxacin) and other pollutants adsorption using aged and modified biochar, Role of biochar in crop growth and production under various climate change conditions, Soil remediation by Biochar.

Ghazala Mustafa is an Associate Professor in the Department of Plant Sciences at Quaid-i-Azam University, Islamabad. Her research focuses on plant proteomics, with an emphasis on understanding the stress response mechanisms of crops under challenging environmental conditions, particularly abiotic stresses such as flooding. Dr. Mustafa is deeply engaged in studying plant-nanoparticle interaction mechanisms at the molecular level. Her expertise spans a wide range of fields, including abiotic stress tolerance, plant physiology, proteomics, molecular biology, biochemistry, and mass spectrometry techniques such as LC-MS/MS. With a robust academic portfolio, Dr. Mustafa has authored numerous research articles and book chapters in esteemed international journals, amassing over 2,000 citations. Her contributions significantly advance the understanding of plant resilience and adaptation in the face of environmental challenges.

Muhammad Muzamil Sattar attained his BS Honors degree in Botany from Quaid-i-Azam University, Islamabad, and subsequently completed his M.Phil. in Plant Sciences at the same institution. He worked on the isolation, culturing and purification of algal species during his research work in Bs honors and he has finalized his MPhil thesis on the synthesis and characterization of Chara-based zinc oxide nanoparticles and their PVP-doped derivatives to improve biological efficacy. He has authored a book chapter and served as a co-author for a research article. His research work, as the primary author, is currently under modification.

P. Selvakumar is currently an Associate Professor in the Department of Science and Humanities at Nehru Institute of Technology, Coimbatore, Tamil Nadu, India. Since 2023, he has served as the Department IQAC Coordinator at the institute. Prior to joining Nehru Institute of Technology, he held positions as an Assistant Professor in the Department of Science and Humanities various engineering colleges in Coimbatore, Tamil Nadu, India, spanning from 2011 to 2023. Dr.P. Selvakumar's academic and research interests lie primarily in the field of phytochemistry. He has made significant contributions to his field, with an impressive record that includes 30 journal publications, 12 patents, and 20 conference papers. Additionally, he has organized 5 seminars and conferences, further demonstrating his commitment to academic and scientific exchange. Throughout his career, Dr. P.Selvakumar has been recognized for his dedication and excellence in teaching, having been honored with several awards, including the Best Teacher Awards. He holds a prominent position in the academic community, contributing actively to research.

Najeeba Parre Paker, a distinguished scholar in Plant Sciences, completed her M.Sc., M.Phil., and Ph.D. from Quaid-i-Azam University, Islamabad, Pakistan, earning distinction in her M.Phil. research. Currently, she is a Commonwealth Split-Site Scholar and Visiting Research Scholar at the University of York, UK, while continuing her Ph.D. at Quaid-i-Azam University. Paker's research bridges plant-microbe interactions and soil science, focusing on the role of Plant Growth-Promoting Rhizobacteria (PGPR) like Pseudomonas and Bacillus spp. in mitigating abiotic and biotic stresses. Her work innovates in areas such as pesticide assessment, organophosphate degradation, and plant health enhancement, advancing knowledge on nitrogen cycling through biological nitrogen fixation and nitrification. Passionate about microbial ecology, biocontrol strategies, and sustainable agriculture, Paker is dedicated to developing science-driven solutions for global agricultural and environmental challenges.

Saira Qadir has obtained her BS, M.Phil and Ph.D degree in Botany (Plant Sciences) from Federal Urdu University of Arts, Sciences and Technology, Karachi, Pakistan with a distinction in BS (Silver Medalist). Dr. Qadir is currently working as an Assistant Professor at Department of Botany, Rawalpindi women University, Rawalpindi, Pakistan. Her research group is using phytochemicals from different weeds in sustainable agriculture to promote bio fertilizer. currently involved in investing the nutritional value of halophytes that can replace glucophytes to reduce food scarcity. She is also a certified trainer of mushroom cultivation. Furthermore, our research group is also working on phytochemistry of medicinal plants, transcriptional regulation, and abiotic stresses and transformation of plants to combat various biotic and abiotic stresses. She has published several research and review articles in various journals of international repute and is a co-author of several book chapters. She is also the editorial board member and reviewer of many international journals of high repute.

Ijaz Ahmed Qureshi has obtained his B.S M.Phil and Ph.D degree in Botany (Plant Sciences) from Federal Urdu University of Arts, Sciences and Technology karachi, Pakistan. Dr. Ijaz iscurrently working as an Examiner PBRR at Ministry of National Food security and Research, Pakistan. His research group is working on invasiveness and allelopathic effect of inter topping and intra cropping and their drawbacks to yield and economy. Further, our research group is also working on phytochemistry of medicinal plants, transcriptional regulation, seasonal variation in seaweeds in coastal line of Sindh, and abiotic stresses and transformation of plants to combat various biotic and abiotic stresses. He has published several research and review articles in various journals of international repute and is a co-author of several book chapters. He is also the editorial board member and reviewer of scientific reports, European journal of nutrition & food safety, international journals of high repute.

Zarqa Riaz has obtained her M.Phil. degree in Plant Sciences from Quaid-i-Azam University, Islamabad, Pakistan. Zarqa Riaz is currently working as visiting Lecturer at Department of Botany, Government College Women University Sialkot, Pakistan. Her main areas of research are Proteomic analysis, nanoparticles synthesis, stress physiology specifically biotic & abiotic stress tolerance in crop plants. She has published some research articles and currently working as collaborator with research groups.

Ammara Saleem is currently pursuing a Ph.D. in Botany at the Institute of Botany, University of the Punjab, where she also completed her MPhil degree. Her research interests lie in exploring the impacts of abiotic stress on cereal crops, focusing on a comprehensive range of morpho-physiological and biochemical analyses. Ammara has contributed significantly to her field through her published research papers, which have garnered strong citations in reputable scientific journals. Her scholarly work has made noteworthy contributions to the understanding of plant resilience under abiotic stress. Additionally, she serves as a dedicated reviewer for various scientific journals, leveraging her expertise to support the academic community and ensure the quality of published research.

Junaid Shehzad completed his M.Phil. in the Department of Plant Sciences at Quaid-i-Azam University, Islamabad, where he is currently pursuing a Ph.D. His research focuses on the proteomic responses of crops to abiotic stress, with a particular interest in understanding the mitigative role of nanoparticles under stresses such as heavy metal toxicity. Mr. Shehzad has authored research articles and book chapters published in international journals, contributing to the fields of plant proteomics and stress tolerance mechanisms.

C.A. Sowmiya, an accomplished scholar in the field of Genetics and Plant Breeding, hails from Kanyakumari, Tamil Nadu. She holds a Ph.D. in Genetics and Plant Breeding from Tamil Nadu Agricultural University, Coimbatore, where she achieved a commendable OGPA of 9.13. Her academic journey includes an M.Sc. in the same specialization from Annamalai University, where she was a top-ranking student with an OGPA of 9.64, and a B.Sc. in Agriculture from the Agricultural College and Research Institute, Killikulam. Dr. Sowmiya has made significant contributions to rice genetics, including the successful introgression of blast and bacterial blight resistance genes into rice genotypes through marker-assisted backcross breeding. Her research is well-documented in various esteemed journals. She is a recipient of several prestigious awards, including the INSPIRE Fellowship and numerous medals for her academic excellence. Dr. Sowmiya's academic excellence has been recognized through numerous accolades, including: - Certificate of Merit for 'Second Rank' in Ph.D. from the Indian Society of Plant Breeders (ISPB). - "Srilochani Varadarajulu Prize" Medal for securing the top rank in M.Sc. (Ag.) Genetics and Plant Breeding. - "Dr. M.S. Swaminathan Prize" for the highest marks in M.Sc. Genetics and Plant Breeding. - "Dr. C.N. Sambandam Prize" for excellence in Genetics subjects during her M.Sc. - "Thiru. V.S. Ramalingam Pillai Prize" for the highest marks in Plant Breeding during M.Sc. - "Dr. V. Sivasubramanian Prize" for the best thesis in M.Sc. Genetics and Plant Breeding. - "T. Vivekanandan Memorial Award" for achieving the second-highest OGPA among female students during her B.Sc. Beyond academics, Dr. Sowmiya has hands-on teaching experience as an Assistant Professor, mentoring students in plant genetics. She has also coordinated multiple institutional initiatives. Passionate about plant genetics, she aspires to inspire the next generation of agricultural scientists while contributing to sustainable crop improvement.

Zakir Ullah is a PhD student of plant Biochemistry and Molecular Biology at Quaid-I-Azam University Islamabad, Pakistan. He has published more than 60 research articles in various international journals. He has expertise in the synthesis of novel nanomaterials, cancer cell lines, formulations, photochemistry, plant virology. He has published his research in well-reputed journals.

Rooma Waqar has obtained her M.Sc., M.Phil., and Ph.D. degrees in Plant Sciences from Quaid-i-Azam University in Islamabad, Pakistan. His extensive education has provided with a deep understanding of phycology and its various applications. Her research group leads a research group focused on eco-friendly and sustainable practices in the fields of phycology, bioremediation, and transcriptomics. Our work in phycology examines the ecological roles and potential applications of algae, particularly in addressing environmental challenges. Through bioremediation, our group explores innovative methods to use living organisms for detoxifying polluted environments, promoting sustainable ecological practices. Additionally, our team, research on transcriptomics investigates gene expression in response to environmental stimuli, which can inform strategies for enhancing plant resilience and productivity. She has published articles and book chapters in various international repute. She is also a reviewer of many international journals.

Tabassum Yaseen obtained his M.Sc. and M.Phil and Ph.D degree in Botany from Department of Botany University of Peshawar, Pakistan. Dr. Yaseen is currently working as an Assistant Professor and Chairperson of the Department at Bacha Khan University, Charsadda, Pakistan. Her research group is involved in vesicular arbuscular mycorrhizal fungi, synthesis of nanoparticles from medicinal plants for biomedical and environmental applications. She has published more than 83 research articles in various journals of international repute, and is a co-author of several book chapters

Asma Zulfiqar is an Associate Professor in the Institute of Botany, University of the Punjab, where she serves as the head of the Molecular Plant Physiology Lab. She has guided numerous research theses at various academic levels, including PhD, MPhil, MS, MSc, and BS students, showcasing her dedication to mentoring and academic excellence. Dr. Zulfiqar is highly regarded for her expertise in plant physiology and related fields, contributing significantly to advancing plant science research. Her scholarly work includes the publication of numerous research articles in reputable, high-impact journals, demonstrating her commitment to scientific progress. In addition to her research and teaching responsibilities, she has also served as the MSc program coordinator, further contributing to academic program development within the department. Dr. Asma Zulfiqar is an active reviewer for several well-regarded journals, reflecting her standing in the scientific community and her role in maintaining the quality of academic publications.

Index

A

B

C

D

E

F

G

M

O

P

Q

R

S

T

V

www.ingramcontent.com/pod-product-compliance
Lightning Source LLC
LaVergne TN
LVHW082051011026
849162LV00019B/743